Tumor Microenvironment

Tumor Microenvironment

Edited by

Dietmar W. Siemann
University of Florida, Shands Cancer Center

WILEY-BLACKWELL
A John Wiley & Sons, Ltd., Publication

This edition first published 2011 © 2011 John Wiley & Sons, Ltd.

Wiley-Blackwell is an imprint of John Wiley & Sons, formed by the merger of Wiley's global Scientific, Technical and Medical business with Blackwell Publishing.

Registered office: John Wiley & Sons Ltd, The Atrium, Southern Gate, Chichester, West Sussex, PO19 8SQ, UK

Editorial offices: 9600 Garsington Road, Oxford, OX4 2DQ, UK
The Atrium, Southern Gate, Chichester, West Sussex, PO19 8SQ, UK
111 River Street, Hoboken, NJ 07030-5774, USA

For details of our global editorial offices, for customer services and for information about how to apply for permission to reuse the copyright material in this book please see our website at www.wiley.com/wiley-blackwell.

Library of Congress Cataloging-in-Publication Data

Tumor microenvironment / editor, Dietmar Siemann.
p. cm.
Includes index.
ISBN 978-0-470-74996-8 (cloth)
1. Cancer cells. 2. Carcinogenesis. I. Siemann, Dietmar W.
RC269.T86 2011
616.99′4071 – dc22

2010023380

A catalogue record for this book is available from the British Library.

This book is published in the following electronic formats: ePDF 9780470669808; Wiley Online Library 9780470669891

Typeset in 10/12pt Sabon by Laserwords Private Limited, Chennai, India
Printed and bound in Singapore by Markono Print Media Pte Ltd.

First Impression 2011

Every day is a journey,
and the journey itself is home.

– Basho

Contents

Preface

The tumor microenvironment has long been identified as a major factor influencing treatment resistance of cancers to radiotherapy and chemotherapy. In addition, it is now well recognized to play a critical role in neoplastic cell initiation, malignant progression, and the metastatic spread of tumor cells.

The tumor microenvironment, or stroma, consists of cells, extracellular matrix and extracellular molecules. Important cell types that have been identified include fibroblasts, epithelial cells, infiammatory cells, immunocytes, and vascular cells. Research indicates that interactions between the neoplastic cells, the extracellular matrix (ECM), and stromal cells are bidirectional and dynamic. The microenvironment can exert both stimulatory and growth inhibitory effects on tumor cells. Indeed, cancer cells are often critically dependent on the stroma, as is the case between neoplastic and endothelial cells. It is well established that most tumors cannot grow larger than a few millimeters in diameter without inducing their own vascular network, primarily through the process of angiogenesis.

Because stromal elements and their manifestations are essential for all stages of tumor development, and lead to significant negative consequences affecting the management of cancer by conventional anticancer therapies, a number of treatment strategies directed against tumor microenvironmental infiuences are being pursued. These fall into two general categories. The first strategy seeks to target unique conditions of the tumor microenvironment which may be exploited for therapeutic gain. A primary target in this approach has been tumor hypoxia – a consequence of the abnormal and inadequate development of blood vessel networks in tumors. Examples of therapeutic strategies being pursued include hypoxic cell radiosensitizers, bioreductive drugs, HIF-1 targeting agents, and gene therapies. The second approach targets stroma itself. Among the approaches are drugs that induce apoptosis or inhibit the function of the stromal cells, or inhibit the factors secreted by stroma that are required for tumor progression and metastasis. The development of novel targeting therapies which attenuate VEGF and EGF signaling illustrate progress in the latter area.

The goal of this book is to review the importance of the tumor microenvironment in cancer management. Particular emphasis is placed on: (i) the characterization of the unique features of the tumor microenvironment, (ii) the determination of factors that are critical in the progression and metastatic spread of tumor cells, (iii) the identification of tumor cell stem cells and their interactions with stroma at sites of dissemination, (iv) the impact of the tumor microenvironment on the response to conventional anticancer therapies, and (v) the development and assessment of novel therapeutic strategies that target the stromal or neoplastic components of the tumor microenvironment or its functions. Most of all, this book is about researchers' efforts to achieve better treatment for cancer patients.

This book is the culminating effort of a team of international investigators; some of whom are long-time friends, and others who are new collaborators. For many the importance of the tumor microenvironment as it relates to cancer therapy has been a long-standing passion. I feel privileged to have worked on this text with such an outstanding group of co-authors. To them I owe a debt of gratitude, for without their efforts and dedication this book would not have materialized. I thank them for their willingness to participate in this project, their timely contributions and their integrity. It has been a pleasure working with these colleagues. The quality of the book reflects their efforts and is to their credit.

I am also grateful to all the staff at Wiley-Blackwell, especially Rachel Moore (Concept Contact), Fiona Woods (Project Editor), Izzy Canning (Publishing Assistant) and Sarah Abdul Karim (Production Editor), who have worked diligently to bring this book to fruition. Their assistance in all matters has been invaluable and much appreciated.

Too many lives are lost each year to cancer. This book reflects the dedication of researchers throughout the world who, through their tireless pursuit to gain a better understanding of the molecular and biologic underpinnings of cancer, seek to identify the causes of cancer and develop strategies for treatments and cure.

Dietmar W. Siemann
Gainesville, FL

List of Contributors

Kevin L. Bennewith Stanford University School of Medicine, Department of Radiation Oncology, CCSR-South, Room 1255, 269 Campus Drive, Stanford, CA 94305-5152, USA

Allison S. Betof Duke University Medical Center, Durham, NC 27710, USA

Rolf Brekken Hamon Center for Therapeutic Oncology, Simmons Comprehensive Cancer Center, Division of Surgical Oncology, UT Southwestern Medical Center, 5323 Harry Hines Blvd, Dallas, TX 75390-8593, USA

Marie Schluterman Burdine Hamon Center for Therapeutic Oncology, Simmons Comprehensive Cancer Center, Division of Surgical Oncology, UT Southwestern Medical Center, 5323 Harry Hines Blvd, Dallas, TX 75390-8593, USA

Naz Chaudary Ontario Cancer Institute, Princess Margaret Hospital and the Campbell Family Institute for Cancer Research, 610 University Avenue, Toronto ON, Canada, MG52M9

Julia A. D'Ambrosio Department of Pharmacology and Physiology, Drexel University College of Medicine, Philadelphia, PA, USA

Nicholas C. Denko Division of Radiation and Cancer Biology, Dept of Radiation Oncology, Stanford University School of Medicine, 269 Campus Dr. West, CCSR-1245C, Stanford, CA 94305-5152, USA

Mark W. Dewhirst Gustavo S. Montana Professor of Radiation Oncology, Professor of Pathology and Biomedical Engineering, Duke University Medical Center, Durham, NC 27710, USA

Graeme J. Dougherty Department of Radiation Oncology, Arizona Cancer Center, 1501 North Campbell Avenue, Tucson, AZ 85724-5081, USA

Shona T. Dougherty Department of Radiation Oncology, Arizona Cancer Center, 1501 North Campbell Avenue, Tucson, AZ 85724-5081, USA

Janine T. Erler Stanford University School of Medicine, Department of Radiation Oncology, CCSR-South, Room 1255, 269 Campus Drive, Stanford, CA 94305-5152, USA

Alessandro Fatatis Associate Professor, Department of Pharmacology and Physiology, Department of Pathology and Laboratory Medicine, Drexel University College of Medicine, 245 N. 15th Street-MS488, New College Building, Philadelphia, PA 19102, USA

Bernard Gallez Louvain Drug Research Institute, Biomedical Magnetic Resonance Unit, Director, Université catholique de Louvain, REMA, Avenue Mounier 73.40, B-1200 Brussels, Belgium

Amato J. Giaccia Jack, Lulu, and Sam Willson Professor of Cancer Biology, Director Division of Cancer and Radiation Biology, Stanford University School of Medicine, Department of Radiation Oncology, CCSR-South, Room 1255, 269 Campus Drive, Stanford, CA 94305-5152, USA

Tereza Goliasova Division of Radiation and Cancer Biology, Dept of Radiation Oncology, Stanford University School of Medicine, 269 Campus Dr. West, CCSR-1245C, Stanford, CA 94305-5152, USA

Ester M. Hammond Radiation Oncology & Biology, University of Oxford, Old Road Campus Research Building, Off Roosevelt Drive, Churchill Hospital, Oxford, OX3 7DQ, UK

Richard P. Hill Senior Scientist OCI/PMH, Professor Dept of Medical Biophysics, University of Toronto, 610 University Ave, Rm 10-113, Toronto, ON, Canada, M5G2M9

Michael R. Horsman Danish Cancer Society, Department of Experimental Clinical Oncology, Norrebrogade 44, DK-8000 Aarhus C, Denmark

Zhongbo Hu Department of Pediatrics, Division of Hematology/Oncology, University of Florida School of Medicine, Box 100296, Gainesville, FL 32610-0296, USA

Whitney L. Jamieson Department of Pharmacology and Physiology, Drexel University College of Medicine, Philadelphia, PA, USA

Danielle L. Jernigan Department of Pharmacology and Physiology, Drexel University College of Medicine, Philadelphia, PA, USA

Bénédicte F. Jordan Louvain Drug Research Institute, Biomedical Magnetic Resonance Unit, Director, Université catholique de Louvain, REMA, Avenue Mounier 73.40, B-1200 Brussels, Belgium

Albert C. Koong Radiation Oncology, Stanford University School of Medicine, 269 Campus Dr. West, CCSR-1245C, Stanford, CA 94305-5152, USA

Alastair H. Kyle BC Cancer Research Centre, University of British Columbia, 675 West 10th Avenue, Vancouver, BC, Canada, V5Z1L3

Giovanni Melillo Head, Tumor Hypoxia Laboratory, Developmental Therapeutics Program, SAIC Frederick, Inc. Bldg 432, Rm 218, National Cancer Institute at Frederick, Frederick, MD 21702-1201, USA

Michael F Milosevic Department of Radiation Oncology, University of Toronto, Princess Margaret Hospital, 5th Floor Rm 634, 610 University Ave, Toronto, ON, Canada, M5G2M9

Andrew I. Minchinton BC Cancer Research Centre, University of British Columbia, 675 West 10th Avenue, Vancouver, BC, Canada, V5Z1L3

Jens Overgaard Danish Cancer Society, Department of Experimental Clinical Oncology, Norrebrogade 44, DK-8000 Aarhus C, Denmark

Nicole N. Parker Department of Radiation Oncology, University of Florida Shands Cancer Center, 2000 SW Archer Road, Gainesville, FL 32610, USA

Isabel M. Pires Radiation Oncology & Biology, University of Oxford, Old Road Campus Research Building, Off Roosevelt Drive, Churchill Hospital, Oxford, OX3 7DQ, UK

Rachel Poole Radiation Oncology & Biology, University of Oxford, Old Road Campus Research Building, Off Roosevelt Drive, Churchill Hospital, Oxford, OX3 7DQ, UK

Chiara Porta Istituto Clinico Humanitas, 20089 Rozzano, Milan, Italy

Annamaria Rapisarda Tumor Hypoxia Laboratory, Developmental Therapeutics Program, SAIC Frederick, Inc. Bldg 432, Rm 218, National Cancer Institute at Frederick, Frederick, MD 21702-1201, USA

Amyn M. Rojiani Department of Pathology-BF-104, Medical College of Georgia, 1120 Fifteenth Street, Augusta, GA 30912-3600

Mumtaz V. Rojiani Department of Pathology-BF-104, University of South Florida, Tampa, FL, USA

Mike R. Russell Department of Pharmacology and Physiology, Drexel University College of Medicine, Philadelphia, PA, USA

Antonio Sica Istituto Clinico Humanitas, 20089 Rozzano, Milan, Italy

Dietmar W. Siemann John P. Cofrin Professor for Cancer Research, Associate Chair, Department of Radiation Oncology, University of Florida Shands Cancer Center, 2000 SW Archer Road, Gainesville, FL 32610, USA

William B. Slayton Department of Pediatrics, Division of Hematology/Oncology, University of Florida School of Medicine, Box 100296, Gainesville, FL 32610-0296, USA

Marzenna Wiranowska Department of Pathology-BF-104, University of South Florida, Tampa, FL, USA

Jing Zhang Radiation Oncology, Stanford University School of Medicine, 269 Campus Dr. West, CCSR-1245C, Stanford, CA 94305-5152, USA

1

The Microenvironment in Cancer

Nicole N. Parker and Dietmar W. Siemann

Department of Radiation Oncology and University of Florida Shands Cancer Center, University of Florida, Gainesville, Florida, USA

1.1 Introduction

In the development of a cancer, the transformation of epithelial cells into a neoplastic and progressively invasive tumor occurs though the acquisition of several procancer characteristics that can take years or decades to develop. The particular stages of transformation have been established and a general consensus exists about the properties of a successful malignancy. While many therapeutics have been developed to combat these properties, these therapies are not universally successful, and their efficacy depends on the type and site of the primary tumor, its degree of vascularization, the proliferative compartment of the tumor, and in particular, the tumor microenvironment. The latter is the key support system of a cancer, and is an important source of critical protumorigenic factors that facilitate growth, invasion, angiogenesis, and metastatic ability. The focus of this book is to examine how the reliance of tumors on their microenvironments for development and preservation of key cellular functions is now recognized not only as a major contributor to cancer aggression and treatment resistance but also as a potential target for novel therapeutic intervention strategies.

1.2 A highly selective process is required to obtain the cancer phenotype

Many studies have focused on the predetermining factors that cause hyperplasia, or the hyperproliferation of cells within their normal environment. On the path to a cancer, regions of hyperplasticity must subsequently become dysplastic, or display a highly disordered pattern of proliferation with little or no growth regulation. Predisposing genomic lesions in various genes of these dysplastic cells

Tumor Microenvironment Edited by Dietmar W. Siemann

confer a proliferative advantage over normal cellular counterparts. Therefore, the out-proliferation exhibited by dysplastic populations requires additional genetic instability or genomic modifications, and these cells are considered neoplastic: they have an advantageous rate of proliferation with a lack of regulation, and possess other procancerous features at the time of transformation. Oncogenic lesions, coupled with inhibition of tumor suppressors, together contribute to cellular transformation. Genomic, proteomic, post-translational, and epigenetic mutations are responsible for activating oncogenes and inhibiting tumor suppressor genes.

1.3 The cancer phenotype

Although a multitude of potential cancer etiologies may occur for cancer development, several essential characteristics are present in malignancies (Hanahan and Weinberg, 2000). A key element of malignant transformation is the loss of regulatory control mechanisms present in normal somatic non-stem cells that are growth arrested and do not divide. Cancer cells not only possess heightened rates of cell proliferation and aberrant cell cycle checkpoints, but also lose contact-inhibited growth regulation. As a result, unlike detached normal cells which die by anoikis or cell detachment-induced apoptosis, cancer cells continue to grow unabated, breaking through basement membranes and invading extracellular spaces around tissues and organs.

Extracellular matrix remodeling and cellular changes in adhesion molecules are both required for a cancer cell to become more motile. Rearrangement of the actin cytoskeleton facilitates cell motility and plasticity, as does downregulation of adhesion proteins that bind tightly to the extracellular matrix (Chapter 3). A cancer cell modifies its adhesive properties and implements a program of non-adhesion through multiple modifications.

The progressive growth of a tumor ultimately results in an inability of normal tissue blood vessels to oxygenate and provide nutrients to tumor cells most distal to the blood supply. As a consequence oxygen-deficient (hypoxic) regions develop within the tumor (Chapters 2 and 9). The ability of transformed cells to survive hypoxic conditions requires a switch from aerobic to anaerobic glycolysis, a major approach by which cancer cells circumvent the cytotoxic effects of oxygen deprivation (Gatenby and Gawlinski, 2003). As a result of glycolysis, lactic acid byproducts accumulate in cells undergoing this process. Although this acidification is generally toxic to cells, cancer cells upregulate acid transporter proteins and efficiently secrete acid products into the surrounding environment. A side effect of acid secretion is an increase in local extracellular acidity, fluid retention, and subsequently, an increase in interstitial pressure (Chapter 9). Still, the ability of cancer cells to metabolically adapt by preferentially undergoing glycolysis even in the presence of oxygen not only provides a survival advantage over non-transformed cells but also ensures the persistence of only the most successful cancer cells (Gatenby and Gillies, 2004).

The outgrowth of a tumor that is beyond the diffusion limits of nearby blood vessels, which supply nutrients and oxygen, leads to another critical phenotypic advantage of cancer cells: the ability to induce angiogenesis, or the process of developing new blood vessels from existing vascular structures. Cancer cells accomplish this through the upregulation and release of proangiogenic factors that can destabilize endothelial cells and induce vascular outgrowth from normal blood vessels. Endothelial cells proliferate toward the source of the chemoattractant angiogenic factors to form a new capillary network for the tumor mass. However, unlike normal vasculature, which is extremely ordered, this newly developed tumor neovasculature is highly aberrant in structure, lacking organization and vessel integrity. Although many proangiogenic factors have now been identified, vascular endothelial growth factor (VEGF) is believed to be the major inducer of tumor angiogenesis. It has not only been implicated in many cancer types (Fukumura *et al.*, 1998) but importantly, VEGF expression has been shown to correlate with tumor angiogenesis and aggression, poor patient outcome, and is a predictor for metastasis and high tumor grade in multiple cancer types (Brychtova *et al.*, 2008). Interestingly, some studies have demonstrated that acidic microenvironments can induce vesicle lysing, thereby secreting VEGF into the tumor microenvironment and contributing to a feed-forward mechanism in which tumor-induced hypoxia and cellular acidification lead to the formation of neovasculature.

It is generally believed that tumors cannot grow to a size larger than a few cubic millimeters without inducing a neovasculature. Once cancer cells induce revascularization, thereby ensuring a more constant nutrient supply, this growth restriction is effectively removed. The requirement for additional tumor space necessitates the ability of cancer cells to invade into surrounding tissue. It is most advantageous for cancer cells to digest adjacent extracellular matrix and force the local reorganization of normal epithelia and surrounding stromal elements.

1.4 The extracellular matrix

The extracellular matrix is comprised of various cell types and secreted proteins that help maintain the organization of higher-order cellular structures. In addition to containing various cell types, the matrix is deposited as a mix of such proteins as collagens, fibronectin, laminins, hyoluronan, plasminogens, proteases, and numerous others, which collectively form an inflexible scaffold to which cells attach. In addition, other secreted cellular proteins such as cytokines and extracellular matrix remodeling proteins normally reside in the extracellular matrix (Chapters 3 and 4). These proteins are released when the matrix is degraded, and upon their release become activated due to proteases and other activating enzymes present in the extracellular environment, further contributing to the regulation of extracellular matrix turnover.

Many cell types are present in the extracellular matrix and the tumor milieu. Examples include fibroblasts, which are an integral inducer of matrix remodeling, as well as endothelial cells, hematopoietic-derived cells, and immune cells, which normally monitor this environment for foreign (i.e., non-host) bodies (Chapter 5). In cancer, particularly at the later stages of transformation and invasion, normal immune functions are subverted, leading to recognition of the tumor as part of the host, rather than as an invading foreign entity.

1.5 Motility, invasion, and metastatic ability

Successful and evolutionarily adapted cancer cells are motile, have no major attachments to extracellular substrata, and can more easily move through the extracellular and intracellular space due to decreased cell adhesion and an increase in factors which facilitate extracellular matrix degradation and remodeling. A natural effect of motility is that cancer cells invade into surrounding tissue, colonize and populate the area given a favorable microenvironment. Further, cancer cell motility facilitates the movement of the cancer cell through layers of endothelial cells surrounding blood vessels, enabled in part through cancer cell secretion of vascular destabilizing factors. As a result the vasculature is perturbated and cancer cells gain access to the circulation, which is the major mode of transport for cancer cells to reach distant organs.

In the metastatic cascade, the tumor microenvironments of both the primary tumor and the target sites colonized by cells shed from the primary tumor are of critical importance to the successful spread of neoplastic cells (Joyce and Pollard, 2009) (Chapters 6–8 and 12). The classic 'seed and soil' mechanism describes a situation in which only permissive target microenvironments enable the attachment and subsequent proliferation by a metastatic cell. In addition, cells in the primary site of a cancer shed factors and progrowth signals that contribute to the tumor microenvironment at the secondary sites of tumor formation (Chapter 8). In this way, metastatic tumor cells establishing new colonies continue to receive progrowth support signals while they are colonizing the secondary site and during subsequent phases of secondary tumor growth.

In addition, the evolution of the microenvironment can impact premetastatic cells in a manner that leads to an invasive, advanced, and evolutionarily favored metastatic phenotype that can survive extravasation, intravasation, and can establish new tumors at distant sites. Accumulating evidence suggests that the hypoxic conditions that select for successful tumor types also contribute to the metastatic potential of that tumor (Chapter 14). Therefore, eradication of the hypoxic regions of a cancer has short-term and long-term benefits (Chapters 16–18), in that both tumor bulk and metastatic capability are reduced.

1.6 Impact of the tumor microenvironment on the control of cancer

The tumor microenvironment is a growing target for consideration of cancer therapeutics due to its varied influence on the cells and on the physical aspects of chemotherapeutic delivery (Chapters 2 and 15). Several drawbacks to traditional chemotherapies that do not account for the microenvironment are: the tumor vasculature, which is highly disordered and leaky; tumor core hypoxia, which confers radiation resistance on tumor cells in this state; cells furthest from blood vessels become growth-arrested, preventing efficacious chemotherapeutic inhibition of proliferating cells; and the upregulation of acid transporter and other transporter proteins, which efficiently excrete chemotherapeutics from cancer cells and hinder successful cancer treatment (Chapters 2, 9, and 15). Importantly, offspring of chemotherapeutic survivors can pass this genetic property to daughter cells, making subsequent populations of tumor cells highly resistant to subsequent therapy. One further consideration regarding the tumor microenvironment is the stem cell population, a slow-growing subset of cancer cells, which is inherently resistant to therapies targeting cells that are actively cycling. Stem- or stem-like cancer cells are pluripotent, highly plastic, and dedifferentiated entities that easily and steadily repopulate tumors following therapy. Considering these scenarios, targeting the tumor microenvironment becomes an increasingly logical and attractive therapeutic option in cancer management.

1.7 Targeting the tumor microenvironment

Classical anticancer therapies including radiotherapy and chemotherapy are toxic to cancer cells but such treatments are typically also associated with inadvertent damage to critical normal tissues. Newer and more specific therapies have become more prevalent in the treatment of specific cancers as the molecular mechanisms of carcinogenesis become better characterized. The approach of uncovering molecular etiologies of cancer coupled with the development of targeted therapies that exploit essential signaling pathways (Chapters 10, 13 and 17) will undoubtedly contribute to the future arsenal of anticancer therapeutics.

Because the microenvironment of tumors not only severely impairs the treatment efficacy of conventional anticancer therapies but also differs significantly from those found in normal tissues, research is beginning to focus on the tumor microenvironment as a separate cancer-associated entity that may be targeted (Chapters 10, 13 and 17). Indeed, several strategies have already been identified that exert an anticancer effect through the specific targeting of the tumor microenvironment. Oxygen-poor cells display greater resistance to radiotherapy, and methods for reversing the radioprotective effects of hypoxia in order to enhance the treatment efficacy of radiotherapy have received considerable attention (Chapters 9 and 16). The other compartment of the tumor microenvironment that has been extensively targeted is the tumor vasculature (Chapter 18). As an essential part of tumor survival, such a strategy seeks to deprive the tumor of critical nutrients and means to spread. The use of antiangiogenic and vascular disruptive therapies provide powerful adjuncts to conventional anticancer treatments. All such potential therapeutic interventions will be critically dependent upon the establishment of novel approaches to non-invasive imaging of the tumor microenvironment (Chapter 11).

1.8 Summary

The microenvironment of tumors creates a significant hindrance to the control of cancers by conventional anticancer therapies. The physical conditions present are imposing and manifold, and include elevated interstitial pressure, localized extracellular acidity, regions of oxygen and nutrient deprivation, and contraction of the extracellular matrix. No less important are the functional consequences experienced by the tumor cells residing in such environments: adaptation to hypoxia, cell quiescence, modulation of transporters, and enhanced metastatic potential. Together these factors lead to therapeutic barriers that may render the chance of tumor elimination as minimal.

However, the aberrant nature of the tumor microenvironments also offers unique therapeutic opportunities. Reducing tumor hypoxia can improve drug delivery and enhance radiotherapy. The inhibition of fibroblasts and other cell types exploits the tumor's reliance on the microenvironment for various factors and properties necessary for its survival. Targeting the tumor vasculature would destroy the nutritional support network of the tumor. These approaches and many others directed against the tumor microenvironment are under active investigation in the laboratory and the clinic.

Because the molecular underpinnings of cancer development are becoming increasingly well characterized, future studies will undoubtedly identify distinct molecular markings that are characteristic of the tumor microenvironment. Such advances will lead to the development of targeted therapies that will selectively impair the neoplastic cell populations residing in these environments. Ultimately, by combining such therapies with conventional anticancer treatments it may be possible to bring cancer growth, invasion, and metastasis to a halt.

References

Brychtova, S., Bezdekova, M., Brychta, T., and Tichy, M. (2008) The role of vascular endothelial growth factors and their receptors in malignant melanomas. *Neoplasma*, **55**, 273–279.

Fukumura, D., Xavier, R., Sugiura, T. *et al.* (1998) Tumor induction of VEGF promoter activity in stromal cells. *Cell*, **94**, 715–725.

Gatenby, R.A. and Gawlinski, E.T. (2003) The glycolytic phenotype in carcinogenesis and tumor invasion: insights through mathematical models. *Cancer Research*, **63**, 3847–3854.

Gatenby, R.A. and Gillies, R.J. (2004) Why do cancers have high aerobic glycolysis? *Nature Reviews Cancer*, **4**, 891–899.

Hanahan, D. and Weinberg, R.A. (2000) The hallmarks of cancer. *Cell*, **100**, 57–70.

Joyce, J.A. and Pollard, J.W. (2009) Microenvironmental regulation of metastasis. *Nature Reviews Cancer*, **9**, 239–252.

2

Establishing the Tumor Microenvironment

Allison S. Betof[1] and Mark W. Dewhirst[2]

[1]*Department of Pathology, Duke University Medical Center, Durham, USA*
[2]*Department of Radiation Oncology, Duke University Medical Center, Durham, USA*

2.1 Introduction

For years cancer researchers focused on a seemingly obvious target: tumor cells. While this research yielded invaluable knowledge, a comprehensive understanding of tumor behavior remains elusive. Just as children are influenced by their environment as they grow, so too are developing tumors. Tumors do not grow in a vacuum. Instead, they are surrounded by the extracellular matrix, blood vessels, immune cells, and other supporting structures that make up human organs. Together this support system comprises a tumor's microenvironment. In recent years, cancer research has experienced a paradigm shift toward trying to understand the role of the microenvironment in tumor development, growth, metastasis, and treatment. This chapter will focus on the establishment of that environment.

There are two principal fields of microenvironmental research. One focuses on cellular components, cell–cell interactions, and the matrix that makes up the tumor parenchyma in addition to tumor cells. The other component is the physiological microenvironment. This involves the exchange of oxygen, nutrients, and waste products through tumor vasculature. While they are often discussed separately, these two components of the microenvironment do not exist in isolation. Though each component of the microenvironment will be addressed separately, this chapter will also review areas of overlap and interaction between the physical and physiological microenvironments.

National Institutes of Health/National Cancer Institute
NIH/NCI CA40355-25

Tumor Microenvironment Edited by Dietmar W. Siemann

2.2 From cancerous cells to a tumor

The complex, heterogeneous structures we know as tumors arise from cancer cells that proliferate within the space dedicated to another organ. The presence of malignant cells within normal tissue alters the environment and interrupts tissue homeostasis. Reciprocal interactions with the host stroma facilitate tumor proliferation, invasion, and metastasis. However, the question of which of these changes occurs first has become a central issue. Do cancer cells transform and then change the microenvironment where they are located to suit their needs? Or are there changes in the previously normal tissue that promote transformation and tumorigenesis in a particular location?

The answer, it now seems, is that both of these alterations can result in the formation of a tumor. For evidence that stromal changes can result in tumorigenesis in the absence of cancer cells, the reader is directed to one of several excellent reviews (Ariztia *et al.*, 2006; Bhowmick, Neilson, and Moses, 2004b). However, for most tumors that are epithelial in origin, the initiating event is usually an epithelial mutation followed by key paracrine signals from stromal fibroblasts that promote neoplastic progression (Lengauer, Kinzler, and Vogelstein, 1998; Bhowmick, Neilson, and Moses, 2004b). Reciprocal interactions between differentiating epithelia and adjacent stromal cells are essential to reaching proper tissue maturation in normal tissue (Bhowmick, Neilson, and Moses, 2004b). In contrast, malignant transformation of epithelial cells is preceded by or occurs simultaneously with host stromal activation, inducing angiogenesis, fibroblast proliferation, and recruitment of inflammatory mediators (Tomakidi *et al.*, 1999; Ronnov-Jessen, Petersen, and Bissell, 1996; Tlsty and Hein, 2001). Stromal fibroblasts respond to tumor-derived pathological signals by producing collagen and activating myofibroblast-associated proteins such as α-smooth muscle actin, vimentin, and desmin (Tlsty and Hein, 2001). In addition, activated fibroblasts proliferate and migrate more *in vitro* than fibroblasts from benign tissue (Schor, Schor, and Rushton, 1988; Schor *et al.*, 1985).

Cancer-associated fibroblasts (CAFs) are stromal fibroblasts with key roles in transformation, proliferation, and invasion, three of the hallmarks of cancer. These fibroblasts secrete growth factors and chemokines that signal critical changes in the extracellular matrix (ECM) and contribute oncogenic signals that increase proliferation and invasion (Kalluri and Zeisberg, 2006). For example, CAFs are able to stimulate both growth and transformation of immortalized prostate epithelial cells *in vitro* and *in vivo*, whereas normal fibroblasts are not (Olumi *et al.*, 1999). In addition, Parrott *et al.* used an ovarian cancer model to show that tumor fibroblasts may not just assist in tumorigenesis, but may actually be necessary to form a tumor from cancer cells (Parrott *et al.*, 2001). This group then used histology to show that the ovarian cancer cells recruited adjacent stroma from the normal mouse tissue to form the stroma of the tumor. Thus, the CAFs play a crucial role in establishing the physical microenvironment of a growing tumor.

One possible mechanism for the tumor promoting effects of CAFs comes from their overexpression of fibroblast secreted protein-1 (FSP1) (Grum-Schwensen *et al.*,

2005). When FSP1 knockout mice are injected with carcinoma cells, tumor formation is decreased compared to wild-type mice and the tumors that do form do not metastasize. Furthermore, adding FSP1$^{+/+}$ fibroblasts in addition to the carcinoma cells restores the wild-type phenotype. A second possible mechanism is that stromal CAFs secrete stromal cell-derived factor 1α (SDF-1α, also known as the cytokine CXCL12), which promotes proliferation of mammary carcinoma cells when it binds to the receptor CXCR4 on the cancer cells (Orimo *et al.*, 2005). These signaling molecules clearly contribute to the developing tumor microenvironment, but their precise effects and the contributions of other factors remain unknown.

2.3 A tumor is more than cancer cells and fibroblasts

In addition to the tumor cells and CAFs already discussed, the microenvironment consists of normal parenchymal and epithelial cells, ECM, normal stromal fibroblasts, soluble growth factors and cytokines, components of the immune system, nerves, and blood and lymphatic vessels. However, this is more than just a physical space. Implied within the term 'microenvironment' is a venue for these factors to interact.

2.3.1 The extracellular matrix

Within a carcinoma, epithelial cells are supported by a three-dimensional structure known as the ECM. The proteins that make up the ECM are produced by fibroblasts. The stroma is separated from the epithelium and endothelium by the basement membrane, a specialized type of ECM composed of collagen IV, laminin, and heparan sulfate proteoglycans (Kalluri, 2003).

There are considerable differences in matrix composition between tumor stroma and the stroma associated with normal tissue. Cells from the tumor secrete a variety of proteins into the ECM that are involved in cell adhesion, motility, communication, and invasion (Mbeunkui and Johann, 2009). Notably, some of these molecules are involved in degrading the ECM. This degradation occurs near the tumor-host interface, where the tumor-derived proteases overwhelm the host's endogenous inhibitors leading to extensive remodeling and stimulating alternative signals from the cell surface (Handsley and Edwards, 2005). Remodeling is achieved by combined efforts of secreted and membrane-anchored matrix metalloproteinases (MMPs), adamalysin-related membrane proteases, bone morphogenic protein-1-type metalloproteinases, endoglycosidases, and tissue serine proteases including tissue plasminogen activator, urokinase, thrombin, and plasmin (Carmeliet, 2003; Werb, 1997). Interestingly, most of these factors involved in remodeling the tumor-host interface originate from host cells, not from the growing carcinoma (Bowden *et al.*, 1999; Coussens *et al.*, 2000; Nakahara *et al.*, 1997; Sternlicht *et al.*, 1999).

Infiltrating inflammatory cells release MMP-9, MMP-12, and MMP-8 from intracellular stores, but they also release cytokines including IL-1β and tumor necrosis factor-α (TNF-α) that stimulate fibroblasts to produce more MMPs (Van Kempen *et al.*, 2003). In addition to fibroblasts and inflammatory cells, paracrine signals stimulate other stromal cells to be the predominant source of microenvironmental MMPs during tumorigenesis (Van Kempen *et al.*, 2003; Coussens and Werb, 2002).

Secreted MMPs degrade both ECM and other proteins in the microenvironment, including growth factors, cytokines, and receptors. Therefore, the effects of individual MMPs on the microenvironment are diverse, but they are critical players in establishing the environment surrounding tumor cells. One potent example is MMP-7, which is expressed by malignant breast epithelial cells. MMP-7 degrades the ECM, disrupts the basement membrane, and cleaves E-cadherin, weakening the connection between breast epithelial cells (Fingleton *et al.*, 2001; Noe *et al.*, 2001). MMPs are found in the microenvironment in zymogen form, and they require activation by other MMPs and related molecules. The reader is cautioned that MMP overexpression alone, which has often been reported in tumors, does not explain MMP activity in the microenvironment, since immunohistochemistry often does not distinguish between the zymogen and active forms (Coussens, Fingleton, and Matrisian, 2002). Under normal conditions MMPs are also regulated by endogenous tissue inhibitors of metalloproteinases (TIMPs), so the balance between TIMPs and MMPs critically affects the microenvironment (Mbeunkui and Johann, 2009).

To balance these remodeling and degradative enzymes, synthesis of matrix components is upregulated by both the tumor and host. Activated fibroblasts synthesize collagen type I and fibronectin in response to the binding of mast cell tryptase to protease-activated receptor-2 (Coussens and Werb, 2002; Frungieri *et al.*, 2002). Macrophages also contribute interleukin (IL)-1β and nitric oxide synthase, both of which augment type I collagen synthesis (Van Kempen *et al.*, 2003). The newly synthesized collagen that forms the peritumoral stroma is loosely woven and disorganized, contributing to the overall disorder surrounding a developing tumor (Ruiter *et al.*, 2002).

2.3.2 Immune cells in the microenvironment

The immune system makes an invaluable contribution to the tumor microenvironment. Inflammatory cells secrete growth factors, cytokines, and chemokines that stimulate epithelial proliferation and generate reactive oxygen species (ROS) that damage DNA, promoting tumor initiation and progression (Coussens and Werb, 2002). They also release proteolytic enzymes, leading to matrix remodeling and angiogenesis. The first cells to respond to tumor growth factors are mast cells, which are attracted by stem cell factor (SCF), vascular endothelial growth factor (VEGF), epidermal growth factor (EGF), basic fibroblast growth factor (bFGF), and platelet-derived growth factor (PDGF) (Hiromatsu and Toda, 2003). Upon arrival, mast cells degranulate to release VEGF (Norrby, 2002), serine proteases (Yong, 1997), and MMP-9 (Fang *et al.*, 1997).

Macrophages are typically the next inflammatory cells to infiltrate the growing microenvironment. Once activated by tumor-derived transforming growth factor beta-1 (TGF-β1), these tumor-associated macrophages (TAMs) secrete TNF-α, IL-1, IL-6, IL-8, and/or bFGF (Ono *et al.*, 1999; Varney *et al.*, 2002). TAMs are thought to play an integral role in signaling angiogenesis, which is necessary for continued tumor growth (Van Kempen *et al.*, 2003). Other inflammatory cells also make invaluable contributions that are beyond the scope of this chapter, but many excellent reviews on the subject are available.

2.4 Communication between the tumor cells and stroma

It is clear that many components go into making up a heterogeneous tumor that is capable of surviving, growing, and invading other tissues. However, this complex system requires organization. Communication between the tumor cells, host stroma, and tumor stroma is mediated by soluble autocrine and paracrine signaling molecules. The most common of these factors are members of the FGF, insulin-like growth factor (IGF), EGF, hepatocyte growth factor (HGF), TGF-β, and PDGF families (Ariztia *et al.*, 2006). The majority of these fibroblast-derived factors enhance proliferation and tumorigenesis (Bhowmick, Neilson, and Moses, 2004b). However, TGF-β exhibits different properties.

Secretion of TGF-β regulates cell proliferation and normal fibroblast transformation (Moses *et al.*, 1981; Roberts and Wakefield, 2003). Early studies showed that TGF-β inhibits the growth of epithelial cells, and subsequent transgenic mouse experiments reinforced the hypothesis that TGF-β is a tumor suppressor (Tucker *et al.*, 1984; Pierce *et al.*, 1995; Bottinger *et al.*, 1997; Gorska *et al.*, 2003). However, TGF-β is also known to both stimulate tumor progression by inducing epithelial-to-mesenchymal transition (EMT) in a variety of carcinomas and sarcomas and promote angiogenesis (Akhurst and Derynck, 2001). Therefore, TGF-β exhibits both tumor suppressive and tumor promoting behavior. At least some of the protumor effects of TGF-β appear to be mediated by HGF (Bhowmick *et al.*, 2004a). For a more comprehensive discussion on the role of growth factors in the tumor microenvironment, the reader is directed to an excellent review (Bhowmick, Neilson, and Moses, 2004b).

In addition to these growth factors, cytokines and chemokines play an essential signaling role in the microenvironmental milieu. Analysis of the gene-expression profiles of cells from normal breast and breast carcinomas showed that expression of the cytokines CXCL14 and CXCL12 by myoepithelial cells and myofibroblasts augments epithelial cell proliferation and invasion (Allinen *et al.*, 2004). CXCL12, in addition to promoting proliferation, enhances recruitment of endothelial progenitor cells (EPCs), thereby supporting angiogenesis (Orimo *et al.*, 2005). The proangiogenic effects of CXCL12 may be mediated by MMP-9, since the cytokine is known to activate MMP-9 in bone marrow cells and MMP-9 knockout mice are unable to respond to CXCL12-triggered EPC recruitment (Heissig *et al.*, 2002). CXCL12 also affects inflammatory cells in the tumor microenvironment, and these effects are likely synergistic with those of TGF-β (Littlepage, Egeblad, and Werb, 2005).

2.5 Hypoxia and angiogenesis

Among the most important components of the tumor microenvironment are the growing and developing blood vessels. Both rapidly dividing tumor cells and the influx of many microenvironmental cells increase the demand for oxygen greatly. As the oxygenation state of a tumor is determined by both oxygen supply and demand, tissue oxygenation will fall as a result of either diminished arteriolar supply or increased oxygen consumption. Interestingly, our group has shown that the amount of hypoxia in a tumor depends more heavily on oxygen consumption than either flow rate or arteriolar pO_2 (Secomb *et al.*, 1995). Thus, the growing and developing microenvironment creates conditions that require increases in oxygen supply to maintain normoxia and support continued growth. This is the physiological microenvironment that will be the focus of the remainder of this chapter.

Before delving into the details of tumor hypoxia and angiogenesis, it is important to highlight ways in which the physical and physiological microenvironments overlap and interact. Few reviews have addressed this topic. Table 2.1 summarizes what is known about the effects of hypoxia on the various components of the physical microenvironment discussed above. Carbonic anhydrase IX (CA IX) is an endogenous hypoxia marker expressed by a variety of cells. Carcinoma-associated fibroblasts are known to express CA IX, and expression by CAFs correlates with poor outcomes (Nakao *et al.*, 2009). Furthermore, hypoxia seems to affect collagen deposition, ECM degradation, and connections between the ECM and tumor cells in various types of tumors (Erkan *et al.*, 2009; Higgins *et al.*, 2007; Hull and Warfel, 1986; Leask and Abraham, 2006; Wu *et al.*, 2010). Various MMPs are affected by hypoxia, including MT1-MMP, MMP-2, MMP-7, MMP-9, and MMP-13 (Miyoshi *et al.*, 2006; Munoz-Najar *et al.*, 2006; Osinsky *et al.*, 2005; Swinson *et al.*, 2004). This will affect matrix remodeling, leading to changes in the physical environment surrounding and supporting the growing tumor.

Most of the research on interactions between the physical environment and hypoxia has focused on signaling. IL-8 is a chemokine with important functions in immune response. Expression of IL-8 is increased in hypoxia, and this is effect is inhibited by the free radical scavenger *N*-acetyl-L-cysteine (Galindo *et al.*, 2001; Wysoczynski *et al.*, 2010). This is particularly interesting because free radicals and ROS are found in high concentrations in hypoxic human tumors. In addition, the stromal-derived chemokine CXCL12 and its receptor CXCR4, discussed above for their roles in inflammation, proliferation, and migration, are both known to be upregulated in a variety of hypoxic tumor cells (Greijer *et al.*, 2008; Kim *et al.*, 2009; Komatani *et al.*, 2009; Liu *et al.*, 2006; Marchesi *et al.*, 2004; Nomura, Yoshida, and Teramoto, 2009; Pan *et al.*, 2006; Schioppa *et al.*, 2003; Zagzag *et al.*, 2006). Based on these observations, even though the physiologic and physical microenvironments are often studied in isolation, it is clear that there are important connections between these systems that require further investigation.

In their often-cited review on 'The Hallmarks of Cancer,' Hanahan and Weinberg highlighted six key features of tumors: including self-sufficiency in growth signals, insensitivity to antigrowth signals, tissue invasion and metastasis, limitless replicative potential, sustained angiogenesis, and evading apoptosis (Hanahan and

Table 2.1 Effects of hypoxia on components of the microenvironment.

Component	Observed effect of hypoxia	Reference
CAFs	Expression of CA IX by CAFs is a prognostic indicator	Nakao *et al.* (2009)
ECM	Abnormal collagen fibrils	Hull and Warfel (1986)
	MMP-9 increased in non small cell lung cancer	Swinson *et al.* (2004)
	Increased MMP-2 and MMP-9 in Lewis lung carcinoma	Osinsky *et al.* (2005)
	Changes in CCN protein expression, which affects cell surface connections to the ECM	Leask and Abraham (2006)
	MMP-7 and MT1-MMP upregulated in hepatoma cell lines	Miyoshi *et al.* (2006)
	In pancreatic cancer, hypoxia-fibrosis cycles are mediated by stellate cells	Erkan *et al.* (2009)
	Activates heparanase via NF-kappaB in pancreatic cancer cells	Wu *et al.* (2010)
Signaling	IGFBP-3, MMP-13, and fibroblast growth factor-3 all increased	Koong (2000)
	CXCR4 increased in human monocytes, HUVEC, ovarian cancer cells, breast cancer cells	Schioppa *et al.* (2003)
	CXCR4 increased in human pancreatic tumor cells	Marchesi *et al.* (2004)
	Increases in CXCR4/CXCL12 in glioblastoma multiforme	Zagzag *et al.* (2006), Komatani *et al.* (2009)
	CXCR4 increased in renal cell carcinoma cells, and expression correlates with metastatic ability	Pan *et al.* (2006)
	CXCR4 levels and CXCR4-mediated metastasis increase in non small cell lung cancer	Liu *et al.* (2006)
	Increased expression of MT-MMP and MMP-2 in breast tumors, increased invasion	Munoz-Najar *et al.* (2006)
	Increased HGF/c-Met signaling, involved in tumor–stromal interactions	Ide (2006)
	CXCR4 and CXCL12 increased in microvascular endothelial cells, CXCR4 increased in melanoma	Schutyser (2007)
	Increased GLUT1 and CXCL12 in colorectal carcinomas	Greijer *et al.* (2008)
	Increases CXCL12 and induces recruitment of bone marrow-derived CD45+ cells that secrete MMP-9	Du (2008)
	CXCR4 increased in myeloma cells	Kim *et al.* (2009)
	Bone morphogenetic protein 4 induced in hepatocellular carcinoma	Maegdefrau (2009)
	IL-8 increases independent of HIF-1α, seemingly through the MAPK pathway	Wysoczynski *et al.* (2010)

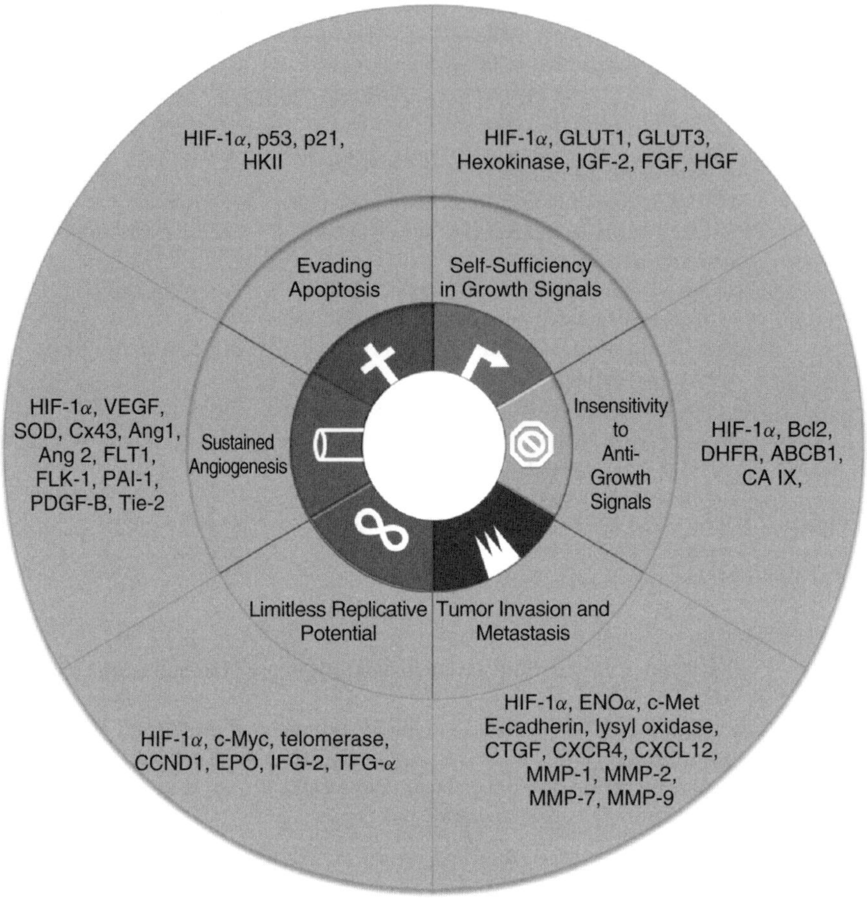

Figure 2.1 The effects of hypoxia on the six 'hallmarks of cancer' as described by Hanahan and Weinberg (2000). While some of these effects are clearly mediated by hypoxia-inducible factor-1α (HIF-1α), the mechanisms underlying many of these observed changes are unresolved. It is clear, however, that HIF-1α plays an integral role in the effects of hypoxia on all of these tumor characteristics. Furthermore, this figure highlights the areas of overlap between the physical and physiological microenvironments that require further investigation.

Weinberg, 2000). While there has been some research on the effects of hypoxia on these processes, our understanding is incomplete. Figure 2.1 summarizes what is currently known about the effects of hypoxia on each of these hallmarks of cancer.

2.5.1 Introduction to angiogenesis

In 1908, Goldman first hypothesized that angiogenesis is intimately related to tumor growth, and Folkman subsequently popularized the notion that angiogenesis is necessary for tumor growth, raising the possibility of modifying this process for therapeutic gain (Goldmann, 1908; Folkman, 1971). We now understand that if tumors did not form new vessels, they could not grow beyond a very small size and would be unable to metastasize. In fact, if angiogenesis is inhibited at the beginning of tumor formation using a truncated form of the soluble VEGF receptor, tumor growth is inhibited (Li *et al.*, 2000). Thus, it appears that angiogenesis is essential to the developing microenvironment.

In the mid-nineteenth century, Virchow was the first to recognize that tumor vessels have an abnormal architecture (David, 1988). Tumor vascular networks are collections of heterogeneous vessels (Decking, 2002; Pries, Secomb, and Gaehtgens, 1995a, 1995b; Pries and Secomb, 2009). There are long and short flow pathways, and the number of component vessels is highly varied. The topological heterogeneity is somewhat tempered by maturation of the neovasculature and pruning of unnecessary and non-functional vessels. Nonetheless, what remains is a disorganized network. This abnormal structure of individual vessels and larger networks leads to poor function, causing regions of inadequate oxygen and nutrient delivery.

Hypoxia refers to a below-normal oxygenation state in a tissue. This deficiency triggers subsequent angiogenesis, which is known to be abnormal in tumors. Thus, hypoxia and angiogenesis are tightly interrelated components of the microenvironment. As such, we will consider them together and discuss the interactions between these processes.

2.5.2 The importance of hypoxia-inducible factor-1

The oxygen tension in normal tissues generally exceeds 20 mmHg (Braun *et al.*, 2001). When the pO_2 dips below 10 mmHg, the hypoxic tissue begins to produce specific proteins mediated by hypoxia-inducible factor-1 (HIF-1), which has been referred to as the master regulator of oxygen homeostasis' (Semenza, 2003). In addition to oxygen tension, free radicals, especially oxygen containing radicals such as superoxide (O_2^-), modulate HIF-1 levels (Dewhirst, Cao, and Moeller, 2008). HIF-1 is a heterodimeric transcription factor that increases the expression of genes involved in angiogenesis, adaptation to hypoxia, invasiveness, and resistance to oxidative stress (Semenza and Wang, 1992; Wang and Semenza, 1993; Semenza, 2003).

Accumulation of HIF-1 in a tumor mediates and integrates four major aspects of tumor biology: mutations and metabolism in addition to hypoxia and angiogenesis (Semenza, 2008). The most notable and well studied of the effects of HIF-1 are the transcriptional activation of genes involved in angiogenesis, most notably VEGF. VEGF is particularly important in the context of this chapter because it is a key regulator of vascular adaptation and angiogenesis. The HIF-1 complex binds to the hypoxia-responsive element upstream of the VEGF gene, directly activating VEGF transcription (Forsythe *et al.*, 1996). Other HIF-1 activated genes involved in angiogenesis include angiopoietin 1 and 2, placental growth factor, and PDGF B (Pugh and Ratcliffe, 2003).

2.5.3 Causes of hypoxia

At its core, hypoxia results from an imbalance between oxygen delivery and consumption. Our laboratory recently described the eight key features of hypoxia relevant to the developing and evolving tumor microenvironment (Dewhirst, Cao, and Moeller, 2008). These features are all depicted in Figure 2.2, an image of regions of differential hemoglobin saturation in a mouse mammary tumor. First, the tumor receives an inadequate amount of oxygenated blood due to a limited arteriolar supply, which causes small tumor vessels far from the originating arterioles to contain very low levels of oxygen (Dewhirst *et al.*, 1992, 1999; Sorg *et al.*, 2008). Second, tumor vessels are oriented in such a way that some regions are well perfused (or even overly perfused) and others do not have enough vascularity (Secomb *et al.*, 1993). Also related to vessel distribution and orientation, the third feature is that the center of a tumor tends to have fewer vessels than the tumor periphery.

Some tumor microvessels contain many red blood cells, whereas others contain few or none at all (Dewhirst *et al.*, 1996). Thus, the fourth feature is that tumors show wide variations in red blood cell flux, defined as the number of red blood cells that traverse a particular microvessel per unit time. All of these cause an imbalance between oxygen supply and demand, which is the fifth aspect of tumor hypoxia (Secomb *et al.*, 1995).

The remaining features of tumor hypoxia relate to blood flow patterns within the tumor. For example, hypoxic red blood cells shrink and stiffen, increasing blood viscosity (Kavanagh *et al.*, 1993). The resulting decrease in flow velocity alters the distribution of erythrocytes at vascular bifurcations. In addition, vascular shunts have been observed that redirect arteriolar blood into draining veins, bypassing parts of the tumor (Eddy and Casarett, 1973). The last feature of tumor hypoxia is that it is temporally unstable. Changes in microvessel red blood cell flux cause intermittent periods of hypoxia, a phenomenon that has come to be known as cycling hypoxia because the variations exhibit periodicity in a variety of tumors (Dewhirst, Cao, and Moeller, 2008; Dewhirst, 2007).

2.5.3.1 Cycling hypoxia

Chaplin and Durand first reported that tumor oxygenation varies with the frequency of a few cycles per hour, and they later showed that this happens in large regions of the tumor over prolonged periods of time (Durand and Aquino-Parsons, 2001; Trotter *et al.*, 1989; Bennewith and Durand, 2004; Bennewith, Raleigh, and Durand, 2002). Based on these studies, it is now understood that up to one-fifth of tumor cells may experience cycling hypoxia, and these cells are not typically adjacent to tumor microvessels. Our laboratory has used phosphorescence lifetime imaging in experimental tumors grown in dorsal skinfold window chambers to show that there are significant inter- and intratumoral differences in the spatial relationships between these fluctuations (Cardenas-Navia *et al.*, 2008). Studies using blood oxygen level dependent functional MRI, an imaging technique that differentiates between

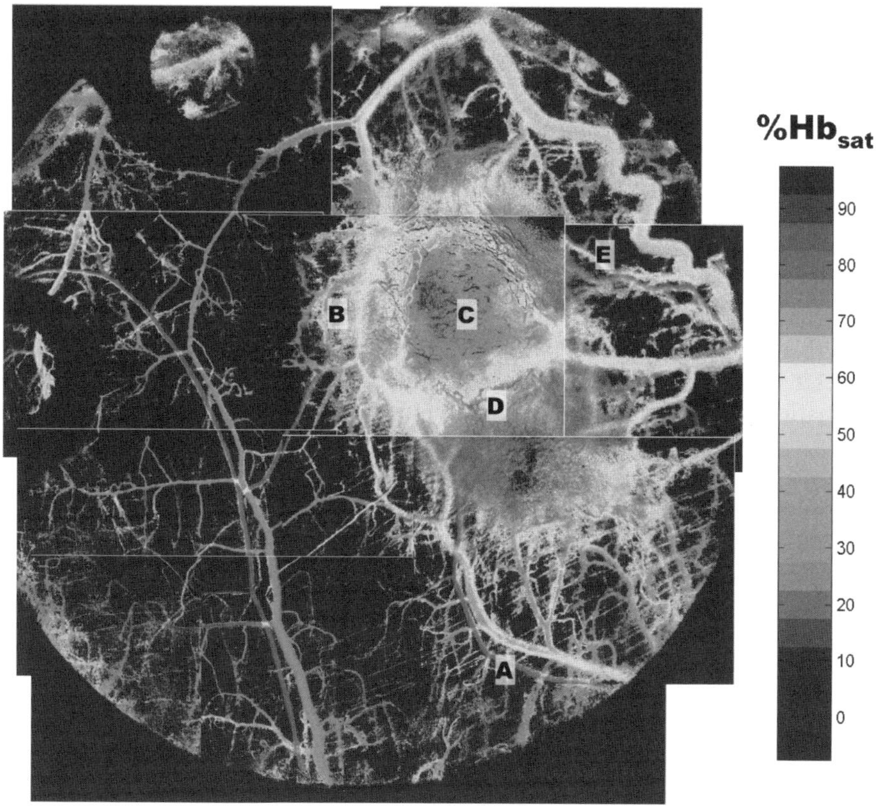

Figure 2.2 This mouse mammary tumor grown in a dorsal skin-fold window chamber was analyzed for regions of different hemoglobin saturations. This image shows many of the key features of tumor hypoxia. Areas of low hemoglobin saturation depict hypoxia and areas of high hemoglobin saturation represent normoxic regions. (A) Low arteriolar supply. This is one of the very few arterioles feeding this tumor. (B) Area of lower oxygenation because this area is poorly vascularized. (C) This region shows a steep longitudinal oxygen gradient, beginning with a moderately oxygenated arteriole leading to a hypoxic region, because the vessel does not contain enough oxygen to deliver to the tissue. This ends in a region of intravascular hypoxia. Furthermore, region C shows that the center of the tumor has fewer vessels than the periphery. (D). By comparing region C with region D, it is clear that there are mismatches in oxygen supply and demand. Some areas receive enough oxygen to remain normoxic, while others experience hypoxia because of inadequate oxygen supply. (E). Vascular shunt that redirects oxygenated arteriolar blood into the draining veins, bypassing parts of the tumor. The remaining features of tumor hypoxia, variations in flux and temporal instability, require serial imaging and therefore cannot be visualized in this figure. (This figure was reproduced with permission from Hardee, M.E., Dewhirst, M.W., Agarwal, N. and Sorg, B.S. (2009) Novel imaging provides new insights into mechanisms of oxygen transport in tumors. *Curr Mol Med*, **9**, 435–441.) A full color version of this figure can be found in the color plate section.

oxygenated and deoxygenated hemoglobin, have suggested that immature vessels may be more likely to experience cycling hypoxia (Baudelet *et al.*, 2004, 2006).

Fourier transform analysis, a mathematical method used to identify frequencies in complex data sets, has shown dominant fluctuations in tumor oxygenation of less than two cycles per minute in a variety of preclinical tumor models (Sorg *et al.*, 2008; Brurberg *et al.*, 2004, 2005; Brurberg, Graff, and Rofstad, 2003; Cardenas-Navia *et al.*, 2004). Measurements of flow using laser Doppler flowmetry and fiberoptic oxygen sensors have reported similar results in human tumors, but Fourier analysis has not been used in the clinical setting (Pigott *et al.*, 1996; Brurberg *et al.*, 2005). Other studies have demonstrated slower variations in oxygen tension within the tumor, on the magnitude of hours measured over a 24, 36, or 96-hour period (Brown, 1979; Bennewith and Durand, 2004; Skala *et al.*, 2009; Skala *et al.*, in press). There is no consensus on the kinetics of cycling hypoxia, and further investigation is necessary to understand these periodic fluctuations in oxygenation.

2.5.3.1.1 *Mechanisms of cycling hypoxia*

Based on experiments correlating microvessel red blood cell flux and interstitial pO_2, it appears that cycling hypoxia results from changes in the flow of red blood cells through tumor microvessels (Kimura *et al.*, 1996; Lanzen *et al.*, 2006). As a natural extension of this observation, the question arises as to what causes these variations in erythrocyte flux. One possible source, though it likely does not explain all of the variations in red blood cell flow, is a change in the vasomotor activity of the tumor's arteriolar supply (Dewhirst *et al.*, 1996; Baudelet *et al.*, 2006). A more likely explanation comes from understanding vascular remodeling. Changes in flow greatly affect the hemodynamics of a vascular network through processes including vascular pruning, formation of new vascular connections, and intussusception (Kiani *et al.*, 1994; Patan, Munn, and Jain, 1996; Zakrzewicz, Secomb, and Pries, 2002). With respect to cycling hypoxia, intussusception is of particular interest. This process involves splitting of microvessels into smaller parallel vessels over a period of minutes (Patan, Munn, and Jain, 1996). Since flow resistance is inversely proportional to vessel radius to the fourth power, small changes in microvessel size lead to large redistributions in flow (Chien, Usami, and Skalak, 1984; Kiani *et al.*, 1994; Pries *et al.*, 1997). Thus, vascular remodeling resulting in changes in flow resistance likely account for the majority of erythrocyte flux causing cycling hypoxia.

2.5.3.1.2 *Interactions between chronic and cycling hypoxia*

The overall oxygenation state of a tumor region has profound effects on whether that region will experience cycling hypoxia, so chronic hypoxia and cycling hypoxia are inextricably linked phenomena. For example, poorly perfused areas that are far from microvessels are unlikely to be affected by changes in red cell flux, and the pO_2 in this region is unlikely to change much because it is already quite low. Alternatively, regions that are perfused by an abundance of vessels are equally unlikely to experience large changes in oxygen tension due to changes in flow through only one of the feeding microvessels. Our laboratory has likened this

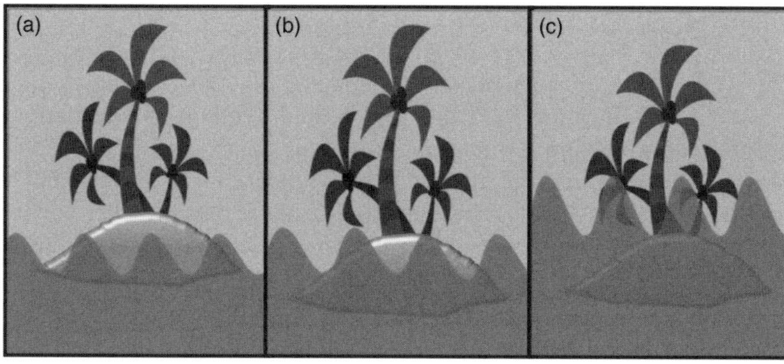

Figure 2.3 The height of tides and waves on an island can be used as an analogy to understand cycling hypoxia. Tides refer to the overall oxygenation state of a tumor, and waves depict temporal variations in red blood cell flux. The exposed island is a representation of the amount of hypoxia, which is also depicted by the background color, with red indicating well-oxygenated areas and blue indicating hypoxia. (a) A region of chronic hypoxia. In this region, the low overall oxygenation state cannot be overcome by the height of individual waves. In (b), individual waves determine whether the island is oxygenated at a particular time. Therefore, this panel represents a region of cycling hypoxia. (c) an area that is so well-oxygenated that individual waves do not cause hypoxia. Therefore, this panel depicts a tumor region that is normoxic.

situation to the effects of tides and waves on an island in the ocean, as depicted in Figure 2.3 (Dewhirst, 2009). When tides are high, the island may be completely awash and individual waves do not affect the amount of beach that is exposed. However, at low tide, when the water no longer covers the entire island, the height of individual waves determines how much of the beach is exposed at a given time. We liken the tides to overall oxygenation state, waves to episodes of cycling hypoxia, and the amount of beach exposed to the level of hypoxia at a particular time. If the overall oxygenation of a tumor region is high (high tide), the individual waves do not cause significant hypoxia. However, if the overall oxygenation is lower, individual fluctuations (waves) cause large changes in tissue oxygenation.

While this analogy provides a framework for understanding these phenomena, it clearly does not explain the whole story. The temporal variations between different reports of cycling hypoxia make the two processes difficult to differentiate. For example, while variations on the order of seconds or minutes seem clearly to be cycling hypoxia, reports of variations on the order of hours to days blur the distinction (Skala *et al.*, 2009; Skala *et al.*, in press).

It is well known that chronic hypoxia has important physiological consequences for tumors that lead to treatment resistance, but cycling hypoxia appears to have effects that are distinct from those of chronic hypoxia. Richard Hill's group has shown that metastatic frequency is dependent on the degree of tumor hypoxia and that cycling hypoxia increases metastatic frequency over that of chronic hypoxia (Cairns and Hill, 2004; Cairns, Kalliomaki, and Hill, 2001; De Jaeger, Kavanagh, and Hill, 2001; Rofstad *et al.*, 2007). Tumor-bearing mice subjected to cycling hypoxia also showed increased expression of genes associated with metastasis, including CXCR4, uPAR, VEGF, and osteopontin (Chaudary and Hill, 2009).

The effects of cycling hypoxia are not limited to metastasis. Cycling hypoxia appears to influence HIF-1α protein levels and transcriptional activity more than chronic hypoxia (Martinive *et al.*, 2006; Ning *et al.*, 2007; Peng *et al.*, 2006). Cycling hypoxia may increase ROS due to repeated hypoxia-reoxygenation cycles, and a recent study involving a transgenic model of breast cancer showed evidence of significant oxidative damage to DNA and lipids caused by cycling hypoxia (Kalliomaki *et al.*, 2008, 2009). Furthermore, it is clear that the microenvironment greatly affects mammalian target of rapamycin (mTOR) activity, and cycling hypoxia seems to have opposite effects from chronic hypoxia on mTOR function and interaction with HIF-1 (Dunlop and Tee, 2009; Hudson *et al.*, 2002; Yuan *et al.*, 2008). This is significant because mTOR is an integral part of the mTOR complex that is involved in modifying response to changes in nutritional and energy status and oxidative stress. Hypoxia and oxidative stress both induce the unfolded protein response (UPR), which alters protein expression, metabolism, and cell death in response to stress (Wouters and Koritzinsky, 2008). It seems likely that cycling hypoxia will affect the UPR, since genes controlled by HIF-1 are often contained in the stress granules formed by the UPR and cycling hypoxia increases oxidative stress (Moeller *et al.*, 2004). However, further investigation into these changes is needed to better understand the pathophysiological responses to cycling hypoxia.

In accordance with the clear effects on the environment of the tumor, hypoxia, and cycling hypoxia have important implications for tumor treatment (Brown and Wilson, 2004). In the setting of radiotherapy, oxygen helps to stabilize treatment-induced DNA damage such that the DNA cannot be effectively repaired. Therefore, chronic hypoxia decreases the cytotoxic effects of radiation. Furthermore, the delivery and activity of chemotherapeutic agents is often decreased under hypoxic conditions. In addition to the chemo- and radioresistance conferred on tumor cells by chronic hypoxia, cycling hypoxia is known to cause resistance to radiation therapy (Brown, 1979; Yamaura and Matsuzawa, 1979). Therefore, these microenvironmental effects are critical to understand in order to design better targeted therapeutic strategies.

2.5.4 Interactions between hypoxia and angiogenesis

The term "angiogenic switch" refers to the balance of pro- and antiangiogenic factors, leading to the initiation of angiogenesis. Low tissue oxygenation is generally considered to be the predominant weight that tips the scales in favor of angiogenesis, since hypoxia enhances HIF-1 protein levels and activity, directly upregulating VEGF (Diaz-Gonzalez *et al.*, 2005; Laderoute *et al.*, 2000; Mazure *et al.*, 1996). Other microenvironmental components, such as oncogenes and growth factors, can act via the PI3K-AKT pathway to increase expression of HIF-1 to the point that it overcomes oxygen-mediated degradation (Feldkamp *et al.*, 1999; Jiang and Liu, 2008).

For primary tumors to grow beyond a few millimeters in diameter, angiogenesis is generally required. In early malignant breast tumors, HIF-1 expression correlates with VEGF levels and angiogenesis (Bos *et al.*, 2001). This is consistent with a model of HIF-1 causing VEGF production and angiogenesis, but it does not answer the question of whether hypoxia-mediated stabilization of HIF-1 is the principal underlying cause of neovascularization. Using a rat glioma model, Holash *et al.* showed that vessel cooption occurs in the first week after tumor transplantation, followed by evidence of vessel regression in week 2 and angiogenesis by week 4 (Holash *et al.*, 1999). In this experiment, increased expression of angiopoietin 2 was hypothesized to cause vessel regression, leading to the 'hypoxic crisis' that triggered the onset of angiogenesis.

In our laboratory, using fluorescent labels for blood vessels and hypoxic regions in a dorsal skin fold window chamber tumor model, we observed that angiogenesis precedes hypoxia (Cao *et al.*, 2005). There was no evidence of vascular stasis before angiogenesis, though angiogenesis was enhanced by HIF-1α upregulation. This led us to put forth the 'acceleration model,' suggesting that HIF-1α is not necessary for angiogenic initiation but instead accelerates neovascularization. This would mimic the role of angiogenesis in wound healing (Haroon *et al.*, 2000). However, more evidence is needed to better understand the temporal and functional relationships between vessel cooption and angiogenesis.

It is not only the onset of angiogenesis that is of interest. Ongoing angiogenesis affects delivery of oxygen, nutrients, and therapeutic agents to the tumor. Using a fluorescent reporter for HIF-1 expression, it is clear that as tumors grow, some well-oxygenated microvessels are found in regions with high expression of HIF-1 (Sorg *et al.*, 2005). It is possible that ROS and/or NO may stabilize HIF-1 in these regions, leading to VEGF expression in the absence of hypoxia and causing vascular remodeling and angiogenesis.

2.5.5 Communication in the vasculature

The development and remodeling of vessels is a complex process that requires coordination not only between the tumor and the vessels but also among and within the vessels themselves. A variety of microenvironmental signaling molecules, including NO, modulate vascular tone, and sustained changes in vascular tone affect the structure of tumor vessels (Cosby *et al.*, 2003; Bakker *et al.*, 2002; Martinez-Lemus, Hill, and Meininger, 2009). We are beginning to understand the mechanisms by which these vessels communicate to enact these structural modifications.

Communication is essential not only between vessels, but also within an individual vessel. Gap junctions are plasma membrane channels that connect cells, and they play a key role in intercellular communication in vessel walls. These channels enable small molecules (less than 100 Da), including ions, amino acids, and second messengers, to pass between adjacent cells. Transmembrane proteins known as connexins (Cxs) are essential components of these junctions. Radially aligned connexins form hexameric complexes known as connexons. Channels are made of two connexons, one from each cell. Connexin proteins are composed of four hydrophobic transmembrane domains, two extra-cellular domains involved in connecting to an adjacent connexon, and three cytoplasmic domains (Willecke *et al.*, 2002).

Four connexin isoforms have been identified in the vascular wall: Cx37, Cx40, Cx43, and Cx45 (Brisset, Isakson, and Kwak, 2009). These isoforms form connections between endothelial cells or between smooth muscle cells, and to a lesser extent, allow communication between endothelial and smooth muscle cells. While knockout experiments indicate that both Cx40 and Cx43 are involved in vascular tone, Cx43 is the isoform most likely involved in the myoendothelial junction (Isakson *et al.*, 2006). Therefore, most research on vascular connexins has focussed on Cx43.

The permeability of a gap junction is determined by the connexin isoform it is composed of, and channel opening and closing is controlled by environmental signals (Lampe and Lau, 2004). The short half-life of Cx43 in cultured cells and tissues, ranging from 1 to 3 hours, suggests tight post-translational regulation. Serine phosphorylation is the predominant modification to Cx43, and these phosphorylation events appear to correlate with changes in the formation and degradation of gap junctions (Musil *et al.*, 1990; Cooper *et al.*, 2000; Lampe and Lau, 2004). Additionally, phosphorylation of tyrosine residues on Cx43 is correlated with significant decreases in intercellular communication (Filson *et al.*, 1990). Phosphorylation of Y247 by $pp60^{v\text{-}src}$ leads to closure of the gap junction (Lin *et al.*, 2001). It seems likely that phosphorylation of Y265 stabilizes the interaction between Cx43 and v-Src, thereby enabling the phosphorylation of Y247 that leads to closure of the channel (Lampe and Lau, 2004). Therefore, phosphorylation of Cx43 could greatly affect communication between endothelial cells.

A variety of microenvironmental factors cause differential phosphorylation of Cx43. Binding of the growth factors EGF and PDGF to their receptors leads to serine phosphorylation (Lampe and Lau, 2004). Treatment of endothelial cells with VEGF is associated with Cx43 phosphorylation and reduction in endothelial cell–cell communication as mediated by both the c-Src tyrosine kinase and MAP kinase pathways (Suarez and Ballmer-Hofer, 2001).

2.5.5.1.1 *The role of connexin-43 in hypoxia and angiogenesis*

While it is known that there is a considerable amount of heterogeneity in tumor microvascular networks, the mechanisms underlying the structural and functional abnormalities are poorly understood. Recent computer modeling of tumor vessels revealed that reduced communication along vessels is a leading cause of aberrant tumor vessel structure and function (Pries *et al.*, 2009). Similar models revealed that information must be transferred along the length of vessels in order for functional networks to form, and impaired communication leads to shunting of blood flow, one of the eight causes of tumor hypoxia addressed above (Pries, Reglin, and Secomb, 2001; Pries, Secomb, and Gaehtgens, 1998). Furthermore, red blood cell flux is disturbed by a lack of cell–cell communication, which could explain the changes in erythrocyte flow that result in cycling hypoxia.

While normal vessels coordinate blood flow through endothelial cell and smooth muscle cell gap junctions, little is known about the expression or function of connexins in tumor vasculature (Segal, 1994). Tumor vessels lack the regular organization of endothelial cells, basement membrane, and smooth muscle (Carmeliet and Jain, 2000). They are often composed of just endothelial cells or even of tumor cells themselves (Di Tomaso *et al.*, 2005). Since tumor vessel composition is abnormal, it

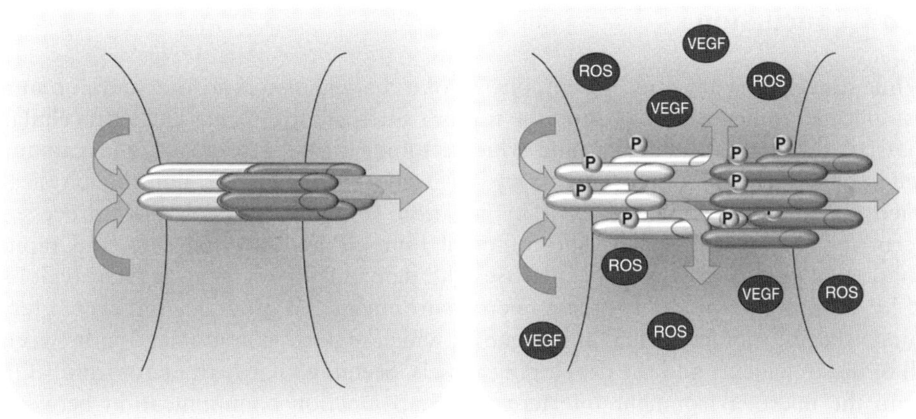

Figure 2.4 Gap junction connecting two endothelial cells in a tumor vessel. The junction is composed of a hexameric complex of connexin 43 (Cx43), which forms a channel to allow the passage of small signaling molecules between adjacent cells. The panel on the left shows a well-oxygenated region in which the Cx43 is not phosphorylated. This leads to a functional junction and enables intercellular communication. The blue panel on the right depicts a hypoxic region containing reactive oxygen species (ROS) and vascular endothelial growth factor (VEGF). These microenvironmental factors lead to phosphorylation of Cx43, disrupting the gap junction. As a result, the passage of materials between adjacent tumor endothelial cells is ineffective.

follows that communication between these cells is likely disrupted. Figure 2.4 is a schematic diagram of an abnormal gap junction in tumor vasculature.

Of the vascular connexins, Cx43 has been implicated in dysfunctional tumor vasculature, because its expression and function are altered by the microenvironment (Errede *et al.*, 2002). Some groups have reported that decreases in Cx43 expression in breast tumor models are associated with a more aggressive tumor phenotype (Shao *et al.*, 2005; McLachlan *et al.*, 2006; Pollmann *et al.*, 2005). While some studies have shown downregulation of Cx43 in tumors, another reports upregulation of the phosphorylated form of the protein (Gould *et al.*, 2005; Laird *et al.*, 1999; Van Beijnum *et al.*, 2006). Phosphorylation of Cx43 in breast carcinomas is associated with decreased gap junction formation (Gould *et al.*, 2005).

As VEGF is known to cause Cx43 phosphorylation and a transient disruption of endothelial cell communication, tumor angiogenesis driven by VEGF is likely to be plagued by miscommunication between contributing endothelial cells. Furthermore, ROS, which are known to be present in the tumor microenvironment, are known to cause hyperphosphorylation of Cx43 in liver cells (Upham *et al.*, 1997). Since hypoxia is known to be associated with increased oxidative stress and VEGF expression, and both ROS and VEGF are correlated with increased phosphorylation of Cx43, it follows that areas with dysfunctional vasculature will likely have hyperphosphorylated forms of Cx43. Further investigation is clearly warranted, but it seems likely that Cx43 phosphorylation decreases communication in developing and remodeling tumor vessels, leading to diminished functionality in delivering adequate amounts of oxygen, nutrients, and chemotherapeutic agents.

2.6 Conclusion

While cancer cells are critical components of a malignant tumor, they develop in a complicated milieu that makes up the tumor's microenvironment. The extracellular matrix provides a structure around which a tumor grows. Fibroblasts and immune cells contribute key signals necessary for tumor propagation. But the physical microenvironment does not exist in isolation from the physiological processes occurring within this space. The influx of immune and stromal cells and rapid division of tumor cells greatly increases the demand for oxygen, causing regions of hypoxia in the rapidly evolving microenvironment. To grow beyond even a few hundred cells, tumors initiate angiogenesis, which requires communication between all of these elements and the developing vessels. Seemingly, growth factors and ROS from the microenvironment interfere with gap junction communication between endothelial cells in the neovasculature, which likely induces more hypoxia and aberrant angiogenesis. Only by understanding tumors in the context of the complex interactions between these components can we hope to find effective ways to kill tumors and treat human malignancies.

Acknowledgements

This work was supported by a grant from the NIH/NCI CA40355-25. The authors would like to thank Dr Tim Secomb and Dr Axel Pries for their helpful discussions on gap junctions. We would also like to thank Caroline Hadley for her assistance in reviewing and editing this chapter.

References

Akhurst, R.J. and Derynck, R. (2001) TGF-beta signaling in cancer – a double-edged sword. *Trends in Cell Biology*, **11**, S44–S51.

Allinen, M., Beroukhim, R., Cai, L. *et al.* (2004) Molecular characterization of the tumor microenvironment in breast cancer. *Cancer Cell*, **6**, 17–32.

Ariztia, E.V., Lee, C.J., Gogoi, R., and Fishman, D.A. (2006) The tumor microenvironment: key to early detection. *Critical Reviews in Clinical Laboratory Sciences*, **43**, 393–425.

Bakker, E.N., Van Der Meulen, E.T., Van Den Berg, B.M. *et al.* (2002) Inward remodeling follows chronic vasoconstriction in isolated resistance arteries. *Journal of Vascular Research*, **39**, 12–20.

Baudelet, C., Ansiaux, R., Jordan, B.F. *et al.* (2004) Physiological noise in murine solid tumours using T2*-weighted gradient-echo imaging: a marker of tumour acute hypoxia? *Physics in Medicine and Biology*, **49**, 3389–3411.

Baudelet, C., Cron, G.O., Ansiaux, R. *et al.* (2006) The role of vessel maturation and vessel functionality in spontaneous fluctuations of T2*-weighted GRE signal within tumors. *NMR in Biomedicine*, **19**, 69–76.

Bennewith, K.L. and Durand, R.E. (2004) Quantifying transient hypoxia in human tumor xenografts by flow cytometry. *Cancer Research*, **64**, 6183–6189.

Bennewith, K.L., Raleigh, J.A., and Durand, R.E. (2002) Orally administered pimonidazole to label hypoxic tumor cells. *Cancer Research*, **62**, 6827–6830.

Bhowmick, N.A., Chytil, A., Plieth, D. *et al.* (2004a) TGF-beta signaling in fibroblasts modulates the oncogenic potential of adjacent epithelia. *Science*, **303**, 848–851.

Bhowmick, N.A., Neilson, E.G., and Moses, H.L. (2004b) Stromal fibroblasts in cancer initiation and progression. *Nature*, **432**, 332–337.

Bos, R., Zhong, H., Hanrahan, C.F. *et al.* (2001) Levels of hypoxia-inducible factor-1 alpha during breast carcinogenesis. *Journal of National Cancer Institute*, **93**, 309–314.

Bottinger, E.P., Jakubczak, J.L., Haines, D.C. *et al.* (1997) Transgenic mice overexpressing a dominant-negative mutant type II transforming growth factor beta receptor show enhanced tumorigenesis in the mammary gland and lung in response to the carcinogen 7,12-dimethylbenz-[a]-anthracene. *Cancer Research*, **57**, 5564–5570.

Bowden, E.T., Barth, M., Thomas, D. *et al.* (1999) An invasion-related complex of cortactin, paxillin and PKCmu associates with invadopodia at sites of extracellular matrix degradation. *Oncogene*, **18**, 4440–4449.

Braun, R.D., Lanzen, J.L., Snyder, S.A., and Dewhirst, M.W. (2001) Comparison of tumor and normal tissue oxygen tension measurements using OxyLite or microelectrodes in rodents. *American Journal of Physiology: Heart and Circulatory Physiology*, **280**, H2533–H2544.

Brisset, A.C., Isakson, B.E., and Kwak, B.R. (2009) Connexins in vascular physiology and pathology. *Antioxidants and Redox Signaling*, **11**, 267–282.

Brown, J.M. (1979) Evidence for acutely hypoxic cells in mouse tumours, and a possible mechanism of reoxygenation. *British Journal of Radiology*, **52**, 650–656.

Brown, J.M. and Wilson, W.R. (2004) Exploiting tumour hypoxia in cancer treatment. *Nature Reviews Cancer*, **4**, 437–447.

Brurberg, K.G., Graff, B.A., Olsen, D.R., and Rofstad, E.K. (2004) Tumor-line specific pO(2) fluctuations in human melanoma xenografts. *International Journal of Radiation Oncology, Biology, Physics*, **58**, 403–409.

Brurberg, K.G., Graff, B.A., and Rofstad, E.K. (2003) Temporal heterogeneity in oxygen tension in human melanoma xenografts. *British Journal of Cancer*, **89**, 350–356.

Brurberg, K.G., Skogmo, H.K., Graff, B.A. *et al.* (2005) Fluctuations in pO2 in poorly and well-oxygenated spontaneous canine tumors before and during fractionated radiation therapy. *Radiotherapy and Oncology*, **77**, 220–226.

Cairns, R.A. and Hill, R.P. (2004) Acute hypoxia enhances spontaneous lymph node metastasis in an orthotopic murine model of human cervical carcinoma. *Cancer Research*, **64**, 2054–2061.

Cairns, R.A., Kalliomaki, T., and Hill, R.P. (2001) Acute (cyclic) hypoxia enhances spontaneous metastasis of KHT murine tumors. *Cancer Research*, **61**, 8903–8908.

Cao, Y., Li, C.Y., Moeller, B.J. *et al.* (2005) Observation of incipient tumor angiogenesis that is independent of hypoxia and hypoxia inducible factor-1 activation. *Cancer Research*, **65**, 5498–5505.

Cardenas-Navia, L.I., Mace, D., Richardson, R.A. *et al.* (2008) The pervasive presence of fluctuating oxygenation in tumors. *Cancer Research*, **68**, 5812–5819.

Cardenas-Navia, L.I., Yu, D., Braun, R.D. *et al.* (2004) Tumor-dependent kinetics of partial pressure of oxygen fluctuations during air and oxygen breathing. *Cancer Research*, **64**, 6010–6017.

Carmeliet, P. (2003) Angiogenesis in health and disease. *Nature Medicine*, **9**, 653–660.

Carmeliet, P. and Jain, R.K. (2000) Angiogenesis in cancer and other diseases. *Nature*, **407**, 249–257.

Chaudary, N. and Hill, R.P. (2009) Increased expression of metastasis-related genes in hypoxic cells sorted from cervical and lymph nodal xenograft tumors. *Laboratory Investigation*, **89**, 587–596.

Chien, S., Usami, S., and Skalak, R. (1984) in *Handbook of Physiology* (eds E. Renkin, C. Michel, and S. Geiger), American Physiological Society, Bethesda.
Cooper, C.D., Solan, J.L., Dolejsi, M.K., and Lampe, P.D. (2000) Analysis of connexin phosphorylation sites. *Methods*, **20**, 196–204.
Cosby, K., Partovi, K.S., Crawford, J.H. *et al.* (2003) Nitrite reduction to nitric oxide by deoxyhemoglobin vasodilates the human circulation. *Nature Medicine*, **9**, 1498–1505.
Coussens, L.M., Fingleton, B., and Matrisian, L.M. (2002) Matrix metalloproteinase inhibitors and cancer: trials and tribulations. *Science*, **295**, 2387–2392.
Coussens, L.M., Tinkle, C.L., Hanahan, D., and Werb, Z. (2000) MMP-9 supplied by bone marrow-derived cells contributes to skin carcinogenesis. *Cell*, **103**, 481–490.
Coussens, L.M. and Werb, Z. (2002) Inflammation and cancer. *Nature*, **420**, 860–867.
David, H. (1988) Rudolf Virchow and modern aspects of tumor pathology. *Pathology, Research and Practice*, **183**, 356–364.
Decking, U.K. (2002) Spatial heterogeneity in the heart: recent insights and open questions. *News in Physiological Sciences*, **17**, 246–250.
De Jaeger, K., Kavanagh, M.C., and Hill, R.P. (2001) Relationship of hypoxia to metastatic ability in rodent tumours. *British Journal of Cancer*, **84**, 1280–1285.
Dewhirst, M.W. (2007) Intermittent hypoxia furthers the rationale for hypoxia-inducible factor-1 targeting. *Cancer Research*, **67**, 854–855.
Dewhirst, M.W. (2009) Relationships between cycling hypoxia, HIF-1, angiogenesis and oxidative stress. *Radiation Research*, **172**, 653–665.
Dewhirst, M.W., Cao, Y., and Moeller, B. (2008) Cycling hypoxia and free radicals regulate angiogenesis and radiotherapy response. *Nature Reviews Cancer*, **8**, 425–437.
Dewhirst, M.W., Kimura, H., Rehmus, S.W. *et al.* (1996) Microvascular studies on the origins of perfusion-limited hypoxia. *British Journal of Cancer Supplement*, **27**, S247–S251.
Dewhirst, M.W., Ong, E.T., Braun, R.D. *et al.* (1999) Quantification of longitudinal tissue pO2 gradients in window chamber tumours: impact on tumour hypoxia. *British Journal of Cancer*, **79**, 1717–1722.
Dewhirst, M.W., Ong, E.T., Klitzman, B. *et al.* (1992) Perivascular oxygen tensions in a transplantable mammary tumor growing in a dorsal flap window chamber. *Radiation Research*, **130**, 171–182.
Diaz-Gonzalez, J.A., Russell, J., Rouzaut, A. *et al.* (2005) Targeting hypoxia and angiogenesis through HIF-1alpha inhibition. *Cancer Biology and Therapy*, **4**, 1055–1062.
Di Tomaso, E., Capen, D., Haskell, A. *et al.* (2005) Mosaic tumor vessels: cellular basis and ultrastructure of focal regions lacking endothelial cell markers. *Cancer Research*, **65**, 5740–5749.
Du, R., Lu, K.V., Petritsch, C., *et al.* (2008) HIF1alpha induces the recruitment of bone marrow-derived vascular modulatory cells to regulate tumor angiogenesis and invasion. *Cancer Cell*, **13** (3), 206–220.
Dunlop, E.A. and Tee, A.R. (2009) Mammalian target of rapamycin complex 1: signalling inputs, substrates and feedback mechanisms. *Cell Signaling*, **21**, 827–835.
Durand, R.E. and Aquino-Parsons, C. (2001) Non-constant tumour blood flow–implications for therapy. *Acta Oncologica*, **40**, 862–869.
Eddy, H.A. and Casarett, G.W. (1973) Development of the vascular system in the hamster malignant neurilemmoma. *Microvascular Research*, **6**, 63–82.
Erkan, M., Reiser-Erkan, C., Michalski, C.W. *et al.* (2009) Cancer-stellate cell interactions perpetuate the hypoxia-fibrosis cycle in pancreatic ductal adenocarcinoma. *Neoplasia*, **11**, 497–508.
Errede, M., Benagiano, V., Girolamo, F. *et al.* (2002) Differential expression of connexin43 in foetal, adult and tumour-associated human brain endothelial cells. *Histochemical Journal*, 34, 265–271.

Fang, K.C., Raymond, W.W., Blount, J.L., and Caughey, G.H. (1997) Dog mast cell alpha-chymase activates progelatinase B by cleaving the Phe88-Gln89 and Phe91-Glu92 bonds of the catalytic domain. *The Journal of Biological Chemistry*, **272**, 25628–25635.

Feldkamp, M.M., Lau, N., Rak, J. *et al.* (1999) Normoxic and hypoxic regulation of vascular endothelial growth factor (VEGF) by astrocytoma cells is mediated by Ras. *International Journal of Cancer*, **81**, 118–124.

Filson, A.J., Azarnia, R., Beyer, E.C. *et al.* (1990) Tyrosine phosphorylation of a gap junction protein correlates with inhibition of cell-to-cell communication. *Cell Growth and Differentiation*, **1**, 661–668.

Fingleton, B., Vargo-Gogola, T., Crawford, H.C., and Matrisian, L.M. (2001) Matrilysin [MMP-7] expression selects for cells with reduced sensitivity to apoptosis. *Neoplasia*, **3**, 459–468.

Folkman, J. (1971) Tumor angiogenesis: therapeutic implications. *New England Journal of Medicine*, **285**, 1182–1186.

Forsythe, J.A., Jiang, B.H., Iyer, N.V. *et al.* (1996) Activation of vascular endothelial growth factor gene transcription by hypoxia-inducible factor 1. *Molecular and Cellular Biology*, **16**, 4604–4613.

Frungieri, M.B., Weidinger, S., Meineke, V. *et al.* (2002) Proliferative action of mast-cell tryptase is mediated by PAR2, COX2, prostaglandins, and PPARgamma: possible relevance to human fibrotic disorders. *Proceedings of the National Academy of Sciences of the United States of America*, **99**, 15072–15077.

Galindo, M., Santiago, B., Alcami, J. *et al.* (2001) Hypoxia induces expression of the chemokines monocyte chemoattractant protein-1 (MCP-1) and IL-8 in human dermal fibroblasts. *Clinical and Experimental Immunology*, **123**, 36–41.

Goldmann, E. (1908) The growth of malignant disease in man and the lower animals, with special reference to the vascular system. *Proceedings of the Royal Society of Medicine*, **1**, 1–13.

Gorska, A.E., Jensen, R.A., Shyr, Y. *et al.* (2003) Transgenic mice expressing a dominant-negative mutant type II transforming growth factor-beta receptor exhibit impaired mammary development and enhanced mammary tumor formation. *American Journal of Pathology*, **163**, 1539–1549.

Gould, V.E., Mosquera, J.M., Leykauf, K. *et al.* (2005) The phosphorylated form of connexin43 is up-regulated in breast hyperplasias and carcinomas and in their neoformed capillaries. *Human Pathology*, **36**, 536–545.

Greijer, A.E., Delis-Van Diemen, P.M., Fijneman, R.J. *et al.* (2008) Presence of HIF-1 and related genes in normal mucosa, adenomas and carcinomas of the colorectum. *Virchows Archiv*, **452**, 535–544.

Grum-Schwensen, B., Klingelhofer, J., Berg, C.H. *et al.* (2005) Suppression of tumor development and metastasis formation in mice lacking the S100A4(mts1) gene. *Cancer Research*, **65**, 3772–3780.

Hanahan, D. and Weinberg, R.A. (2000) The hallmarks of cancer. *Cell*, **100**, 57–70.

Handsley, M.M. and Edwards, D.R. (2005) Metalloproteinases and their inhibitors in tumor angiogenesis. *International Journal of Cancer*, **115**, 849–860.

Haroon, Z.A., Raleigh, J.A., Greenberg, C.S., and Dewhirst, M.W. (2000) Early wound healing exhibits cytokine surge without evidence of hypoxia. *Annals of Surgery*, **231**, 137–147.

Heissig, B., Hattori, K., Dias, S. *et al.* (2002) Recruitment of stem and progenitor cells from the bone marrow niche requires MMP-9 mediated release of kit-ligand. *Cell*, **109**, 625–637.

Higgins, D.F., Kimura, K., Bernhardt, W.M. *et al.* (2007) Hypoxia promotes fibrogenesis *in vivo* via HIF-1 stimulation of epithelial-to-mesenchymal transition. *Journal of Clinical Investigation*, **117**, 3810–3820.

Hiromatsu, Y. and Toda, S. (2003) Mast cells and angiogenesis. *Microscopy Research and Technique*, **60**, 64–69.

Holash, J., Maisonpierre, P.C., Compton, D. *et al.* (1999) Vessel cooption, regression, and growth in tumors mediated by angiopoietins and VEGF. *Science*, **284**, 1994–1998.

Hudson, C.C., Liu, M., Chiang, G.G. *et al.* (2002) Regulation of hypoxia-inducible factor 1alpha expression and function by the mammalian target of rapamycin. *Molecular Cell and Biology*, **22**, 7004–7014.

Hull, M.T. and Warfel, K.A. (1986) Ultrastructure of abnormal collagen in human tumors. *Ultrastructure Pathology*, **10**, 293–301.

Ide, T., Kitajima, Y., Miyoshi, A. *et al.* (2006) Tumor-stromal cell interaction under hypoxia increases the invasiveness of pancreatic cancer cells through the hepatocyte growth factor/c-Met pathway. *International Journal of Cancer*, **119** (12), 2750–1759.

Isakson, B.E., Damon, D.N., Day, K.H. *et al.* (2006) Connexin40 and connexin43 in mouse aortic endothelium: evidence for coordinated regulation. *American Journal of Physiology: Heart and Circulatory Physiology*, **290**, H1199–H1205.

Jiang, B.H. and Liu, L.Z. (2008) AKT signaling in regulating angiogenesis. *Current Cancer Drug Targets*, **8**, 19–26.

Kalliomaki, T.M., Mccallum, G., Lunt, S.J. *et al.* (2008) Analysis of the effects of exposure to acute hypoxia on oxidative lesions and tumour progression in a transgenic mouse breast cancer model. *BMC Cancer*, **8**, 151.

Kalliomaki, T.M., Mccallum, G., Wells, P.G., and Hill, R.P. (2009) Progression and metastasis in a transgenic mouse breast cancer model: effects of exposure to *in vivo* hypoxia. *Cancer Letters*, **282**, 98–108.

Kalluri, R. (2003) Basement membranes: structure, assembly and role in tumour angiogenesis. *Nature Reviews Cancer*, **3**, 422–433.

Kalluri, R. and Zeisberg, M. (2006) Fibroblasts in cancer. *Nature Reviews Cancer*, **6**, 392–401.

Kavanagh, B.D., Coffey, B.E., Needham, D. *et al.* (1993) The effect of flunarizine on erythrocyte suspension viscosity under conditions of extreme hypoxia, low pH, and lactate treatment. *British Journal of Cancer*, **67**, 734–741.

Kiani, M.F., Pries, A.R., Hsu, L.L. *et al.* (1994) Fluctuations in microvascular blood flow parameters caused by hemodynamic mechanisms. *American Journal of Physiology*, **266**, H1822–H1828.

Kim, S.W., Kim, H.Y., Lee, H.J. *et al.* (2009) Dexamethasone and hypoxia upregulate CXCR4 expression in myeloma cells. *Leukemia and Lymphoma*, **50**, 1163–1173.

Kimura, H., Braun, R.D., Ong, E.T. *et al.* (1996) Fluctuations in red cell flux in tumor microvessels can lead to transient hypoxia and reoxygenation in tumor parenchyma. *Cancer Research*, **56**, 5522–5528.

Komatani, H., Sugita, Y., Arakawa, F. *et al.* (2009) Expression of CXCL12 on pseudopalisading cells and proliferating microvessels in glioblastomas: an accelerated growth factor in glioblastomas. *International Journal of Oncology*, **34**, 665–672.

Koong, A.C., Denko, N.C., Hudson, K.M. *et al.* (2000) Candidate genes for the hypoxic tumor phenotype. *Cancer Research*, **60**, 883–887.

Laderoute, K.R., Alarcon, R.M., Brody, M.D. *et al.* (2000) Opposing effects of hypoxia on expression of the angiogenic inhibitor thrombospondin 1 and the angiogenic inducer vascular endothelial growth factor. *Clinical Cancer Reseach*, **6**, 2941–2950.

Laird, D.W., Fistouris, P., Batist, G. *et al.* (1999) Deficiency of connexin43 gap junctions is an independent marker for breast tumors. *Cancer Research*, **59**, 4104–4110.

Lampe, P.D. and Lau, A.F. (2004) The effects of connexin phosphorylation on gap junctional communication. *International Journal of Biochemistry and Cell Biology*, **36**, 1171–1186.

Lanzen, J., Braun, R.D., Klitzman, B. *et al.* (2006) Direct demonstration of instabilities in oxygen concentrations within the extravascular compartment of an experimental tumor. *Cancer Research*, **66**, 2219–2223.

Leask, A. and Abraham, D.J. (2006) All in the CCN family: essential matricellular signaling modulators emerge from the bunker. *Journal of Cell Science*, **119**, 4803–4810.

Lengauer, C., Kinzler, K.W., and Vogelstein, B. (1998) Genetic instabilities in human cancers. *Nature*, **396**, 643–649.

Li, C.Y., Shan, S., Huang, Q. *et al.* (2000) Initial stages of tumor cell-induced angiogenesis: evaluation via skin window chambers in rodent models. *Journal of National Cancer Institute*, **92**, 143–147.

Lin, R., Warn-Cramer, B.J., Kurata, W.E., and Lau, A.F. (2001) v-Src phosphorylation of connexin43 on Tyr247 and Tyr265 disrupts gap junctional communication. *Journal of Cell Biology*, **154**, 815–827.

Littlepage, L.E., Egeblad, M., and Werb, Z. (2005) Coevolution of cancer and stromal cellular responses. *Cancer Cell*, 7, 499–500.

Liu, Y.L., Yu, J.M., Song, X.R. *et al.* (2006) Regulation of the chemokine receptor CXCR4 and metastasis by hypoxia-inducible factor in non small cell lung cancer cell lines. *Cancer Biology and Therapy*, **5**, 1320–1326.

Maegdefrau, U., Amann, T., Winklmeier, A. *et al.* (2009) Bone morphogenetic protein 4 is induced in hepatocellular carcinoma by hypoxia and promotes tumour progression. *Journal of Pathology*, **218** (4), 520–529.

Marchesi, F., Monti, P., Leone, B.E. *et al.* (2004) Increased survival, proliferation, and migration in metastatic human pancreatic tumor cells expressing functional CXCR4. *Cancer Research*, **64**, 8420–8427.

Martinez-Lemus, L.A., Hill, M.A., and Meininger, G.A. (2009) The plastic nature of the vascular wall: a continuum of remodeling events contributing to control of arteriolar diameter and structure. *Physiology (Bethesda)*, **24**, 45–57.

Martinive, P., Defresne, F., Bouzin, C. *et al.* (2006) Preconditioning of the tumor vasculature and tumor cells by intermittent hypoxia: implications for anticancer therapies. *Cancer Research*, **66**, 11736–11744.

Mazure, N.M., Chen, E.Y., Yeh, P. *et al.* (1996) Oncogenic transformation and hypoxia synergistically act to modulate vascular endothelial growth factor expression. *Cancer Research*, **56**, 3436–3440.

Mbeunkui, F. and Johann, D.J. Jr. (2009) Cancer and the tumor microenvironment: a review of an essential relationship. *Cancer Chemotherapy and Pharmacology*, **63**, 571–582.

Mclachlan, E., Shao, Q., Wang, H.L. *et al.* (2006) Connexins act as tumor suppressors in three-dimensional mammary cell organoids by regulating differentiation and angiogenesis. *Cancer Research*, **66**, 9886–9894.

Miyoshi, A., Kitajima, Y., Ide, T. *et al.* (2006) Hypoxia accelerates cancer invasion of hepatoma cells by upregulating MMP expression in an HIF-1alpha-independent manner. *International Journal of Oncology*, **29**, 1533–1539.

Moeller, B.J., Cao, Y., Li, C.Y., and Dewhirst, M.W. (2004) Radiation activates HIF-1 to regulate vascular radiosensitivity in tumors: role of reoxygenation, free radicals, and stress granules. *Cancer Cell*, **5**, 429–441.

Moses, H.L., Branum, E.L., Proper, J.A., and Robinson, R.A. (1981) Transforming growth factor production by chemically transformed cells. *Cancer Research*, **41**, 2842–2848.

Munoz-Najar, U.M., Neurath, K.M., Vumbaca, F., and Claffey, K.P. (2006) Hypoxia stimulates breast carcinoma cell invasion through MT1-MMP and MMP-2 activation. *Oncogene*, **25**, 2379–2392.

Musil, L.S., Cunningham, B.A., Edelman, G.M., and Goodenough, D.A. (1990) Differential phosphorylation of the gap junction protein connexin43 in junctional communication-competent and -deficient cell lines. *Journal of Cell Biology*, **111**, 2077–2088.

Nakahara, H., Howard, L., Thompson, E.W. *et al.* (1997) Transmembrane/cytoplasmic domain-mediated membrane type 1-matrix metalloprotease docking to invadopodia is required for cell invasion. *Proceedings of the National Academy of Sciences of the United States of America*, **94**, 7959–7964.

Nakao, M., Ishii, G., Nagai, K. *et al.* (2009) Prognostic significance of carbonic anhydrase IX expression by cancer-associated fibroblasts in lung adenocarcinoma. *Cancer*, **115**, 2732–2743.

Ning, X.H., Chen, S.H., Buroker, N.E. *et al.* (2007) Short-cycle hypoxia in the intact heart: hypoxia-inducible factor 1alpha signaling and the relationship to injury threshold. *American Journal of Physiology: Heart and Circulatory Physiology*, **292**, H333–H341.

Noe, V., Fingleton, B., Jacobs, K. *et al.* (2001) Release of an invasion promoter E-cadherin fragment by matrilysin and stromelysin-1. *Journal of Cell Science*, **114**, 111–118.

Nomura, R., Yoshida, D., and Teramoto, A. (2009) Stromal cell-derived factor-1 expression in pituitary adenoma tissues and upregulation in hypoxia. *Journal of Neurooncology*, **94**, 173–181.

Norrby, K. (2002) Mast cells and angiogenesis. *APMIS*, **110**, 355–371.

Olumi, A.F., Grossfeld, G.D., Hayward, S.W. *et al.* (1999) Carcinoma-associated fibroblasts direct tumor progression of initiated human prostatic epithelium. *Cancer Research*, **59**, 5002–5011.

Ono, M., Torisu, H., Fukushi, J. *et al.* (1999) Biological implications of macrophage infiltration in human tumor angiogenesis. *Cancer Chemotherapy and Pharmacology*, **43** (Suppl.), S69–S71.

Orimo, A., Gupta, P.B., Sgroi, D.C. *et al.* (2005) Stromal fibroblasts present in invasive human breast carcinomas promote tumor growth and angiogenesis through elevated SDF-1/CXCL12 secretion. *Cell*, **121**, 335–348.

Osinsky, S.P., Ganusevich, I.I., Bubnovskaya, L.N. *et al.* (2005) Hypoxia level and matrix metalloproteinases-2 and -9 activity in Lewis lung carcinoma: correlation with metastasis. *Experimental Oncology*, **27**, 202–205.

Pan, J., Mestas, J., Burdick, M.D. *et al.* (2006) Stromal derived factor-1 (SDF-1/CXCL12) and CXCR4 in renal cell carcinoma metastasis. *Molecular Cancer*, **5**, 56.

Parrott, J.A., Nilsson, E., Mosher, R. *et al.* (2001) Stromal-epithelial interactions in the progression of ovarian cancer: influence and source of tumor stromal cells. *Molecular and Cellular Endocrinology*, **175**, 29–39.

Patan, S., Munn, L.L., and Jain, R.K. (1996) Intussusceptive microvascular growth in a human colon adenocarcinoma xenograft: a novel mechanism of tumor angiogenesis. *Microvascular Research*, **51**, 260–272.

Peng, Y.J., Yuan, G., Ramakrishnan, D. *et al.* (2006) Heterozygous HIF-1alpha deficiency impairs carotid body-mediated systemic responses and reactive oxygen species generation in mice exposed to intermittent hypoxia. *Journal of Physiology*, **577**, 705–716.

Pierce, D.F. Jr., Gorska, A.E., Chytil, A., *et al.* (1995) Mammary tumor suppression by transforming growth factor beta 1 transgene expression. *Proceedings of the National Academy of Sciences of the United States of America*, **92**, 4254–4258.

Pigott, K.H., Hill, S.A., Chaplin, D.J., and Saunders, M.I. (1996) Microregional fluctuations in perfusion within human tumours detected using laser Doppler flowmetry. *Radiotherapy and Oncology*, **40**, 45–50.

Pollmann, M.A., Shao, Q., Laird, D.W., and Sandig, M. (2005) Connexin43 mediated gap junctional communication enhances breast tumor cell diapedesis in culture. *Breast Cancer Research*, 7, R522–R534.

Pries, A.R. and Secomb, T.W. (2009) Origins of heterogeneity in tissue perfusion and metabolism. *Cardiovascular Research*, **81**, 328–335.

Pries, A.R., Reglin, B., and Secomb, T.W. (2001) Structural adaptation of microvascular networks: functional roles of adaptive responses. *American Journal of Physiology: Heart and Circulatory Physiology*, **281**, H1015–H1025.

Pries, A.R., Cornelissen, A.J., Sloot, A.A. *et al.* (2009) Structural adaptation and heterogeneity of normal and tumor microvascular networks. *PLoS Computational Biology*, **5**, e1000394.

Pries, A.R., Schonfeld, D., Gaehtgens, P. *et al.* (1997) Diameter variability and microvascular flow resistance. *American Journal of Physiology*, **272**, H2716–H2725.

Pries, A.R., Secomb, T.W., and Gaehtgens, P. (1995a) Design principles of vascular beds. *Circulation Research*, **77**, 1017–1023.

Pries, A.R., Secomb, T.W., and Gaehtgens, P. (1995b) Structure and hemodynamics of microvascular networks: heterogeneity and correlations. *American Journal of Physiology*, **269**, H1713–H1722.

Pries, A.R., Secomb, T.W., and Gaehtgens, P. (1998) Structural adaptation and stability of microvascular networks: theory and simulations. *American Journal of Physiology*, **275**, H349–H360.

Pugh, C.W. and Ratcliffe, P.J. (2003) Regulation of angiogenesis by hypoxia: role of the HIF system. *Nature Medicine*, **9**, 677–684.

Roberts, A.B. and Wakefield, L.M. (2003) The two faces of transforming growth factor beta in carcinogenesis. *Proceedings of the National Academy of Sciences of the United States of America*, **100**, 8621–8623.

Rofstad, E.K., Galappathi, K., Mathiesen, B., and Ruud, E.B. (2007) Fluctuating and diffusion-limited hypoxia in hypoxia-induced metastasis. *Clinical Cancer Research*, **13**, 1971–1978.

Ronnov-Jessen, L., Petersen, O.W., and Bissell, M.J. (1996) Cellular changes involved in conversion of normal to malignant breast: importance of the stromal reaction. *Physiological Reviews*, **76**, 69–125.

Ruiter, D., Bogenrieder, T., Elder, D., and Herlyn, M. (2002) Melanoma-stroma interactions: structural and functional aspects. *The Lancet Oncology*, **3**, 35–43.

Schioppa, T., Uranchimeg, B., Saccani, A. *et al.* (2003) Regulation of the chemokine receptor CXCR4 by hypoxia. *Journal of Experimental Medicine*, **198**, 1391–1402.

Schor, S.L., Schor, A.M., and Rushton, G. (1988) Fibroblasts from cancer patients display a mixture of both foetal and adult-like phenotypic characteristics. *Journal of Cell Science*, **90** (Pt 3), 401–407.

Schor, S.L., Schor, A.M., Rushton, G., and Smith, L. (1985) Adult, foetal and transformed fibroblasts display different migratory phenotypes on collagen gels: evidence for an isoformic transition during foetal development. *Journal of Cell Science*, **73**, 221–234.

Schutyser E., Su Y., Yu Y., *et al.* (2007) Hypoxia enhances CXCR4 expression in human microvascular endothelial cells and human melanoma cells. *European Cytokine Network*, **18** (2), 59–70. Epub ahead of print Jun 26.

Secomb, T.W., Hsu, R., Dewhirst, M.W. *et al.* (1993) Analysis of oxygen transport to tumor tissue by microvascular networks. *International Journal of Radiation Oncology, Biology, Physics*, **25**, 481–489.

Secomb, T.W., Hsu, R., Ong, E.T. *et al.* (1995) Analysis of the effects of oxygen supply and demand on hypoxic fraction in tumors. *Acta Oncologica*, **34**, 313–316.

Segal, S.S. (1994) Cell-to-cell communication coordinates blood flow control. *Hypertension*, **23**, 1113–1120.

Semenza, G.L. (2003) Targeting HIF-1 for cancer therapy. *Nature Reviews Cancer*, **3**, 721–732.

Semenza, G.L. (2008) Hypoxia-inducible factor 1 and cancer pathogenesis. *IUBMB Life*, **60**, 591–597.

Semenza, G.L. and Wang, G.L. (1992) A nuclear factor induced by hypoxia via de novo protein synthesis binds to the human erythropoietin gene enhancer at a site required for transcriptional activation. *Molecular and Cellular Biology*, **12**, 5447–5454.

Shao, Q., Wang, H., Mclachlan, E. *et al.* (2005) Down-regulation of Cx43 by retroviral delivery of small interfering RNA promotes an aggressive breast cancer cell phenotype. *Cancer Research*, **65**, 2705–2711.

Skala, M.C., Fontanella, A., Hendargo, H. *et al.* (2009) Combined hyperspectral and spectral domain optical coherence tomography microscope for noninvasive hemodynamic imaging. *Optics Letters*, **34**, 289–291.

Skala, M.C., Fontanella, A., Lan, L., *et al.* Longitudinal optical imaging of tumor metabolism and hemodynamics. *Journal of Biomedical Optics*, **15** (1), 111–112.

Sorg, B.S., Hardee, M.E., Agarwal, N. *et al.* (2008) Spectral imaging facilitates visualization and measurements of unstable and abnormal microvascular oxygen transport in tumors. *Journal of Biomedical Optics*, **13**, 014026.

Sorg, B.S., Moeller, B.J., Donovan, O. *et al.* (2005) Hyperspectral imaging of hemoglobin saturation in tumor microvasculature and tumor hypoxia development. *Journal of Biomedical Optics*, **10**, 44004.

Sternlicht, M.D., Lochter, A., Sympson, C.J. *et al.* (1999) The stromal proteinase MMP3/stromelysin-1 promotes mammary carcinogenesis. *Cell*, **98**, 137–146.

Suarez, S. and Ballmer-Hofer, K. (2001) VEGF transiently disrupts gap junctional communication in endothelial cells. *Journal of Cell Science*, **114**, 1229–1235.

Swinson, D.E., Jones, J.L., Cox, G. *et al.* (2004) Hypoxia-inducible factor-1 alpha in non small cell lung cancer: relation to growth factor, protease and apoptosis pathways. *International Journal of Cancer*, **111**, 43–50.

Tlsty, T.D. and Hein, P.W. (2001) Know thy neighbor: stromal cells can contribute oncogenic signals. *Current Opinion in Genetics and Development*, **11**, 54–59.

Tomakidi, P., Mirancea, N., Fusenig, N.E. *et al.* (1999) Defects of basement membrane and hemidesmosome structure correlate with malignant phenotype and stromal interactions in HaCaT-Ras xenografts. *Differentiation*, **64**, 263–275.

Trotter, M.J., Chaplin, D.J., Durand, R.E., and Olive, P.L. (1989) The use of fluorescent probes to identify regions of transient perfusion in murine tumors. *International Journal of Radiation Oncology, Biology, Physics*, **16**, 931–934.

Tucker, R.F., Shipley, G.D., Moses, H.L., and Holley, R.W. (1984) Growth inhibitor from BSC-1 cells closely related to platelet type beta transforming growth factor. *Science*, **226**, 705–707.

Upham, B.L., Kang, K.S., Cho, H.Y., and Trosko, J.E. (1997) Hydrogen peroxide inhibits gap junctional intercellular communication in glutathione sufficient but not glutathione deficient cells. *Carcinogenesis*, **18**, 37–42.

Van Beijnum, J.R., Dings, R.P., Van Der Linden, E. *et al.* (2006) Gene expression of tumor angiogenesis dissected: specific targeting of colon cancer angiogenic vasculature. *Blood*, **108**, 2339–2348.

Van Kempen, L.C., Ruiter, D.J., Van Muijen, G.N., and Coussens, L.M. (2003) The tumor microenvironment: a critical determinant of neoplastic evolution. *European Journal of Cell Biology*, **82**, 539–548.

Varney, M.L., Olsen, K.J., Mosley, R.L. *et al.* (2002) Monocyte/macrophage recruitment, activation and differentiation modulate interleukin-8 production: a paracrine role of tumor-associated macrophages in tumor angiogenesis. *In vivo*, **16**, 471–477.

Wang, G.L. and Semenza, G.L. (1993) General involvement of hypoxia-inducible factor 1 in transcriptional response to hypoxia. *Proceedings of the National Academy of Sciences of the United States of America*, **90**, 4304–4308.

Werb, Z. (1997) ECM and cell surface proteolysis: regulating cellular ecology. *Cell*, **91**, 439–442.

Willecke, K., Eiberger, J., Degen, J. *et al.* (2002) Structural and functional diversity of connexin genes in the mouse and human genome. *Biological Chemistry*, **383**, 725–737.

Wouters, B.G. and Koritzinsky, M. (2008) Hypoxia signalling through mTOR and the unfolded protein response in cancer. *Nature Reviews Cancer*, **8**, 851–864.

Wu, W., Pan, C., Meng, K. *et al.* (2010) Hypoxia activates heparanase expression in an NF-kappaB dependent manner. *Oncology Reports*, **23**, 255–261.

Wysoczynski, M., Shin, D.M., Kucia, M., and Ratajczak, M.Z. (2010) Selective upregulation of interleukin-8 by human rhabdomyosarcomas in response to hypoxia: therapeutic implications. *International Journal of Cancer*, **126**, 371–381.

Yamaura, H. and Matsuzawa, T. (1979) Tumor regrowth after irradiation; an experimental approach. *International Journal of Radiation Biology and Related Studies in Physics, Chemistry and Medicine*, **35**, 201–219.

Yong, L.C. (1997) The mast cell: origin, morphology, distribution, and function. *Experimental Toxicology and Pathology*, **49**, 409–424.

Yuan, G., Nanduri, J., Khan, S. *et al.* (2008) Induction of HIF-1alpha expression by intermittent hypoxia: involvement of NADPH oxidase, Ca2+ signaling, prolyl hydroxylases, and mTOR. *Journal of Cellular Physiology*, **217**, 674–685.

Zagzag, D., Lukyanov, Y., Lan, L. *et al.* (2006) Hypoxia-inducible factor 1 and VEGF upregulate CXCR4 in glioblastoma: implications for angiogenesis and glioma cell invasion. *Laboratory Investigation*, **86**, 1221–1232.

Zakrzewicz, A., Secomb, T.W., and Pries, A.R. (2002) Angioadaptation: keeping the vascular system in shape. *News in Physiological Sciences*, **17**, 197–201.

Further reading

Dang, C.V., Kim, J.W., Gao, P., and Yustein, J. (2008) The interplay between MYC and HIF in cancer. *Nature Reviews Cancer*, **8**, 51–56.

3

Contributions of the Extracellular Matrix to Tumorigenesis

Marie Schluterman Burdine and Rolf A. Brekken

Hamon Center for Therapeutic Oncology Research, Division of Surgical Oncology and Departments of Surgery and Pharmacology, University of Texas Southwestern Medical Center, Dallas, TX USA

3.1 The extracellular matrix

Within tissues, cells are surrounded by a meshwork of proteins and proteoglycans collectively called the extracellular matrix (ECM), which compartmentalizes tissues. The ECM is divided into two distinct layers: (i) the basement membrane, which is composed of sheet-like layers of ECM and lies under epithelial cells segregating tissues into functionally distinct regions; and (ii) the interstitial matrix, which exists within intercellular space. The ECM serves multiple functions that are critical for embryonic development and wound repair. These functions include providing tissues with shape and flexibility and acting as a cushion to absorb external pressure. The ECM also serves as a base for cell anchorage, which mediates cell polarity, intracellular signaling, and assists in migration. The key to the ECM's function lies in its unique composition and structure. The ECM is constructed in a specific pattern that is critical to its ability to carry out these functions and we will discuss later in this chapter how alterations in the expression level or arrangement of proteins within the ECM can be used to manipulate its function.

The most obvious function of the ECM is to provide structural support, shape, and stability for tissues. It does this by functioning as a base for cell anchorage. This base consists of three main structural components collagen, fibronectin, and elastic fibers, which bind to one another building a protein lattice upon which cells adhere. Collagen is the most abundant structural protein found in the human body. There are approximately 30 different collagen family members with collagen I and II being the most prevalent. Collagen is secreted predominantly by fibroblasts into the ECM as single polypeptide chains. Once secreted, the single chains are

Tumor Microenvironment Edited by Dietmar W. Siemann

crosslinked to other collagen molecules to form collagen fibrils (Kadler, Hill, and Canty-Laird, 2008). These fibrils arrange in specific patterns forming the foundation of the ECM upon which fibronectin and elastic fibers bind. Similar to collagen, fibronectin is produced and secreted by fibroblasts. It is secreted as dimers that initiate fibronectin fibrillogenesis by binding to cell surface receptors called integrins. Integrin binding induces a conformational change that exposes fibronectin binding sites allowing dimers to bind to each other forming mature fibronectin fibrils (Mao and Schwarzbauer, 2005). This interaction with integrins also allows fibronectin to serve as a substrate for cells linking them to the collagen platform (Mao and Schwarzbauer, 2005). Elastic fibers are the largest structure found in the ECM and provide tissues with elasticity. The two main components of elastic fibers are: (i) elastin, which is secreted in a precursor form as tropoelastin and (ii) microfibrils, which serve as a backbone for elastic fiber assembly. Assembly begins by the coacervation of tropoelastin molecules, which are then modified and crosslinked by lysyl oxidase like-1 (LOXL-1). The elastin 'bundle' forms the core of elastic fibers, which are then wrapped by microfibrils forming mature elastic fibers (Wagenseil and Mecham, 2007). These proteins, along with other fibrous matrix proteins including laminin and vitronectin intertwine with collagen fibrils generating a heterogeneous structure upon which cells bind. In addition, the structure is filled with proteoglycans (proteins containing large clusters of carbohydrate chains) such as herapan sulfate that coat structural fibers adding depth and cushion to the ECM (Marastoni *et al.*, 2008).

Cell adherence to the ECM lattice provides cells support needed for cell migration. This is particularly important during embryonic development when cells are required to migrate into surrounding regions and differentiate into specific tissues (Svoboda, Fischman, and Gordon, 2008). Cells adhere to the ECM through cell surface adhesion molecules known as integrins. Integrins are a family of heterodimeric transmembrane proteins that contain α and β subunits. The particular ECM protein an integrin binds to depends on the combination of subunits. For example, $\alpha_1\beta_1$ and $\alpha_2\beta_1$ interact with collagen while $\alpha_5\beta_1$ serves as the primary fibronectin receptor (Mizejewski, 1999). ECM binding results in integrin activation and formation of a focal adhesion complex that links the ECM to actin filaments of the cytoskeleton (Lock, Wehrle-Haller, and Stromblad, 2008). This enables the cell to pull its way through the ECM.

A less obvious yet possibly more important function of the ECM in regards to tissue homeostasis and disease is its ability to mediate intracellular signaling. The ECM affects signaling through three main mechanisms: (i) cell–ECM interaction, (ii) regulation of the bioavailability of growth factors, and (iii) the function of matricellular proteins. Cell–ECM interactions generate outside-in stimuli that initiate cellular activities required for proper cell function (Mizejewski, 1999). For instance, cell attachment to the ECM via integrins induces signaling cascades that promote survival. Loss of cell–ECM contact can result in a form of apoptosis termed anoikis (Giannoni *et al.*, 2008). Anchorage-dependent survival is observed in most cells with the exception of red blood cells and inflammatory cells. However, tumor cells are often resistant to anoikis and can survive without a physical attachment to the ECM allowing them to successfully metastasize to distant tissues (Chiarugi and Giannoni, 2008). The exact pathway initiated by cell–ECM contact and the extent of activation is related directly to the specific combination of integrin and ECM protein involved.

The ECM also affects cellular activity by serving as a reservoir for proteins required for proper tissue function and repair. This includes a plethora of growth factors and proteases. Growth factors such as basic fibroblast growth factor (bFGF), platelet-derived growth factor (PDGF), vascular endothelial growth factor (VEGF), and transforming growth factor β (TGFβ) are involved intricately in development and continued expression of these factors is required to maintain tissue homeostasis during adulthood. These pleiotropic molecules have been shown to robustly affect proliferation, survival, and migration in numerous cell types. Once growth factors are secreted from cells, they often become embedded within the ECM and require ECM degradation by proteases such as elastase to release the active protein allowing it to interact with surrounding cells and transduce downstream signaling. For instance, the ECM serves as a VEGF 'sink.' High levels of VEGF are found incorporated within the ECM lattice shielded from cellular contact. Matrix metalloproteinases (MMPs), a family of proteases, which degrade structural proteins within the ECM, liberate VEGF from the ECM allowing it to bind to receptors on the cell surface and activate downstream pathways (Lee *et al.*, 2005; Bergers *et al.*, 2000). Also, TGFβ is deposited within the ECM in a latent form, which requires proteolytic processing to generate active TGFβ (Jenkins, 2008). The latent form binds to microfibrils of elastic fibers, which prevents cleavage by shielding the protein from proteases. Breakdown of microfibrils by elastase releases latent TGFβ freeing it for cleavage into its active form (Ten Dijke and Arthur, 2007). The ability of the ECM to control the bioavailability of growth factors provides another means of regulating cellular activities and further explains how alterations in the makeup of the ECM as observed in diseases such as cancer affect cell response.

Matricellular proteins also reside in the ECM. They are a unique family of proteins that do not function as structural proteins but rather orchestrate the deposition of the ECM and mediate cell–cell and cell–ECM interactions (for a comprehensive review on matricellular proteins see *Journal of Cell Communication and Signaling* Volume 3, Issues 3–4, 2009). To do this, matricellular proteins interact directly with cell surface receptors, structural proteins, growth factors, and proteases found within the ECM (Framson and Sage, 2004). Their expression is found in every tissue, begins early in development, persists throughout adulthood and is increased during tissue remodeling events. Matricellular proteins are critical regulators of many aspects of cell function including differentiation, survival, proliferation, and migration making them necessary for proper tissue function. Not surprisingly, given their affect on cell–ECM mediated signaling pathways, matricellular proteins have been shown to strongly influence tumor growth. The focus of this chapter will be on reviewing present data concerning the function of matricellular proteins in the context of tumor development.

3.2 Manipulation of the ECM during tumor development

For tumor cells to metastasize, the local ECM must be remodeled to create an environment conducive to tumor survival and progression. This includes altering the architecture and composition of the tumor-associated ECM or tumor microenvironment (TME) to facilitate tumor cell dissemination (Pupa *et al.*, 2002). Changes in ECM architecture are primarily carried out by enzymes such as MMPs which assist in remodeling of the TME by degrading structural proteins such as collagen and fibronectin allowing tumor cells to freely navigate through the surrounding ECM. MMPs and other proteases assist in destruction of the first barrier tumor cells face to successful metastasis, the basement membrane. They degrade the underlying basement membrane allowing tumor cells to escape the primary tumor and invade into surrounding non-neoplastic tissues. MMPs continue to breakdown barriers in the surrounding ECM clearing a path to blood vessels where tumor cells will intravasate into the circulatory system and seed secondary tumors (Hofmann *et al.*, 2005). Destruction of the ECM by proteases also promotes tumor progression by facilitating the release of angiogenic and mitogenic factors bound within the ECM (van Kempen *et al.*, 2003). For instance, the release of VEGF from the ECM promotes the establishment of a tumor vascular system which is required for continued tumor survival. These protumor effects explain why in most human cancers such as breast and colon cancer, elevated levels of MMP expression correlate with a highly invasive, metastatic phenotype, and poor patient prognosis (Deryugina and Quigley, 2006). These findings spurred the development of multiple synthetic MMP inhibitors that in preclinical mouse tumor studies were highly effective at controlling tumor growth and metastasis (Wojtowicz-Praga *et al.*, 1997). Unfortunately, these results were not observed in human clinical trials, forcing a global re-evaluation of the function of MMPs in cancer (Figueira *et al.*, 2009; Wojtowicz-Praga *et al.*, 1997). In a surprising unexpected twist, studies revealed that the breakdown of ECM proteins by MMPs was more complex than anticipated. In fact it is a highly organized process which results in the generation of both protumor and antitumor cleavage products (Lopez-Otin, Palavalli, and Samuels, 2009). For example, cleavage of type IV and type XVIII collagen by MMP-9 releases cryptic protein fragments with antiangiogenic properties such as α3NC1 and endostatin respectively (Pasco *et al.*, 2004; O'Reilly *et al.*, 1997). These endogenous angiogenic inhibitors reduce tumor angiogenesis and growth *in vivo* (Pasco *et al.*, 2004; O'Reilly *et al.*, 1997). It is still unclear how the specific processing of ECM proteins by MMPs is regulated within the TME but it appears that the source of MMP expression (host or tumor derived) contributes to the diverse response.

Presence of the ECM is required for cellular survival therefore; increased degradation of the ECM within the TME must be balanced by an increase in ECM synthesis. The development of a tumor, much like a wound, provokes a robust inflammatory response causing an influx of mast cells, macrophages, and neutrophils into the TME (Wu and Zhou, 2009). These immune cells release cytokines (e.g., interleukin-1α and β, TGFβ1) that stimulate the secretion of structural ECM proteins by tumor-associated fibroblasts. This continuous influx of immune cells and deposition of ECM proteins results in the buildup of large amounts of stroma within tumors (Wu and Zhou, 2009). Desmoplasia, a unique characteristic of ductal carcinomas, refers to a track of dense fibrous tissue composed of collagen, fibronectin, and laminin, that forms around the edge of tumors (Pandol *et al.*, 2009). It is believed that this thick band serves to encapsulate the tumor and prevent cancer cell progression. However, tumor cells have learned to manipulate this response for their own benefit (Hartel *et al.*, 2004). The large track of stroma within tumors are full of growth factors which are co-opted by tumor cells and used to evade apoptotic signals and promote survival and proliferation (Korc, 2007).

In non-neoplastic ECM, structural proteins are deposited in a specific pattern that (i) organizes tissues into specific compartments and (ii) provides correct cell–ECM signaling. This organized pattern is lost in the presence of a tumor. For example, type 1 collagen lies parallel to epithelial cells in non-transformed tissues; however, within the TME, it is deposited in a random, loosely woven and disorganized pattern (van Kempen *et al.*, 2003). Although the pattern may seem obscure, it promotes ECM stiffening, which is beneficial for tumors (Levental *et al.*, 2009). Because of increased deposition of ECM proteins and elevated interstitial pressure, tumors are characteristically stiffer than normal tissues. This characteristic enables ECM proteins to properly align and interact with cell surface receptors to induce signaling pathways critical for tumor progression. In support of this preventing ECM stiffening in tumors has been shown to reduce tumor growth (Erler and Weaver, 2009).

The presence of a tumor also results to dramatic changes in the expression of matricellular proteins. The altered expression of matricellular proteins suggests a possible function for these proteins during tumor growth and metastasis. Therefore, elucidating the function matricellular proteins serve in tumor progression has become a focal point of intense research. What has been discovered so far is the focus of the rest of this chapter.

3.3 Matricellular proteins and their complex effects on tumor development

Matricellular proteins are expressed at sites of tissue remodeling where they coordinate cell–ECM interaction. As such this unique class of proteins is well suited to influence the TME and tumor progression. Much of our understanding about the function of matricellular proteins in tumorigenesis is a result of studies in mice engineered to lack the expression of specific matricellular proteins including *s*ecreted *p*rotein *a*cidic and *r*ich in *c*ysteine (SPARC), thrombospondin-1 (TSP-1), osteopontin (OPN), and fibulin-5 (Fbln5). Although it is clear that these proteins influence tumor growth, the exact effect appears to be dependent on the origin of the protein, whether it is produced by host cells or tumor cells and on the specific type of tumor. These obstacles make understanding the function of matricellular proteins complex and contextually dependent. The rest of this chapter will discuss the best-characterized matricellular proteins with a specific emphasis on Fbln5.

3.3.1 SPARC

SPARC, a founding member of the matricellular protein family, has been shown to modulate multiple cellular functions. It has deadhesive and antiproliferative effects and is critical for proper ECM (collagen) deposition (Framson and Sage, 2004). These effects are attributed to SPARC's ability to bind to ECM proteins such as collagen and vitronectin and to growth factors such as PDGF and bFGF (Hasselaar and Sage, 1992; Raines *et al.*, 1992; Yan and Sage, 1999). SPARC has also been shown to bind to VEGF resulting in the inhibition of endothelial cell proliferation and migration. The effect of SPARC on the development of cancer is dependent on the source of its expression (for a comprehensive review on SPARC and cancer see (Arnold and Brekken, 2009). For example, in relation to breast cancer, forced expression of SPARC by human breast cancer cells, MDA-MB-231, inhibited their invasiveness *in vitro*, indicating a negative relationship between tumor-cell derived SPARC and breast cancer progression (Koblinski *et al.*, 2005). Given these results, it was surprising when reports began to surface showing an increase in SPARC expression within human mammary tumors. Upon closer examination, it was determined that SPARC expression was restricted to the host stroma compartment of these tumors and not expressed by the tumor cells. These reports identified a positive correlation between increased expression of SPARC in the stroma of human breast cancer samples and disease-free survival indicating that SPARC is upregulated by the host to control tumor growth (Beck *et al.*, 2008; Bergamaschi *et al.*, 2008). These findings also held true for other types of cancers. Studies in our laboratory have found that host-derived SPARC is a critical factor in controlling the host response to tumor formation. We observed an increase in the invasiveness of pancreatic tumors in SPARC-*null* animals were host-derived SPARC is absent (Arnold *et al.*, 2008, 2009). Also, the SPARC promoter was found to be hypermethylated in multiple colorectal cancer cell lines as well as pancreatic, lung, and prostate cancer lines (Tai and Tang, 2008). Expressing SPARC in these colorectal cancer cell lines enhanced cell death after treatment with chemotherapeutics indicating that tumor cells purposely repress SPARC production because of its ability to facilitate apoptosis (Tai *et al.*, 2005). This hypothesis was validated when it was shown that SPARC could turn on apoptosis by increasing gene expression of several members of the caspase family including caspase 8 and 10 (Tang and Tai, 2007). These results coincide with expression analysis of SPARC in human colorectal cancer. SPARC was again found to be highly expressed in the tumor stroma but not expressed in tumor cells and patients with higher-expressing SPARC colorectal tumors had better clinical outcome than those with tumors that did not express SPARC (Lussier, Sodek, and Beaulieu, 2001; Yang *et al.*, 2007). It appears from these studies that host cells express SPARC within the TME to inhibit tumor growth while many types of epithelial tumor cells attempt to block these effects by reducing their level of SPARC expression.

3.3.2 Thrombospondin-1

Similar findings have been observed with the matricellular protein, TSP-1. TSP-1 has been well characterized as an antiangiogenic protein that functions via multiple mechanisms. For example, TSP-1 regulates VEGF activity by binding to VEGF and preventing its release from the ECM (Wang-Rodriguez *et al.*, 2003; Greenaway *et al.*, 2007). TSP-1 has also been shown to directly induce apoptosis of endothelial cells by decreasing expression of Bcl-2 and increasing expression of Bax (Jimenez *et al.*, 2000; Nor *et al.*, 2000). Given its effect on angiogenesis, it is not surprising that various groups have shown TSP-1 expression to be regulated by oncogenes as well as tumor suppressor genes. For example, the RAS-oncoprotein modifies TSP-1 expression via a mechanism dependent on Myc-phosphorylation. Tumor cells with high RAS expression express low levels of TSP-1 and have a more aggressive nature when grown *in vivo* than cells with elevated TSP-1 expression (Watnick *et al.*, 2003). It has also been reported that tumor suppressor genes such as p53 turn on expression of TSP-1 to control tumor growth in mice by reducing the level of VEGF present in the TME (Gautam *et al.*, 2002). Human tumor samples with a loss in p53 expression also exhibited a decrease in TSP-1 expression confirming the existence of this relationship in human disease (Kazerounian, Yee, and Lawler, 2008). Much like SPARC, TSP-1 was found expressed highly in the stroma of various tumor types, such as invasive breast carcinoma, but not by the tumor cells suggesting that the host uses TSP-1 as a tool to control tumor growth (Kazerounian, Yee, and Lawler, 2008). This was further supported by tumor studies performed in *TSP-1-null* mice where tumors grew significantly faster than in *wild-type* (*WT*) mice (Lawler *et al.*, 2001).

Although TSP-1 as an angiogenic inhibitor has been well documented, a second function for TSP-1 has been identified that labels TSP-1 as a protumorigenic protein making its effect on tumor growth even more complex. The primary goal of a developing tumor is to invade surrounding tissue and spread beyond the initial site. To accomplish this, two important steps must occur. The surrounding ECM must be degraded to allow tumor cell invasion and tumor cells must be able to bind to components of the TME to migrate through the ECM. Recent reports have identified functions of TSP-1 that promote both of these steps. First, TSP-1 increases the expression of the proteolytic enzyme plasmin, which functions to degrade the ECM providing tumor cells with clear avenues to blood vessels where they are then able to metastasize to secondary sites (Albo *et al.*, 1999). Second, TSP-1 directly anchors tumor cells to structural proteins of the ECM including type IV collagen enabling them to invade and migrate through the TME (Wang *et al.*, 1996). These characteristics help explain why TSP-1 has reportedly been found highly expressed by tumor cells in late stage, highly aggressive breast cancers (Clezardin *et al.*, 1993). It has been hypothesized that in the early stages of tumor development TSP-1 expression is turned off by tumor cells to help create a proangiogenic environment, but turned on by host cells to limit tumor angiogenesis. However, once the tumor adapts to this environment and becomes vascularized, tumors begin to express TSP-1 for its prometastatic functions. Further research is needed to better understand TSP-1 and its effect on cancer. It is yet to be determined how best to target TSP-1 for therapeutic purposes.

3.3.3 Osteopontin

The effects of OPN on tumor development appear to be more straightforward than the aforementioned matricellular proteins. It is a multifunctional protein implicated in promoting a metastatic phenotype in numerous tumor types. OPN affects cell adhesion, survival, proliferation, and migration. It serves as an adhesion molecule by binding to various integrins, thus anchoring cells to the ECM. Its interaction with cell surface integrins induces survival and evasion of apoptosis in various tumor cell lines (Johnston *et al.*, 2008). By inducing migration and the expression of MMPs, OPN expression correlates with a more metastatic tumor phenotype. For example, transformed NIH 3T3 fibroblasts develop a highly invasive, malignant phenotype. This phenotype corresponds with a higher level of OPN expression (Chambers *et al.*, 1992). OPN has also been shown to promote tumor angiogenesis by binding to the vascular integrin $\alpha_V\beta_3$. This interaction induces migration of endothelial cells and allows OPN to protect endothelial cells from apoptosis *in vitro* (Standal, Borset, and Sundan, 2004). *In vivo*, forced overexpression of OPN in SBC-3 lung cancer cells increased tumor growth by augmenting angiogenesis (Cui *et al.*, 2007). This effect was ameliorated by abrogation of the interaction between OPN and $\alpha_V\beta_3$ (Cui *et al.*, 2007). OPN expression has been analyzed in multiple different human carcinomas to determine its relevancy in human disease. OPN mRNA was expressed highly in cancers of the stomach, breast, bladder, pancreas and prostate compared to normal tissues (Brown *et al.*, 1994). In addition, increased expression of OPN correlated with advancing stages of colon cancer. This work was supported by *in vitro* data showing treatment of human colon cancer cells with OPN resulted in increased invasiveness (Irby, Mccarthy, and Yeatman, 2004). OPN now serves as a prognostic marker for several different cancer types including ovarian and malignant glioblastomas (Johnston *et al.*, 2008). Inhibition of OPN as a therapeutic approach is intriguing but to date has not been well established. Preclinical trials using antibodies against OPN showed decreased prostate cancer growth in mice; however, no strategies have advanced to clinical trials (Weber, 2001).

3.3.4 Fibulin-5

Fbln5 is a matricellular protein required for maturation of elastic fibers, which provide elasticity to the blood vessel wall. Therefore Fbln5 has a direct effect on the efficiency of the vasculature but it also has been shown to directly regulate angiogenic development and cell-ECM signaling making it an interesting protein for cancer research.

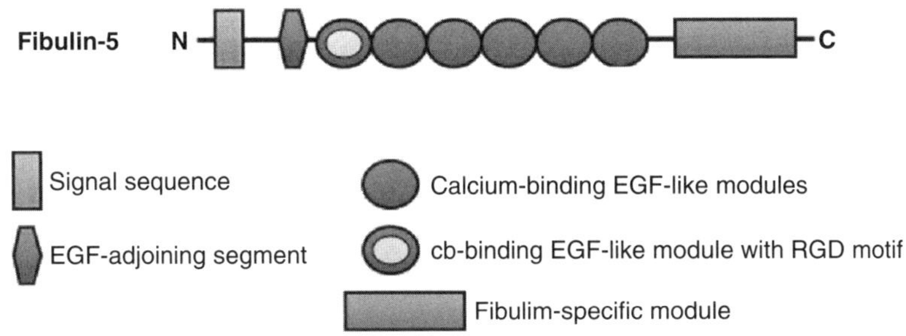

Figure 3.1 A schematic representation of the protein structure of Fbln5. The first cb-binding EGF-like module contains an integrin-binding RGD motif.

Fbln5 (also known as DANCE and EVEC) is a member of the fibulin family of ECM proteins, which all contain a string of calcium-binding epidermal growth factor-like (cbEGF) repeats at the N-terminus followed by the defining globular COOH-terminal fibulin-type module. cbEGF motifs have been shown to be important for proper protein folding and structure stabilization and act as signaling sequences for protein interaction. To date, the function for the fibulin-type module is unknown. Unique to Fbln5 is the insertion of an RGD motif in the first cbEGF repeat. This sequence facilitates binding to RGD-dependent integrins such as $\alpha_5\beta_1$ and the $\alpha_v\beta$ integrins (Figure 3.1).

Originally, fibulins were identified for their involvement in the formation and stabilization of structural components such as collagen and elastic fibers (Timpl *et al.*, 2003; Zheng *et al.*, 2007). However, more recently, fibulins have been identified as mediators of cell–cell or cell–ECM interactions. This function allows them to directly regulate processes such as proliferation, survival, and migration.

A main function for Fbln5 was first identified with the generation of *Fbln5*-deficient ($Fbln5^{-/-}$) mice. $Fbln5^{-/-}$ mice are born at Mendelian ratios, are fertile, and appear relatively normal. However, closer examination revealed that $Fbln5^{-/-}$ mice exhibit fragmented and disorganized elastic fibers, indicating Fbln5's function in the proper assembly of elastic fibers (Yanagisawa *et al.*, 2002). It was shown to bind to tropoelastin, the precursor protein of elastin, but not polymerized elastin. Fbln5 is believed to facilitate the coacervation of tropoelastin molecules preparing them for crosslinking by LOXL1. Furthermore, Fbln5 was shown to interact with fibrillin-1, a molecule expressed highly within microfibrils (Freeman *et al.*, 2005). This interaction assists in the tethering of tropoelastin to microfibrils to form mature elastic fibers (Zheng *et al.*, 2007; Liu *et al.*, 2004). The disruption in elastic fiber deposition in $Fbln5^{-/-}$ mice resulted in loose skin and led to the identification of mutations in human FBLN5 that result in cutis laxa (Loeys *et al.*, 2002). $Fbln5^{-/-}$ mice also present with alveolar defects. Lungs of mutant mice were expanded because of dilated alveoli. By 6 months of age, the lung defect progressed to severe emphysema. $Fbln5^{-/-}$ mice also exhibited vascular anomalies. Aortas of $Fbln5^{-/-}$ mice were distended and tortuous compared to aortas from *WT* mice. The defect in vessel development was also observed in subcutaneous vessels from $Fbln5^{-/-}$ mice. These vessels developed in a disorganized, sinuous pattern, and had an increase in

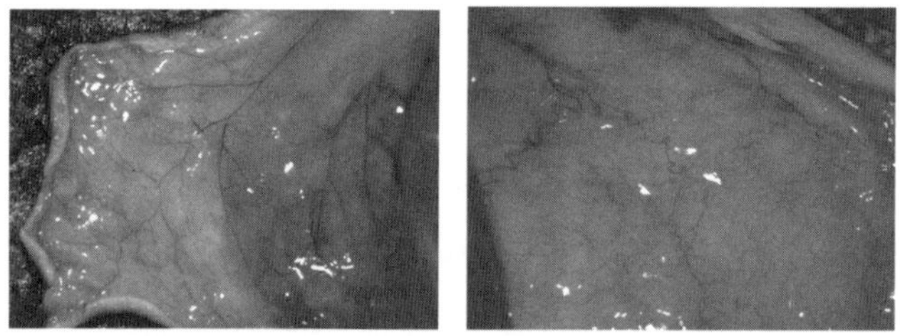

Figure 3.2 Subcutaneous blood vessel defect in Fbln5$^{-/-}$ mice. WT mice exhibit normal blood vessel patterning while blood vessels in Fbln5$^{-/-}$ mice appear tortuous with irregular patterning. Courtesy of Hiromi Yanagiswa, unpublished data.

vessel sprouting (Figure 3.2). The defect in blood vessel patterning was contributed to by fragmented elastic fibers within the elastic laminae decreasing the integrity of the blood vessel wall (Yanagisawa *et al.*, 2002).

Within tissues, Fbln5 colocalizes with elastin, the main protein of elastic fibers. It is expressed predominantly in tissues containing high levels of elastic fibers particularly blood vessels including the great vessels and the aorta. Fbln5 is also present in lungs, skin, and uterus, and it is interesting to note that within these tissues, expression is again mostly localized to blood vessels (Yanagisawa *et al.*, 2002). The level of Fbln5 protein expression is reduced in adult vessels compared to neonatal blood vessels but is elevated in adult mice in response to vascular insult or atherosclerotic plaques and in interstitial fibroblasts during lung injury repair. This suggests a possible function for Fbln5 in the remodeling process that occurs after vascular injury (Kowal *et al.*, 1999; Kuang *et al.*, 2003; Nakamura *et al.*, 1999).

Fbln5 is secreted by fibroblasts, vascular smooth muscle cells (VSMCs), and endothelial cells. Within the ECM, Fbln5 binds to structural proteins such as collagen allowing it to assist in cell adhesion by acting as a bridge between cells and the ECM via its interaction with integrins through its RGD motif (Timpl *et al.*, 2003; Lomas *et al.*, 2007). Fbln5 serves as a substrate for cells through interaction with integrins $\alpha_V\beta_3$, $\alpha_V\beta_5$, and $\alpha_9\beta_1$ (Nakamura *et al.*, 2002). Fbln5 was also shown to bind to smooth muscle cells by interacting with $\alpha_5\beta_1$ and $\alpha_4\beta_1$, the primary fibronectin receptors. The ability of Fbln5 to bind to integrins allows it to directly influence signals transduced by cell-ECM interactions. However, the effect is cell type dependent. For example, overexpression of Fbln5 by retroviral infection in 3T3-L1 fibroblasts increased DNA synthesis and proliferation through a mechanism involving TGFβ. In the same study, researchers showed that increased expression of Fbln5 in an epithelial cell line (Mv1Lu) curtailed proliferation by suppressing cyclin A expression (Schiemann *et al.*, 2002).

Convincing evidence indicates that Fbln5 functions as a blocking peptide by binding to integrins and preventing their interaction with growth factors and other ECM proteins. For instance, VSMCs harvested from *Fbln5*$^{-/-}$ mice exhibited no difference in proliferation rates compared to *WT* VSMCs. However, when treated with PDGF, *Fbln5*$^{-/-}$ VSMCs exhibited a twofold increase in proliferation compared to PDGF-treated *WT* cells. This effect was curtailed by treatment with recombinant Fbln5 protein suggesting that the physical presence of Fbln5 inhibits PDFG cell stimulation (Spencer *et al.*, 2005). In relation to this, Fbln5 binding to integrins $\alpha_5\beta_1$ and $\alpha_4\beta_1$ does not result in integrin activation and cells plated on fibronectin in the presence of Fbln5 do not undergo fibronectin-mediated spreading, migration, or proliferation (Lomas *et al.*, 2007). These results support the hypothesis that Fbln5 functions as a blocking peptide to control integrin-mediated cell signaling.

3.3.4.1 *Fibulin-5 and angiogenesis*

The high level of Fbln5 expression by endothelial cells and the aberrant vessel defect in *Fbln5*$^{-/-}$ mice indicates a critical function for Fbln5 in the vascular environment. Initial *in vitro* studies denoted Fbln5 as an inhibitor of angiogenesis. Treatment with recombinant Fbln5 inhibited the proliferation and invasion of murine brain microvascular endothelial cells through matrigel by antagonizing VEGF activation of the ERK1/ERK2 signaling pathway. Furthermore, Fbln5 was shown to be a target of TGFβ and induced the expression of the antiangiogenic protein TSP-1. In addition, activated endothelial cells undergoing tubulogenesis downregulated expression of Fbln5 (Albig and Schiemann, 2004).

In vivo studies using *Fbln5*$^{-/-}$ mice correlated with the *in vitro* findings and further indicated Fbln5's function as an angiogenic inhibitor. *Fbln5*$^{-/-}$ mice bearing implanted PVA sponges had increased vascular invasion compared to *WT* animals (Sullivan *et al.*, 2007). Expression of proangiogenic proteins such as VEGF and angiopoietin-1, -2, and -3 were increased in sponges removed from *Fbln5*$^{-/-}$ mice compared to those from *WT* mice. *Fbln5*$^{-/-}$ mice also exhibited an increase in vessel branching off the long thoracic artery (Sullivan *et al.*, 2007). Furthermore, studies addressing Fbln5's function during wound repair, a process that requires angiogenesis, revealed increased neovascularization in the skin of *Fbln5*$^{-/-}$ mice before and during wound repair. However, this study did not observe a difference in wound repair rates between *Fbln5*$^{-/-}$ and *WT* mice despite the increase in vessel formation (Lee *et al.*, 2004; Zheng *et al.*, 2006). These studies suggest that in certain physiological circumstances Fbln5 can inhibit angiogenesis through an undefined mechanism. However, Schluterman *et al.* (2010) suggest that Fbln5 controls angiogenic stimulation by limiting the production of reactive oxygen species (ROS), potent proangiogenic molecules. In this study, the loss of Fbln5 expression in endothelial cells resulted in a significant increase in fibronectin-mediated ROS production. ROS production was inhibited when Fbln5 deficient cells were treated with an antibody that blocked the activation of β1 integrins indicating that Fbln5 controlled ROS production by abrogating the activation of β1 integrins by fibronectin (Chiarugi *et al.*, 2003). ROS stimulate expression of VEGF and other angiogenic proteins and the regulation of ROS production by Fbln5 could explain previous findings where the loss of Fbln5 induced an increase in VEGF expression.

3.3.4.2 *Fibulin-5 and tumor development*

Investigation into the effect of Fbln5 on tumor development and progression are still in the early stages and little is currently known. However, the expression of Fbln5 in human cancers has recently been examined in a small subset of tumor types including kidney, breast, lung, ovary, and some gastrointestinal cancers. In this study, Fbln5 mRNA was evaluated using a cDNA microarray coated with matched normal/tumor cDNA from 68 patients with varying cancers. Fbln5 expression was altered in 44 of 68 samples and of those 44 cases, expression was downregulated in 42 and upregulated in only 2 (Schiemann *et al.*, 2002). It is important to note, however, that the samples examined in this study were derived from whole tumors. Therefore, the source of Fbln5 expression was not determined, whether from the stromal/host compartment or tumor cells. As mentioned previously, studies of SPARC and TSP-1 have shown that proteins expressed by stromal cells can have vastly different effects on tumor growth than proteins secreted from tumor cells. A closer look at the expression level of Fbln5 by these two compartments is needed to fully determine how human tumor development effects Fbln5 expression. In a related study (Wlazlinski *et al.*, 2007), analysis of prostate cancer identified an increase in Fbln5 expression compared to benign tissue. Expression did not correlate with advancing stages but instead remained consistent throughout prostate cancer progression. Interestingly, staining of tumor tissue with a Fbln5-specific antibody found Fbln5 expression to be present in the nuclei of tumor cells (Wlazlinski *et al.*, 2007). A function for Fbln5 in the nucleus has not been investigated but might explain how Fbln5 affects gene expression.

As seen with other matricellular proteins, *in vitro* and *in vivo* studies have shown the effect of Fbln5 on tumor growth to be complex and context-dependent. Given that forced expression of Fbln5 in 3T3-L1 fibroblasts enhanced DNA synthesis and proliferation, experiments were performed to determine the effect of Fbln5 on the tumorigencity of fibrosarcoma cells. Although forced expression of Fbln1, a closely related family member, in fibrosarcoma cells inhibited proliferation, expression of Fbln5 increased proliferation and improved invasion (Schiemann *et al.*, 2002). Fbln1 does not contain an RGD domain; therefore it is possible that the different effects observed in these cells was the result of Fbln5's ability to modify integrin signaling. In stark contrast to the *in vitro* findings, fibrosarcoma cells modified to increase expression of Fbln5 and injected into genetically normal mice developed smaller, slower tumors. Tumors over expressing Fbln5 had impaired angiogenesis, which is believed to be the cause for the decreased tumor growth (Albig, Neil, and Schiemann, 2006). Furthermore, histamine, an inflammatory molecule, was shown to heighten B16-F10 melanoma growth in mice. A study exploring this effect identified Fbln5 as an indirect target for histamine. Melanomas secreting high levels of histamine had significantly reduced expression of Fbln5 compared to tumors not expressing histamine. Histamine secretion also decreased insulin-like growth factor II receptor (IGF-IIR) expression, which is required for activation of TGF-β. Since Fbln5 expression had previously been shown to be induced by TGF-β, it was hypothesized that histamine indirectly regulates the level of Fbln5 expression by reducing IGF-IIR expression thus aiding in the promotion of tumor growth (Pos *et al.*, 2008). Unfortunately, angiogenesis was not evaluated in these tumors.

Although over expression of exogenous Fbln5 protein by tumor cells had a negative impact on tumor growth, loss of endogenous host-derived Fbln5 resulted in reduced pancreatic tumor growth (Schluterman *et al.*, 2010). Pancreatic tumors grown in $Fbln5^{-/-}$ mice grew slower and exhibited a decrease in blood vessel density compared to tumors grown in *WT* littermates. Tumors from $Fbln5^{-/-}$ mice also contained elevated levels of ROS, which caused chronic oxidative damage within the tumor. Similar tumor growth was observed when tumors were grown in transgenic knockin mice containing an altered RGD motif. In these mice, the aspartic acid of the integrin-binding RGD sequence of Fbln5 was changed to glutamic acid (RGE) rendering the protein unable to bind to integrins. Tumors from $Fbln5^{RGE/RGE}$ had elevated ROS confirming that Fbln5 binding to integrins via the RGD domain was critical to its regulation of ROS production. Elevated ROS levels were proven to be the cause for the tumor and blood vessel growth defect in these mice because when tumor experiments were repeated in $Fbln5^{RGE/RGE}$ mice in the presence of an antioxidant to reduce the level of ROS, tumor angiogenesis, and growth was restored to *WT* levels (Schluterman *et al.*, 2010). These results indicated that the loss of Fbln5's regulation of integrin-induced ROS production created a TME that could no longer support endothelial cell survival and therefore was detrimental to tumor growth. Increasing ROS within the TME has long been used as a means of controlling tumor growth. It has therefore been suggested that altering ROS levels by synthetically inhibiting Fbln5's interaction with β1 integrins could be an effective approach to cancer therapy.

3.4 Conclusion

Understanding the many facets of the ECM is daunting. Not only is the ECM critical for maintaining tissue integrity and stability, but it also modulates seemingly every aspect of cell activity including survival, polarity, proliferation, transcription, and migration. Numerous studies have shown that tissue architecture created by the specific deposition of the structural ECM proteins including collagen, fibronectin, and elastic fibers strongly influence the invasive capabilities of tumor cells. Alterations in the structure of the ECM by changes in ECM synthesis and degradation have dramatic effects on tumor outcome, but these effects are highly context dependent making the outcomes unpredictable and hard to target for therapeutic purposes. The effects of matricellular proteins on tumor cells are equally unpredictable. The source of matricellular proteins whether tumor or host-derived is critical and seem to dictate how matricellular proteins will affect tumor growth. Although complex and often times contradictory reports about the ECM make designing therapeutic strategies difficult, it is clear that to be truly effective, the next generation of cancer therapy must be multidimensional using a combination of drugs that target components of both tumor cells and the TME.

References

Albig, A.R., Neil, J.R., and Schiemann, W.P. (2006) Fibulins 3 and 5 antagonize tumor angiogenesis in vivo. *Cancer Research*, **66**, 2621–2629.

Albig, A.R. and Schiemann, W.P. (2004) Fibulin-5 antagonizes vascular endothelial growth factor (VEGF) signaling and angiogenic sprouting by endothelial cells. *DNA and Cell Biology*, **23**, 367–379.

Albo, D., Berger, D.H., Vogel, J., and Tuszynski, G.P. (1999) Thrombospondin-1 and transforming growth factor beta-1 upregulate plasminogen activator inhibitor type 1 in pancreatic cancer. *Journal of Gastrointestinal Surgery*, **3**, 411–417.

Arnold, S.A. and Brekken, R.A. (2009) SPARC: a matricellular regulator of tumorigenesis. *Journal of Cell Communication and Signaling*, **3** (3–4), 255–273.

Arnold, S., Mira, E., Muneer, S. *et al.* (2008) Forced expression of MMP9 rescues the loss of angiogenesis and abrogates metastasis of pancreatic tumors triggered by the absence of host SPARC. *Experimental Biology and Medicine (Maywood)*, **233**, 860–873.

Arnold, S.A., Rivera, L.B., Miller, A.F., *et al.* (2009) Lack of host SPARC enhances vascular function and tumor spread in an orthotopic murine model of pancreatic carcinoma. *Disease Models and Mechanisms*, **3**, 57–72.

Beck, A.H., Espinosa, I., Gilks, C.B. *et al.* (2008) The fibromatosis signature defines a robust stromal response in breast carcinoma. *Laboratory Investigations*, **88**, 591–601.

Bergamaschi, A., Tagliabue, E., Sorlie, T. *et al.* (2008) Extracellular matrix signature identifies breast cancer subgroups with different clinical outcome. *Journal of Pathology*, **214**, 357–367.

Bergers, G., Brekken, R., Mcmahon, G. *et al.* (2000) Matrix metalloproteinase-9 triggers the angiogenic switch during carcinogenesis. *Nature Cell Biology*, **2**, 737–744.

Brown, L.F., Papadopoulos-Sergiou, A., Berse, B. *et al.* (1994) Osteopontin expression and distribution in human carcinomas. *American Journal of Pathology*, **145**, 610–623.

Chambers, A.F., Behrend, E.I., Wilson, S.M., and Denhardt, D.T. (1992) Induction of expression of osteopontin (OPN; secreted phosphoprotein) in metastatic, ras-transformed NIH 3T3 cells. *Anticancer Research*, **12**, 43–47.

Chiarugi, P. and Giannoni, E. (2008) Anoikis: a necessary death program for anchorage-dependent cells. *Biochemical Pharmacology*, **76**, 1352–1364.

Chiarugi, P., Pani, G., Giannoni, E. *et al.* (2003) Reactive oxygen species as essential mediators of cell adhesion: the oxidative inhibition of a FAK tyrosine phosphatase is required for cell adhesion. *Journal of Cell Biology*, **161**, 933–944.

Clezardin, P., Frappart, L., Clerget, M. *et al.* (1993) Expression of thrombospondin (TSP1) and its receptors (CD36 and CD51) in normal, hyperplastic, and neoplastic human breast. *Cancer Research*, **53**, 1421–1430.

Cui, R., Takahashi, F., Ohashi, R. *et al.* (2007) Abrogation of the interaction between osteopontin and alphavbeta3 integrin reduces tumor growth of human lung cancer cells in mice. *Lung Cancer*, **57**, 302–310.

Deryugina, E.I. and Quigley, J.P. (2006) Matrix metalloproteinases and tumor metastasis. *Cancer and Metastasis Reviews*, **25**, 9–34.

Erler, J.T. and Weaver, V.M. (2009) Three-dimensional context regulation of metastasis. *Clinical and Experimental Metastasis*, **26**, 35–49.

Figueira, R.C., Gomes, L.R., Neto, J.S. *et al.* (2009) Correlation between MMPs and their inhibitors in breast cancer tumor tissue specimens and in cell lines with different metastatic potential. *BMC Cancer*, **9**, 20.

Framson, P.E. and Sage, E.H. (2004) SPARC and tumor growth: where the seed meets the soil. *Journal of Cellular Biochemistry*, **92**, 679–690.

Freeman, L.J., Lomas, A., Hodson, N. *et al.* (2005) Fibulin-5 interacts with fibrillin-1 molecules and microfibrils. *Biochemical Journal*, **388**, 1–5.

Gautam, A., Densmore, C.L., Melton, S. *et al.* (2002) Aerosol delivery of PEIp53 complexes inhibits B16-F10 lung metastases through regulation of angiogenesis. *Cancer Gene Therapy*, **9**, 28–36.

Giannoni, E., Buricchi, F., Grimaldi, G. *et al.* (2008) Redox regulation of anoikis: reactive oxygen species as essential mediators of cell survival. *Cell Death and Differentiation*, **15**, 867–878.

Greenaway, J., Lawler, J., Moorehead, R. *et al.* (2007) Thrombospondin-1 inhibits VEGF levels in the ovary directly by binding and internalization via the low density lipoprotein receptor-related protein-1 (LRP-1). *Journal of Cellular Physiology*, **210**, 807–818.

Hartel, M., Di Mola, F.F., Gardini, A. *et al.* (2004) Desmoplastic reaction influences pancreatic cancer growth behavior. *World Journal of Surgery*, **28**, 818–825.

Hasselaar, P. and Sage, E.H. (1992) SPARC antagonizes the effect of basic fibroblast growth factor on the migration of bovine aortic endothelial cells. *Journal of Cellular Biochemistry*, **49**, 272–283.

Hofmann, U.B., Houben, R., Brocker, E.B., and Becker, J.C. (2005) Role of matrix metalloproteinases in melanoma cell invasion. *Biochimie*, **87**, 307–314.

Irby, R.B., Mccarthy, S.M., and Yeatman, T.J. (2004) Osteopontin regulates multiple functions contributing to human colon cancer development and progression. *Clinical and Experimental Metastasis*, **21**, 515–523.

Jenkins, G. (2008) The role of proteases in transforming growth factor-beta activation. *International Journal of Biochemistry and Cell Biology*, **40**, 1068–1078.

Jimenez, B., Volpert, O.V., Crawford, S.E. *et al.* (2000) Signals leading to apoptosis-dependent inhibition of neovascularization by thrombospondin-1. *Nature Medicine*, **6**, 41–48.

Johnston, N.I., Gunasekharan, V.K., Ravindranath, A. *et al.* (2008) Osteopontin as a target for cancer therapy. *Frontiers in Bioscience*, **13**, 4361–4372.

Kadler, K.E., Hill, A., and Canty-Laird, E.G. (2008) Collagen fibrillogenesis: fibronectin, integrins, and minor collagens as organizers and nucleators. *Current Opinion in Cell Biology*, **20**, 495–501.

Kazerounian, S., Yee, K.O., and Lawler, J. (2008) Thrombospondins in cancer. *Cellular and Molecular Life Sciences*, **65**, 700–712.

van Kempen, L.C., Ruiter, D.J., Van Muijen, G.N., and Coussens, L.M. (2003) The tumor microenvironment: a critical determinant of neoplastic evolution. *European Journal of Cell Biology*, **82**, 539–548.

Koblinski, J.E., Kaplan-Singer, B.R., Vanosdol, S.J. *et al.* (2005) Endogenous osteonectin/SPARC/BM-40 expression inhibits MDA-MB-231 breast cancer cell metastasis. *Cancer Research*, **65**, 7370–7377.

Korc, M. (2007) Pancreatic cancer-associated stroma production. *American Journal of Surgery*, **194**, S84–S86.

Kowal, R.C., Richardson, J.A., Miano, J.M., and Olson, E.N. (1999) EVEC, a novel epidermal growth factor-like repeat-containing protein upregulated in embryonic and diseased adult vasculature. *Circulation Research*, **84**, 1166–1176.

Kuang, P.P., Goldstein, R.H., Liu, Y. *et al.* (2003) Coordinate expression of fibulin-5/DANCE and elastin during lung injury repair. *American Journal of Physiology: Lung Cellular and Molecular Physiology*, **285**, L1147–L1152.

Lawler, J., Miao, W.M., Duquette, M. *et al.* (2001) Thrombospondin-1 gene expression affects survival and tumor spectrum of p53-deficient mice. *American Journal of Pathology*, **159**, 1949–1956.

Lee, S., Jilani, S.M., Nikolova, G.V. *et al.* (2005) Processing of VEGF-A by matrix metalloproteinases regulates bioavailability and vascular patterning in tumors. *Journal of Cell Biology*, **169**, 681–691.

Lee, M.J., Roy, N.K., Mogford, J.E. *et al.* (2004) Fibulin-5 promotes wound healing in vivo. *Journal of the American College of Surgeons*, **199**, 403–410.

Levental, K.R., Yu, H., Kass, L. *et al.* (2009) Matrix crosslinking forces tumor progression by enhancing integrin signaling. *Cell*, **139**, 891–906.

Liu, X., Zhao, Y., Gao, J. *et al.* (2004) Elastic fiber homeostasis requires lysyl oxidase-like 1 protein. *Nature Genetics*, **36**, 178–182.

Lock, J.G., Wehrle-Haller, B., and Stromblad, S. (2008) Cell-matrix adhesion complexes: master control machinery of cell migration. *Seminars in Cancer Biology*, **18**, 65–76.

Loeys, B., Van Maldergem, L., Mortier, G. *et al.* (2002) Homozygosity for a missense mutation in fibulin-5 (FBLN5) results in a severe form of cutis laxa. *Human Molecular Genetics*, **11**, 2113–2118.

Lomas, A.C., Mellody, K.T., Freeman, L.J. *et al.* (2007) Fibulin-5 binds human smooth-muscle cells through alpha5beta1 and alpha4beta1 integrins, but does not support receptor activation. *Biochemical Journal*, **405**, 417–428.

Lopez-Otin, C., Palavalli, L.H., and Samuels, Y. (2009) Protective roles of matrix metalloproteinases: from mouse models to human cancer. *Cell Cycle*, **8**, 3657–3662.

Lussier, C., Sodek, J., and Beaulieu, J.F. (2001) Expression of SPARC/osteonectin/BM4O in the human gut: predominance in the stroma of the remodeling distal intestine. *Journal of Cellular Biochemistry*, **81**, 463–476.

Mao, Y. and Schwarzbauer, J.E. (2005) Fibronectin fibrillogenesis, a cell-mediated matrix assembly process. *Matrix Biology*, **24**, 389–399.

Marastoni, S., Ligresti, G., Lorenzon, E. *et al.* (2008) Extracellular matrix: a matter of life and death. *Connective Tissue Research*, **49**, 203–206.

Mizejewski, G.J. (1999) Role of integrins in cancer: survey of expression patterns. *Proceedings of the Society for Experimental Biology and Medicine*, **222**, 124–138.

Nakamura, T., Lozano, P.R., Ikeda, Y. *et al.* (2002) Fibulin-5/DANCE is essential for elastogenesis in vivo. *Nature*, **415**, 171–175.

Nakamura, T., Ruiz-Lozano, P., Lindner, V. *et al.* (1999) DANCE, a novel secreted RGD protein expressed in developing, atherosclerotic, and balloon-injured arteries. *Journal of Biological Chemistry*, **274**, 22476–22483.

Nor, J.E., Mitra, R.S., Sutorik, M.M. *et al.* (2000) Thrombospondin-1 induces endothelial cell apoptosis and inhibits angiogenesis by activating the caspase death pathway. *Journal of Vascular Research*, **37**, 209–218.

O'Reilly, M.S., Boehm, T., Shing, Y. *et al.* (1997) Endostatin: an endogenous inhibitor of angiogenesis and tumor growth. *Cell*, **88**, 277–285.

Pandol, S., Edderkaoui, M., Gukovsky, I. *et al.* (2009) Desmoplasia of pancreatic ductal adenocarcinoma. *Clinical Gastroenterology and Hepatology*, **7**, S44–S47.

Pasco, S., Ramont, L., Venteo, L. *et al.* (2004) In vivo overexpression of tumstatin domains by tumor cells inhibits their invasive properties in a mouse melanoma model. *Experimental Cell Research*, **301**, 251–265.

Pos, Z., Wiener, Z., Pocza, P. *et al.* (2008) Histamine suppresses fibulin-5 and insulin-like growth factor-II receptor expression in melanoma. *Cancer Research*, **68**, 1997–2005.

Pupa, S.M., Menard, S., Forti, S., and Tagliabue, E. (2002) New insights into the role of extracellular matrix during tumor onset and progression. *Journal of Cellular Physiology*, **192**, 259–267.

Raines, E.W., Lane, T.F., Iruela-Arispe, M.L. *et al.* (1992) The extracellular glycoprotein SPARC interacts with platelet-derived growth factor (PDGF)-AB and -BB and inhibits the binding of PDGF to its receptors. *Proceedings of the National Academy of Sciences of the United States of America*, **89**, 1281–1285.

Schiemann, W.P., Blobe, G.C., Kalume, D.E. *et al.* (2002) Context-specific effects of fibulin-5 (DANCE/EVEC) on cell proliferation, motility, and invasion. Fibulin-5 is induced by transforming growth factor-beta and affects protein kinase cascades. *Journal of Biological Chemistry*, **277**, 27367–27377.

Schluterman, M.K., Chapman, S.L., Korpanty, G. *et al.* (2010) Loss of fibulin-5 binding to {beta}1 integrins inhibits tumor growth by increasing the level of ROS. *Disease Models & Mechanisms*, **3** (5–6), 333–342.

Spencer, J.A., Hacker, S.L., Davis, E.C. *et al.* (2005) Altered vascular remodeling in fibulin-5-deficient mice reveals a role of fibulin-5 in smooth muscle cell proliferation and migration. *Proceedings of the National Academy of Sciences of the United States of America*, **102**, 2946–2951.

Standal, T., Borset, M., and Sundan, A. (2004) Role of osteopontin in adhesion, migration, cell survival and bone remodeling. *Experimental Oncology*, **26**, 179–184.

Sullivan, K.M., Bissonnette, R., Yanagisawa, H. *et al.* (2007) Fibulin-5 functions as an endogenous angiogenesis inhibitor. *Laboratory Investigation*, **87**, 818–827.

Svoboda, K.K., Fischman, D.A., and Gordon, M.K. (2008) Embryonic chick corneal epithelium: a model system for exploring cell-matrix interactions. *Developmental Dynamics*, **237**, 2667–2675.

Tai, I.T., Dai, M., Owen, D.A., and Chen, L.B. (2005) Genome-wide expression analysis of therapy-resistant tumors reveals SPARC as a novel target for cancer therapy. *Journal of Clinical Investigation*, **115**, 1492–1502.

Tai, I.T. and Tang, M.J. (2008) SPARC in cancer biology: Its role in cancer progression and potential for therapy. *Drug Resistance Updates,* **11** (6), 231–246.

Tang, M.J. and Tai, I.T. (2007) A novel interaction between procaspase 8 and SPARC enhances apoptosis and potentiates chemotherapy sensitivity in colorectal cancers. *Journal of Biological Chemistry*, **282**, 34457–34467.

Ten Dijke, P. and Arthur, H.M. (2007) Extracellular control of TGFbeta signalling in vascular development and disease. *Nature Reviews Molecular Cell Biology*, **8**, 857–869.

Timpl, R., Sasaki, T., Kostka, G., and Chu, M.L. (2003) Fibulins: a versatile family of extracellular matrix proteins. *Nature Reviews Molecular Cell Biology*, **4**, 479–489.

Wagenseil, J.E. and Mecham, R.P. (2007) New insights into elastic fiber assembly. *Birth Defects Research Part C: Embryo Today*, **81**, 229–240.

Wang, T.N., Qian, X., Granick, M.S. *et al.* (1996) Thrombospondin-1 (TSP-1) promotes the invasive properties of human breast cancer. *Journal of Surgical Research*, **63**, 39–43.

Wang-Rodriguez, J., Urquidi, V., Rivard, A., and Goodison, S. (2003) Elevated osteopontin and thrombospondin expression identifies malignant human breast carcinoma but is not indicative of metastatic status. *Breast Cancer Research*, **5**, R136–R143.

Watnick, R.S., Cheng, Y.N., Rangarajan, A. *et al.* (2003) Ras modulates Myc activity to repress thrombospondin-1 expression and increase tumor angiogenesis. *Cancer Cell*, **3**, 219–231.

Weber, G.F. (2001) The metastasis gene osteopontin: a candidate target for cancer therapy. *Biochimica et Biophysica Acta*, **1552**, 61–85.

Wlazlinski, A., Engers, R., Hoffmann, M.J. *et al.* (2007) Downregulation of several fibulin genes in prostate cancer. *The Prostate*, **67**, 1770–1780.

Wojtowicz-Praga, S.M., Dickson, R.B., and Hawkins, M.J. (1997) Matrix metalloproteinase inhibitors. *Investigational New Drugs*, **15**, 61–75.

Wu, Y. and Zhou, B.P. (2009) Inflammation: a driving force speeds cancer metastasis. *Cell Cycle*, **8**, 3267–3273.

Yan, Q. and Sage, E.H. (1999) SPARC, a matricellular glycoprotein with important biological functions. *Journal of Histochemistry and Cytochemistry*, **47**, 1495–1506.

Yanagisawa, H., Davis, E.C., Starcher, B.C. *et al.* (2002) Fibulin-5 is an elastin-binding protein essential for elastic fibre development in vivo. *Nature*, **415**, 168–171.

Yang, E., Kang, H.J., Koh, K.H. *et al.* (2007) Frequent inactivation of SPARC by promoter hypermethylation in colon cancers. *International Journal of Cancer*, **121**, 567–575.

Zheng, Q., Choi, J., Rouleau, L. *et al.* (2006) Normal wound healing in mice deficient for fibulin-5, an elastin binding protein essential for dermal elastic fiber assembly. *Journal of Investigative Dermatology*, **126**, 2707–2714.

Zheng, Q., Davis, E.C., Richardson, J.A. *et al.* (2007) Molecular analysis of fibulin-5 function during de novo synthesis of elastic fibers. *Molecular and Cellular Biology*, **27**, 1083–1095.

4

Matrix Metalloproteinases and Their Inhibitors – Friend or Foe

Mumtaz V. Rojiani[1], Marzenna Wiranowska[1] and Amyn M. Rojiani[2]

[1]*Department of Pathology and Cell Biology, College of Medicine, University of South Florida, Tampa, FL, USA*
[2]*Department of Pathology, Medical College of Georgia, Augusta GA, USA*

4.1 Introduction

The invasive and metastatic activities of cancer cells remain by far the most significant contributors to morbidity and mortality. Interactions within the tumor microenvironment, extracellular matrix (ECM) including, but not limited to proteolytic enzymes, their inhibitors, and a wide range of cytokines play a fundamental role in this process. The ECM is a complex network of highly specialized assemblies of structural and signaling macromolecules and molecules that provide the cohesiveness and dynamic physiochemical properties of a tissue or organ. ECM structural components include: (i) fibrous elements, for example, collagen, (ii) link proteins, for example, fibronectin or laminin, and (iii) space-filling molecules, for example, glycosaminoglycans (GAGs). In addition, the ECM and the extracellular space house various signaling molecules such as ions, peptides, cytokines, growth factors, hormones, metabolites. The primary elements that modulate change within this component of the physiologic and tumor microenvironment include matrix metalloproteinases (MMPs) and tissue inhibitors of matrix metalloproteinases (TIMPs). The individual and balanced activity of both MMP and TIMP impact a wide range of functions including but not limited to ECM turnover and remodeling, tissue growth, angiogenesis, and programmed cell death. Each of these physiologic activities takes on additional significance in the context of the tumor microenvironment and consequently therapeutic interventions for tumor control must be based on a more complete understanding of the functions and mechanisms by which these molecules act.

Tumor Microenvironment Edited by Dietmar W. Siemann

4.2 Matrix metalloproteinases

The collagenolytic activity of endopeptidases found in tadpoles undergoing metamorphosis, initially observed by Gross and Lapiere (1962) was an early step in the investigations into the family of endopeptidases that were later named MMPs. Since then this rapidly expanding field, evaluating the enzymes in plants and animals, has revealed the highly pleiotropic nature of these molecules, both in health and disease. It became known that the human MMP family comprising of over 23 members, is capable of degrading and processing almost all components of the ECM including the basement membrane, as well as a wide variety of non-matrix substrates important in tumor microenvironment and at all stages of tumor progression (Martin and Matrisian, 2007; Page-Mccaw, Ewald, and Werb, 2007; Rodriguez, Morrison, and Overall, 2009).

4.2.1 Structure and classification of matrix metalloproteinases

Members of the MMP family are synthesized as zymogens and either secreted from the cell into ECM or anchored to the plasma membrane, or some are even found as intracellular proteins. MMP expression is mainly regulated at the transcriptional level, but recent reports suggest that post-transcriptional events may play a role (Fanjul-Fernandez *et al.*, 2010). MMPs share a highly conserved zinc binding site in their catalytic domain and their classification is based on their substrate specificity and domain organization (Fanjul-Fernandez *et al.*, 2010; Martin and Matrisian, 2007). Based primarily on their domain organization the following groups of MMPs are distinguished (Figure 4.1): (i) archetypal MMPs, (ii) matrilysins, (iii) gelatinases, and (iv) furin-activatable MMPs (Fanjul-Fernandez *et al.*, 2010). MMPs which play an essential role in the tumor microenvironment can be found in each of these groups.

4.2.1.1 The archetypal MMPs

This includes three subgroups based on substrate specificity:

Collagenases, which include three enzymes found in mammals: MMP-1, MMP-8, and MMP-13 (also known as collagenases 1, 2, and 3) all of which can cleave collagen and are able to process other ECM molecules and bioactive molecules such as cytokines and growth factors.

Stromelysins, including MMP-3 and MMP-10 (known as Stromelysins 1 and 2) can process many ECM components as well as some growth factors, cytokines, and adhesion molecules, except for the native collagen. Stromelysins can also participate in activation of some proMMPs to generate active forms of these MMPs, such as MMP-1, -8, -13, and -9. Stromelysins are expressed by fibroblast and epithelial cells and are secreted into the extracellular space.

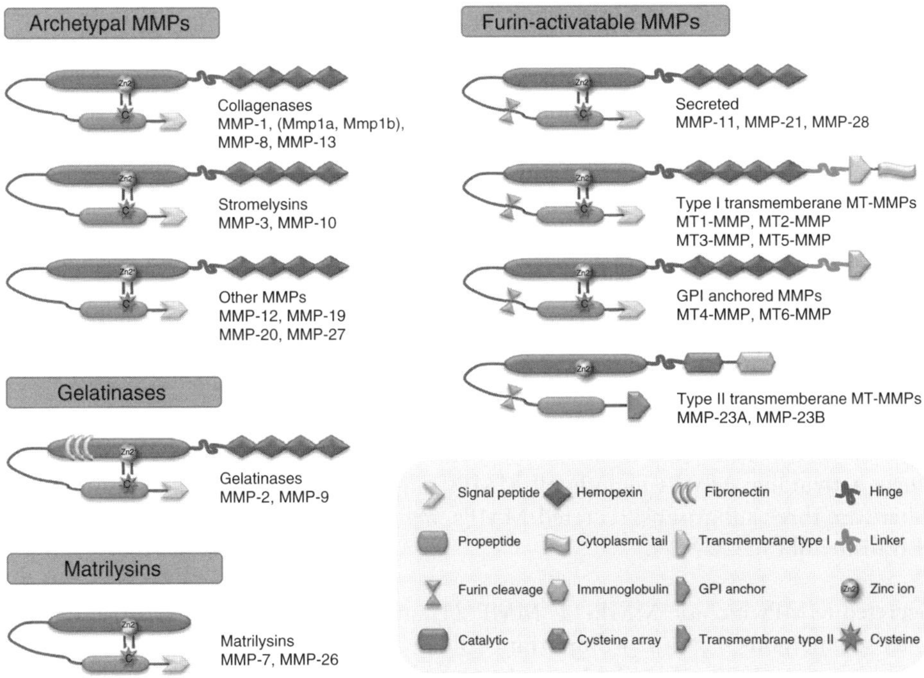

Figure 4.1 The mammalian family of matrix metalloproteinases. Structural classification of MMPs based on their domain arrangement. (Reproduced with permission from Fanjul-Fernandez, M., Folgueras, A.R., Cabrera, S. and Lopez-Otin, C. (2010) Matrix metalloproteinases: evolution gene regulation and functional analysis in mouse models. *Biochimica et Biophysica Acta*, **1803**, 3–19.)

Other archetypal MMPs such as MMP-12, -19, -20, and -27, all of which have specific sequences and substrates. For example, MMP-12 (metalloelastase) is a potent elastolytic enzyme, which also degrades many other ECM components and is mainly expressed by macrophages. MMP-19, expressed by many human tissues, is associated with inflammation and has potent degrading activity against basement membrane components. Finally, MMP-20 is involved in tooth enamel formation while MMP-27 (cloned from chicken embryo fibroblasts) is highly expressed by human B-lymphocytes but little is known about its activity in humans.

4.2.1.2 *Matrilysins*

Matrilysins include MMP-7 and MMP-26 (also known as matrilysins 1 and 2) and are both expressed under physiologic conditions as well as associated with several types of cancer. They play important roles in degradation and processing of ECM and non-ECM proteins. In addition MMP-26 is known to activate proMMP-9 under pathological conditions.

4.2.1.3 Gelatinases

Gelatinases include MMP-2 and MMP-9 (also known as gelatinase-A and -B) and are constitutively expressed by many cell types and associated with many pathologic conditions including cancer. MMP-2 is expressed by fibroblasts, endothelial cells, monocytes, keratinocytes, and chondrocytes while MMP-9 is expressed by alveolar macrophages, polymorphonuclear leukocytes, and osteoclasts. They play an important role in the remodeling of collagenous ECM, also targeting other ECM and non-ECM molecules such as progrowth factors, growth factors, procytokines, and some chemoattractants. They are also capable of releasing or generating several factors with pro- or antiangiogenic properties.

4.2.1.4 Furin-activatable MMPs

Furin-activatable MMPs include all MMPs that contain a furin recognition motif. There are three subgroups: secreted-MMPs, membrane-type (MT) MMPs, and type II transmembrane MMPs.

Secreted MMPs include MMP-11, MMP-21, and MMP-28. MMP-11 (also known as stromelysin-3) plays an important role in normal tissues as well as in cancer, linking it to obesity. MMP-11 was shown to be induced in adipose tissue by cancer cells and is responsible for tumor progression through the degradation of collage VI. Besides its physiologic activities, MMP-21 is also linked to various epithelial cancers. The role of MMP-28 which is also produced by several carcinomas is not yet well defined, but it is believed to play a role in several diseases of the central nervous system (CNS) including multiple sclerosis.

Membrane-type (MT) MMPs include MMP-14, -15, -16, -17, -24, and -25, all of which are located on the cell surface and therefore able to control the local environment of normal and tumor cells. They are divided into two subgroups depending on their attachment to the plasma membrane: the first subgroup is type I transmembrane and includes MMP-14, 15, 16, and 24 (also known as MT1-, MT2-, MT3-, and MT5-MMP respectively); the second subgroup is glycosylphosphatidylinositol (GPI) MT-MMPs and includes MMP-17 and MMP-25 (also known as MT4- and MT-6-MMP). MT-MMPs are expressed under normal conditions by various tissues. They are the main activators of proMMP-2 and are involved in blood vessel formation. In addition, they are upregulated in tumors and in some cases associated with poor prognosis of several types of cancer.

Type II transmembrane MMPs includes MMP-23A and MMP-23B which have an identical amino acid sequence but they are encoded by distinct genes in the human genome. They are produced by various normal tissues but their precise roles have not yet been elucidated (Fanjul-Fernandez *et al.*, 2010).

4.2.1.5 *Disintegrin/Metalloproteinase*

The MMP family also encompasses a disintegrin and metalloproteinases (ADAMs) and the a disintegrin and metalloproteinase with thrombospondin (ADAMTS) motif (Malemud, 2006). An interaction between MMPs and ADAMTS was reported with MT4-MMP contributing to activation of ADAMTS-4 (Gao *et al.*, 2004). MMPs' enzymatic activity can be inhibited by naturally occurring specific endogenous tissue inhibitors of metalloproteinases or TIMPs.

4.2.2 Cellular sources and substrate/targets of MMP

4.2.2.1 *Cellular sources of MMPs in the tumor microenvironment*

4.2.2.1.1 *Tumor cells*

These are known sources of MMPs, detected both *in vivo* and *in vitro*. Various proinflammatory factors, including cytokines and MMPs are produced by the tumor as well as by the stroma. Together, these can enhance turnover of ECM and tumor cell migration. On the other hand, some cytokines can also have an inhibitory effect. For example, interferon (IFN) alpha/beta inhibited tumor cells secretion of active form of MMP-2 and chondrotin sulfate (CS) proteoglycan, an ECM component, as well as tumor cell migration (Wiranowska and Naidu, 1994; Wiranowska, Tresser, and Saporta, 1998, Wiranowska *et al.*, 2000. Stromal cells such as tumor-infiltrating leukocytes including mononuclear cells (monocytes and macrophages), granulocytes (neutrophils, eosinophils, basophils/mast cells), and lymphocytes are known to be the main producers of MMPs . In addition other stromal cells such as fibroblasts, endothelial cells, and vascular pericytes can also express MMPs (Deryugina and Quigley, 2006, 2010).

4.2.2.1.2 *Mononuclear cells (monocytes and macrophages)*

They can produce wide variety of MMPs (MMP-1, -2, -3, -7, -9, -10, -12, -13, -14, -19) capable of degrading ECM and inducing tumor angiogenesis. MMP-9, which is directly involved in angiogenesis, is released mainly by macrophages at the tumor site. Interestingly, the spectrum of MMPs and MMP inhibitors produced by these cells can vary depending on the stage of differentiation or tissue localization (Lewis and Pollard, 2006).

4.2.2.1.3 *Granulocytes (neutrophils, eosinophils, basophils/mast cells)*

Granulocytes, in general, release their MMPs on demand. MMPs released by neutrophils are MMP-8 and MMP-9 with latter shown to be highly potent proangiogenic enzyme. Tumor-infiltrating neutrophils localize predominantly in the tumor interior, in contrast to monocytes/macrophages, which are found at the tumor periphery or tumor/stroma border. Mast cells, like basophils, are able to release the vasoactive agents from their granules and produce MMP-9 and MMP-2.

4.2.2.1.4 *Lymphocytes (T and B cells)*

They express a variety of MMPs such as, MMP-1, -2, -9, and -14, but the role of lymphocyte-derived MMPs in tumor growth is not yet well understood.

4.2.2.1.5 *Fibroblasts*

Tumor-associated fibroblasts have been recognized for their role in cancer progression and angiogenesis. The cross-talk between tumor cells and normal stromal cells can trigger fibroblasts to express MMPs, for example, MMP-9. Other known fibroblast-derived MMPs include MMP-1, -7, and -14.

4.2.2.1.6 *Endothelial cells and perivascular pericytes*

Activated endothelial cells of the capillary network overexpress several MMPs (MMP-1, -9, and -14) during sprouting and formation of lumina-containing tubules. However, in the quiescent state, endothelial cells are relatively deficient in MMP expression. Perivascular cells such as pericytes and smooth muscle cells have been implicated in tumor angiogenesis. However, only pericytes were shown to express MMP-9. It is also believed that upon maturation of newly formed blood vessels, the interaction between pericytes and endothelial cells can lead to the silencing of MMPs (van Hinsbergh and Koolwijk, 2008; Deryugina and Quigley, 2010).

These complex interactions involving many cell types producing a range of MMPs within the tumor and the bordering stroma, as well as interaction of MMPs with their stimulators, inhibitors, and substrates become even more complex at the cellular level. For example, depending on the cellular source, the activity of a single MMP could have either inhibitory or stimulatory effect on the tumor growth.

4.2.2.2 *Targets and substrates of MMP action: cellular or extracellular*

The generation of mouse models has facilitated the identification of some *in vivo* substrates for MMPs. In addition to the ECM molecules first identified as the substrates for MMPs, many more non-ECM bioactive molecules have been identified: for example, growth factor receptors, adhesion molecules, cytokines, chemokines, angiogenic factors, apoptotic ligands, and so on (Yu and Stamenkovic, 1999; Egeblad and Werb, 2002).

4.2.2.2.1 *ECM substrates*

MMP-14 is one of the most potent modifiers of extracellular tumor environment contributing to matrix proteolysis and leading to migration and invasion of tumor and endothelial cells in this modified ECM (Itoh and Seiki, 2006; Packard *et al.*, 2009; Sabeh, Shimizu-Hirota, and Weiss, 2009). In addition, MMP-14 participates in the angiogenic process in growing tumors (Genis *et al.*, 2006; Itoh and Seiki, 2006; van Beem *et al.*, 2009; van Hinsbergh and Koolwijk, 2008). Remodeling of the basement membrane matrix and in particular the basement membrane of the endothelial cells, initiates the angiogenesis process. In that process the basement membrane, rich in laminins and collagen IV, undergoes degradation and reassembly (Davis and Senger, 2008), with MMP-2, -9, and -14 playing a role in remodeling of the basement membrane *in vivo* (Oh *et al.*, 2001).

4.2.2.2.2 Non-ECM substrates

MMPs can also proteolytically modify cell surface molecules. Both modification of ECM or cell surface by MMPs leads to the release of various proangiogenic factors or inhibitors of angiogenesis (Kalluri, 2003). An example of non-ECM substrates are cell surface adhesion molecules, for example, CD44 molecules. The transmembrane proteoglycan CD44, which is found at the leading edge of the invasive tumor (Wiranowska *et al.*, 2006), was shown to bind through its extracellular portion to several MMPs including MMP-2. As a result of this interaction, the MMPs were localized at the migrating front of the tumor. CD44 therefore serves as an MMP substrate and its proteolytic degradation promotes cell mobility (Seiki, 2002). In addition it was also found that binding of MMP-9 to CD44 promotes MMP activation (Yu and Stamenkovic, 1999).

4.2.3 Impact of MMP on tumor microenvironment and growth

MMPs function not only as effectors of tissue remodeling but also interact with and are modulated by a network of cytokines and growth factors. Both the ECM and basement membrane serve as depots for cytokines and growth factors bound to proteoglycans. Therefore, enzymatic degradation of the ECM results in the release and diffusion of cytokines and growth factors as well as activation of ECM molecules important in tissue pathology. Additionally, MMPs act as sheddases or convertases, as they transform membrane-bound cytokines, cytokine receptors, and adhesion molecules, into their soluble forms (Leppert *et al.*, 2001; Wiranowska *et al.*, 2008). The proteolytic action of MMPs affects basic cellular events such as cell proliferation, migration, adhesion, and also physiological processes related to ECM remodeling such as angiogenesis. In the tumor microenvironment, the proteolytic modification of many complex fibrillar proteins by MMPs facilitates protease-dependent tumor cell migration and tumor angiogenesis. Therefore, traditionally, the upregulation of MMPs has been associated with many pathological processes including inflammation and cancer.

MMPs *in vitro* show overlapping affinities for different ECM and non-ECM substrates and many MMPs can degrade and process several different classes of ECM proteins, for example, MMP-2 and MMP-9 can degrade collagens I, IV, V, VII, and X, gelatin, elastin, fibronectin, and proteoglycans.

4.2.3.1 *Tumor cell proliferation and inhibition of apoptosis*

MMPs are involved in the earliest phase of tumor progression, that is, the growth of tumor cells at the primary site. They control cell division and proliferation through regulation of growth factor availability and activation and inactivation of growth factor receptors, for example, their proteolytic activity contributes to the release and processing of factors such as fibroblast growth factor (FGF). MMPs can also inhibit apoptosis, as in the case of MMP-7, which can trigger an intracellular signaling pathway to promote cell survival (Martin and Matrisian, 2007; Rodriguez, Morrison, and Overall, 2009).

4.2.3.2 *Tumor migration, invasion, metastasis*

The earliest MMPs identified in tumor invasion were MMP-2 and MMP-9, both of which were capable of degrading type IV collagen and disrupting the integrity of the basement membrane. Studies of MMP-9 and MMP-2 null mice provided evidence of the role of these MMPs in metastasis (Martin and Matrisian, 2007).

A large body of information was generated with regard to direct visualization of MMP activity *in vitro* and *in vivo* and localization of their proteolytic activity during migration and invasion of the tumor cells (Sloane *et al.*, 2006; Hobson *et al.*, 2006). In that respect, several studies evaluated MMP-14 in tumor cells (Wolf *et al.*, 2003, 2007; Packard *et al.*, 2009), and found protease activity of MMP-14 to be localized at the polarized leading edge of the tumor cell (Packard *et al.*, 2009) facilitating forward movement of the cell. However, controversy still exists as to extent of the role of MMPs in protease-dependent versus protease-independent amoeboid movement of tumor cells (Wolf *et al.*, 2003, 2007; Sabeh, Shimizu-Hirota, and Weiss, 2009).

4.2.3.3 *Angiogenesis and vasculogenesis*

Angiogenesis, the formation of new blood vessels within the tumor, is initiated when the tumor has reached a critical size. It involves MMPs' degradation of ECM and non-ECM substrates and formation of new blood vessels from the pre-existing vascular network. In contrast, the *de novo* formation of blood vessels (Deryugina and Quigley, 2010), which involves the development of endothelial cell networks by recruitment of circulating progenitors of the endothelial cells, is called vasculogenesis. During tumor progression the 'angiogenic switch' occurs when the balance between the proangiogenic and the antiangiogenic factors tilts toward a proangiogenic outcome. Some MMPs play an important role in the angiogenic process, MMP-9 is particularly known as a critical mediator of the angiogenic switch (Martin and Matrisian, 2007; Baeriswyl and Christofori, 2009). Additional details on MMP and angiogenesis are also explored in other sections.

4.2.3.4 Pleiotropic activities of MMPs in the tumor microenvironment

Based on the observations that MMPs play an essential role in tumor cell proliferation, migration, invasion, metastasis, and angiogenesis, it seemed very plausible that inhibiting MMPs at the tumor site would be a viable therapeutic target. The early transgenic mouse models overexpressing various MMPs supported the notion that MMPs contribute to tumor progression (Ha *et al.*, 2001). In addition, the observation that high levels of MMPs correlated with poor prognosis of cancer patients, paved the way for the clinical trials using inhibitors of MMPs. The outcomes of these trials were disappointing showing that indiscriminate targeting and broad-spectrum inhibition of MMPs did not result in the anticipated inhibitory effect on the tumor growth. On the contrary, in some studies inhibition of MMPs or MMP-induced molecules resulted in promoting tumor growth (Coussens, Fingleton, and Matrisian, 2002; Fingleton, 2008). These unexpected findings led to the extensive studies using new mouse models of MMP knock-outs in which the generation of gain or loss of function revealed the highly complex and pleiotropic nature of MMPs and their function (Kruger, 2009). Currently available MMP knock-out mouse models only encompass 17 out of 23 murine MMP genes. Additional MMP knock-out mouse models are necessary in order to better understand MMP functions in human malignancies. In addition, generation of double or even triple knock-out mouse models may be necessary to minimize existing functional redundancy or compensatory mechanisms between various members of the MMP family (Fanjul-Fernandez *et al.*, 2010).

4.2.3.5 Inhibitory and stimulatory effects of MMPs on tumor growth

There are number of MMPs that can effect tumor growth in a stimulatory or inhibitory way, but some can exhibit both of these activities. In addition, remodeling of cell surface, basement membrane, and ECM by MMP's, leads to the release of several membrane- or matrix-bound growth factors and cytokines. This includes, for example, positive and negative regulators of angiogenesis that impact tumor growth.

4.2.3.5.1 Stimulatory activities of MMPs

Several MMPs were shown to stimulate tumor growth, for example, MMP-1. However, indirect stimulatory activities of MMPs on tumor progression are known as well. For example, one of the most studied positive regulators of angiogenesis released from ECM by MMP proteolysis is vascular endothelial growth factor (VEGF). In this regard, numerous studies have reported the effect of MMP-9 derived from inflammatory leukocytes such as macrophages and neutrophils (Bergers *et al.*, 2000) as well as tumor cells (Belotti *et al.*, 2003) on VEGF release. Also, MMP-2 and MMP-14 were shown to be involved in mobilization and upregulation of VEGF

(Dean and Overall, 2007; Deryugina, Soroceanu, and Strongin, 2002). In addition, MMP-3, -7, and -19 cleave matrix-bound isoforms of VEGF (Lee *et al.*, 2005).

FGF-2, another molecule with stimulatory activity on angiogenesis is released from the ECM as result of proteolytic activity of MMPs (Presta *et al.*, 2005; Ardi *et al.*, 2009). Endothelial basement membrane-bound FGF-2 resides in the ECM as an inactive form and needs to be proteolytically cleaved and released from the ECM in order to be biologically active (Iozzo and San Antonio, 2001; Sanderson *et al.*, 2005). MMP-9 was shown to induce the release of FGF-2 (Ardi *et al.*, 2009).

4.2.3.5.2 Inhibitory activities of MMPs

The generation of new genetically modified animal models demonstrated that several MMPs, such as MMP-8 and MMP-12 have inhibitory activities on tumor growth. The expression of MMP-8 derived from neutrophils was shown to be elevated in a non-metastatic cell line and correlated with its protective effect on tumor cell invasion and metastasis (Lopez-Otin and Matrisian, 2007; Martin and Matrisian, 2007; Gutierrez-Fernandez *et al.*, 2008). Most recently, tumor-suppressive functions of MMP-9 were shown in colitis-associated cancer (Garg *et al.*, 2010) and in colorectal cancer (Bendardaf *et al.*, 2010).

In addition, inhibitors of angiogenesis are also released indirectly during ECM remodeling by MMPs (Kalluri, 2003). These inhibitors include angiostatin, endostatin, and tumstatin (Folkman, 2004). The MMPs capable of contributing to the production of angiostatin include MMP-2, -7, -9, and -12 with MMP-12 being the most efficient inducer of angiostatin resulting in inhibition of angiogenesis (Nyberg, Xie, and Kalluri, 2005; Gorrin-Rivas *et al.*, 2000). Endostatin, another inhibitor of angiogenesis, was reported to be produced by cleavage from collagen type XVIII of basement membrane by MMP-3, -7, -9, -12, and -20. Also some select MMPs, for example, MMP-9 were implicated in production of tumstatin, another inhibitor of angiogenesis, from collagen type IV (Nyberg, Xie, and Kalluri, 2005).

4.2.3.6 Dual action of MMPs and MMP-induced molecules as biological stimulators and inhibitors

As mentioned earlier, clinical studies using inhibitors of MMPs failed to show desired, expected antitumor effect (Coussens, Fingleton, and Matrisian, 2002; Fingleton, 2008). It has been reported that several MMPs can play dual roles as tumor stimulators or inhibitors depending on the type of tissue and progression of the disease. The enzymes found to have this dual role include MMP-3, -9, and -11 (Andarawewa *et al.*, 2003; Jost *et al.*, 2006; Martin and Matrisian, 2007; Fanjul-Fernandez *et al.*, 2010).

In addition, it was found that depending on the cellular source of the same MMP, the activity may be contradictory; for example, MMP-12 could be protumorigenic when derived from tumor cells, but when derived from tumor-associated macrophages, it has a protective effect, leading to better differentiation of tumor cells and better outcome of the disease (Kerkela *et al.*, 2002).

A similar finding related to MMP-induced molecules such as VEGF involved in tumor angiogenesis and known to be sequestered in the ECM. Targeting and inhibition of the VEGF pathway *in vivo* in pancreatic carcinoma and glioblastoma mouse models resulted in worsening of the disease process by increasing tumor cell invasiveness and metastasis (Paez-Ribes *et al.*, 2009). Similarly it was shown that deletion of VEGF in myeloid cells accelerated tumorigenesis (Stockmann *et al.*, 2008). Yet, another paradox was reported relating to MMP-9, which is known to be associated with the production of tumstatin, an inhibitor of angiogenesis. It is known that MMP-9 activity is associated either with promoting or decreasing angiogenesis (Deryugina and Quigley, 2010).

4.3 Tissue inhibitors of matrix metalloproteinases

4.3.1 Introduction

TIMPs are endogenous endoproteinases that are known for their inhibition of MMP activity. The balance between TIMPs and MMPs regulates the turnover and remodeling that occurs in both normal development as well as variety of pathologic processes. Initial studies in this area suggested that TIMPs may be particularly important in inhibiting tumorigenesis and subsequent malignant progression. The earlier inhibitory effects of TIMPs on this processes was shown by Albini *et al.* (1991) and Baker *et al.* (1999) whereby TIMP overexpression in tumor cells achieved an inhibitory effect. However, the effects of TIMPs on tumorigenesis and metastasis are both multifunctional and often paradoxical. We will examine some of the data with regard to TIMPs having growth stimulatory and anti-apoptotic effects as well as the effect of TIMPs on tumor angiogenesis.

4.3.1.1 The TIMP family

The TIMP gene family has been identified in species ranging from C. elegans to drosophila, and humans (Lambert *et al.*, 2004; Crocker, Pagenstecher, and Campbell, 2004; Brew, Dinakarpandian, and Nagase, 2000). The TIMP gene family is composed of four members in most species including humans. They are secreted proteins with a molecular weight ranging from 21 to 28 kDa. There are several features that are common to all four members of the TIMP family and these are responsible for differences in their functions. TIMPs have a two-domain structure with N- and C-terminal regions, each of which contains six cysteine residues, forming three disulfide bonds (Murphy *et al.*, 1991; Williamson *et al.*, 1990). There is significant homology in the N-terminal domain, whereas the C-terminal portion displays prominent variability. This structural variation supports the contention that the N-terminal domain is responsible for the MMP inhibitory functions of TIMP. On the other hand variability within the C-terminal domain speaks for the different functions specific to each of the TIMPs on an individual level.

4.3.1.2 *Inhibition of tumorigenesis and metastasis by TIMPs*

The interactions of TIMPs and MMPs are responsible for regulation of MMP activity. Increased production or activity of MMPs has very frequently been linked to malignancy.

Some of the earliest evidence supporting a tumor inhibitory function of TIMPs was obtained in Swiss3T3 fibroblasts in which downregulation of TIMP-1 resulted in increased tumorigenesis and metastatic potential of these cells (Khokha *et al.*, 1989). In various other investigations, overexpression of TIMP-1–4 in both animal and human cell lines has been shown to decrease their ability to degrade the ECM and to invade *in vitro*. Upregulating the expression of TIMP does indeed result in inhibition or tumor invasion or metastases (Albini *et al.*, 1991; Baker *et al.*, 1999; Valente *et al.*, 1998; Matsuzawa *et al.*, 1996; Wang *et al.*, 1997). Using poorly differentiated human pancreatic adenocarcinoma cells, we have previously shown that there is inhibition of tumor growth and that the TIMP-1 and MMP-2 identified in this model originated from the implanted tumor and not the host (Bloomston *et al.*, 2002). The use of recombinant TIMP-1 and TIMP-2 has been shown to inhibit B16 melanoma cells' metastasis to the lung (Schultz *et al.*, 1988; Alvarez, Carmichael, and Declerck, 1990).

Inhibition of tumor growth and metastasis of melanoma cells (Khokha, 1994) was shown to be a result of overexpression of TIMP-1. Similarly, the metastatic ability of human gastric cancer cells was reduced, as was that of oral squamous cell carcinoma (Watanabe *et al.*, 1996; Nii *et al.*, 2000). It is important to note that it is not just tumor cell invasion and metastasis but also inhibition of primary tumor growth that is accomplished by overexpression of TIMP (Wang *et al.*, 1997; Khokha, 1994; Brand *et al.*, 2000). The mechanism by which this tumor inhibitory function is achieved is multifold. The suppression of MMPs by TIMPs may be the primary player, in that MMPs have been recognized to have a role in early stages of primary tumor growth. Similar functions have been attributed to other MMPs as well. For example, MMP-3 and MMP-9 have been shown to be involved in mammary carcinogenesis (Sternlicht *et al.*, 1999) and skin carcinogenesis (Coussens *et al.*, 2000).

The second suggested paradigm in this context was the effect of tumor-induced angiogensis. It is well recognized that MMPs play critical role in this process (Bergers *et al.*, 2000). TIMP-1 inhibition of MMPs and hence of angiogenesis may therefore play a significant role in reducing tumor growth. These concepts however, may also be challenged by more recent evidence indicating a direct (non MMP-dependent) mechanism for TIMP activity.

4.3.1.3 *Tumor-promoting activities of TIMPs*

When initially identified, TIMP-1 was noted to be identical to the erythroid-potentiating activity (EPA) protein that was responsible for stimulating growth and differentiation of murine erythroid precursors and various leukemic cell lines (Docherty *et al.*, 1985; Niskanen *et al.*, 1988; Murate *et al.*, 1993). Similar growth-promoting capabilities were also identified for TIMP2 (Stetler-Stevenson, Bersch, and Golde, 1992). Subsequently many investigations have demonstrated growth promoting activities of TIMP-1 and TIMP-2 in a wide array of cell types. TIMP-1 and TIMP-2 promote cell division and tumor growth in epithelial, mesenchymal, and other tumor cell lines (Bertaux *et al.*, 1991; Corcoran and Stetler-Stevenson, 1995; Hayakawa *et al.*, 1994; Nemeth and Goolsby, 1993). TIMP-3 and TIMP-4 also have growth promoting activities (Yang and Hawkes, 1992; Celiker *et al.*, 2001). An important observation in this context is that synthetic MMP inhibitors do not demonstrate similar activities. Additionally the use of reductive alkylation or mutations that eliminate or disrupt the MMP inhibitory activity of TIMP1 do not impact the growth stimulatory activities of TIMPs (Hayakawa *et al.*, 1994; Chesler *et al.*, 1995). The growth-promoting activity of TIMP, however, is eliminated by complex formation with either pro- or active MMPs (Hayakawa *et al.*, 1994). This observation suggests that free TIMP is a requirement for this function, and that TIMPs act directly on cell growth through binding to cell surface proteins, that is, possibly through a cell surface receptor mechanism. TIMP-4 has also been reported to promote tumorigenesis of human mammary carcinoma cells in nude mice (Jiang *et al.*, 2001).

Increased TIMP expression has been demonstrated in a range of different tumors such as breast, colorectal, gastric, and lung cancer (Jones, Glynn, and Walker, 1999; Yoshiji, Gomez, and Thorgeirsson, 1996; Powe *et al.*, 1997; Zeng *et al.*, 1995; Joo *et al.*, 2000; Michael *et al.*, 1999). If indeed the predominant function of TIMPs was restricted to inhibition of MMPs it is logical to assume then that increased TIMP levels would inhibit tumor invasion and metastasis, thereby improving prognosis.

Paradoxically, however, elevated TIMP levels have often been associated with poor prognosis in many solid tumors. This has, for instance, been documented in colorectal cancer, breast cancer, gastric cancer, lymphoma, prostate, and lung cancer (Zeng *et al.*, 1995; Rowe *et al.*, 1997; Hewitt *et al.*, 2000; Yoshiji, Gomez, and Thorgeirsson, 1996; Mccarthy *et al.*, 1999; Joo *et al.*, 2000; Kossakowska, Urbanski, and Edwards, 1991; Still *et al.*, 2000; Fong *et al.*, 1996). Elevated TIMP-1 mRNA in colorectal cancer stroma correlates with lymphoid metastasis (Zeng *et al.*, 1995) whereas patients with prostate cancer metastasis have significantly higher levels of TIMP-1 in plasma. In patients with cervical and breast carcinoma, high levels of TIMP-2 have been correlated with poor prognosis (Davidson *et al.*, 1999; Visscher *et al.*, 1994). Jiang *et al.* (2001) have also immunohistochemically demonstrated high levels of TIMP-4 expression in breast carcinoma cells than in normal breast epithelial cells. Our studies have examined the effects of overexpression of TIMP-1 in a CNS model of metastasis, focussing primarily on the interaction of TIMP-1 in the CNS microenvironment, particularly its impact on the implantation and growth of tumor. Following implantation of lung adenocarcinoma cells transfected to overexpress human TIMP-1, we demonstrated both increased tumor size as well as more aggressive tumor growth patterns with multiplicity of tumors and increased invasion (Rojiani and Rojiani, 2005; DeMers *et al.*, 2010). While the possibility that TIMP overexpression in malignancies may simply be a response to elevated MMP levels during continued tumorigenesis, a more plausible alternative is that this high level of TIMP expression is itself responsible for the tumor promoting and apoptotic regulatory functions described by others.

4.3.1.4 TIMPs in apoptosis

The regulation of programmed cell death (apoptosis) by TIMPs has been shown to utilize both MMP-dependent and MMP-independent pathways. TIMP-1 and to a lesser extent TIMP-2 inhibit apoptosis, most frequently by MMP-independent mechanisms, whereas data for TIMP-3 suggest that a proapoptotic mechanism is at play.

Release of anchorage-dependent cells from the ECM, as mediated by specific MMPs will result in programmed cell death (Frisch and Francis, 1994; Boudreau, Werb, and Bissell, 1996). MMP-3 overexpression in mammary epithelial cells, both *in vitro* and *in vivo*, induced apoptosis that was significantly inhibited by TIMP-1. This antiapoptotic activity of TIMP-1 was dependent on MMP inhibition (Alexander *et al.*, 1996).

Enhanced cell survival resulting from TIMP-1's inhibition of apoptosis in normal lymphoid cells, as well as Burkitt's lymphoma cells occurs primarily through induction of the anti-apoptotic gene bcl-x_L. TIMP-1 expression has been inversely correlated with induction of programmed cell death in various lymphoma cell lines (Guedez, Courtemanch, and Stetler-Stevenson, 1998a). In these studies it was also demonstrated through forced expression of TIMP-1 or following treatment with recombinant TIMP-1 that there was antiapoptotic activity, affecting both intrinsic and extrinsic pathways, with suppression of caspase activity. Using the MMP inhibitor BB-94 as well as an alklylated form of TIMP-1 (which lacks MMP inhibitory activity) it was confirmed that TIMP-1 exerted its antiapoptotic activity by MMP-independent mechanisms (Guedez *et al.*, 1998b, 2001). TIMP-1 has also been shown to have an anti-apoptotic effect on MCF10A human breast epithelial cells via both intrinsic and extrinsic pathways (Li, Fridman, and Kim, 1999; Liu *et al.*, 2003, 2005). These studies also confirmed that regulation of apoptosis by TIMP-1 was not a function of its MMP inhibitory activity. There are data to suggest that TIMP-4 will inhibit apoptosis in human breast cancer cells, *in vitro* and mammary tumors *in vivo*.

TIMP-1 functions downstream of Bcl-2 (Li, Fridman, and Kim, 1999) and also upregulates bcl-X_L in B cells (Guedez *et al.*, 1998b). The translocation of GFP-TIMP-1 construct into the nucleus of MCF-7 cells would suggest that TIMP-1 may also function in the nucleus (Ritter, Garfield, and Thorgeirsson, 1999). The data also suggest that this regulation of apoptosis by TIMPs may be a result of targeting unidentified TIMP-binding proteins on the cell membrane and initiating signaling pathways.

The apoptotic inhibitory effects of TIMPs may also be the result of inhibition of MMPs. It is recognized that cell – matrix interactions will influence viability. Once tumor cells dissociate from the ECM, apoptotic cell death will result. Turnover of the ECM as initiated by MMPs has been shown to modulate cell survival (Lukashev and Werb, 1998). In contrast TIMP-3 has been shown to have a predominantly proapoptotic mode of action in MCF cells, melanoma, smooth muscle, and various other cell lines. It is most likely that TIMP-3 acts primarily by MMP-dependent mechanisms

Collectively the data on the role that TIMPs play in the context of apoptosis is significant, supporting an antiapoptotic role by both MMP-dependent and -independent mechanisms for TIMP-1, and to a limited extent for TIMP-2 and TIMP-4; while defining a proapoptotic function for TIMP-3.

4.3.1.5 TIMPs and angiogenesis

TIMPs have antiangiogenic activity by virtue of their ability to inhibit the activity of MMPs (Fernandez *et al.*, 1999). MMPs have been shown to promote endothelial cell migration and trigger the angiogenic switch. MMP-9 brings about the release of VEGF from the matrix and it has been suggested that MMP-2 activity was necessary for the switch to an angiogenic phenotype (Bergers *et al.*, 2000; Fang *et al.*, 2000). In a study examining vascular proliferation in brain metastasis of lung adenocarcinoma, it was noted that MMP-2-expressing tumors displayed increased angiogenesis at the tumor–brain interface (Rojiani *et al.*, 2001). On the other hand MMPs have well defined antiangiogenic properties, primarily by the generation of potent angiogenic inhibitors. The conversion of plasminogen to angiostatin, a well-recognized angiogenesis inhibitor has been defined (Patterson and Sang, 1997; Cornelius *et al.*, 1998; O'Reilly *et al.*, 1994; Pozzi *et al.*, 2000). Similarly MMPs and elastase contribute to the formation of endostatin, the C-terminal fragment of the basement membrane protein collagen XVIII, another potent angiogenisis inhibitor (Wen *et al.*, 1999). Inhibition of MMP activity therefore may have a positive impact on angiogenesis. In the final analysis the inhibition of MMPs by TIMPs may yield paradoxical effects on the angiogenic profile.

The direct contribution of TIMPs on endothelial cells and angiogenesis continue to be elaborated. MMP-independent mechanisms using TIMP-interacting cell surface proteins that regulate angiogenesis have also been described. The interaction of TIMP-2 with $\alpha_3\beta1$ culminates in cell cycle arrest of endothelial cells and inhibition of angiogenesis. TIMP-3 has the ability to bind VEGF receptor2, competing with VEGF for this site (Qi *et al.*, 2003). Both of the above mentioned cell-surface protein interactions were shown to be independent of their MMP-related functions. In studies inducing overexpression of TIMP-1, VEGF expression in mammary carcinoma is increased (Yoshiji *et al.*, 1998). VEGF-induced neovascularization in the retina was also enhanced under similar experimental conditions (Yamada *et al.*, 2001; Murphy *et al.*, 1991). Murphy, Unsworth, and Stetler-Stevenson (1993) have described decreased growth of basic FGF-stimulated endothelial cells, whereas TIMP-3 negatively impacts both endothelial cell motility and proliferation. In our model of TIMP-1 overexpressing lung carcinoma metastases to the CNS, we describe increased angiogenesis. Both *in vivo* and *in vitro* analysis of vascular patterns either in tumors from these cells or when serum-free medium from TIMP-1 overexpressing clones was used, enhanced vascular profiles were identified. Gene expression profiling of TIMP-1 overexpressing clones exhibited significant reduction of thrombospondin-1 (TSP-1). TSP-1 is a well-documented inhibitor of angiogenesis. These studies further confirm the role of TIMP-1 to promote tumor growth and suggest yet another possible mechanism for its interaction in the host microenvironment (DeMers *et al.*, 2010; Rojiani and Rojiani, 2006).

A clear understanding of the proangiogenic properties and antiangiogenic effects mediated by MMP and TIMP continues to elude us and demands more complete investigation if it is to be used to therapeutic advantage.

4.3.1.6 *TIMPs in cell signaling*

Over the years evidence has been mounting about the signaling capacities of TIMPs that are independent of their role in MMP inhibition. Numerous *in vitro* reports had indicated that TIMP-1 and TIMP-2 bind to the surface of many normal and malignant cell with high affinity and stimulate cell growth leading to speculations of putative TIMP receptors on cell surfaces. As pleiotropic roles of TIMPs began to emerge, for example, in cell growth, angiogenesis, and apoptosis, several studies documented downstream activation of cell signaling pathways.

TIMP-2 had been known to inhibit endothelial cell proliferation (Murphy, Unsworth, and Stetler-Stevenson, 1993). Subsequently, $\alpha_3\beta_1$ was identified as the bona fide TIMP-2 receptor mediating this effect via the tyrosine kinase receptor signal transduction pathway (Seo *et al.*, 2003).

TIMP-3 has been documented as a pro-apoptotic molecule that inhibits the shedding of death receptors mediated by tumor necrosis factor (TNF)-alpha-converting enzyme (Amour *et al.*, 1998) as well as a potent inhibitor of angiogenesis (Anand-Apte *et al.*, 1997). Recently, it has been shown that TIMP-3 interacts with VEGF-R2, a principal receptor involved in endothelial cell proliferation, thus inhibiting VEGF-A binding and hence angiogenesis. Interestingly, TIMP-3 is the only member of this family described as a tumor suppressor.

A substantial amount of data defining the multipotential activities of TIMP-1 and its effect on intracellular signaling pathways had prompted a search for TIMP-1-specific receptors. Several investigations had documented molecular weights and K_d values of putative receptors of TIMP-1, indicative of intense research in this area. The role of TIMP-1 in extrinsic and intrinsic apoptotic pathways was shown to involve the focal adhesion kinase (FAK), phosphotidylinositol 3-kinase, and ERKs. These studies have led investigators in search of TIMP-1-specific receptors and hence identification of CD63 along with integrin β_1 subunit as TIMP-1 receptor (Jung *et al.*, 2006). The authors found that TIMP-1 expression level corresponded with integrin β_1 levels and this synergy was CD63-dependent. Finally, Lambert *et al.* have shown that TIMP-1 has an MMP-independent, antiapoptotic effect on UT-7 erythroid cells. Their results indicate that this antiapoptotic signaling of TIMP-1 is dependent upon interactions of TIMP-1, proMMP-9, and CD44 yielding a complex on the UT-7 erythroid cell surface (Lambert *et al.*, 2009). Investigations thus continue into determining the precise mechanisms by which many of the MMP-independent functions of TIMPs manifest their effects.

4.4 Concluding comments

The above description of the highly pleiotropic nature of MMPs further illustrates the diversity of their functions in cancer progression. Clearly, MMPs can orchestrate complex signaling pathways through their proteolytic processing of ECM and non-ECM molecules. Therefore, considering the above multiple aspects of MMPs' role in the tumor microenvironment, the use of broad-spectrum inhibitors interfering with the function of many of these enzymes, led to unforeseen and often undesirable biological consequences and resulted in disappointing outcomes of clinical studies.

In a manner similar to the way in which earlier paradigms of the functions and mechanisms of TIMP activity have been challenged, current understanding of the effects of TIMP on everything from angiogenesis to tumor growth or inhibition, as well as apoptosis have also been revisited. The typical assumptions of TIMP activity have centered on their inhibition of MMP; however, MMP-independent pathways have clearly been identified and continue to assume equal if not greater significance. Thus, even after decades of investigations, we continue to unravel the complexities of MMP and TIMP. Despite limited success with the broad-spectrum inhibitor clinical trials, inhibitors of MMP remain a viable therapeutic target in oncology. It will however require a clear understanding of the pleiotropic activities of TIMPs, particularly at the molecular level, thereby facilitating development of targeted, mechanism-based pharmacologic, therapeutic interventions.

References

Albini, A., Melchiori, A., Santi, L. *et al.* (1991) Tumor cell invasion inhibited by TIMP-2. *Journal of National Cancer Institute*, **83**, 775–779.

Alexander, C.M., Howard, E.W., Bissell, M.J., and Werb, Z. (1996) Rescue of mammary epithelial cell apoptosis and entactin degradation by a tissue inhibitor of metalloproteinases-1 transgene. *Journal of Cell Biology*, **135**, 1669–1677.

Alvarez, O.A., Carmichael, D.F., and Declerck, Y.A. (1990) Inhibition of collagenolytic activity and metastasis of tumor cells by a recombinant human tissue inhibitor of metalloproteinases. *Journal of National Cancer Institute*, **82**, 589–595.

Amour, A., Slocombe, P.M., Webster, A. *et al.* (1998) TNF-alpha converting enzyme (TACE) is inhibited by TIMP-3. *FEBS Letters*, **435**, 39–44.

Anand-Apte, B., Pepper, M.S., Voest, E. *et al.* (1997) Inhibition of angiogenesis by tissue inhibitor of metalloproteinase-3. *Investigative Ophthalmology and Visual Science*, **38**, 817–823.

Andarawewa, K.L., Boulay, A., Masson, R. *et al.* (2003) Dual stromelysin-3 function during natural mouse mammary tumor virus-ras tumor progression. *Cancer Research*, **63**, 5844–5849.

Ardi, V.C., Van Den Steen, P.E., Opdenakker, G. *et al.* (2009) Neutrophil MMP-9 proenzyme, unencumbered by TIMP-1, undergoes efficient activation in vivo and catalytically induces angiogenesis via a basic fibroblast growth factor (FGF-2)/FGFR-2 pathway. *The Journal of Biological Chemistry*, **284**, 25854–25866.

Baeriswyl, V. and Christofori, G. (2009) The angiogenic switch in carcinogenesis. *Seminars in Cancer Biology*, **19**, 329–337.

Baker, A.H., George, S.J., Zaltsman, A.B. *et al.* (1999) Inhibition of invasion and induction of apoptotic cell death of cancer cell lines by overexpression of TIMP-3. *British Journal of Cancer*, **79**, 1347–1355.

Belotti, D., Paganoni, P., Manenti, L. *et al.* (2003) Matrix metalloproteinases (MMP9 and MMP2) induce the release of vascular endothelial growth factor (VEGF) by ovarian carcinoma cells: implications for ascites formation. *Cancer Research*, **63**, 5224–5229.

Bendardaf, R., Buhmeida, A., Hilska, M. *et al.* (2010) MMP-9 (gelatinase B) expression is associated with disease-free survival and disease-specific survival in colorectal cancer patients. *Cancer Investigation*, **28**, 38–43.

Bergers, G., Brekken, R., Mcmahon, G. *et al.* (2000) Matrix metalloproteinase-9 triggers the angiogenic switch during carcinogenesis. *Nature Cell Biology*, **2**, 737–744.

Bertaux, B., Hornebeck, W., Eisen, A.Z., and Dubertret, L. (1991) Growth stimulation of human keratinocytes by tissue inhibitor of metalloproteinases. *Journal of Investigative Dermatology*, **97**, 679–685.

Bloomston, M., Shafii, A., Zervos, E.E. *et al.* (2002) MMP-2 and TIMP-1 are derived from, not in response to, pancreatic cancer. *Journal of Surgical Research*, **102**, 35–38.

Boudreau, N., Werb, Z., and Bissell, M.J. (1996) Suppression of apoptosis by basement membrane requires three-dimensional tissue organization and withdrawal from the cell cycle. *Proceedings of the National Academy of Sciences of the United States of America*, **93**, 3509–3513.

Brand, K., Baker, A.H., Perez-Canto, A. *et al.* (2000) Treatment of colorectal liver metastases by adenoviral transfer of tissue inhibitor of metalloproteinases-2 into the liver tissue. *Cancer Research*, **60**, 5723–5730.

Brew, K., Dinakarpandian, D., and Nagase, H. (2000) Tissue inhibitors of metalloproteinases: evolution, structure and function. *Biochimica et Biophysica Acta*, **1477**, 267–283.

Celiker, M.Y., Wang, M., Atsidaftos, E. *et al.* (2001) Inhibition of Wilms' tumor growth by intramuscular administration of tissue inhibitor of metalloproteinases-4 plasmid DNA. *Oncogene*, **20**, 4337–4343.

Chesler, L., Golde, D.W., Bersch, N., and Johnson, M.D. (1995) Metalloproteinase inhibition and erythroid potentiation are independent activities of tissue inhibitor of metalloproteinases-1. *Blood*, **86**, 4506–4515.

Corcoran, M.L. and Stetler-Stevenson, W.G. (1995) Tissue inhibitor of metalloproteinase-2 stimulates fibroblast proliferation via a cAMP-dependent mechanism. *The Journal of Biological Chemistry*, **270**, 13453–13459.

Cornelius, L.A., Nehring, L.C., Harding, E. *et al.* (1998) Matrix metalloproteinases generate angiostatin: effects on neovascularization. *Journal of Immunology*, **161**, 6845–6852.

Coussens, L.M., Fingleton, B., and Matrisian, L.M. (2002) Matrix metalloproteinase inhibitors and cancer: trials and tribulations. *Science*, **295**, 2387–2392.

Coussens, L.M., Tinkle, C.L., Hanahan, D., and Werb, Z. (2000) MMP-9 supplied by bone marrow-derived cells contributes to skin carcinogenesis. *Cell*, **103**, 481–490.

Crocker, S.J., Pagenstecher, A., and Campbell, I.L. (2004) The TIMPs tango with MMPs and more in the central nervous system. *Journal of Neuroscience Research*, **75**, 1–11.

Davidson, B., Goldberg, I., Kopolovic, J. *et al.* (1999) MMP-2 and TIMP-2 expression correlates with poor prognosis in cervical carcinoma – a clinicopathologic study using immunohistochemistry and mRNA in situ hybridization. *Gynecologic Oncology*, **73**, 372–382.

Davis, G.E. and Senger, D.R. (2008) Extracellular matrix mediates a molecular balance between vascular morphogenesis and regression. *Current Opinion in Hematology*, **15**, 197–203.

Dean, R.A. and Overall, C.M. (2007) Proteomics discovery of metalloproteinase substrates in the cellular context by iTRAQ labeling reveals a diverse MMP-2 substrate degradome. *Molecular and Cellular Proteomics*, **6**, 611–623.

DeMers, N., Rojiani, A.M., Brem, S., and Rojiani, M.V. (eds) (2010) *Molecular Mechanisms of Metastasis to the CNS*, in press.

Deryugina, E.I. and Quigley, J.P. (2006) Matrix metalloproteinases and tumor metastasis. *Cancer and Metastasis Reviews*, **25**, 9–34.

Deryugina, E.I. and Quigley, J.P. (2010) Pleiotropic roles of matrix metalloproteinases in tumor angiogenesis: contrasting, overlapping and compensatory functions. *Biochimica et Biophysica Acta*, **1803** (1), 103–20.

Deryugina, E.I., Soroceanu, L., and Strongin, A.Y. (2002) Up-regulation of vascular endothelial growth factor by membrane-type 1 matrix metalloproteinase stimulates human glioma xenograft growth and angiogenesis. *Cancer Research*, **62**, 580–588.

Docherty, A.J., Lyons, A., Smith, B.J. *et al.* (1985) Sequence of human tissue inhibitor of metalloproteinases and its identity to erythroid-potentiating activity. *Nature*, **318**, 66–69.

Egeblad, M. and Werb, Z. (2002) New functions for the matrix metalloproteinases in cancer progression. *Nature Reviews Cancer*, **2**, 161–174.

Fang, J., Shing, Y., Wiederschain, D. *et al.* (2000) Matrix metalloproteinase-2 is required for the switch to the angiogenic phenotype in a tumor model. *Proceedings of the National Academy of Sciences of the United States of America*, **97**, 3884–3889.

Fanjul-Fernandez, M., Folgueras, A.R., Cabrera, S., and Lopez-Otin, C. (2010) Matrix metalloproteinases: evolution, gene regulation and functional analysis in mouse models. *Biochimica et Biophysica Acta*, **1803** (1), 3–19.

Fernandez, H.A., Kallenbach, K., Seghezzi, G. *et al.* (1999) Inhibition of endothelial cell migration by gene transfer of tissue inhibitor of metalloproteinases-1. *Journal of Surgical Research*, **82**, 156–162.

Fingleton, B. (2008) MMP inhibitor clinical trials – the past, present, and future, in *The Cancer Degradome* (eds D. Edwards, G. Hoyer-Hansen, F. Blasi and B.F. Sloane), Springer, New York, pp. 759–785.

Folkman, J. (2004) Endogenous angiogenesis inhibitors. *APMIS*, **112**, 496–507.

Fong, K.M., Kida, Y., Zimmerman, P.V., and Smith, P.J. (1996) TIMP1 and adverse prognosis in non-small cell lung cancer. *Clinical Cancer Research*, **2**, 1369–1372.

Frisch, S.M. and Francis, H. (1994) Disruption of epithelial cell-matrix interactions induces apoptosis. *Journal of Cell Biology*, **124**, 619–626.

Gao, G., Plaas, A., Thompson, V.P. *et al.* (2004) ADAMTS4 (aggrecanase-1) activation on the cell surface involves C-terminal cleavage by glycosylphosphatidyl inositol-anchored membrane type 4-matrix metalloproteinase and binding of the activated proteinase to chondroitin sulfate and heparan sulfate on syndecan-1. *The Journal of Biological Chemistry*, **279**, 10042–10051.

Garg, P., Sarma, D., Jeppsson, S. *et al.* (2010) Matrix metalloproteinase-9 functions as a tumor suppressor in colitis-associated cancer. *Cancer Research*, **70**, 792–801.

Genis, L., Galvez, B.G., Gonzalo, P., and Arroyo, A.G. (2006) MT1-MMP: universal or particular player in angiogenesis? *Cancer and Metastasis Reviews*, **25**, 77–86.

Gorrin-Rivas, M.J., Arii, S., Furutani, M. *et al.* (2000) Mouse macrophage metalloelastase gene transfer into a murine melanoma suppresses primary tumor growth by halting angiogenesis. *Clinical Cancer Research*, **6**, 1647–1654.

Gross, J. and Lapiere, C.M. (1962) Collagenolytic activity in amphibian tissues: a tissue culture assay. *Proceedings of the National Academy of Sciences of the United States of America*, **48**, 1014–1022.

Guedez, L., Courtemanch, L., and Stetler-Stevenson, M. (1998a) Tissue inhibitor of metalloproteinase (TIMP)-1 induces differentiation and an antiapoptotic phenotype in germinal center B cells. *Blood*, **92**, 1342–1349.

Guedez, L., Stetler-Stevenson, W.G., Wolff, L. *et al.* (1998b) In vitro suppression of programmed cell death of B cells by tissue inhibitor of metalloproteinases-1. *Journal of Clinical Investigation*, **102**, 2002–2010.

Guedez, L., Mcmarlin, A.J., Kingma, D.W. *et al.* (2001) Tissue inhibitor of metalloproteinase-1 alters the tumorigenicity of Burkitt's lymphoma via divergent effects on tumor growth and angiogenesis. *American Journal of Pathology*, **158**, 1207–1215.

Gutierrez-Fernandez, A., Fueyo, A., Folgueras, A.R. *et al.* (2008) Matrix metalloproteinase-8 functions as a metastasis suppressor through modulation of tumor cell adhesion and invasion. *Cancer Research*, **68**, 2755–2763.

Ha, H.Y., Moon, H.B., Nam, M.S. *et al.* (2001) Overexpression of membrane-type matrix metalloproteinase-1 gene induces mammary gland abnormalities and adenocarcinoma in transgenic mice. *Cancer Research*, **61**, 984–990.

Hayakawa, T., Yamashita, K., Ohuchi, E., and Shinagawa, A. (1994) Cell growth-promoting activity of tissue inhibitor of metalloproteinases-2 (TIMP-2). *Journal of Cell Science*, **107** (Pt 9), 2373–2379.

Hewitt, R.E., Brown, K.E., Corcoran, M., and Stetler-Stevenson, W.G. (2000) Increased expression of tissue inhibitor of metalloproteinases type 1 (TIMP-1) in a more tumourigenic colon cancer cell line. *Journal of Pathology*, **192**, 455–459.

Hobson, J.P., Liu, S., Rono, B. *et al.* (2006) Imaging specific cell-surface proteolytic activity in single living cells. *Nature Methods*, **3**, 259–261.

Iozzo, R.V. and San Antonio, J.D. (2001) Heparan sulfate proteoglycans: heavy hitters in the angiogenesis arena. *Journal of Clinical Investigation*, **108**, 349–355.

Itoh, Y. and Seiki, M. (2006) MT1-MMP: a potent modifier of pericellular microenvironment. *Journal of Cellular Physiology*, **206**, 1–8.

Jiang, Y., Wang, M., Celiker, M.Y. *et al.* (2001) Stimulation of mammary tumorigenesis by systemic tissue inhibitor of matrix metalloproteinase 4 gene delivery. *Cancer Research*, **61**, 2365–2370.

Jones, J.L., Glynn, P., and Walker, R.A. (1999) Expression of MMP-2 and MMP-9, their inhibitors, and the activator MT1-MMP in primary breast carcinomas. *Journal of Pathology*, **189**, 161–168.

Joo, Y.E., Seo, K.S., Kim, H.S. *et al.* (2000) Expression of tissue inhibitors of metalloproteinases (TIMPs) in gastric cancer. *Digestive Diseases and Science*, **45**, 114–121.

Jost, M., Folgueras, A.R., Frerart, F. *et al.* (2006) Earlier onset of tumoral angiogenesis in matrix metalloproteinase-19-deficient mice. *Cancer Research*, **66**, 5234–5241.

Jung, K.K., Liu, X.W., Chirco, R. *et al.* (2006) Identification of CD63 as a tissue inhibitor of metalloproteinase-1 interacting cell surface protein. *EMBO Journal*, **25**, 3934–3942.

Kalluri, R. (2003) Basement membranes: structure, assembly and role in tumour angiogenesis. *Nature Reviews Cancer*, **3**, 422–433.

Kerkela, E., Ala-Aho, R., Klemi, P. *et al.* (2002) Metalloelastase (MMP-12) expression by tumour cells in squamous cell carcinoma of the vulva correlates with invasiveness, while that by macrophages predicts better outcome. *Journal of Pathology*, **198**, 258–269.

Khokha, R. (1994) Suppression of the tumorigenic and metastatic abilities of murine B16-F10 melanoma cells in vivo by the overexpression of the tissue inhibitor of the metalloproteinases-1. *Journal of the National Cancer Institute*, **86**, 299–304.

Khokha, R., Waterhouse, P., Yagel, S. *et al.* (1989) Antisense RNA-induced reduction in murine TIMP levels confers oncogenicity on Swiss 3T3 cells. *Science*, **243**, 947–950.

Kossakowska, A.E., Urbanski, S.J., and Edwards, D.R. (1991) Tissue inhibitor of metalloproteinases-1 (TIMP-1) RNA is expressed at elevated levels in malignant non-Hodgkin's lymphomas. *Blood*, 77, 2475–2481.

Kruger, A. (2009) Functional genetic mouse models: promising tools for investigation of the proteolytic internet. *Biological Chemistry*, **390**, 91–97.

Lambert, E., Bridoux, L., Devy, J. *et al.* (2009) TIMP-1 binding to proMMP-9/CD44 complex localized at the cell surface promotes erythroid cell survival. *International Journal of Biochemistry and Cell Biology*, **41**, 1102–1115.

Lambert, E., Dassé, E., Haye, B., and Petitfrère, E. (2004) TIMPs as multifacial proteins. *Critical Reviews in Oncology/Hematology*, **49**, 187–198.

Lee, S., Jilani, S.M., Nikolova, G.V. *et al.* (2005) Processing of VEGF-A by matrix metalloproteinases regulates bioavailability and vascular patterning in tumors. *Journal of Cell Biology*, **169**, 681–691.

Leppert, D., Lindberg, R.L., Kappos, L., and Leib, S.L. (2001) Matrix metalloproteinases: multifunctional effectors of inflammation in multiple sclerosis and bacterial meningitis. *Brain Research: Brain Research Reviews*, **36**, 249–257.

Lewis, C.E. and Pollard, J.W. (2006) Distinct role of macrophages in different tumor microenvironments. *Cancer Research*, **66**, 605–612.

Li, G., Fridman, R., and Kim, H.R. (1999) Tissue inhibitor of metalloproteinase-1 inhibits apoptosis of human breast epithelial cells. *Cancer Research*, **59**, 6267–6275.

Liu, X.W., Bernardo, M.M., Fridman, R., and Kim, H.R. (2003) Tissue inhibitor of metalloproteinase-1 protects human breast epithelial cells against intrinsic apoptotic cell death via the focal adhesion kinase/phosphatidylinositol 3-kinase and MAPK signaling pathway. *The Journal of Biological Chemistry*, **278**, 40364–40372.

Liu, X.W., Taube, M.E., Jung, K.K. *et al.* (2005) Tissue inhibitor of metalloproteinase-1 protects human breast epithelial cells from extrinsic cell death: a potential oncogenic activity of tissue inhibitor of metalloproteinase-1. *Cancer Research*, **65**, 898–906.

Lopez-Otin, C. and Matrisian, L.M. (2007) Emerging roles of proteases in tumour suppression. *Nature Reviews Cancer*, **7**, 800–808.

Lukashev, M.E. and Werb, Z. (1998) ECM signalling: orchestrating cell behaviour and misbehaviour. *Trends in Cell Biology*, **8**, 437–441.

Malemud, C.J. (2006) Matrix metalloproteinases (MMPs) in health and disease: an overview. *Frontiers in Bioscience*, **11**, 1696–1701.

Martin, M.D. and Matrisian, L.M. (2007) The other side of MMPs: protective roles in tumor progression. *Cancer and Metastasis Reviews*, **26**, 717–724.

Matsuzawa, K., Fukuyama, K., Hubbard, S.L. *et al.* (1996) Transfection of an invasive human astrocytoma cell line with a TIMP-1 cDNA: modulation of astrocytoma invasive potential. *Journal of Neuropathology and Experimental Neurology*, **55**, 88–96.

Mccarthy, K., Maguire, T., Mcgreal, G. *et al.* (1999) High levels of tissue inhibitor of metalloproteinase-1 predict poor outcome in patients with breast cancer. *International Journal of Cancer*, **84**, 44–48.

Michael, M., Babic, B., Khokha, R. *et al.* (1999) Expression and prognostic significance of metalloproteinases and their tissue inhibitors in patients with small-cell lung cancer. *Journal of Clinical Oncology*, **17**, 1802–1808.

Murate, T., Yamashita, K., Ohashi, H. *et al.* (1993) Erythroid potentiating activity of tissue inhibitor of metalloproteinases on the differentiation of erythropoietin-responsive mouse erythroleukemia cell line, ELM-I-1-3, is closely related to its cell growth potentiating activity. *Experimental Hematology*, **21**, 169–176.

Murphy, A.N., Unsworth, E.J., and Stetler-Stevenson, W.G. (1993) Tissue inhibitor of metalloproteinases-2 inhibits bFGF-induced human microvascular endothelial cell proliferation. *Journal of Cellular Physiology*, **157**, 351–358.

Murphy, G., Houbrechts, A., Cockett, M.I. *et al.* (1991) The N-terminal domain of tissue inhibitor of metalloproteinases retains metalloproteinase inhibitory activity. *Biochemistry*, **30**, 8097–8102.

Nemeth, J.A. and Goolsby, C.L. (1993) TIMP-2, a growth-stimulatory protein from SV40-transformed human fibroblasts. *Experimental Cell Research*, **207**, 376–382.

Nii, M., Kayada, Y., Yoshiga, K. *et al.* (2000) Suppression of metastasis by tissue inhibitor of metalloproteinase-1 in a newly established human oral squamous cell carcinoma cell line. *International Journal of Oncology*, **16**, 119–124.

Niskanen, E., Gasson, J.C., Teates, C.D., and Golde, D.W. (1988) In vivo effect of human erythroid-potentiating activity on hematopoiesis in mice. *Blood*, **72**, 806–810.

Nyberg, P., Xie, L., and Kalluri, R. (2005) Endogenous inhibitors of angiogenesis. *Cancer Research*, **65**, 3967–3979.

O'Reilly, M.S., Holmgren, L., Shing, Y. *et al.* (1994) Angiostatin: a novel angiogenesis inhibitor that mediates the suppression of metastases by a Lewis lung carcinoma. *Cell*, **79**, 315–328.

Oh, J., Takahashi, R., Kondo, S. *et al.* (2001) The membrane-anchored MMP inhibitor RECK is a key regulator of extracellular matrix integrity and angiogenesis. *Cell*, **107**, 789–800.

Packard, B.Z., Artym, V.V., Komoriya, A., and Yamada, K.M. (2009) Direct visualization of protease activity on cells migrating in three-dimensions. *Matrix Biology*, **28**, 3–10.

Paez-Ribes, M., Allen, E., Hudock, J. *et al.* (2009) Antiangiogenic therapy elicits malignant progression of tumors to increased local invasion and distant metastasis. *Cancer Cell*, **15**, 220–231.

Page-Mccaw, A., Ewald, A.J., and Werb, Z. (2007) Matrix metalloproteinases and the regulation of tissue remodelling. *Nature Reviews: Molecular and Cellular Biology*, **8**, 221–233.

Patterson, B.C. and Sang, Q.A. (1997) Angiostatin-converting enzyme activities of human matrilysin (MMP-7) and gelatinase B/type IV collagenase (MMP-9). *The Journal of Biological Chemistry*, **272**, 28823–28825.

Powe, D.G., Brough, J.L., Carter, G.I. *et al.* (1997) TIMP-3 mRNA expression is regionally increased in moderately and poorly differentiated colorectal adenocarcinoma. *British Journal of Cancer*, **75**, 1678–1683.

Pozzi, A., Moberg, P.E., Miles, L.A. *et al.* (2000) Elevated matrix metalloprotease and angiostatin levels in integrin alpha 1 knockout mice cause reduced tumor vascularization. *Proceedings of the National Academy of Sciences of the United States of America*, **97**, 2202–2207.

Presta, M., Dell'era, P., Mitola, S. *et al.* (2005) Fibroblast growth factor/fibroblast growth factor receptor system in angiogenesis. *Cytokine Growth Factor Reviews*, **16**, 159–178.

Qi, J.H., Ebrahem, Q., Moore, N. *et al.* (2003) A novel function for tissue inhibitor of metalloproteinases-3 (TIMP3): inhibition of angiogenesis by blockage of VEGF binding to VEGF receptor-2. *Nature Medicine*, **9**, 407–415.

Ritter, L.M., Garfield, S.H., and Thorgeirsson, U.P. (1999) Tissue inhibitor of metalloproteinases-1 (TIMP-1) binds to the cell surface and translocates to the nucleus of human MCF-7 breast carcinoma cells. *Biochemical and Biophysical Research Communications*, **257**, 494–499.

Rodriguez, D., Morrison, C.J., and Overall, C.M. (2010) Matrix metalloproteinases: What do they not do? New substrates and biological roles identified by murine models and proteomics. *Biochimica et Biophysica Acta*, **1803** (1), 39–54.

Rojiani, A.M., Lin, J.T., Korabathina, K. *et al.* (2001) Expression of MMP-2 Correlates With Increased Angiogenesis at CNS/Metastatic Carcinoma Interface. *Journal of Neuropathology and Experimental Neurology*, **60**, 444.

Rojiani, M.V. and Rojiani, A.M. (2005) Aggressive growth pattern of H2009 lung carcinoma following TIMP-1 overexpression in CNS metastasis model. *AACR Meeting Abstracts*, Apr 2005, **38**.

Rojiani, M.V. and Rojiani, A.M. (2006) Down-regulation of Thrombospondin-1 following TIMP-1 overexpression in lung carcinoma enhances angiogenesis. *AACR Meeting Abstracts*, Apr 2006, 655–656.

Rowe, T.F., King, L.A., Macdonald, P.C., and Casey, M.L. (1997) Tissue inhibitor of metalloproteinase-1 and tissue inhibitor of metalloproteinase-2 expression in human amnion mesenchymal and epithelial cells. *American Journal of Obstetrics and Gynecology*, **176**, 915–921.

Sabeh, F., Shimizu-Hirota, R., and Weiss, S.J. (2009) Protease-dependent versus -independent cancer cell invasion programs: three-dimensional amoeboid movement revisited. *Journal of Cell Biology*, **185**, 11–19.

Sanderson, R.D., Yang, Y., Kelly, T. *et al.* (2005) Enzymatic remodeling of heparan sulfate proteoglycans within the tumor microenvironment: growth regulation and the prospect of new cancer therapies. *Journal of Cellular Biochemistry*, **96**, 897–905.

Schultz, R.M., Silberman, S., Persky, B. *et al.* (1988) Inhibition by human recombinant tissue inhibitor of metalloproteinases of human amnion invasion and lung colonization by murine B16-F10 melanoma cells. *Cancer Research*, **48**, 5539–5545.

Seiki, M. (2002) The cell surface: the stage for matrix metalloproteinase regulation of migration. *Current Opinion in Cell Biology*, **14**, 624–632.

Seo, D.W., Li, H., Guedez, L. *et al.* (2003) TIMP-2 mediated inhibition of angiogenesis: an MMP-independent mechanism. *Cell*, **114**, 171–180.

Sloane, B.F., Sameni, M., Podgorski, I. *et al.* (2006) Functional imaging of tumor proteolysis. *Annual Review of Pharmacology and Toxicology*, **46**, 301–315.

Sternlicht, M.D., Lochter, A., Sympson, C.J. *et al.* (1999) The stromal proteinase MMP3/stromelysin-1 promotes mammary carcinogenesis. *Cell*, **98**, 137–146.

Stetler-Stevenson, W.G., Bersch, N., and Golde, D.W. (1992) Tissue inhibitor of metalloproteinase-2 (TIMP-2) has erythroid-potentiating activity. *FEBS Letters*, **296**, 231–234.

Still, K., Robson, C.N., Autzen, P. *et al.* (2000) Localization and quantification of mRNA for matrix metalloproteinase-2 (MMP-2) and tissue inhibitor of matrix metalloproteinase-2 (TIMP-2) in human benign and malignant prostatic tissue. *Prostate*, **42**, 18–25.

Stockmann, C., Doedens, A., Weidemann, A. *et al.* (2008) Deletion of vascular endothelial growth factor in myeloid cells accelerates tumorigenesis. *Nature*, **456**, 814–818.

Valente, P., Fassina, G., Melchiori, A. *et al.* (1998) TIMP-2 over-expression reduces invasion and angiogenesis and protects B16F10 melanoma cells from apoptosis. *International Journal of Cancer*, **75**, 246–253.

Van Beem, R.T., Verloop, R.E., Kleijer, M. *et al.* (2009) Blood outgrowth endothelial cells from cord blood and peripheral blood: angiogenesis-related characteristics in vitro. *Journal of Thrombosis and Haemostasis*, **7**, 217–226.

Van Hinsbergh, V.W. and Koolwijk, P. (2008) Endothelial sprouting and angiogenesis: matrix metalloproteinases in the lead. *Cardiovascular Research*, **78**, 203–212.

Visscher, D.W., Hoyhtya, M., Ottosen, S.K. *et al.* (1994) Enhanced expression of tissue inhibitor of metalloproteinase-2 (TIMP-2) in the stroma of breast carcinomas correlates with tumor recurrence. *International Journal of Cancer*, **59**, 339–344.

Wang, M., Liu, Y.E., Greene, J. *et al.* (1997) Inhibition of tumor growth and metastasis of human breast cancer cells transfected with tissue inhibitor of metalloproteinase 4. *Oncogene*, **14**, 2767–2774.

Watanabe, M., Takahashi, Y., Ohta, T. *et al.* (1996) Inhibition of metastasis in human gastric cancer cells transfected with tissue inhibitor of metalloproteinase 1 gene in nude mice. *Cancer*, 77, 1676–1680.

Wen, W., Moses, M.A., Wiederschain, D. *et al.* (1999) The generation of endostatin is mediated by elastase. *Cancer Research*, **59**, 6052–6056.

Williamson, R.A., Marston, F.A., Angal, S. *et al.* (1990) Disulphide bond assignment in human tissue inhibitor of metalloproteinases (TIMP). *Biochemical Journal*, **268**, 267–274.

Wiranowska, M. and Naidu, A.K. (1994) Interferon effect on glycosaminoglycans in mouse glioma in vitro. *Journal of Neurooncology*, **18**, 9–17.

Wiranowska, M., Plaas, A. (2008) Cytokines and extracellular matrix remodeling in the central nervous system, in *NeuroImmune Biology* (eds C. Phelps and E.A. Korneva), Elsevier, Edinburgh, pp. 167–197.

Wiranowska, M., Tresser, N., and Saporta, S. (1998) The effect of interferon and anti-CD44 antibody on mouse glioma invasiveness in vitro. *Anticancer Research*, **18**, 3331–3338.

Wiranowska, M., Rojiani, A.M., Gottschall, P., Moscinski, L., Johnson, J., and Saporta, S. (2000). CD44 expression and MMP-2 secretion by mouse glioma cells: effect of interferon and anti-CD44 antibody. *Anticancer Research*, **20**, 4301-4306.

Wiranowska, M., Ladd, S., Smith, S.R., and Gottschall, P.E. (2006) CD44 adhesion molecule and neuro-glial proteoglycan NG2 as invasive markers of glioma. *Brain Cell Biology*, **35**, 159–172.

Wolf, K., Mazo, I., Leung, H. *et al.* (2003) Compensation mechanism in tumor cell migration: mesenchymal-amoeboid transition after blocking of pericellular proteolysis. *Journal of Cell Biology*, **160**, 267–277.

Wolf, K., Wu, Y.I., Liu, Y. *et al.* (2007) Multi-step pericellular proteolysis controls the transition from individual to collective cancer cell invasion. *Nature Cell Biology*, **9**, 893–904.

Yamada, E., Tobe, T., Yamada, H. *et al.* (2001) TIMP-1 promotes VEGF-induced neovascularization in the retina. *Histology and Histopathology*, **16**, 87–97.

Yang, T.T. and Hawkes, S.P. (1992) Role of the 21-kDa protein TIMP-3 in oncogenic transformation of cultured chicken embryo fibroblasts. *Proceedings of the National Academy of Sciences of the United States of America*, **89**, 10676–10680.

Yoshiji, H., Gomez, D.E., and Thorgeirsson, U.P. (1996) Enhanced RNA expression of tissue inhibitor of metalloproteinases-1 (TIMP-1) in human breast cancer. *International Journal of Cancer*, **69**, 131–134.

Yoshiji, H., Harris, S.R., Raso, E. *et al.* (1998) Mammary carcinoma cells over-expressing tissue inhibitor of metalloproteinases-1 show enhanced vascular endothelial growth factor expression. *International Journal of Cancer*, **75**, 81–87.

Yu, Q. and Stamenkovic, I. (1999) Localization of matrix metalloproteinase 9 to the cell surface provides a mechanism for CD44-mediated tumor invasion. *Genes Development*, **13**, 35–48.

Zeng, Z.S., Cohen, A.M., Zhang, Z.F. *et al.* (1995) Elevated tissue inhibitor of metalloproteinase 1 RNA in colorectal cancer stroma correlates with lymph node and distant metastases. *Clinical Cancer Research*, **1**, 899–906.

5

Role of Tumor-Associated Macrophages (TAM) in Cancer Related Inflammation

Antonio Sica[1,2] and Chiara Porta[1,2]

[1]*DISCAFF, Università del Piemonte Orientale, Novara, Italy*
[2]*Istituto Clinico Humanitas, Rozzano (Milan), Italy*

5.1 Introduction

Macrophages are a fundamental part of the innate defense mechanisms, which can promote specific immunity by inducing T-cell recruitment and activation. Despite this, their presence within the tumor microenvironment has been associated with enhanced tumor progression and shown to promote cancer cell growth and spread, angiogenesis, and immunosuppression. This paradoxical role of macrophages in cancer finds an explanation in their functional plasticity, that may result in the polarized expression of either pro- or antitumoral functions. Key players in the setting of their functional phenotype are the microenvironmental signals to which macrophages are exposed, which selectively tune their functions within a functional spectrum encompassing the M1 and M2 extremes. Here we discuss recent findings suggesting the central role of tumor-associated macrophages (TAMs) in driving cancer-related inflammation.

5.2 Functional plasticity of macrophages

Macrophages are released from the bone marrow in the blood stream as immature monocytes that migrate into tissues and rapidly differentiate into distinct, mature resident macrophage populations, including Kupffer cells in the liver, alveolar macrophages in the lung, and osteoclasts in the bone. Versatility is a hallmark shared

Tumor Microenvironment Edited by Dietmar W. Siemann

by both immature monocytes and mature macrophages; it accounts for phagocytes ability to coordinate the host response to a broad range of stimuli and consequently for their crucial role in both host defense and homeostasis (Martinez, Helming, and Gordon, 2009; Adams and Hamilton, 1984; Nathan and Hibbs, 1991). In primitive organisms, macrophages represent the host defense system, in that they are responsible for both recognition and destruction of threats (Podolsky and Tauber, 1996). In higher organisms, such as humans, macrophages play a key role in the innate and adaptive immune responses to pathogens, and are crucial mediators in the orchestration and resolution of inflammation. Macrophages contribute to the balance between antigen availability and clearance through phagocytosis and subsequent degradation of senescent or apoptotic cells, microbes, and possibly neoplastic cells. Their role is essential for triggering, instructing, and terminating the adaptive immune response. Macrophages collaborate with T and B cells, through both cellcell interactions and fluid-phase mediated mechanisms, based on the release of cytokines, chemokines, enzymes, arachidonic acid metabolites, and reactive radicals (Lawrence, Willoughby, and Gilroy, 2002). Macrophage activation can be either proinflammatory or anti-inflammatory, thus contributing to tissue cell destruction or to tissue regeneration and wound healing. These polar phenotypes are not expressed simultaneously but regulated in such manner that macrophages display a balanced, integrated pattern of functions. In analogy with the T helper 1 (Th1) and Th2 dichotomy, the extremes of the different form of macrophage-polarized activation are called classic or M1 and alternative or M2 (Gordon, 2003; Mantovani, Sica, and Locati, 2005; Mosmann *et al.*, 1986; Mantovani *et al.*, 2004). Classic or M1 macrophage activation in response to microbial products or interferon-γ is characterized by: high capacity to present antigen; high interleukin-12 (IL-12) and -23 (IL-23) production and consequent activation of a polarized type I T-cell response (Mantovani *et al.*, 2002a; Gordon and Taylor, 2005). M1 macrophages have cytotoxic activity toward tumor cells and ingested intracellular microorganisms, by expressing high levels of toxic intermediates, including nitric oxide (NO), reactive oxygen intermediates (ROIs), and tumor necrosis factor alpha (TNFα) (Mantovani *et al.*, 2002a; Gordon and Taylor, 2005). In contrast, alternative or M2 activation of macrophages is promoted by various signals (e.g., IL-4, IL-13, glucocorticoids, IL-10, immunoglobulin complexes/TLR ligands), which elicit different M2 forms sharing a phenotype characterized by an IL-12^{low} IL-10^{high} IL-1 decoyR^{High}, IL-1ra^{high} expression, along with high expression of scavenger and mannose receptors (Mantovani *et al.*, 2002a; Gordon and Taylor, 2005). Further, M2 macrophages express a distinct chemokines expression pattern (e.g., CCL17, CCL22) and have poor antigen presenting capability. Macrophage polarization is also characterized by profound changes in various metabolic pathways, such as arginine and iron metabolism (Martinez *et al.*, 2006). In M1 macrophages, arginine metabolism is oriented toward the production of citrulline and citotoxic NO, while the M2 counterpart metabolizes arginine to give ornitine and polyamines, which promote cell proliferation and fibrosis.

Further, changes in iron (Fe) metabolism is one of the hallmarks of M1–M2 macrophage polarization. Iron metabolism is modulated by polarizing cytokines and M1 cells present increased iron intake and cellular iron retention in order to reduce

Fe available for bacterial and tumor growth. In contrast, M2 macrophages tend to release Fe in the tumor microenvironment (Ludwiczek *et al.*, 2003).

Overall the various forms of M2-activated macrophages are oriented to tune M1 inflammation, promoting adaptive Th2 immunity, scavenge debris, angiogenesis, tissue remodelling, and repair. Thus M2-polarized macrophages protect the host by promoting the killing and encapsulation of parasites and by supporting healing of wounded/damage tissue but they may also coordinate an inflammatory response adverse for the host such as in allergy and in tumors.

5.3 Macrophages as key orchestrators of cancer-related inflammation

The association between cancer and inflammation dates back to 1863, when Rudolf Virchow noticed the presence of leukocytes in neoplastic tissues. Although this suggestion waned for a long time, much epidemiological, and experimental evidence accumulated during recent years demonstrate that inflammatory cells and circuits characterize the tumor microenvironment and represent crucial players in tumor development and progression (Mantovani *et al.*, 2008b). Cancer-related inflammation can be promoted by two major distinct pathways: an extrinsic pathway driven by inflammatory signals (e.g., infections) and autoimmune conditions (e.g., inflammatory bowel disease), and an intrinsic pathway driven by genetic alterations that cause both inflammation and neoplasia (Mantovani *et al.*, 2008b) (Figure 5.1). Thus, irrespective of the trigger for the development, an inflammatory component is present in the microenvironment of the most of if not all tumors, including those that are not causally related to an obvious inflammatory process (e.g., breast cancer). Whereas the leukocyte infiltrate varies in size and composition during cancer progression and differs among different type of tumors, several studies have highlighted macrophages as the key orchestrators of cancer-related inflammation (Mantovani *et al.*, 1992; Coussens *et al.*, 2000; Lin *et al.*, 2001).

The tumor-promoting role of TAM is suggested by the association between high frequency of infiltrating TAM and the poor prognosis for many different human tumors such as lymphoma, cervix, bladder, breast, and lung cancers (Bingle, Brown, and Lewis, 2002). According with these findings, in the post-genomic era, genes associated to leukocyte or macrophage infiltration (e.g., CD68) were identified as a part of the molecule signatures that herald a poor prognosis in lymphomas and breast carcinoma (Paik *et al.*, 2004). Recent studies have also highlighted a positive correlation between TAM abundance in primary tumors and the development of cancer metastasis (Condeelis and Pollard, 2006; Pollard, 2008). The protumorigenic activity of TAM is also supported by studies with genetically modified mice, as well as by cell transfer experiments. Low macrophage infiltration into the tumor mass correlated with the inhibition of tumor growth and metastasis development, in different animal models (Lin *et al.*, 2001, 2006; Hiraoka *et al.*, 2008; Zeisberger *et al.*, 2006). Lin and colleagues demonstrated that crossing of MMTV-PyMT mice, which spontaneously develop mammary tumors, with op/op mice, which lack monocytes/macrophages, inhibits tumor growth and spread (Lin *et al.*, 2001).

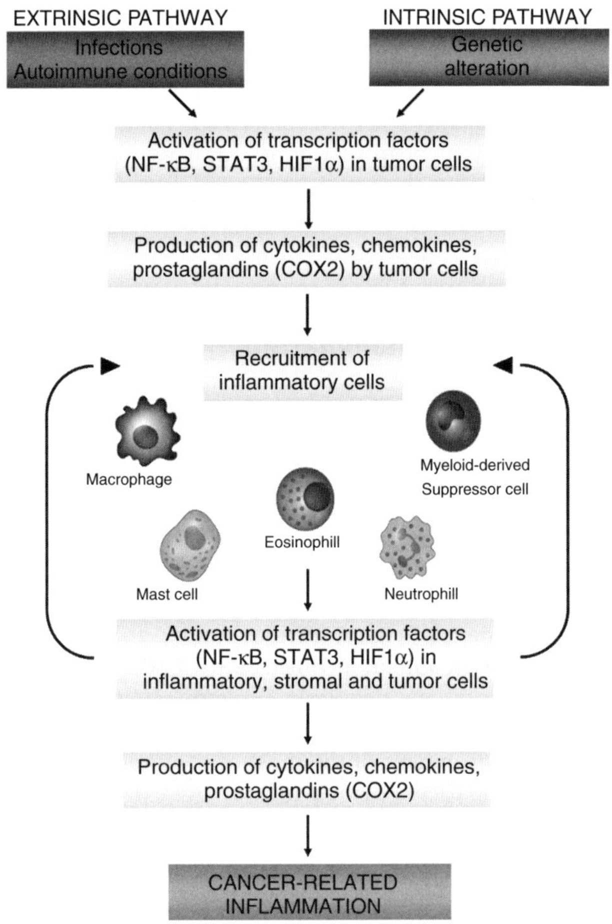

Figure 5.1 Pathways promoting cancer and inflammation: the intrinsic pathway and the extrinsic pathway. The intrinsic pathway includes the activation of various types of oncogene by mutation, chromosomal rearrangement or amplification, and the inactivation of tumor-suppressor genes. Cells that are transformed in this manner produce inflammatory mediators, thereby generating an inflammatory microenvironment in tumors for which there is no underlying inflammatory condition (for example, breast tumors). By contrast, in the extrinsic pathway, inflammatory or infectious conditions augment the risk of developing cancer at certain anatomical sites (for example, the colon, prostate and pancreas). The two pathways converge, resulting in the activation of transcription factors, mainly nuclear factor-κB (NF-κB), signal transducer and activator of transcription 3 (STAT3) and hypoxia-inducible factor 1α (HIF1α), in tumor cells. These transcription factors coordinate the production of inflammatory mediators, including cytokines and chemokines, as well as the production of cyclooxygenase 2 (COX-2) (which, in turn, results in the production of prostaglandins). These factors recruit and activate various leukocytes, most notably cells of the myelomonocytic lineage. The cytokines activate the same key transcription factors in inflammatory cells, stromal cells and tumor cells, resulting in more inflammatory mediators being produced and a cancer-related inflammatory microenvironment being generated.

In this mouse model, they also showed that TAMs play a crucial role in the 'angiogenic switch' when hyperplastic lesions develop into early stage mammary carcinomas (Lin *et al.*, 2006).

5.4 Recruitment and differentiation of TAM

A strict correlation between increased numbers and/or density of macrophages and poor prognosis has been observed. Based on this, both recruitment and activation of TAM are regarded as pivotal events in tumor progression and TAM considered putative targets for therapeutic interventions (Sica *et al.*, 2007). TAM derive from blood monocytes, which are recruited in neoplastic tissue by different chemotactic factors. The first tumor-derived chemotactic factor identified was CCL2 (Bottazzi *et al.*, 1983, 1992), whose level of expression was later correlated with TAM density in various tumors (Mantovani *et al.*, 2002a).

In addition to recruiting monocytes, chemokines, that act in concert with adhesion molecules (Marttila-Ichihara *et al.*, 2009), have an important role in tumor progression by directly stimulating neoplastic growth, promoting inflammation, and inducing angiogenesis (Balkwill, 2003; Sica *et al.*, 2008).

Most recently, the role of chemokines in carcinogenesis has received new confirmation. The missing genetic link has been obtained taking advantage of the D6 atypical chemokine receptor, which acts as a decoy and scavenger for most inflammatory CC chemokines (Biswas *et al.*, 2006). D6-deficient mice show increased susceptibility to skin carcinogenesis (Nibbs *et al.*, 2007) and colitis-associated cancer, the latter being representative of a clinical paradigm of the inflammationcancer connection (Vetrano *et al.*, 2009). Thus, D6-targeted mice provide unequivocal genetic evidence that inflammatory CC chemokines are more than an epiphenomenon in clinically relevant carcinogenesis.

It has become also clear that molecules other than chemokines can promote recruitment of TAM. These factors are produced either by tumor cells or host stromal elements (Sica *et al.*, 2006) and include tyrosine-kinase receptor ligands, such as colony-stimulating factor-1 (CSF-1) (Joyce and Pollard, 2009) and vascular endothelial growth factor (VEGF) (Sica *et al.*, 2007). Moreover, the C5a complement component, which interacts with a G-protein coupled receptor as do chemokines, has also been recently shown to play a role in leukocyte recruitment in cancer-related inflammation (Markiewski and Lambris, 2009). Thus, members of different molecular families contribute to the shaping of cancer-related inflammation. Some of these factors (e.g., IL-10, IL-6), expressed in the tumor microenvironment, also inhibit differentiation of monocytes into dendritic cells (DCs), by activating signal transducer and activator of transcription 3 (STAT3)-dependent signalling (Gabrilovich, 2004; Mellman and Steinman, 2001).

TAM preferentially localize in poorly vascularized regions of tumors (Lewis and Pollard, 2006; Leek *et al.*, 2002). This environment promotes the metabolic adaptation of TAM to hypoxia, through the activation of the hypoxia-inducible factor-1 (HIF-1) and HIF-2 (Leek *et al.*, 2002). We recently have shown that HIF-1α activated in TAM by hypoxia influences the positioning and function of tumor cells, stromal cells and TAM, by selectively upregulating their expression of CXC-chemokine receptor 4 (CXCR4) (Schioppa *et al.*, 2003). Moreover, HIF-1 activation can have a role in the induction of the CXCR4 ligand, CXC-chemokine ligand 12 (CXCL12) (Ceradini *et al.*, 2004), a chemokine involved in cancer metastasis (Muller *et al.*, 2001). Together, these data suggest that oxygen availability has a role in guiding the microanatomical localization and function of TAM. Moreover, hypoxia can also have important consequences on L-Arg metabolism, since it can induce nitric oxide synthase (NOS2) and ARG expression and consequently inhibit adaptive immunity (Sica and Bronte, 2007).

Despite a poor knowledge over the microenvironmental signals driving the TAM phenotype, M-CSF appears to play an important M2 orienting activity (Martinez *et al.*, 2006). Along with M-CSF, IL-10, TGFβ, prostaglandin E2 (PGE2), and IL-6 have also M2-promoting properties (Saccani *et al.*, 2006). Thus, incoming monocytes are exposed to a cocktail of tumor derived factors which strongly polarize their M2 functions (Sica *et al.*, 2006; Mantovani, Allavena, and Sica, 2004; Biswas *et al.*, 2006). IL-10 also favors macrophage differentiation and inhibits DC development (Allavena *et al.*, 1998) thus contributing to TAM expansion.

Adaptive immune responses can also shape the infiltration and function of TAM. In a primary mammary carcinoma model, $CD4^+$ T cells promote metastasis by causing M2 activation via IL-4 (DeNardo *et al.*, 2009). Interestingly, the same group, by pursuing previous studies on multistage epithelial carcinogenesis in the skin driven by transgenic expression of human papillomavirus 16 (HPV16), identified a different pathway of orchestration of inflammation-mediated tumor promotion by adaptive immune responses. Here, T-cell-dependent humoral antibody production against extracellular matrix components orchestrates cancer-related inflammation (de Visser, Korets, and Coussens, 2005). Autoantibodies act via FcγR and regulate recruitment and M2-like polarization of myelomonocytic cells which promote tumor progression (Andreu *et al.*, in press).

Transcriptional analysis of TAM has revealed that these cells express low levels of inflammatory mediators (e.g., IL-12, IL-1, TNF, IL-6, NO, ROIs) (Mantovani, Allavena, and Sica, 2004; Biswas *et al.*, 2006; Dinapoli, Calderon, and Lopez, 1996; Klimp *et al.*, 2001), along with unexpected high levels of Interferon-inducible chemokines (CXCL9, CXCL10) (Biswas *et al.*, 2006). Further, TAM express high levels of immunosuppressive cytokines (e.g., IL-10, TGFβ) (Biswas *et al.*, 2006), scavenger receptors (e.g., SR-A and mannose receptor) and chemokines (CCL17, CCL18, CCL22), which preferentially recruit naïve, T regulatory (Treg), and Th2 lymphocytes. This pattern of gene expression confirmed that TAM represent a distinct M2 macrophage population (Biswas *et al.*, 2006).

The importance of an M2 signature for cancer progression is demonstrated by several evidence. Of note, SH2-containing inositol-5′-phosphatase 1 (SHIP1)-deficient mice, which exhibit a spontaneous macrophage drift toward the M2 polarization, display increased growth of transplanted tumors (Rauh *et al.*, 2005), while p50 nuclear factor-kappaB (NF-κB)-deficient mice, which showed a defective capacity to mount an M2 macrophage polarization (Porta *et al.*, 2009), displayed an increased tumor resistance (Saccani *et al.*, 2006). Further, clinical evidence suggest that the type of immunological profile expressed at the tumor site represents an independent prognostic factor. In particular, an established M2 or type-2 'suppressive' immunological profile correlates with poor prognosis, as shown in colorectal, hepatocellular, and pancreatic carcinomas (Budhu *et al.*, 2006; Galon *et al.*, 2006; Kurahara *et al.*, 2009).

5.5 Protumoral functions of TAM

Several evidence have demonstrated that tumor cells secrete factors able to sustain myelopoiesis, accumulation, and functional differentiation of myelomonocytic cells, which in turn elicit T-cell dysfunction and provide essential supports for the angiogenesis and the stroma remodelling required for tumor growth (Figure 5.2).

5.5.1 Immunosuppression

The molecular basis of TAM-mediated immunosuppression include the inhibition of T-cell activation along with the release of chemokines that preferentially attract T-cell subsets devoid of cytotoxic functions (Mantovani *et al.*, 2002a). In agreement, it has been recently demonstrated that depletion of TAM is associated with increased HPV16 E7(49-57)-specific CD8 lymphocytes infiltrated tumors as well as with tumor growth inhibition (Lepique *et al.*, 2009). In particular TAM express low levels of costimulatory molecules (e.g., CD86) and are poor producers of immunostimulatory cytokines (such as IL-12, IL-1, and TNF) (Sica *et al.*, 2000). In contrast, TAM produce high constitutive levels of the immunosuppressive cytokines (IL-10 and TGFβ) (Sica *et al.*, 2000). These immunosuppressive cytokines tune inflammation and can promote Treg differentiation (Mantovani *et al.*, 2002a). TAM also secrete high levels of CCL17 and CCL22 which preferentially recruit Th2 and Treg (Balkwill, 2004; Mantovani *et al.*, 2004). As previously observed in the ascitic fluid of human ovarian carcinoma, TAM are important producers of CCL18, which attracts naïve T cells (Schutyser *et al.*, 2002). The recruitment of these undifferentiated lymphocytes in a tumor microenvironment is likely to contribute to T-cell anergy (Mantovani *et al.*, 2002a). In conjunction with TAM, an immature population of myeloid cells (myeloid-derived suppressor cells (MDSCs)), with immunosuppressive activities, accumulates in tumor bearers and inhibits the adaptive antitumor response (Bronte *et al.*, 2003). MDSC are defined by the expression of both CD11b and Gr1 markers (Gallina *et al.*, 2006) and similarly to TAM show an M2-skewed gene expression profile (Biswas *et al.*, 2006; Sica and Bronte, 2007; Martinez, Helming, and Gordon, 2009). MDSCs preferentially accumulate in blood and spleen of tumor bearers but they are also recruited into the tumor site where they can express the macrophages marker F4/80 (Kusmartsev and Gabrilovich, 2005).

MDSC use two enzymes involved in the arginine metabolism to control the T-cell response: inducile NOS2 and arginase-1 (Arg-1), which deplete the milieu of arginine, causing peroxynitrite generation, as well as lack of CD3ζ chain expression and T-cell apoptosis. In prostate cancer, selective antagonists of these two enzymes were proved beneficial in restoring T-cell-mediated cytotoxicity (Bronte *et al.*, 2005).

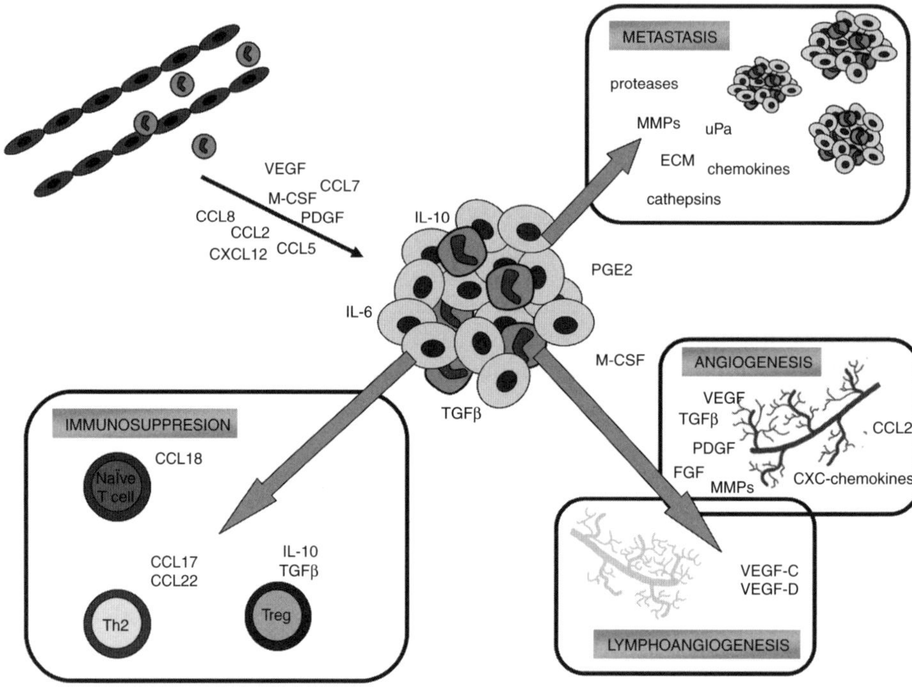

Figure 5.2 Mechanisms of TAM recruitment and their protumoral functions. Chemokines (e.g., CCL2, CCL5, CCL7, CCL8, CXCL12) and growth factors (e.g., M-CSF, PDGF, and VEGF) actively recruit blood monocytes at the tumor site, where they differentiate in TAM. Tumor microenvironmental signals, such as IL-10, PGE2, M-CSF, TGFβ, and IL-6 promote TAM polarization toward the M2 phenotype. TAM promote tumor growth and progression through different mechanisms. TAM inhibit the anti-tumor responses by the secretion of immunosuppressive cytokines, like IL-10 and TGFβ, and by selective recruitment of naıve T cells, via the chemokine CCL18, Th2 and Treg, via the chemokines CCL17 and CCL22. TAM also contribute to the both blood and lymphatic neovascular formation, by releasing angiogenic (VEGF, FGF, TGFβ, PDGF, MMPs, and chemokines) and lymphoangiogenic (VEGF-C, VEGF-D) factors. Further TAM promote tumor invasion and metastasis formation through the secretion of matrix proteins (ECM), matrix remodeling enzymes (MMPs, proteases, cathepsins, uPa), and their activators (chemokines).

5.5.2 Angiogenesis

The generation of new blood vessels in response to increasing demand for nutrients and oxygen experienced by proliferating tumor cells is essential for cancer growth and progression (Carmeliet and Jain, 2000). Experimental murine tumor models have highlighted the importance of CSF-1 in TAM-mediated angiogenesis. In particular, using a MMTV-PyMT model, Lin and coworkers have demonstrated that CSF-1 overexpression increases TAM recruitment, which in turns promotes tumor development by inducing the 'angiogenic switch' (Lin *et al.*, 2006). Further, Abraham and colleagues have recently demonstrated that intratumoral injections of small interfering RNAs directed against CSF-1 results in significant inhibition of neuroblastoma growth associated with the reduction of TAM infiltration, matrix metalloproteinase (MMP)-12 expression and angiogenesis (Abraham *et al.*, 2009). The proangiogenic role of TAM was also highlighted in clinical studies that demonstrated the existence of a positive correlation between TAM density and vascular grade of different types of human tumors, such as glioma, squamous cell carcinoma of esophagus, breast, bladder prostate, and renal cell carcinoma (Leek *et al.*, 1996; Nishie *et al.*, 1999; Hanada *et al.*, 2000; Koide *et al.*, 2004; Toge *et al.*, 2009). The molecular basis of TAM proangiogenic activities are partially elucidated: TAM preferentially accumulate in tumor regions characterized by low oxygen tension and hypoxia is able to triggers a proangiogenic program in these cells. Low oxygen conditions promote HIF-1 and HIF-2 expression, which activate the transcription of several proangiogenic molecules such as VEGF, βFGF, and CXCL8 (Lewis and Murdoch, 2005). In addiction to CXCL8, TAM secrete several other proangiogenic chemokines, such as CCL2, CXCL5, CXCL1, CXCL13, and CXCL12, which play an important role in tumor neovascularization. For example, CXCL5 and CXCL8 levels are associated with increased angiogenesis and decreased survival of small cell lung cancer patients (Strieter *et al.*, 2004). The CXCL12-CXCR4 axis plays a key role in angiogenesis, vasculogenesis, and metastasis formation of different type of human tumors (Kryczek *et al.*, 2007).

TAM secrete several growth factors (e.g., VEGF, platelet-derived growth factor (PDGF), TGFβ, and members of the FGF family) (Mantovani *et al.*, 2002a; Leek *et al.*, 1996; Bingle *et al.*, 2006) as well as matrix remodeling enzymes (e.g., uPa and MMPs) that directly activate endothelial cell proliferation and facilitate their migration within the extracellular matrix (Coussens *et al.*, 2000; Pollard, 2004).

In the epidermal squamous cell carcinoma model, K14-HPV16, TAM produce the potent pro-angiogenic enzyme, MMP-9, which degrades extracellular matrix releasing other proangiogenic growth factors (Coussens *et al.*, 2000). Similarly, a study examining the importance of MMP-9 and VEGF-A in RIP1-Tag2 mice highlighted that inflammatory cells are the main producers of these two proangiogenic molecules in pancreatic tumors (Bergers *et al.*, 2000; Nozawa, Chiu, and Hanahan, 2006; Shojaei *et al.*, 2008). Sierra and coworkers demonstrated that TAM are the principal producers of the pro-angiogenic Semaphorin 4D (SemaD4), which plays a critical role in tumor angiogenesis and blood vessel maturation (Sierra *et al.*, 2008).

In addiction to TAM, a distinct subset of monocytes characterized by the expression of the Tie-2 receptor has been identified in hypoxic areas of solid tumors, in close proximity to nascent tumor vessels. These Tie-2-expressing monocytes (TEMs) show proangiogenic activities in different tumor models (e.g., mouse subcutaneous tumors, human glioblastoma xenografts, and spontaneous pancreatic tumors) (De Palma *et al.*, 2003, 2005; Venneri *et al.*, 2007). Strikingly, analysis of TEM's gene expression profile has highlighted that TEMs are highly related to TAM but express a distinct M2 signature, suggesting that this subset of monocytes is a distinct lineage committed to execute physiologic proangiogenic and tissue-remodeling programs, which can be co-opted by tumors (Pucci *et al.*, 2009).

In addiction to vasculogenesis, the development of the lymphatic vascular networks is a crucial event for tumor progression and dissemination. In human ovarian tumor bearer mice, TAM stimulate lymphoangiogenesis through the secretion of

VEGF-C and VEGF-D, which activate the lymphatic vessel specific receptor VEGF receptor-3 (Jeon *et al.*, 2008). The importance of TAM-derived VEGF-C in the development of peritumoral lymphoangiogenesis is also reported in human cervical, pancreatic, and prostate cancers (Zumsteg *et al.*, 2009).

5.5.3 Metastasis formation

TAM promotes cancer metastasis through several mechanisms, including promotion of angiogenesis and lymphoangiogenesis, induction of tumor growth and enhancement of tumor cells migration, and invasion. Accumulating evidence has highlighted the importance of the cross-talk between TAM and cancer cells for tumor cells migration. Goswami and colleagues reported that epidermal growth factor (EGF) release by TAM and CSF-1 secreted by breast cancer cells acts on the reciprocal cell type to stimulate the migration of both tumor cells and TAM (Goswami *et al.*, 2005). In this scenario, interrupting either of these signals results in decreased tumor cell motility. Using multiphoton-based intravital imaging, these authors also showed that tumor cell intravasation occurs in association with perivascular macrophages (Wyckoff *et al.*, 2007) and suggested a tripartite arrangement of an invasive carcinoma cell, a macrophage, and an endothelial cell as a potential tumor microenvironment of metastasis (TMEM). Recently, they found a significant association between high density of TMEM in human breast carcinoma samples and the development of systemic, hematogenous metastasis (Robinson *et al.*, 2009), suggesting TMEMs density as a new prognostic marker for breast cancer patients.

TAM may promote invasion of neoplastic cells and facilitate their movement in the extracellular matrix. For example, human macrophage-derived TNFα enhances the invasiveness of breast and ovarian cancer cell lines and induces the expression of both macrophage inhibitory factor (MIF) and extracellular MMPs. TAM are key modulators of extracellular matrix remodeling, through the secretion of both matrix-degrading enzymes and matrix proteins. In particular, TAM produce serine proteases, MMPs (e.g., MMP-2, MMP-9), and cathepsins, which act on cell–cell junctions, modify the extracellular matrix composition and promote basal membrane destruction (Egeblad and Werb, 2002; Lynch and Matrisian, 2002; Mohamed and Sloane, 2006). Further, TAM produce chemokines that are able to induce both the expression and the activation of several MMPs, in particular MMP-9 and uPA receptor (Locati *et al.*, 2002). Recently, the importance of both the macrophage-derived MMP-2, MMP-9, and the uPA system in the enhancement of invasive properties observed in B16 melanoma cells stimulated by activated macrophages was demonstrated in *in vitro* coculture (Marconi *et al.*, 2008). *In vivo* evidence on the contribution of hematopoietic cell-derived MMP-9 in carcinogenesis has been observed in a spontaneous model of skin carcinogenesis (Coussens *et al.*, 2000). Recent evidence also indicates that MMP7 enhances release and activation of RANKL, which is a potent prometastatic factor (Lynch *et al.*, 2005; Luo *et al.*, 2007). Since TAM upregulate MMP7 in hypoxic tumor regions they could also promote RANKL-driven tumor cell motility. Other studies have also convincingly illustrated how macrophages play a role in precondition a specific region to metastasic growth.

Using VEGFR1- and MMP-9-deficient mice, Hiratsuka *et al.* demonstrated that tumor cell colonization of the lungs is dependent on VEGF-induced expression of MMP-9, by tissue resident macrophages and endothelial cells (Hiratsuka *et al.*, 2002). TAM also contribute in the built of extracellular matrix architecture by the production of matrix proteins such as secreted protein acidic and rich in cysteine (SPARC), which modulate collagen density, leukocyte, and blood vessel infiltration (Sangaletti *et al.*, 2003).

5.6 Molecular determinants of TAM functions

As mentioned above, TAM generally have phenotype and functions similar to alternative or M2 macrophages (Biswas *et al.*, 2006; Mantovani *et al.*, 2008a). For example, TAM express low levels of the major histocompatibility complex class II and reduced antimicrobial and tumoricidal activity, while increasing production of mediators that promote angiogenesis, such as VEGF and cyclo-oxygenase-2 (COX-2)–derived prostaglandin E2, as well as the anti-inflammatory cytokine IL-10 (Mantovani *et al.*, 2008a). Another hallmark feature of alternative activation expressed by TAM is the low expression of IL-12 and upregulated levels of M2-specific genes, such as Arg-1, macrophage galactose-type C-type lectin-2 (Mgl2), Fizz1, and Ym1 (Biswas *et al.*, 2006; Ghassabeh *et al.*, 2006). Recent additions to the molecular repertoire of TAM include Sema4D and Gas6, which are respectively involved in promoting tumor angiogenesis (Sierra *et al.*, 2008) and cancer cell proliferation (Loges *et al.*, in press). Analysis of the molecular basis of the TAM phenotype has identified the transcriptional factors NF-κB and HIF-1 as master regulators of their transcriptional programs and indicates these factors as central regulators of tumor progression and metastasis (Sica and Bronte, 2007).

NF-κB induces several cellular alterations associated with tumorigenesis and more aggressive phenotypes, including: self-sufficiency in growth signals; insensitivity to growth inhibition; resistance to apoptotic signals; immortalization; angiogenesis; tissue invasion; and metastasis (Karin, 2006). Constitutive NF-κB activation often observed in cancer cells may be either promoted by genetic alterations or by microenvironmental signals, including cytokines, hypoxia, and ROI (Karin, 2006). In particular, proinflammatory cytokines (e.g., IL-1 and TNF), expressed by infiltrating leukocytes, can activate NF-κB in cancer cells and contribute to their proliferation and survival (Karin, 2006).

However, to the extent they have been investigated, TAM display a defective NF-κB activation in response to different proinflammatory signals (Allavena *et al.*, 2008), suggesting their tolerant phenotype. Inhibition of NF-κB activation in TAM correlates with impaired expression of NF-κB-dependent inflammatory functions (e.g., expression of cytotoxic mediators, such as NO, and cytokines (TNFα, IL-1, and IL-12) (Sica *et al.*, 2008; Mantovani *et al.*, 2008a). Hence, there is an apparent contrast with inflammatory functions expressed by TAM during early steps of carcinogenesis. Possible explanations for this discrepancy are based on the functional feature of macrophages, which are considered highly plastic cells able to finely modulate their programs in response to different microenvironmental conditions

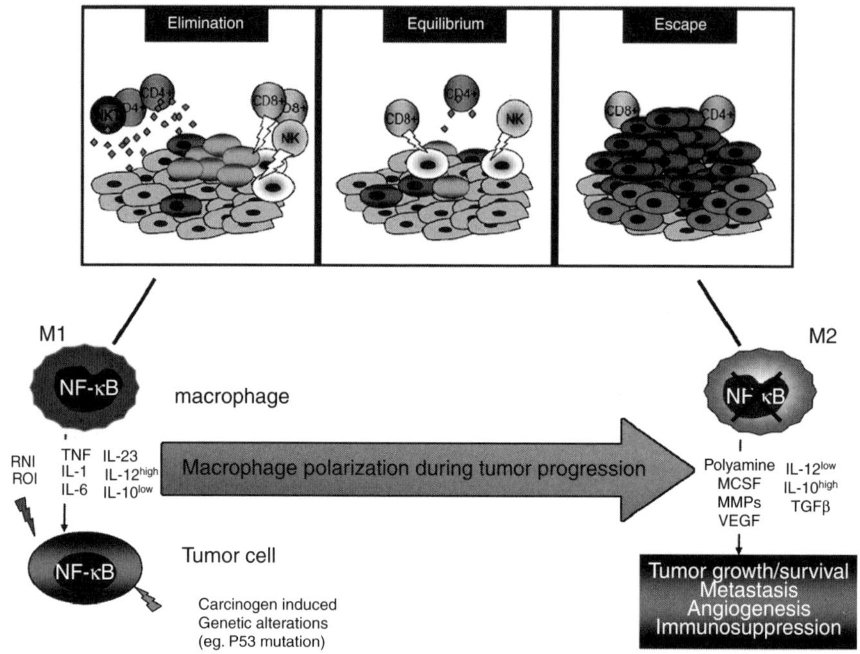

Figure 5.3 Gradual switching of TAM polarization, from M1 to M2, is paralleled by the gradual inhibition of NF-κB, during different stages of tumor progression. Early in carcinogenesis T cell driven M1 activated macrophages may contribute to elimination. Ensuing regulatory mechanisms of results in M2 polarized TAM which orchestrate smoldering, non-resolving tumor-promoting inflammation.

(Sica *et al.*, 2008). It has been speculated that dynamic changes of the tumor microenvironment may occur during the transition from early neoplastic events toward advanced tumor stages (Sica and Bronte, 2007). These events would drive an M1 toward M2 switch of TAM functions (Figure 5.3) (Sica and Bronte, 2007) and are likely connected to the profound changes occurring in the tumor microphysiology (e.g., hypoxia, glucose levels, pH) (Vaupel, 2008). Thus, while full activation of NF-κB in leukocytes would favor M1 inflammation and tumorigenesis, tumor growth, and progression may drive inhibition of NF-κB in infiltrating leukocytes, as reported in both myeloid and lymphoid cells associated with advanced tumor stages (Sica and Bronte, 2007).

By investigating the molecular basis of this phenomenon, TAM were found to be characterized by massive nuclear accumulation of the inhibitory p50 NF-κB homodimer (Saccani *et al.*, 2006), a phenotype also observed in lipopolysaccharide-tolerant macrophages (Porta *et al.*, 2009). Interestingly, TAM from $p50^{-/-}$ tumor-bearing mice express cytokines characteristic of M1 macrophages (e.g., $IL\text{-}12^{high}/IL\text{-}10^{low}$) and their splenocytes produce increased levels of Th1 cytokines (e.g., IFN-γ), which are associated with a delay in tumor growth (Saccani *et al.*, 2006). A detailed analysis of the role of p50 NF-κB homodimer in macrophage functions revealed that its nuclear accumulation, both in TAM and LPS-tolerant macrophages, not only mediates a status of unresponsiveness (tolerance) toward pro-inflammatory signals, but actually plays as key regulator of M2-driven inflammatory reactions, acting through inhibition of NF-κB-driven M1-polarizing IFNβ production and STAT1 phosphorylation (Porta *et al.*, 2009).

A recent study by Hagemann *et al.* demonstrated the requirement for IKKβ to maintain the IL-10^{high}/IL-12^{low} phenotype of TAM in a mouse model of ovarian cancer (Hagemann *et al.*, 2008). These authors showed that targeted deletion or inhibition of IKKβ in TAM increased their tumoricidal activity through elevated NOS2 expression and IL-12-dependent recruitment and activation of natural killer (NK) cells. Because increased expression of IL-12, NOS2, and enhanced tumoricidal activity are associated with M1 characteristics, these data imply that inhibition of the IKKβ/NF-κB pathway promotes an M1-like phenotype in TAM. Additional efforts are required to clarify the role of NF-κB in TAM polarization, but both studies indicate that inhibition of either p50 NF-κB or IKKβ result in restoration of STAT1 activity, which appears a convergent and necessary function to promote M1 macrophage polarization (Porta *et al.*, 2009; Hagemann *et al.*, 2008). An early IKKβ dependent NF-κB activation may trigger cancer-related inflammation and the p50-dependent regulatory pathway may tune and promote M2-associated smoldering inflammation.

A microenvironmental condition that appears to impact on NF-κB signaling in TAM is hypoxia (low oxygen tension). The presence of many areas of hypoxia is a hallmark feature of most forms of solid tumor (Vaupel, 2008) and TAM have been shown to accumulate in these areas where hypoxia promotes their protumor phenotype (Murdoch *et al.*, 2008). HIF-1 has been shown to control the cellular response to hypoxia. Hypoxia stabilizes HIF-1α, preventing posttranslational hydroxylation and subsequent degradation via the proteasome. More recently, short-term exposure of murine bone marrow-derived macrophages to hypoxia has been shown to upregulate NF-κB activity, which in turn upregulates HIF-1α levels (Rius *et al.*, 2008). This study used macrophages from IKKβ$^{-/-}$ mice to show that NF-κB is a critical transcriptional activator of HIF-1α and that basal NF-κB activity is required for HIF-1α protein accumulation under hypoxia. Overall, this evidence indicates a considerable plasticity in NF-κB in TAM and suggests that modulation of its activity in these cells maintains their immunosuppressive, tumor-promoting phenotype. Further studies addressing the relative contribution of individual NF-κB members (p65, c-Rel, p50, BCL3) and their combinatorial transcriptional partners, such as STATs and IRF3 (Kawai and Akira, 2007), will likely contribute to fully clarify its role in cancer-related inflammation.

5.7 Therapeutic targeting of TAM

Restoration of NF-κB activity in TAM is a potential strategy to restore M1 inflammation and intratumoral cytotoxicity (Sica, Allavena, and Mantovani, 2008). In agreement, accumulating evidence indicate that restoration of an M1 phenotype in TAM may provide therapeutic benefit in tumor-bearing mice. In particular, combination of CpG plus an anti-IL-10 receptor antibody switched infiltrating macrophages from M2 to M1 and triggered innate responses, debulking large tumors within

16 hours (Guiducci *et al.*, 2005). In line with this study, several other studies in preclinical models of cancers have confirmed the antitumor properties of TLR9 agonists which are currently used to treat solid and hematologic malignancies (Krieg, 2008). In addiction to TLR agonists, it has been shown that other approaches based on bacteriophages will be able to recover M1 macrophage activation as well as inhibition of tumor growth and metastasis (Eriksson *et al.*, 2009). It is likely that the restoration of NF-κB activation in TAM is the common molecular mechanism of action of these immune adjuvants.

In spontaneous breast cancer model, *in vivo* treatment with zoledronic acid, a well-known antitumor drug, is able to revert TAM polarization from the M2 to M1 phenotype, as well as to inhibit mammary carcinogenesis (Coscia *et al.*, 2009). These data support that switching the TAM phenotype from M2 to M1 during tumor progression may promote antitumor response. Recently it has been reported that IFNγ can restore an M1 phenotype in TAM purified from human ovarian ascites, thus suggesting the local use of this cytokines to potentiate the efficacy of antitumor immunotherapies (Duluc *et al.*, 2009).

Efforts were made in the identification of the key molecules driving the switching from M1 to M2 phenotype during tumor progression. In this regards TAM from $STAT6^{-/-}$ tumor-bearing mice display an M1 phenotype, with a low level of Arg and a high level of NO. As a result, these mice immunologically rejected spontaneous mammary carcinoma (Sinha *et al.*, 2005).

Along with STAT6, STAT3 activation is associated with M2 macrophage polarization (Mantovani *et al.*, 2002a, 2002b; O'Shea *et al.*, 2004). STAT3 is constitutively activated in tumor cells (Wang *et al.*, 2004) and in diverse tumor-infiltrating immune cells, including TAM (Kortylewski *et al.*, 2005), leading to inhibition of proinflammatory cytokine and chemokine production and to the release of factors that suppress DC maturation. Ablating STAT3 in hematopoietic cells triggers an intrinsic immune surveillance system that inhibits tumor growth and metastasis and that is associated with enhanced functional activity of DCs, T cells, NK cells, and neutrophils (Kortylewski *et al.*, 2005).

It was shown that SHIP1 phosphatase plays a critical role in programming macrophage M1 versus M2 functions. Mice deficient for SHIP1 display a skewed development away from M1 macrophages (which have high inducible NOS2 levels and produce NO), toward M2 macrophages (which have high Arg levels and produce ornithine) associated with enhance growth of transplanted tumors (Rauh *et al.*, 2005). We have previously demonstrated that nuclear accumulation p50 NF-κB homodimers in TAM is associated with defective NF-κB activation in response to inflammatory stimulus (Saccani *et al.*, 2006). Ablation of p50 is associated with restoration of NF-κB dependent M1 inflammatory response capable to inhibit growth (Saccani *et al.*, 2006). This data have suggested p50 NF-κB as an important molecule for the expression of an M2-polarized TAM phenotype. Accordingly, we have recently shown that the p50 subunit of NF-κB is an essential orchestrator of M2 driven inflammatory reactions, *in vitro* and *in vivo* (Porta *et al.*, 2009). As a consequence, drugs able to selectively target p50 homodimers could represent a valuable approach to switch TAM polarization from the protumoral M2 to the anti-tumoral M1 phenotype. In this context, it is interesting to note that PGE2, which selectively induces p50 homodimers, has M2-orienting properties (Eruslanov *et al.*, 2009). Tumors produce copious amount of PGE2 as result of activation of COX-2 and down-regulation of enzymes (e.g., 15-PGDH) involved in PGE2 catabolism. Recently, Eruslanov and coworkers have demonstrated that adenovirus-mediated delivery of 15-PGDH resulted in restoration of M1 antitumor response *in vivo*, associated with tumor inhibition and long-term survival even in mice with pre-established tumors (Eruslanov *et al.*, 2009).

In addition to TAM, MDSC are a myeloid M2-biased cell population present in lymphoid organs and peripheral tissues of tumor-bearing hosts and potently contributing to T cell immunosuppression, through the release of NOS2 and Arg1 (Bronte *et al.*, 2005). In prostate cancer, selective antagonists of these two enzymes were proved beneficial in restoring T-cell-mediated cytotoxicity (Bronte *et al.*, 2005).

The IFN-γ-inducible enzyme indoleamine 2,3-dioxygenase (IDO) is a well known suppressor of T-cell activation. It catalyzes the initial rate-limiting step in tryptophan catabolism, which leads to the biosynthesis of nicotinamide adenine dinucleotide. By depleting tryptophan from local microenvironment, IDO blocks activation of T lymphocytes. Recently it was shown that inhibition of IDO, which is highly expressed by macrophages, may cooperate with cytotoxic agents to elicit regression of established tumors (Uyttenhove *et al.*, 2003) and may increase the efficacy of cancer immunotherapy (Muller *et al.*, 2005).

5.8 Conclusions

Recent findings suggest that polarized inflammation plays different roles during tumor progression. Within this scenario, experimental and clinical evidence support a dual effect of inflammatory cells, macrophages in particular, in early as opposed to late phases of cancer development. While high production of M1 inflammatory mediators (e.g., TNF, ROIs) by inflammatory cells appears to support neoplastic transformation (Pikarsky *et al.*, 2004; Greten *et al.*, 2004; Balkwill and Coussens, 2004), phenotypic characterization of TAM in established cancers has invariably revealed their lack of M1 functions, along with the predominant expression of M2 properties (Dinapoli, Calderon, and Lopez, 1996; Sica *et al.*, 2000; Saccani *et al.*, 2006). While the production of M1 mediators, including cytokines (e.g., IL-12), by innate immune cells provides the functional bridge promoting specific antitumor immunity, progressive M2 skewing of macrophage functions in cancer bearers correlates with impaired antitumor Th1 cell responses (immunosuppression), as well as with poor prognosis (Galon *et al.*, 2006; Budhu *et al.*, 2006). Emerging evidence indicate the gradual inhibition of the NF-κB activity and increased STAT3 activation in hematopoietic cells, including macrophages, may represent a tumor-mediated mechanism triggering their M1/M2 skewing (Saccani *et al.*, 2006). In advanced cancers, strategies targeting mechanisms and molecules (e.g., p50 NF-κB, STAT3) supporting M2 inflammatory programs, in combination with active (vaccines) and passive immunotherapy (adoptive transfer of *ex vivo* expanded tumor-infiltrating T cells), are likely to provide therapeutic benefit.

References

Abraham, D., Zins, K., Sioud, M. *et al.* (2009) Stromal cell-derived CSF-1 blockade prolongs xenograft survival of CSF-1-negative neuroblastoma. *International Journal of Cancer*, **126**, 1339–1352.

Adams, D.O. and Hamilton, T.A. (1984) The cell biology of macrophage activation. *Annual Review of Immunology*, **2**, 283–318.

Allavena, P., Piemonti, L., Longoni, D. *et al.* (1998) IL-10 prevents the differentiation of monocytes to dendritic cells but promotes their maturation to macrophages. *European Journal of Immunology*, **28**, 359–369.

Allavena, P., Garlanda, C., Borrello, M.G. *et al.* (2008) Pathways connecting inflammation and cancer. *Current Opinion in Genetics and Development*, **18**, 3–10.

Andreu, P., Johansson, M., Affara, N.I. *et al.* FcRgamma activation regulates inflammation-associated squamous carcinogenesis. *Cancer Cell*, **17**, 121–134.

Balkwill, F. (2003) Chemokine biology in cancer. *Seminars in Immunology*, **15**, 49–55.

Balkwill, F. (2004) Cancer and the chemokine network. *Nature Reviews Cancer*, **4**, 540–550.

Balkwill, F. and Coussens, L.M. (2004) Cancer: an inflammatory link. *Nature*, **431**, 405–406.

Bergers, G., Brekken, R., McMahon, G. *et al.* (2000) Matrix metalloproteinase-9 triggers the angiogenic switch during carcinogenesis. *Nature Cell Biology*, **2**, 737–744.

Bingle, L., Brown, N.J., and Lewis, C.E. (2002) The role of tumour-associated macrophages in tumour progression: implications for new anticancer therapies. *Journal of Pathology*, **196**, 254–265.

Bingle, L., Lewis, C.E., Corke, K.P. *et al.* (2006) Macrophages promote angiogenesis in human breast tumour spheroids *in vivo*. *British Journal of Cancer*, **94**, 101–107.

Biswas, S.K., Gangi, L., Paul, S. *et al.* (2006) A distinct and unique transcriptional programme expressed by tumor-associated macrophages: defective NF-kB and enhanced IRF-3/STAT1 activation. *Blood*, **107**, 2112–2122.

Bottazzi, B., Polentarutti, N., Acero, R. *et al.* (1983) Regulation of the macrophage content of neoplasms by chemoattractants. Science, **220**, 210–212.

Bottazzi, B., Walter, S., Govoni, D. *et al.* (1992) Monocyte chemotactic cytokine gene transfer modulates macrophage infiltration, growth, and susceptibility to IL-2 therapy of a murine melanoma. *Journal of Immunology*, **148**, 1280–1285.

Bronte, V., Serafini, P., Mazzoni, A. *et al.* (2003) L-arginine metabolism in myeloid cells controls T-lymphocyte functions. *Trends in Immunology*, **24**, 302–306.

Bronte, V., Kasic, T., Gri, G. *et al.* (2005) Boosting antitumor responses of T lymphocytes infiltrating human prostate cancers. *Journal of Experimental Medicine*, **201**, 1257–1268.

Budhu, A., Forgues, M., Ye, Q.H. *et al.* (2006) Prediction of venous metastases, recurrence, and prognosis in hepatocellular carcinoma based on a unique immune response signature of the liver microenvironment. *Cancer Cell*, **10**, 99–111.

Carmeliet, P. and Jain, R.K. (2000) Angiogenesis in cancer and other diseases. *Nature*, **407**, 249–257.

Ceradini, D.J., Kulkarni, A.R., Callaghan, M.J. *et al.* (2004) Progenitor cell trafficking is regulated by hypoxic gradients through HIF-1 induction of SDF-1. *Nature Medicine*, **10**, 858–864.

Condeelis, J. and Pollard, J.W. (2006) Macrophages: obligate partners for tumor cell migration, invasion, and metastasis. *Cell*, **124**, 263–266.

Coscia, M., Quaglino, E., Iezzi, M. *et al.* (2009) Zoledronic acid repolarizes tumor-associated macrophages and inhibits mammary carcinogenesis by targeting the mevalonate pathway. *Journal of Cellular and Molecular Medicine*, ePub ahead of print.

Coussens, L.M., Tinkle, C.L., Hanahan, D., and Werb, Z. (2000) MMP-9 supplied by bone marrow-derived cells contributes to skin carcinogenesis. *Cell*, **103**, 481–490.

DeNardo, D.G., Barreto, J.B., Andreu, P. *et al.* (2009) CD4(+) T cells regulate pulmonary metastasis of mammary carcinomas by enhancing protumor properties of macrophages. *Cancer Cell*, **16**, 91–102.

De Palma, M., Venneri, M.A., Roca, C., and Naldini, L. (2003) Targeting exogenous genes to tumor angiogenesis by transplantation of genetically modified hematopoietic stem cells. *Nature Medicine*, **9**, 789–795.

De Palma, M., Venneri, M.A., Galli, R. *et al.* (2005) Tie2 identifies a hematopoietic lineage of proangiogenic monocytes required for tumor vessel formation and a mesenchymal population of pericyte progenitors. *Cancer Cell*, **8**, 211–226.

Dinapoli, M.R., Calderon, C.L., and Lopez, D.M. (1996) The altered tumoricidal capacity of macrophages isolated from tumor-bearing mice is related to reduce expression of the inducible nitric oxide synthase gene. *Journal of Experimental Medicine*, **183**, 1323–1329.

Duluc, D., Corvaisier, M., Blanchard, S. *et al.* (2009) Interferon-gamma reverses the immunosuppressive and protumoral properties and prevents the generation of human tumor-associated macrophages. *International Journal of Cancer*, **125**, 367–373.

Egeblad, M. and Werb, Z. (2002) New functions for the matrix metalloproteinases in cancer progression. *Nature Reviews Cancer*, **2**, 161–174.

Eriksson, F., Tsagozis, P., Lundberg, K. *et al.* (2009) Tumor-specific bacteriophages induce tumor destruction through activation of tumor-associated macrophages. *Journal of Immunology*, **182**, 3105–3111.

Eruslanov, E., Kaliberov, S., Daurkin, I. *et al.* (2009) Altered expression of 15-hydroxyprostaglandin dehydrogenase in tumor-infiltrated CD11b myeloid cells: a mechanism for immune evasion in cancer. *Journal of Immunology*, **182**, 7548–7557.

Gabrilovich, D. (2004) Mechanisms and functional significance of tumour-induced dendritic-cell defects. *Nature Reviews Immunology*, **4**, 941–952.

Gallina, G., Dolcetti, L., Serafini, P. *et al.* (2006) Tumors induce a subset of inflammatory monocytes with immunosuppressive activity on CD8+ T cells. *Journal of Clinical Investigation*, **116**, 2777–2790.

Galon, J., Costes, A., Sanchez-Cabo, F. *et al.* (2006) Type, density, and location of immune cells within human colorectal tumors predict clinical outcome. *Science*, **313**, 1960–1964.

Ghassabeh, G.H., De Baetselier, P., Brys, L. *et al.* (2006) Identification of a common gene signature for type II cytokine-associated myeloid cells elicited *in vivo* in different pathologic conditions. *Blood*, **108**, 575–583.

Gordon, S. (2003) Alternative activation of macrophages. *Nature Reviews Immunology*, **3**, 23–35.

Gordon, S. and Taylor, P.R. (2005) Monocyte and macrophage heterogeneity. *Nature Reviews Immunology*, **5**, 953–964.

Goswami, S., Sahai, E., Wyckoff, J.B. *et al.* (2005) Macrophages promote the invasion of breast carcinoma cells via a colony-stimulating factor-1/epidermal growth factor paracrine loop. *Cancer Research*, **65**, 5278–5283.

Greten, F.R., Eckmann, L., Greten, T.F. *et al.* (2004) IKKbeta links inflammation and tumorigenesis in a mouse model of colitis-associated cancer. *Cell*, **118**, 285–296.

Guiducci, C., Vicari, A.P., Sangaletti, S. *et al.* (2005) Redirecting *in vivo* elicited tumor infiltrating macrophages and dendritic cells towards tumor rejection. *Cancer Research*, **65**, 3437–3446.

Hagemann, T., Lawrence, T., McNeish, I. *et al.* (2008) 'Re-educating' tumor-associated macrophages by targeting NF-kappaB. *Journal of Experimental Medicine*, **205**, 1261–1268.

Hanada, T., Nakagawa, M., Emoto, A. *et al.* (2000) Prognostic value of tumor-associated macrophage count in human bladder cancer. *International Journal of Urology*, **7**, 263–269.

Hiraoka, K., Zenmyo, M., Watari, K. *et al.* (2008) Inhibition of bone and muscle metastases of lung cancer cells by a decrease in the number of monocytes/macrophages. *Cancer Science*, **99**, 1595–1602.

Hiratsuka, S., Nakamura, K., Iwai, S. *et al.* (2002) MMP9 induction by vascular endothelial growth factor receptor-1 is involved in lung-specific metastasis. *Cancer Cell*, **2**, 289–300.

Jeon, B.H., Jang, C., Han, J. *et al.* (2008) Profound but dysfunctional lymphangiogenesis via vascular endothelial growth factor ligands from CD11b+ macrophages in advanced ovarian cancer. *Cancer Research*, **68**, 1100–1109.

Joyce, J.A. and Pollard, J.W. (2009) Microenvironmental regulation of metastasis. *Nature Reviews Cancer*, **9**, 239–252.

Karin, M. (2006) Nuclear factor-kappaB in cancer development and progression. *Nature*, **441**, 431–436.

Kawai, T. and Akira, S. (2007) Signaling to NF-kappaB by Toll-like receptors. *Trends in Molecular Medicine*, **13**, 460–469.

Klimp, A.H., Hollema, H., Kempinga, C. *et al.* (2001) Expression of cyclooxygenase-2 and inducible nitric oxide synthase in human ovarian tumors and tumor-associated macrophages. *Cancer Research*, **61**, 7305–7309.

Koide, N., Nishio, A., Sato, T. *et al.* (2004) Significance of macrophage chemoattractant protein-1 expression and macrophage infiltration in squamous cell carcinoma of the esophagus. *American Journal of Gastroenterology*, **99**, 1667–1674.

Kortylewski, M., Kujawski, M., Wang, T. *et al.* (2005) Inhibiting Stat3 signaling in the hematopoietic system elicits multicomponent antitumor immunity. *Nature Medicine*, **11**, 1314–1321.

Krieg, A.M. (2008) Toll-like receptor 9 (TLR9) agonists in the treatment of cancer. *Oncogene*, **27**, 161–167.

Kryczek, I., Wei, S., Keller, E. *et al.* (2007) Stroma-derived factor (SDF-1/CXCL12) and human tumor pathogenesis. *American Journal of Physiology: Cell Physiology*, **292**, C987–C995.

Kurahara, H., Shinchi, H., Mataki, Y. *et al.* (2009) Significance of M2-Polarized Tumor-Associated Macrophage in Pancreatic Cancer. *Journal of Surgical Research*.

Kusmartsev, S. and Gabrilovich, D.I. (2005) STAT1 signaling regulates tumor-associated macrophage-mediated T cell deletion. *Journal of Immunology*, **174**, 4880–4891.

Lawrence, T., Willoughby, D.A., and Gilroy, D.W. (2002) Anti-inflammatory lipid mediators and insights into the resolution of inflammation. *Nature Reviews Immunology*, **2**, 787–795.

Leek, R.D., Lewis, C.E., Whitehouse, R. *et al.* (1996) Association of macrophage infiltration with angiogenesis and prognosis in invasive breast carcinoma. *Cancer Research*, **56**, 4625–4629.

Leek, R.D., Talks, K.L., Pezzella, F. *et al.* (2002) Relation of hypoxia-inducible factor-2 alpha (HIF-2 alpha) expression in tumor-infiltrative macrophages to tumor angiogenesis and the oxidative thymidine phosphorylase pathway in human breast cancer. *Cancer Research*, **62**, 1326–1329.

Lepique, A.P., Daghastanli, K.R., Cuccovia, I.M., and Villa, L.L. (2009) HPV16 tumor associated macrophages suppress antitumor T cell responses. *Clinical Cancer Research*, **15**, 4391–4400.

Lewis, C. and Murdoch, C. (2005) Macrophage responses to hypoxia: implications for tumor progression and anti-cancer therapies. *American Journal of Pathology*, **167**, 627–635.

Lewis, C.E. and Pollard, J.W. (2006) Distinct role of macrophages in different tumor microenvironments. *Cancer Research*, **66**, 605–612.

Lin, E.Y., Li, J.F., Gnatovskiy, L. *et al.* (2006) Macrophages regulate the angiogenic switch in a mouse model of breast cancer. *Cancer Research*, **66**, 11238–11246.

Lin, E.Y., Nguyen, A.V., Russell, R.G., and Pollard, J.W. (2001) Colony-stimulating factor 1 promotes progression of mammary tumors to malignancy. *Journal of Experimental Medicine*, **193**, 727–740.

Locati, M., Deuschle, U., Massardi, M.L. *et al.* (2002) Analysis of the gene expression profile activated by the CC chemokine ligand 5/RANTES and by lipopolysaccharide in human monocytes. *Journal of Immunology*, **168**, 3557–3562.

Loges, S., Schmidt, T., Tjwa, M. *et al.* Malignant cells fuel tumor growth by educating infiltrating leukocytes to produce the mitogen Gas6. *Blood,* **115**, 2264–2273.

Ludwiczek, S., Aigner, E., Theurl, I., and Weiss, G. (2003) Cytokine-mediated regulation of iron transport in human monocytic cells. *Blood*, **101**, 4148–4154.

Luo, J.L., Tan, W., Ricono, J.M. *et al.* (2007) Nuclear cytokine-activated IKKalpha controls prostate cancer metastasis by repressing Maspin. *Nature*, **446**, 690–694.

Lynch, C.C., Hikosaka, A., Acuff, H.B. *et al.* (2005) MMP-7 promotes prostate cancer-induced osteolysis via the solubilization of RANKL. *Cancer Cell*, **7**, 485–496.

Lynch, C.C. and Matrisian, L.M. (2002) Matrix metalloproteinases in tumor-host cell communication. *Differentiation*, **70**, 561–573.

Mantovani, A., Allavena, P., and Sica, A. (2004) Tumour-associated macrophages as a prototypic type II polarised phagocyte population: role in tumour progression. *Europena Journal of Cancer*, **40**, 1660–1667.

Mantovani, A., Sica, A., and Locati, M. (2005) Macrophage polarization comes of age. *Immunity*, **23**, 344–346.

Mantovani, A., Sica, A., Sozzani, S. *et al.* (2004) The chemokine system in diverse forms of macrophage activation and polarization. *Trends in Immunology*, **25**, 677–686.

Mantovani, A., Allavena, P., Sica, A., and Balkwill, F. (2008a) Cancer-related inflammation. *Nature*, **454**, 436–444.

Mantovani, A., Allavena, P., Sica, A., and Balkwill, F. (2008b) Cancer-related inflammation. *Nature*, **454**, 436–444.

Mantovani, A., Bottazzi, B., Colotta, F. *et al.* (1992) The origin and function of tumor-associated macrophages. *Immunology Today*, **13**, 265–270.

Mantovani, A., Sozzani, S., Locati, M. *et al.* (2002a) Macrophage polarization: tumor-associated macrophages as a paradigm for polarized M2 mononuclear phagocytes. *Trends in Immunology*, **23**, 549–555.

Mantovani, A., Sozzani, S., Locati, M. *et al.* (2002b) Macrophage polarization: tumor-associated macrophages as a paradigm for polarized M2 mononuclear phagocytes. *Trends in Immunology*, **23**, 549–555.

Marconi, C., Bianchini, F., Mannini, A. *et al.* (2008) Tumoral and macrophage uPAR and MMP-9 contribute to the invasiveness of B16 murine melanoma cells. *Clinical and Experimental Metastasis*, **25**, 225–231.

Markiewski, M.M. and Lambris, J.D. (2009) Is complement good or bad for cancer patients? A new perspective on an old dilemma. *Trends in Immunology*, **30**, 286–292.

Martinez, F.O., Gordon, S., Locati, M., and Mantovani, A. (2006) Transcriptional profiling of the human monocyte-to-macrophage differentiation and polarization: new molecules and patterns of gene expression. *Journal of Immunology*, **177**, 7303–7311.

Martinez, F.O., Helming, L., and Gordon, S. (2009) Alternative activation of macrophages: an immunologic functional perspective. *Annual Review of Immunology*, **27**, 451–483.

Marttila-Ichihara, F., Auvinen, K., Elima, K. *et al.* (2009) Vascular adhesion protein-1 enhances tumor growth by supporting recruitment of Gr-1+CD11b+ myeloid cells into tumors. *Cancer Research*, **69**, 7875–7883.

Mellman, I. and Steinman, R.M. (2001) Dendritic cells: specialized and regulated antigen processing machines. *Cell*, **106**, 255–258.

Mohamed, M.M. and Sloane, B.F. (2006) Cysteine cathepsins: multifunctional enzymes in cancer. *Nature Reviews Cancer*, **6**, 764–775.

Mosmann, T.R., Cherwinski, H., Bond, M.W. *et al.* (1986) Two types of murine helper T cell clone. I. Definition according to profiles of lymphokine activities and secreted proteins. *Journal of Immunology*, **136**, 2348–2357.

Muller, A., Homey, B., Soto, H. *et al.* (2001) Involvement of chemokine receptors in breast cancer metastasis. *Nature*, **410**, 50–56.

Muller, A.J., DuHadaway, J.B., Donover, P.S. *et al.* (2005) Inhibition of indoleamine 2,3-dioxygenase, an immunoregulatory target of the cancer suppression gene Bin1, potentiates cancer chemotherapy. *Nature Medicine*, **11**, 312–319.

Murdoch, C., Muthana, M., Coffelt, S.B., and Lewis, C.E. (2008) The role of myeloid cells in the promotion of tumour angiogenesis. *Nature Reviews Cancer*, **8**, 618–631.

Nathan, C.F. and Hibbs, J.B. Jr. (1991) Role of nitric oxide synthesis in macrophage antimicrobial activity. *Current Opinion in Immunology*, **3**, 65–70.

Nibbs, R.J., Gilchrist, D.S., King, V. *et al.* (2007) The atypical chemokine receptor D6 suppresses the development of chemically induced skin tumors. *Journal of Clinical Investigation*, **117**, 1884–1892.

Nishie, A., Ono, M., Shono, T. *et al.* (1999) Macrophage infiltration and heme oxygenase-1 expression correlate with angiogenesis in human gliomas. *Clinical Cancer Research*, **5**, 1107–1113.

Nozawa, H., Chiu, C., and Hanahan, D. (2006) Infiltrating neutrophils mediate the initial angiogenic switch in a mouse model of multistage carcinogenesis. *Proceedings of the National Academy of Sciences of the United States of America*, **103**, 12493–12498.

O'Shea, J.J., Pesu, M., Borie, D.C., and Changelian, P.S. (2004) A new modality for immunosuppression: targeting the JAK/STAT pathway. *Nature Reviews Drug Discovery*, **3**, 555–564.

Paik, S., Shak, S., Tang, G. *et al.* (2004) A multigene assay to predict recurrence of tamoxifen-treated, node-negative breast cancer. *New England Journal of Medicine*, **351**, 2817–2826.

Pikarsky, E., Porat, R.M., Stein, I. *et al.* (2004) NF-kappaB functions as a tumour promoter in inflammation-associated cancer. *Nature*, **431**, 461–466.

Podolsky, S.H. and Tauber, A.I. (1996) Darwinism and antibody diversity: a historical perspective. *Research in Immunology*, **147**, 199–202.

Pollard, J.W. (2004) Tumour-educated macrophages promote tumour progression and metastasis. *Nature Reviews Cancer*, **4**, 71–78.

Pollard, J.W. (2008) Macrophages define the invasive microenvironment in breast cancer. *Journal of Leukocyte Biology*, **84**, 623–630.

Porta, C., Rimoldi, M., Raes, G. *et al.* (2009) Tolerance and M2 (alternative) macrophage polarization are related processes orchestrated by p50 nuclear factor kappaB. *Proceedings of the National Academy of Sciences of the United States of America*, **106**, 14978–14983.

Pucci, F., Venneri, M.A., Biziato, D. *et al.* (2009) A distinguishing gene signature shared by tumor-infiltrating Tie2-expressing monocytes, blood 'resident' monocytes, and embryonic macrophages suggests common functions and developmental relationships. *Blood*, **114**, 901–914.

Rauh, M.J., Ho, V., Pereira, C. *et al.* (2005) SHIP represses the generation of alternatively activated macrophages. *Immunity*, **23**, 361–374.

Rius, J., Guma, M., Schachtrup, C. *et al.* (2008) NF-kappaB links innate immunity to the hypoxic response through transcriptional regulation of HIF-1alpha. *Nature*, **453**, 807–811.

Robinson, B.D., Sica, G.L., Liu, Y.F. *et al.* (2009) Tumor microenvironment of metastasis in human breast carcinoma: a potential prognostic marker linked to hematogenous dissemination. *Clinical Cancer Research*, **15**, 2433–2441.

Saccani, A., Schioppa, T., Porta, C. *et al.* (2006) p50 nuclear factor-kappaB overexpression in tumor-associated macrophages inhibits M1 inflammatory responses and antitumor resistance. *Cancer Research*, **66**, 11432–11440.

Sangaletti, S., Stoppacciaro, A., Guiducci, C. *et al.* (2003) Leukocyte, rather than tumor-produced SPARC, determines stroma and collagen type IV deposition in mammary carcinoma. *Journal of Experimental Medicine*, **198**, 1475–1485.

Schioppa, T., Uranchimeg, B., Saccani, A. *et al.* (2003) Regulation of the chemokine receptor CXCR4 by hypoxia. *Journal of Experimental Medicine*, **198**, 1391–1402.

Schutyser, E., Struyf, S., Proost, P. *et al.* (2002) Identification of biologically active chemokine isoforms from ascitic fluid and elevated levels of CCL18/pulmonary and activation-regulated chemokine in ovarian carcinoma. *The Journal of Biological Chemistry*, **277**, 24584–24593.

Shojaei, F., Singh, M., Thompson, J.D., and Ferrara, N. (2008) Role of Bv8 in neutrophil-dependent angiogenesis in a transgenic model of cancer progression. *Proceedings of the National Academy of Sciences of the United States of America*, **105**, 2640–2645.

Sica, A. and Bronte, V. (2007) Altered macrophage differentiation and immune dysfunction in tumor development. *Journal of Clinical Investigation*, **117**, 1155–1166.

Sica, A., Allavena, P., and Mantovani, A. (2008) Cancer related inflammation: the macrophage connection. *Cancer Letters*, **267**, 204–215.

Sica, A., Larghi, P., Mancino, A. *et al.* (2008) Macrophage polarization in tumour progression. *Seminars in Cancer Biology*, **18**, 349–355.

Sica, A., Rubino, L., Mancino, A. *et al.* (2007) Targeting tumour-associated macrophages. *Expert Opinion in Therapy Targets*, **11**, 1219–1229.

Sica, A., Saccani, A., Bottazzi, B. *et al.* (2000) Autocrine production of IL-10 mediates defective IL-12 production and NF-kappa B activation in tumor-associated macrophages. *Journal of Immunology*, **164**, 762–767.

Sica, A., Schioppa, T., Mantovani, A., and Allavena, P. (2006) Tumour-associated macrophages are a distinct M2 polarised population promoting tumour progression: potential targets of anti-cancer therapy. *European Journal of Cancer*, **42**, 717–727.

Sierra, J.R., Corso, S., Caione, L. *et al.* (2008) Tumor angiogenesis and progression are enhanced by Sema4D produced by tumor-associated macrophages. *Journal of Experimental Medicine*, **205**, 1673–1685.

Sinha, P., Clements, V.K., Miller, S., and Ostrand-Rosenberg, S. (2005) Tumor immunity: a balancing act between T cell activation, macrophage activation and tumor-induced immune suppression. *Cancer Immunology and Immunotherapy*, **54**, 1137–1142.

Strieter, R.M., Belperio, J.A., Burdick, M.D. *et al.* (2004) CXC chemokines: angiogenesis, immunoangiostasis, and metastases in lung cancer. *Annals of the New York Academy of Science*, **1028**, 351–360.

Toge, H., Inagaki, T., Kojimoto, Y. *et al.* (2009) Angiogenesis in renal cell carcinoma: the role of tumor-associated macrophages. *International Journal of Urology*, **16**, 801–807.

Uyttenhove, C., Pilotte, L., Theate, I. *et al.* (2003) Evidence for a tumoral immune resistance mechanism based on tryptophan degradation by indoleamine 2,3-dioxygenase. *Nature Medicine*, **9**, 1269–1274.

Vaupel, P. (2008) Hypoxia and aggressive tumor phenotype: implications for therapy and prognosis. *Oncologist*, **13** (Suppl. 3), 21–26.

Venneri, M.A., De Palma, M., Ponzoni, M. *et al.* (2007) Identification of proangiogenic TIE2-expressing monocytes (TEMs) in human peripheral blood and cancer. *Blood*, **109**, 5276–5285.

Vetrano, S., Borroni, E.M., Sarukhan, A., *et al.* (2009) The lymphatic system controls intestinal inflammation and inflammation-associated colon cancer through the chemokine decoy receptor D6. *Gut*.

de Visser, K.E., Korets, L.V., and Coussens, L.M. (2005) De novo carcinogenesis promoted by chronic inflammation is B lymphocyte dependent. *Cancer Cell*, **7**, 411–423.

Wang, T., Niu, G., Kortylewski, M. *et al.* (2004) Regulation of the innate and adaptive immune responses by Stat-3 signaling in tumor cells. *Nature Medicine*, **10**, 48–54.

Wyckoff, J.B., Wang, Y., Lin, E.Y. *et al.* (2007) Direct visualization of macrophage-assisted tumor cell intravasation in mammary tumors. *Cancer Research*, **67**, 2649–2656.

Zeisberger, S.M., Odermatt, B., Marty, C. *et al.* (2006) Clodronate-liposome-mediated depletion of tumour-associated macrophages: a new and highly effective antiangiogenic therapy approach. *British Journal of Cancer*, **95**, 272–281.

Zumsteg, A., Baeriswyl, V., Imaizumi, N. *et al.* (2009) Myeloid cells contribute to tumor lymphangiogenesis. *PLoS One*, **4**, e7067.

6

Bone Marrow Stroma and the Leukemic Microenvironment

William B. Slayton and Zhongbo Hu
Division of Hematology/Oncology, Department of Pediatrics, University of Florida College of Medicine, Gainesville, FL, USA

6.1 Introduction

Bone marrow is a special place where immature and mature blood cells are interspersed with structural stromal cells. This spongy, fatty tissue is located inside bones, such as the human skull, sternum, ribs, pelvis, and femurs. Bone marrow is nourished and connected to the circulation through specialized blood vessels, including specialized fenestrated capillaries called sinusoids. These sinusoids penetrate the sponge-like extracellular matrix (ECM) produced by reticular fibroblasts, and these cells and proteins organize the hematopoietic cells into discrete compartments. Within the bone marrow are specialized microenvironments, comprised of hematopoietic cells, stromal cells, ECM and liquid containing cytokines, growth factors, and chemokines. Within these microenvironments, developing hematopoietic cells receive signals to survive, differentiate, and proliferate. The bone marrow microenvironment also plays an important role in the development and progression of leukemia and other types of cancer. In many ways, the bone marrow provides an ideal microenvironment for malignant cells to grow.

6.2 Components and function of the normal bone marrow microenvironment

Bone marrow cell types can be functionally categorized as hematopoietic, parenchymal and stromal, or mesenchymal. Normal hematopoiesis occurs in a complex marrow microenvironment consisting of hematopoietic cells, stromal cells, and

Tumor Microenvironment Edited by Dietmar W. Siemann

ECM components. Hematopoietic progenitor proliferation and differentiation are regulated by receptor-mediated signals initiated by cytokines or interactions between progenitors and stromal cells or ECM components within this microenvironment.

6.2.1 Stromal cells

Bone marrow stroma is a connective tissue that is not directly involved in hematopoiesis. However, mesenchymal cells within the stroma provide a structural microenvironment that facilitates hematopoiesis. Mesenchymal cells produce cytokines, growth factors, and ECM proteins. Mesenchymal cells promote hematopoiesis by modulating stem cell quiescence and self-renewal, but also provide signals that lead to the proliferation, maturation, and apoptosis of more mature hematopoietic cells. Mesenchymal cell types that constitute the bone marrow stroma include: mesenchymal stem cells, fibroblasts, advential reticular cells, adipocytes, osteoblasts, osteoclasts, endothelial cells, and endothelial progenitor cells (see Figure 6.1).

6.2.1.1 *Mesenchymal stem cells (MSCs)*

Mesenchymal stem cells are non-hematopoietic stem cells present in the bone marrow defined by Pittenger *et al.* (1999). Before that, researchers used various terms for this population, such as colony-forming unit fibroblast (CFU-F), stromal stem cell, bone marrow stromal stem cell, marrow stromal cells, mesodermal progenitor cells (Reyes *et al.*, 2001), marrow isolated adult multilineage inducible cells (D'Ippolito *et al.*, 2004), multipotent stromal cells, or skeletal stem cells (Prockop, 1997; Kuznetsov *et al.*, 2001). Under certain conditions, MSC give rise to fibroblasts, osteoblasts, osteocytes, chondrocytes, adipocytes, adventitial reticular cells, smooth muscle cells, endothelial cells, myelosupportive stroma, and nerve tissue (Pittenger *et al.*, 1999; Bianco, Robey, and Simmons, 2008). MSC can be identified by expression of certain markers, such as Stro-1, Thy-1, vascular cell adhesion molecule-1, endoglin, α1 integrin, and MUC-18/CD146 (Pittenger *et al.*, 1999; Dennis and Charbord, 2002). Human marrow stromal cell (hMSC) can be identified by the monoclonal antibodies SH-2, SH-3, and SH-4 and lack of expression of CD14, CD34, and CD45 when analyzed by flow cytometry. MSCs produce large number of cytokines and chemokines (see Table 6.1). hMSCs are estimated to be present at a frequency of 1 in 10^5 bone marrow cells. Like hematopoietic stem cells (HSC), MSC are able to home to the bone marrow (Devine *et al.*, 2001).

6.2.1.2 *Fibroblasts*

Fibroblasts are the best-studied of the marrow stromal cells. Marrow fibroblasts can produce a large variety of cytokines and growth factors, including interleukin (IL)-6, granulocyte-macrophage colony-stimulating factor (GM-CSF) (Guba *et al.*, 1992), granulocyte colony-stimulating factor (G-CSF) (Fibbe *et al.*, 1988a; Kaushansky,

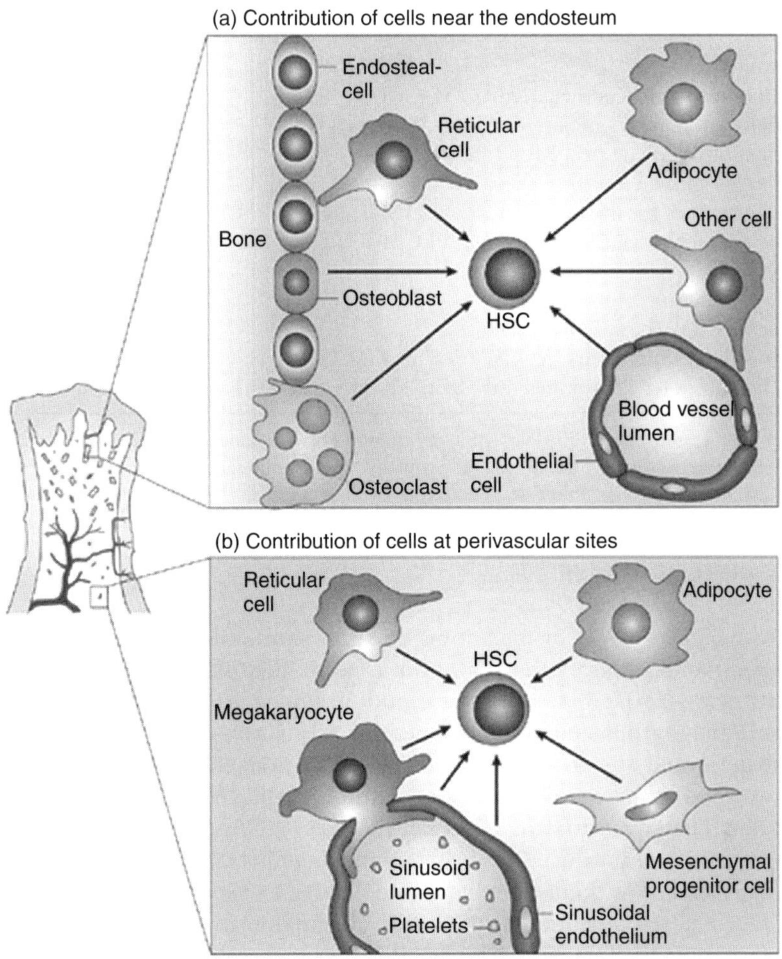

Figure 6.1 (a,b) Stromal cells in the bone marrow microenvironment. From Keil, M.J. and Morrison, S.J. (2008) Uncertainty in the niches that maintain haematopoietic stem cells. *Nature Reviews Immunology*, **8**, 290–301. © 2008 Reproduced by permission of Nature Publishing Group.

Lin, and Adamson, 1988), thrombopoietin (TPO) (Guerriero *et al.*, 1997), stem cell factor (SCF). They also synthesize type I and III procollagen with a I/III ratio of 1 : 3 (Bainton *et al.*, 1986). Fibroblasts produce a structural scaffolding to which hematopoietic cells adhere. Fibroblasts express adhesion molecules, such as vascular cell adhesion molecule (VCAM)-1, a ligand for $\alpha 4\beta 1$, to which hematopoietic stem cells bind (Avraham *et al.*, 1992).

6.2.1.3 *Adventitial reticular cells*

Adventitial reticular cells are stromal cells that reside immediately adjacent to and outside of bone marrow sinusoids (Bianco, Robey, and Simmons, 2008). CD146,

Table 6.1 Characterization of human mesenchymal stem cells.

Flow cytometric
Hematopoietic markers (negative): CD14, CD34, CD45
Mesenchymal markers (positive): SH-2, SH-3, SH-4
Integrins (positive): α1, α2, β1, β2, β3
Integrins (negative): α4, αL, Cβ2
Matrix receptors (positive): ICAM-1, VCAM-1, LFA-3, L-selectin
Matrix receptors (negative): ICAM-3, PECAM-1, E-selectin
P-selectin
Cytokine production
Interleukins: 1a, 1b, 6, 7, 8, 11, 14, 15
Colony-stimulating factors: M-CSF, G-CSF, GM-CSF
Other hematopoietic cytokines: LIF, SCF, Flt-3 ligand, TPO

CFU-F, colony-forming units fibroblast; Flt, fetal liver tyrosine kinase; G-CSF, granulocyte colony-stimulating factor; GM-CSF, granulocyte-macrophage colony-stimulating factor; ICAM, intercellular adhesion molecule; LFA, lymphocyte function-associated antigen; LIF, leukemia inhibitory factor; M-CSF, macrophage colony-stimulating factor; PECAM, platelet/endothelial cell adhesion molecule; SCF, stem cell factor; TPO, thrombopoietin; VCAM, vascular cell adhesion molecule.
Cited from Devine and Hoffman, 2000.

also known as Mel-CAM or melanoma cell adhesion molecule (MCAM), MUC18, A32 antigen, and S-Endo-1, have all been used to identify adventitial reticular cells (Sacchetti *et al.*, 2007; Bianco, Robey, and Simmons, 2008; Shih, 1999). CD146 is a membrane glycoprotein, which functions as a Ca^{2+}-independent cell adhesion molecule involved in heterophilic cell to cell interactions (Shih, 1999). $CD146^+$ cells express angiopoietin-1 and have features of mural cells, 'niche' cells, and osteogenic progenitors. The adventitial reticular cells express VCAM-1, alkaline phosphatase and alpha actin and produce CXC chemokine ligand (CXCL)12 (Porter and Calvi, 2008), IL-7, and SCF (Funk, Stephan, and Witte, 1995). IL-7 plays an important role in B-cell development. CXCL12 is a chemokine involved (with its receptor CXC receptor (CXCR)4) in retaining HSC within supportive niches provided by adventitial reticular cells (Porter and Calvi, 2008). Adventitial reticular cells also function to direct vasculogenesis and stabilize blood vessels within the bone marrow (Moore and Lemischka, 2006).

6.2.1.4 Adipocytes

Bone marrow adipocytes are derived from MSCs. Adipocyte progenitors, called preadipocytes, provide some of the cytokines and ECM proteins required for the maturation and proliferation of hematopoietic cells. Bone marrow adipocytes provide a localized energy reservoir for emergency situations requiring increased hematopoiesis (blood loss) or osteogenesis (Gimble *et al.*, 1996). Within the bone marrow, the differentiation of MSCs into adipocytes or osteoblasts is competitively balanced by cross talk between complex signaling pathways, including those derived from bone morphogenic proteins (BMPs), wingless type mouse mammary tumor virus integration site proteins (Wnt), hedgehog proteins, delta/jagged proteins, fibroblastic growth factors (FGFs), insulin, insulin-like growth factors (IGFs), and transcriptional regulators of adipocyte and osteoblast differentiation, including peroxisome proliferator-activated receptor-γ (PPARγ) and runt-related transcription factor 2 (Runx2) (Muruganandan, Roman, and Sinal, 2009; Nuttall and Gimble, 2004). Adipocytes are less supportive of hematopoiesis *in vitro* than their undifferentiated stromal or preadipocytic counterparts, in part because of reduced production of growth factors such as GM-CSF and G-CSF (Nishikawa *et al.*, 1993; Corre *et al.*, 2006). Moreover, adipose tissue secretes neuropilin-1 (Belaid-Choucair *et al.*, 2008), lipocalin 2 (Miharada *et al.*, 2008; Yan *et al.*, 2007), adiponectin (Yokota *et al.*, 2000) and tumor necrosis factor-α (TNF-α) (Zhang *et al.*, 1995; Hotamisligil, Shargill, and Spiegelman, 1993), each of which inhibit hematopoietic proliferation. Of note, TNF-α and adiponectin inhibit progenitor activity while positively regulating the most primitive HSCs for optimal proliferation (Zhang *et al.*, 1995; Dimascio *et al.*, 2007), suggesting that adipocytes prevent hematopoietic progenitor expansion while preserving the hematopoietic stem-cell pool. Recently, Naveiras *et al.* (2009) used flow cytometry, colony-forming activity, and competitive repopulation assays to analyze murine vertebrae and found adipocyte-rich marrow harbors a decreased frequency of progenitors and proportionally more quiescent stem cells. They used lipoatrophic A-ZIP/F1 'fatless' mice, which are genetically incapable of forming adipocytes, and mice treated with the PPARγ inhibitor bisphenol, a diglycidyl ether, which inhibits adipogenesis, as recipients in bone marrow transplants. Engraftment after irradiation was found to be accelerated relative to wild-type or untreated mice. This demonstrates that adipocytes act as a negative regulator of hematopoietic proliferation in the bone marrow microenvironment, balancing the supportive role of the osteoblasts in the HSC niche.

6.2.1.5 Osteoblasts

Osteoblasts are also derived from mesenchymal cells (Mackie, 2003). Ostoeblasts are located on the surface of the endosteum of trabecular bone. Osteoblasts provide factors to maintain the quiescent status of HSC, and these factors establish an osteoblastic niche (Taichman and Emerson, 1998; Arai and Suda, 2007). Osteoblasts produce G-CSF, macrophage colony-stimulating factor (M-CSF), GM-CSF, IL-1, IL-6, leukemia inhibitory factor (LIF), TNF-α, vascular endothelial growth factor (VEGF), transforming growth factor (TGF)-β1, and angiopoietin-1 to regulate hematopoiesis (Arai *et al.*, 2004; Taichman and Emerson, 1998). Osteoblasts also secrete several matrix proteins, such as osteopontin, that negatively regulate HSC number (Taichman and Emerson, 1998). Osteoblasts interact with hematopoietic cells through a number of receptor-ligand cell adhesion molecules, including the cadherins, immunoglobulins, integrins, and selectins, and the ECM. These receptor-ligand adhesion molecules are described later in this chapter.

6.2.1.6 *Osteoclasts*

Osteoclasts have a hematopoietic origin and are derived from colony-forming unit granulocyte-macrophage (CFU-GM). Osteoclasts express tartrate-resistant acid phosphatase (TRAP), matrix metalloproteinase-9 (MMP-9), IL-6, IL-1, TGF-β to regulate bone resorption and HSC homeostasis. Osteoclast function is controlled by cross-talk with MSCs and osteoblasts through expression of many cytokines and hormones (Roodman, 1999; Reddy and Roodman, 1998).

6.2.1.7 *Endothelial cells*

Endothelial cells line blood vessels that carry blood through the bone marrow. Bone marrow nutrient arteries deliver blood to enervated arterioles that control the flow of blood into regions of the bone marrow. Arterioles deliver oxygenated blood into sinusoids, whose walls consisted of a single layer of endothelial cells. Sinusoids form a complex three-dimensional structure. Sinusoidal endothelium not only acts as a gatekeeper controlling the trafficking and homing of hematopoietic progenitors, but also provides cellular contact and secretes cytokines that promote various aspects of hematopoietic maturation. Sinusoidal endothelial cells produce lineage-specific cytokines such as G-CSF, GM-CSF, M-CSF, Kit-ligand, IL-6, fetal liver kinase-2 ligand, and LIF. Direct cellular contact between hematopoietic progenitor cells and bone marrow endothelial cells (BMECs) monolayers plays a critical role in trafficking and possibly proliferation of hematopoietic stem cells. This contact is mediated through specific adhesion molecules including β1 integrins, such as very late antigen-4 (VLA-4), β2 integrins, and selectins (Rafii *et al.*, 1997; Shiozawa *et al.*, 2008). BMECs also express the chemokine receptor CXCR4, which binds to CXCL12 leading to the homing of human $CD34^+$ hematopoietic stem and progenitors to the bone marrow (Dar *et al.*, 2005). BMECs also play a critical role in thrombopoiesis. CXCL12 and FGF-4 induce the localization and adhesion of megakaryocytes to BMECs. The direct contact between megakaryocytes and BMECs promotes survival, maturation, and release of platelets through vascular endothelial (VE)-cadherin and/or the VLA-4/VCAM-1 axis (Shiozawa *et al.*, 2008).

6.2.1.8 *Endothelial progenitor cells*

The definitions and names of endothelial precursor populations are not consistently used leading to confusion and lack of clarity within the field. One commonly held model of the origin and immunophenotypic characteristics of distinct subpopulations of circulating endothelial cells (CECs) in peripheral blood is described by Steurer *et al.* (2008). Circulating endothelial precursors or progenitor cells (CEPs) are endothelial precursor or progenitor cells (EPCs) that circulate in human peripheral blood. CEPs are derived from bone marrow EPCs that contribute to postnatal vasculogenesis. Cell surface proteins used to identify CEPs are not uniform. CEPs are characterized by the ability to bind lectins, such as *Ulex europeus* lectin 1, and the uptake of acetylated low density lipoproteins (Ac-LDL). More primitive endothelial progenitors express

CD34, CD133, and vascular endothelial growth factor receptor-2 (VEGFR2, kinase domain receptor), whereas more mature cells are characterized by the absence of CD133. During maturation, cells upregulate the expression of endothelial markers, such as CD31 (platelet/endoethelial cell adhesion molecule), CD144 (VE-Cadherin), and von Willebrand factor (vWF) (Untergasser *et al.*, 2006). The pluripotent stem cells in the bone marrow give rise to 'hemangioblasts' that have the capacity to differentiate into hematopoietic progenitor cells (HPCs) or EPC. EPC differentiate into CEPs and CECs. HPC differentiate into myeloid cells such as monocytes, which can transdifferentiate into myeloid EC. Mature EC shed from the vessel wall can enter the circulation (Steurer *et al.*, 2008). Other work suggests that bone marrow derived progenitor cells provide structures that support vasculogenesis by producing inductive cytokines rather than producing endothelium that incorporates into the vessel wall (Slayton *et al.*, 2007; Timmermans *et al.*, 2009). For instance, marrow sinusoids injured by radiation are repaired by the proliferation of endothelial cells themselves rather than by circulating progenitors (Li *et al.*, 2008).

6.2.2 Other microenvironmental cells

6.2.2.1 Macrophages

Macrophages that reside in the bone marrow are called fixed macrophages. Macrophages are mononuclear phagocytes responsible for numerous homeostatic, immunological, and inflammatory processes. Macrophages (including osteoclasts) are the only microenvironmental cell population derived from the hematopoietic stem cells rather than mesenchymal stem cells. Macrophages can be distinguished from other microenvironmental cells based on expression of the pan-hematopoeitic marker, CD45.

Macrophages regulate angiogenesis, especially during tumorigenesis. Macrophages promote angiogenesis through secretory products. Macrophages secrete or produce several factors that induce mitoses of capillary endothelium. Whereas the mitogenic effects of basic fibroblast growth factor-B (bFGF), TGF-α, GM-CSF, M-CSF, VEGF, vascular permeability factor (VPF), IL-8, and substance P (SP) are well characterized, the proliferative actions of IGF-I, somatomedin C, and platelet-derived growth factor (PDGF) on endothelial cells may need further confirmation (Ribatti *et al.*, 2007). Activated macrophages are the only blood cells besides platelets that produce PDGF. Macrophages also release several factors that inhibit migration or mitosis of endothelial cells, such as monocyte-derived endothelial cell inhibitory factor (MECIF), macrophage-derived endothelial cell inhibitor (MD-ECI), thrombospondin I, interferon (IFN)-α, and IFN-γ (Lamagna, Aurrand-Lions, and Imhof, 2006).

In addition to their important role in tumor angiogenesis, macrophages play an important role in tumor metastasis. Some solid tumors contain substantial numbers of tumor-associated macrophages (TAMs) (Balkwill and Mantovani, 2001). TAM express and release epidermal growth factor (EGF), FGF-2, TGF-α and -β, VEGF, TNF-α, IL-1, IL-6, IL-8, platelet-activating factor, PDGF, G-CSF, and

GM-CSF, thymidine phosphorylase, and chemokines such as CCL2. In addition, TAMs produce angiogenic factors such as thrombospondin-1, IL-12, IL-18, and the MMP -9 through 12. TAMs induce tissue remodeling by producing various proteinase activators and inhibitors, such as MMP-2, MMP-9, MMP-12, and cyclo-oxygenase-2 (COX-2), which disrupt the integrity of the basement membrane and ECM, liberating matrix-bound factors (Ribatti *et al.*, 2007). Macrophages contribute to erythropoiesis by providing iron for hemoglobin production. Macrophage-specific markers include F4/80 in the mouse and CD68 in human and the mouse (Gordon, 2007). Major histocompatibility complex (MHC) class II, CD11c, CD14, macrosialin, sialoadhesin, and α-actinin are non-specific markers of macrophages (Taylor *et al.*, 2003; Khazen *et al.*, 2005).

6.2.2.2 *Megakaryocytes*

Megakaryocytes are polyploid bone marrow cells responsible for the production of platelets. Megakaryocytes are 10 to 15 times larger than a typical erythrocyte, averaging 50–100 μm in diameter. The primary signal for megakaryocyte production is TPO. Other molecular signals that promote megakaryopoiesis include GM-CSF, IL-3, IL-6, IL-11, IL-12, and erythropoietin (Pang, Weiss, and Poncz, 2005). Two chemokines, stromal cell-derived factor-1 (SDF-1; CXCL12) and platelet factor 4 (PF4, CXCL4), also have important effects on megakaryopoiesis and platelet production. Other cytokines, including IL-1α and LIF, modulate megakaryocyte maturation and platelet release. Other than playing an important role in hemostasis by generating platelets, megakaryocytes have other important roles. By expressing and secreting angiopoietin, bFGF, β-thromboglobulin, PF4, EGF, G-CSF, GM-CSF,SCF, IFN-α, TGF-β, TNF-α, IL-1, -3, -4, -6, -7, -8, -9, and -11, LIF, PDGF, MMP-2 and -9, and VEGF, megakaryocytes comprise one of the most important microenvironmental cells in the bone marrow. Megakaryocytes regulate angiogenesis, inflammation, atherosclerosis, antimicrobial host defense, wound healing, bone homeostasis, and malignancy (Blair and Flaumenhaft, 2009; Kacena and Horowitz, 2006). Megakaryocytes play a key role in the regulation of bone remodeling. For instance, megakaryocytes express a number of bone regulatory factors, including the BMP 2, 4, and 6 (Sipe *et al.*, 2004), receptor activator for nuclear factor κ B ligand (Kartsogiannis *et al.*, 1999), *N*-methyl-D-aspartic acid-type glutamate receptors (Genever *et al.*, 1999), calcium-sensing receptors (House *et al.*, 1997), osteonectin, osteocalcin (Kelm *et al.*, 1992; Thiede *et al.*, 1994), bone sialoprotein, osteopontin, osteoprotegerin (Kacena, Gundberg, and Horowitz, 2006), TGF-β and their receptors and estrogen receptors (Bord *et al.*, 2001). TPO inhibits osteoclastogenesis (Wakikawa *et al.*, 1997). Increased megakaryocyte number is associated with marrow fibrosis and osteosclerosis (Poulsen, Melsen, and Bendix, 1998).

6.2.3 Extracellular matrix

The extracellular component of the hematopoietic microenvironment plays an important role in the regulation of cell proliferation and differentiation. In general,

ECMs are composed of collagen, non-collagenous glycoproteins, proteoglycans, and small amounts of other proteins (Gordon, 1988; Kreis and Vale, 1999). Variations in type and amount of these components produce the characteristic properties of ECMs in different tissues. Bone marrow ECM molecules include fibronectin (FN), collagen, hyaluronan, heparin sulfate, tenascin, proteoglycans, osteonectin, osteopontin, vitronectin, laminin, and bone sialoproteins. Interference with cell–matrix interaction or digestion of the ECM leads to mobilization of HSCs by agents such as G-CSF, plerixafor hydrochloride (also known as AMD3100), and IL-8 (Broxmeyer *et al.*, 2005; Laterveer *et al.*, 1995; Link, 2000).

6.2.3.1 *Fibronectin*

FN is a 450-kDa fibril-forming glycoprotein composed of two subunits that is a major component of the bone marrow microenvironment. FN is produced by both marrow stromal (endothelial cells and fibroblasts) and hematopoietic cells, and mediates homing of HSCs to the marrow. FN can bind cells, growth factors and ECM components. Distinct domains of FN interact with different integrins, such as α4β1 (VLA-4) and α5β1 (VLA-5) (see Figure 6.2). Multiple molecular mechanisms have been proposed for the effects of fibronectin on cells expressing integrin receptors, and these interactions serve as a paradigm for the supportive effects of this entire class of microenvironmental signals (Abboud and Lichtman, 2006). Fibronectin contains a specific cell attachment domain with the minimum amino acid sequence arginine-glycine-aspartic acid (RGD the ligand for VLA-5) (Hynes, Schwarzbauer, and Tamkun, 1987). Several studies have indicated that fibronectin is involved in the specific localization of certain types of hematopoietic cells (Schick *et al.*, 1998; Sorrell, 1988).

Integrin engagement by fibronectin triggers a number of intracellular signaling events that affect the cellular cytoskeleton and transcription. Complexes composed of kinases, adaptors, and cytoskeletal components are recruited to sites of integrin engagement, initiated by interactions with integrin cytoplasmic domains. A critical molecule for integrin-based signaling is paxillin, a 68-kDa protein that contains a number of protein–protein binding domains, and which binds to the cytoplasmic

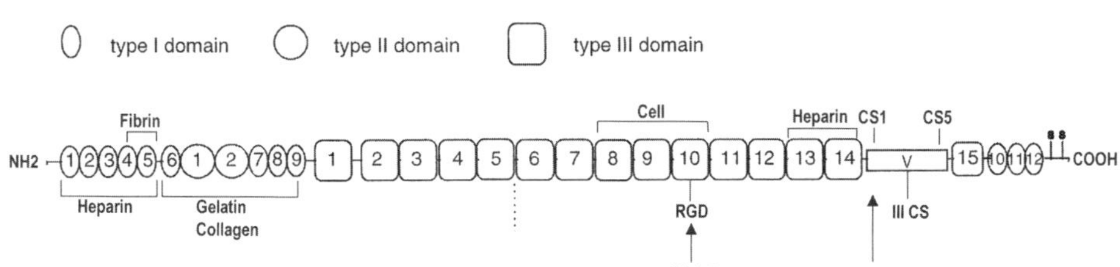

Figure 6.2 Schematic illustration of the modular composition of human fibronectin . From Cryptic self-association sites in type III modules of fibronectin, Kenneth C. Ingham, Shelesa A. Brew, Sheela Huff, and Sergei V. Litvinovich, *Journal of Biological Chemistry* 272 (3), © 1997 Reproduced by permission of The American Society for Biochemistry and Molecular Biology, Inc.

domain of the integrin. Additional binding partners also help trigger intracellular signaling, including focal adhesion kinase (FAK) and the closely related Pyk2 kinase. Upon recruitment, FAK and Pyk2 are activated and initiate tyrosine phosphorylation of paxillin and other associated molecules, creating additional protein binding sites and activating tethered secondary messenger molecules. One vital signaling pathway downstream of FAK and Pyk2 is phosphoinositide 3-kinase (PI3K), which is activated by the association of its regulatory p85 subunit with the adhesion kinases. FAK also directly activates a pathway that results in up-regulation of the cyclin D promoter, affecting cell proliferation. Integrin engagement also leads to Src activation, engagement of Grb2, and activation of Ras, pathways also activated by stem cell factor and TPO. FAK signalling potentially provides a mechanism by which diverse extrinsic stimuli of HSCs may converge (Abboud and Lichtman, 2006). Differentiation of erythroid progenitor cells cultured on stromal cells seems to depend on contact with fibronectin. Also, cultured stromal cells can replace IL-3 and support self-renewal and differentiation by otherwise IL-3-dependent cell lines (Spooncer *et al.*, 1986).

6.2.3.2 *Heparan sulfate*

Long-term cultures that support hematopoiesis develop a heparan sulfate proteoglycan (HSPG) layer. Immunochemical analysis has shown that marrow stromal cell lines synthesize and secrete numerous members of the syndecan family of heparan sulfate, including glypican, betaglycan, and perlecan. Evidence is accumulating that heparan sulfate-containing proteoglycans may be vital components of the stem cell niche. For example, the structure of the heparan sulfate secreted from stromal cell lines that support long-term hematopoiesis is significantly larger and more highly sulfated than heparan sulfate from nonsupportive stromal cell lines. When added to long-term cultures, the larger formers of heparin sulfate support long-term culture-initiating cell whereas the smaller, desulfated form cannot (Gupta *et al.*, 1998).

6.2.3.3 *Tenascins*

Tenascins are large glycoproteins that are upregulated after tissue damage during the process of repair or regeneration. Tenascin-C is composed of six subunits linked like spokes in a wheel by their C-terminal fibrinogen-like domains. Each subunit is composed of multiple EGF-like and fibronectin type III modules. Hematopoietic cells adhere to tenascin-C via the fibrinogen- and fibrinectin-like domains leading to proliferation. Mice lacking tenascin have diminished levels of hematopoietic progenitors that may be due to reduction of fibronectin that is seen in these mice (Abboud and Lichtman, 2006). Little is known about the effect of tenascins on the initiation or maintenance of hematologic malignancies. However, tenascin levels are increased in states of increased megakaryopoiesis such as myelofibrosis (Soini *et al.*, 1993).

6.2.3.4 *Laminin*

Laminins control cell function by binding to integrin and non-integrin receptors. Laminins are heterotrimeric extracellular proteins with characterized five α chains, four β chains, and three γ chains, combining to form at least 15 distinct laminin trimers (Colognato and Yurchenco, 2000). Laminins are named based on the chains that comprise each isoform. For example, laminin-511 contains an $\alpha 5$-chain, a $\beta 1$-chain, and a $\gamma 1$ chain (Aumailley *et al.*, 2005). Laminin production suggests the presence of endothelial cells and has been found in long-term bone marrow cultures (LTBMCs). Northern blot and immunoblot analysis showed that laminin-8/9 ($\alpha 4\beta 1\gamma 1/\alpha 4\beta 2\gamma 1$, 411/421) and laminin-10/11 ($\alpha 5\beta 1\gamma 1/\alpha 5\beta 2\gamma 1$) are the most abundant laminin isoforms synthesized by human bone marrow stromal cells (Siler *et al.*, 2000). Laminin-10/11 maximally binds to $\alpha 6\beta 1$ integrin on primitive hematopoietic cell lines and to primary human $CD34^+/CD38^-$ stem and progenitor cells without integrin activation (Gu *et al.*, 2003). A second, non-integrin laminin receptor also binds laminins, as well as other components of the ECM such as fibronectin, collagen, and elastin, and is composed of an acylated dimer of 32-kDa subunits. Although not an integrin, the laminin receptor associates with integrins (e.g., integrin) to modulate laminin binding. Functionally, laminin-10/11 facilitates SDF-1 stimulated transmigration of $CD34^+$ cells, and causes human hematopoietic cells to undergo mitosis. The non-integrin laminin receptor associates with the GM-CSF-R to modulate its signaling properties, downmodulating receptor signaling in the absence of laminin, and releasing the inhibition when binding its ligand. This arrangement could provide a novel molecular explanation for how laminins affect cell proliferation (Abboud and Lichtman, 2006; Gaynor, 2003).

Relatively little is known about how laminin affects the proliferation or homing of leukemia cells. Laminin 5 (3,2,2) has been identified as a critical motility factor in B-cell chronic lymphocytic leukemia, and is thought to be involved in the migration of these cells throughout involved lymph nodes (Spessotto *et al.*, 2007). Acute myelogenous leukemia with monocytic differentiation often has a high level of expression of the laminin receptor 67LR (Montuori *et al.*, 1999). Exposure of the relatively resistant promyelocytic leukemia cell line NB4 to laminin rendered it more sensitive to the prodifferentation effect of retinoic acid (Becker *et al.*, 1996).

6.2.3.5 *Proteoglycans and glycosaminoglycans*

Proteoglycans are glycoproteins consisting of glycosaminoglycan (GAG) side chains covalently linked to a protein core. GAGs bind to key cytokines, such as GM-CSF and IL-3, and retain them in specific marrow compartments. Proteoglycans with chondroitin and heparan sulfate GAG sidechains are produced in LTBMCs. While most GAGs synthesized in mouse LTBMC are heparin sulfate (Gallagher, Spooncer, and Dexter, 1983), LTBMC of human cells produce mainly chondroitin sulfate (Wight *et al.*, 1986). GAG side chains are though to alter the ability of malignant cells to metastasize. Little is known about specific GAG side chains and leukemia onset or progression. One *in vitro* study demonstrated that heparin derivatives inhibited the leukemia cell line U-937, whereas chondroitin and dermatan sulfate stimulated cell growth (Volpi *et al.*, 1994).

6.2.3.6 *Hyaluronic acid (hyaluronan)*

Hyaluronan (also called hyaluronic acid or hyaluronate) is a proteoglycan associated with the progression of a number of cancer types including leukemia. Hyaluronic acid is a GAG that is a component of hyaluronan. Hematopoietic stem cells express a receptor to hyaluronan called CD44. CD44 has active and inactive conformation. Although most HSCs express CD44, only a fraction of them adhere to hyaluronan. Antibodies to CD44 can block the interaction between HSCs and the marrow stroma. HSCs also produce hyaluronan, which enhances their retention within the marrow and subsequent proliferation (Abboud and Lichtman, 2006).

Interaction between leukemic cells and hylauronan are important in the progression of leukemia. CD 44 and the receptor for hyaluronic acid-mediated motility (RHAMM) are upregulated in chronic lymphocytic leukemia and have been shown to be markers of poor prognosis (Giannopoulos *et al.*, 2009). RHAMM in particular is not expressed on normal hematopoietic stem and progenitor cells, but is differentially expressed on myeloma, chronic lymphocytic leukemia, chronic myelogenous leukemia (CML), and acute myeloid leukemia cells. Vaccines targeting RHAMM have demonstrated clinical activity (Schmitt *et al.*, 2008).

6.2.3.7 *Collagen*

Collagen is a critical structural protein in the bone marrow providing structural support for developing bone as well as other ECM components. Type I collagen predominates in long-term culture, but types III, IV, and VI are present in both culture and in bone marrow sections. Collagen forms long fibrils called reticulin. Reticulin is abundant in normal bone marrow. In contrast, type IV collagen is the major component of basement membranes and forms a structure more similar to a net or mesh. Collagens and laminins interact in the marrow (Klein, 1995). Various hematopoietic cell lines and marrow mononuclear cells, including committed myeloid and erythroid progenitors strongly bind to type I and VI collagen. The integrin receptor $\alpha1\beta1$ and $\alpha2\beta1$, and the non-integrin glycoprotein VI, present predominantly on platelets are major receptors that bind to collagen (Abboud and Lichtman, 2006). Disregulation of collagen production is a major element of myelofibrosis. Myelofibrosis is the result of abnormal megakaryopoiesis seen in myeloproliferative states and is frequently fatal (Thiele and Kvasnicka, 2007).

6.2.4 The role of extracellular matrix in hematopoiesis

6.2.4.1 *Physical support for hematopoietic cells*

The bone marrow produces ECM that is critical for the support of hematopoiesis. Fibronectin is a critical protein that is involved in the compartmentalization and retention of hematopoietic cells in the bone marrow. Hematopoietic cells bind to fibronectin via a specific attachment domain in the fibronectin molecule consisting of arg-gly-asp (RGD) (Hynes, Schwarzbauer, and Tamkun, 1987). Hematopoietic progenitor cells bind to fibronectin through interactions between the integrin receptor VLA-5 and the RGD sequence, and signaling through this receptor accelerates erythrocyte differentiation *in vitro* (Tanaka *et al.*, 2009). Erythroid progenitors also express the integrin receptor VLA-4 which binds to fibronectin but also binds to nursing macrophages expressing VCAM-1 to make up erythroid islands. Signaling through the integrin receptor VLA-4 bound to fibronectin has been shown to be critical for the survival of leukemic blasts after exposure to chemotherapy (Matsunaga *et al.*, 2003). A monoclonal antibody called natalizumab inhibits VLA-4 interactions. Natalizumab is currently in clinical trials for acute myelogenous leukemia.

In contrast to erythrocytic progenitors, granulocyte progenitors have little tendency to bind to fibronectin. Rather, neutrophil progenitors bind to an ECM protein called hemonectin (Campbell, Long, and Wicha, 1987; Sullenbarger *et al.*, 1995; White, Totty, and Panayotou, 1993) Other evidence implicates intercellular adhesion molecule-1 (ICAM-1) the ligand for the integrin receptor leukocyte function antigen-1 (LFA-1) in adhesion of myeloid as well as lymphoid cells (Makgoba *et al.*, 1988). Relative to fibronectin, relatively little is known about the role of hemonectin or ICAM-1 in normal hematopoiesis or leukemia. Studies have demonstrated aberrant expression of cellular adhesion molecules in patients with multiple myeloma including ICAM-1, CD44, and N-CAM (CD56) (Barker *et al.*, 1992).

6.2.4.2 *Growth factor binding*

Interactions between GAGs and cytokines lead to retention of cytokines in specific areas of the bone marrow. Because of their negative charge, GAGs bind and trap growth factors in the marrow microenvironment. Proteoglycans have specific distribution patterns within the bone marrow and these patterns vary based on presence of inflammation, marrow injury, or hematopoietic stress (Borojevic *et al.*, 2003). Specific proteoglycans tend to bind to specific cytokines. For example, betaglycan, ECM and plasma membrane HSPG, and syndecan bind to FGFs, ECM HSPG binds to GM-CSF, IFN-γ, and IL-3; and betaglycan, biglycan, and decorin bind to TGF-β (Parker *et al.*, 1996).

ECM degradation via proteases such as the metalloproteinase MMP-9, or serine proteases, such as the urokinase-type plasminogen activator (uPA)-plasmin system leads to stimulation of hematopoiesis and mobilization of bone marrow components. Matrix-bound bFGF is released by plasmin as a non-covalent complex with HSPG or GAG (Parker *et al.*, 1996). A portion of the cell surface HSPG is anchored via a covalently linked glycosyl-phosphatidylinositol residue, which can be released by treatment with a specific phospholipase C. A fraction of the associated bFGF is also released by this treatment. Plasminogen present in the ECM is thought to activate latent forms of TGF-β.

MMP-9 has been demonstrated to release soluble kit ligand leading to the transfer of hematopoietic stem cells from the quiescent to proliferative niche (Heissig *et al.*,

2002). bFGF and VEGF only bind to specific cell surface receptors if attached either to free to membrane-bound heparan sulfate- or other heparin-like components of the ECM. The presentation of matrix-bound factors may be a mechanism by which factors interact with receptors on nearby cells. Interaction with matrix-bound factors may serve to establish cell-to-cell contacts and induce biological activities in juxtaposed cells (Kreis and Vale, 1993).

6.2.4.3 *HSC self-renewal and differentiation*

Current models suggest that HSCs occupy a specific microenvironment, the niche, which is the product of the ECM and specialized support cells (Li and Xie, 2005). HSC niches are thought to be present along the endosteal surface, the osteoblastic niche, and along the marrow blood vessels, the sinusoids, the vascular niche. The chemokine stromal derived factor-1 (CXCL-12) plays a major role in the homing and retention of hematopoietic stem cells in the niche. Cell surface proteins produced by osteoblasts (Weber and Calvi, 2009), reticular cells (Nagasawa, 2007), or sinusoidal endothelial cells (Kiel *et al.*, 2005) have also been demonstrated to play a role in the spatial distribution and maintenance of HSCs.

Integrins, the major class of cell-ECM adhesion molecules, play a critical role in providing survival and differentiation signals to hematopoietic cells at different stages of maturation. According to current models, migrating HSCs use VLA-4-mediated adhesion to attach to VCAM-1, which is expressed on the endothelial surface of bone marrow sinusoids (Levesque *et al.*, 2001). This binding initiates the transit of the stem cell through the endothelium and into the extravascular bone marrow space. Once in the extravascular space, HSCs find niches through interaction with the chemokine SDF-1 and repopulate the marrow microenvironment (Ellis and Tanentzapf, 2009).

6.2.5 Cytokines

6.2.5.1 *Macrophage colony-stimulating factor*

M-CSF, also known as CSF-1, was the first cytokine discovered and was found to be secreted by the murine clonal stromal cell line H1 (Charbord, 2001). M-CSF is also synthesized by bone marrow macrophages. Different isoforms of M-CSF are generated by alternative splicing. One isoform is slowly released from the membrane. M-CSF can bind to proteoglycans, resulting in a high molecular weight ECM-bound molecule. This membrane-bound molecule has been detected in murine and human stromal cell lines. Membrane-bound M-CSF can serve as an adhesion molecule. TNF-α stimulation can modify the ratio of membrane- to proteoglycan-bound isoforms (Zheng *et al.*, 2000). M-CSF promotes macrophage development and impedes the generation of other lineages. M-CSF may also regulate the production of stromal cells and macrophages by paracrine regulatory loops (Sherr, 1990).

6.2.5.2 *Granulocyte-macrophage colony-stimulating factor*

GM-CSF is found in the conditioned medium of murine stromal cells treated with pokeweed mitogen or lipopolysaccharide (Pluznik and Mergenhagen, 1986). The synthesis and release of GM-CSF is stimulated by IL-1 and TNF-α. GM-CSF regulates microenvironmental cell homeostasis, modulates macrophage to stromal cell ratio and stimulates the generation of marrow fibroblastic colonies (CFU-F) in humans and the growth of murine spleen cell lines (Charbord, 2001; Hamilton and Anderson, 2004).

6.2.5.3 *Granulcyte colony-stimulating factor*

G-CSF was detected in Dexter culture as a constitutive cytokine and in human long-term cultures after induction by IL-1 (Metcalf, 1985). Expression of G-CSF is increased by exposure to pokeweed mitogen or IL-1 and in murine macrophages stimulated by M-CSF or pokeweed mitogen (Souza *et al.*, 1986). G-CSF accounts for colony-forming unit-granulocyte macrophage output in long-term marrow cultures and promotes the proliferation of immature precursors (Charbord, 2001). G-CSF is commonly given to cancer patients in order to boost production of neutrophils during myelosuppressive chemotherapy.

6.2.5.4 *Interleukin-1*

The constitutive or induced expression of IL-1 is crucial for maintenance of hematopoiesis in long-term cultures and *in vivo*. Originally, IL-1 was found to contain two members, IL-1α and IL-1β. But recently more members have been found in the IL-1 superfamily, such as interleukin-1 receptor antagonist (IL-1RA) (Huising *et al.*, 2004). Both IL-1α and IL-1β are produced by macrophages, monocytes, and dendritic cells. IL-1 induces the production of several different hematopoietic growth factors, including GM-CSF, G-CSF, M-CSF, and IL-6 (Fibbe *et al.*, 1988b). In addition, IL-1 acts synergistically with colony-stimulating factors to promote the proliferation of primitive hematopoietic progenitor cells. IL-1 induces mobilization of hematopoietic progenitor cells into the peripheral blood, and these cells may be used to repopulate the bone marrow of lethally irradiated recipients (Fibbe and Willemze, 1991). In circulation, these progenitor cells form an important part of the inflammatory response. The administration of IL-1 also accelerates hematopoietic reconstitution after chemotherapy or radiation induced myelosuppression.

6.2.5.5 *Interleukin-3*

IL-3 is also known as multiple colony-stimulating factor. IL-3 is a multipotential growth factor produced by activated T cells, monocytes/macrophages, and stromal cells. IL-3 stimulates the differentiation of multipotent hematopoietic progenitors into myeloid rather than lymphoid progenitor cells where differentiation is stimulated by IL-7. IL-3 also stimulates proliferation of late myeloid progenitors. IL-3 also provides survival signals. Murine bone marrow-derived cells, dependent on IL-3 for their growth in culture, undergo apoptosis, upon cytokine withdrawal (Collins *et al.*, 1992).

6.2.5.6 *Interleukin-6*

IL-6 is a pleiotropic cytokine that exerts its effect through interaction with the IL-6 receptor complex composed of an α-chain subunit (IL-6Rα/gp80) and a β-chain subunit (gp130) (Heinrich *et al.*, 2003). IL-6 has been detected in cultures containing stromal cells and macrophages. Production can by stimulated by IL-1, M-CSF, IL-3, -4, -7, FGF2, LIF, TNF-α, lipopolysaccharide, or by IL-6 itself. IL-6 is a potent activator of osteoclasts. IL-6 can also act as an angiogenic factor by stimulating vascular endothelial growth factor secretion by plasma cells in multiple myeloma (Cohen *et al.*, 1996; Bellamy *et al.*, 1999; Dankbar *et al.*, 2000; Gupta *et al.*, 2001).

6.2.5.7 *Interleukin-7*

IL-7 is absolutely required for marrow B lymphopoiesis. IL-7 is a hematopoietic growth factor secreted by the bone marrow and thymic stromal cells. IL-7 stimulates differentiation of multipotent (pluripotent) hematopoietic stem cells into lymphoid progenitor cells and the proliferation of all lymphoid lineage cells (B and T lymphocytes and natural killer (NK) cells). IL-7 is important for proliferation during certain stages of B-cell maturation, T and NK cell survival, development, and homeostasis (Tan *et al.*, 2001; von Freeden-Jeffry *et al.*, 1995; Vivien, Benoist, and Mathis, 2001).

6.2.5.8 *Interleukin-10*

IL-10 is also known as human cytokine synthesis inhibitory factor. IL-10 is produced by monocytes and type 2 T helper cells (TH2), mast cells, $CD4^+$ $CD25^+$ $Foxp3^+$ regulatory T cells, and also in a certain subset of activated T cells and B cells. IL-10 enhances B-cell survival, proliferation, and antibody production. This cytokine blocks nuclear factor-κB activity, and is involved in the regulation of the JAK-STAT (Janus kinase/signal transducer and activator of transcription) signaling pathway. IL-10 is also reported to be a negative regulator of osteoclastogenesis (Yong *et al.*, 2009).

6.2.5.9 *Interleukin-11*

IL-11 is a cytokine that signals through the IL-11 receptor, which is a member of the gp130 receptor family. IL-11 stimulates megakaryopoiesis (Paul *et al.*, 1990). It also plays a role in the differentiation of progenitors from more immature precursors.

IL-11 plays a role in regulating stromal formation and phenotype when combined with cytokines such as IL-3, IL-4, IL-7, IL-12, IL-13, SCF, FMS-like tyrosine kinase-3 (Flt3)-L, and GM-CSF (Leng and Elias, 1997).

6.2.5.10 *Leukemia inhibitory factor*

LIF is another cytokine that signals through a member of the gp130 receptor family, LIF-receptor. LIF is constitutively expressed by murine stromal cell lines and is induced by IL-1 and TNF-α, or by LIF itself. LIF-R and other members of the gp130 receptor family play a role in the equilibrium between bone formation and hematopoiesis support by marrow mesenchymal progenitor cells.

6.2.5.11 *Stem cell factor*

SCF, also known as Kit ligand, Steel factor and mast cell growth factor, is detected in bone marrow macrophages and stromal cell cultures. SCF is the ligand for the protein tyrosine kinase receptor c-kit. Stromal cells express both the soluble and transmembrane isoforms of SCF. Several reports suggest that SCF is critical for myelopoiesis but not essential for the survival of the most primitive hematopoietic progenitor cells. SCF is a potent costimulating cytokine. By itself, it has mild stimulating activity for hematopoietic stem and progenitor cells, but when used in combination with a CSF *in vitro*, or with various selected ILs or other cytokines, SCF greatly augments the numbers of detectable stem and progenitor cells and the size of resultant colonies formed (Broxmeyer, 1999). SCF is also a chemotactic agent for HPC and believed to play a role in HSC localization to endosteal niches. SCF promotes cell adhesion by two distinct mechanisms. First, binding membrane-bound SCF on stromal cells to c-Kit expressed by hematopoietic cells leads to attachment to stroma, and second, signaling through c-Kit upregulates the avidity of mast cell and progenitor cell β1 integrins, VLA-4 and VLA-5, for the ECM component, fibronectin. Mobilization by blocking antibodies to VLA-4 or its cellular ligand VCAM-1 required functional c-Kit, indicating integrin/receptor cross-talk and supporting the central role of c-Kit/SCF in mobilization (Ashman, 1999).

6.2.5.12 *FMS-like tyrosine kinase-3 ligand*

The Flt3-ligand, FL belongs to a small family of α-helical cytokines and is structurally related to SCF and CSF. Flt3-ligand is expressed by T lymphocytes and bone marrow stromal fibroblasts as a membrane-bound and a soluble isoform. Treatment with IL-1α increases the expression of Flt-3 ligand in bone marrow fibroblasts. Flt3-ligand stimulates the proliferation and differentiation of primitive hematopoietic cells, for example, early B-cell lineage differentiation and expansion of monocytes and immature dendritic cells (Wodnar-Filipowicz, 2003).

Table 6.2 VEGF family members, receptors, and their functions.

VEGF family members	Receptors	Functions
VEGF (VEGF-A)	VEGFR-1, VEGFR-2, neuropilin-1	Angiogenesis Vascular maintenance
VEGF-B	VEGFR-1	Not established
VEGF-C	VEGFR-2, VEGFR-3	Lymphangiogenesis
VEGF-D	VEGFR-2, VEGFR-3	Lymphangiogenesis
VEGF-E (viral factor)	VEGFR-2	Angiogenesis
Placental growth factor (PlGF)	VEGFR-1, neuropilin-1	Angiogenesis Inflammation

From http://www.researchvegf.com/

6.2.5.13 Vascular endothelial growth factor

VEGFs plays an important role in angiogenesis. As its name suggests, VEGF stimulates vascular endothelial cell growth, survival, and proliferation. VEGF is a member of a family of six structurally related proteins (see Table 6.2) that regulate the growth and differentiation of multiple components of the vascular system, especially blood and lymph vessels.

There are four major isoforms of VEGFA, named based on the size of the VEGF gene. These isoforms are VEGF121, VEGF165, VEGF189, and VEGF206. Although these isoforms behave identically in solution, they differ in their ability to bind heparin and the ECM. The production of VEGF is stimulated by upstream activators, including environmental cues, growth factors, oncogenes, cytokines, and hormones. The binding of VEGF to its receptors on the surface of endothelial cells activates intracellular tyrosine kinases, triggering multiple downstream signals that promote angiogenesis. Although there are multiple variants of both the VEGF ligand and its receptor, the angiogenic effects of this pathway are primarily mediated through the interaction of VEGF-A (the most common variant, often referred to simply as VEGF) with VEGFR-2 (Ferrara, 2004). Other non-VEGF factors are thought to play a secondary role in angiogenesis. VEGF ligands mediate angiogenic effects by binding to specific VEGF receptors, leading to receptor dimerization and subsequent signal transduction. VEGF ligands bind to three primary receptors and two coreceptors. Of the primary receptors, VEGFR-1 and VEGFR-2 are mainly associated with angiogenesis. The third primary receptor, VEGFR-3, is associated with lymphangiogenesis.

Both bone marrow hematopoietic stem/progenitor cells and mesenchymal stem cells may secret of VEGF in response to hypoxia or TNF by stimulating the hypoxia inducible factor-1α pathway (Wang *et al.*, 2006, 2007). Myeloid cells, megakaryocytes, bone marrow stromal cells, and different types of tumor cells also produce VEGF.

6.2.5.14 Stromal derived factor-1α

SDF-1α is small cytokine belonging to the chemokine family that is also called chemokine (C-X-C motif) ligand 12 (CXCL12). In the bone marrow, SDF-1 is produced by immature osteoblasts in the endosteal region and also by adventitial reticular cells. SDF-1 is also produced by vascular endothelium, megakaryocytes, and platelets. Mesenchymal adipocyte cell lines produce SDF-1 in vitro. SDF-1 is produced in two forms, SDF-1α/CXCL12a and SDF-1β/CXCL12b, by alternate splicing of the same gene. SDF-1 is a major factor in regulating stem cell niches. Gradients of SDF-1 regulate migration of megakaryocyte progenitors to inductive niches adjacent to bone marrow sinusoids (Avecilla *et al.*, 2004). SDF-1/CXCR4 interactions are required for homing and engraftment of hematopoietic stem cells during transplantation (Lapidot and Kollet, 2002).

6.2.5.15 *Thrombopoietin*

TPO is the ligand for the proto-oncogene encoded receptor c-mpl. TPO is an essential growth factor for stem cell maintenance and proliferation. TPO is mainly synthesized in the liver, and its levels are regulated by clearance through platelet binding. TPO's major role is to stimulate megakaryocytic activity and enhance platelet production and function, together with IL-6 and IL-11. Although not the only cytokine that has this function, TPO presently seems to be the most potent platelet inducer. TPO plays a role in regulating hematopoietic stem cell proliferation. TPO also acts as a survival factor, an action mediated by the anti-apoptotic effects of upregulation of the promoter comformation of the tumor suppressor gene encoded protein, p53 (Ritchie *et al.*, 1997). TPO may also be involved in migration, homing, and mobilization of stem/progenitor cells. TPO can activate integrins, especially very late antigen VLA-4 and VLA-5, and enhances adherence of progenitor cells to the stromal cell component fibronectin (Broxmeyer, 1999). Mutations in the TPO receptor, c-mpl, are associated with amegakaryocytic thrombocytopenia and a predisposition to aplastic anemia and myeloid leukemia (Kaplan and Bussel, 2004).

6.2.5.16 *Fibroblast growth factor-B*

bFGF is produced by bone marrow stromal cells and several mature peripheral blood lineages. It is released and stored in the bone marrow ECM. Fibroblastic growth factor-receptors (FGF-Rs) are expressed on nearly every hematopoietic cell type. bFGF promotes hematopoiesis, by acting on various cellular targets, including stromal cells, early and committed hematopoietic progenitors, and mature blood cells. bFGF synergizes with hematopoietic cytokines, or antagonizes the negative regulatory effects of another factor, TGF-β, thus potentially playing a central role in hematopoiesis (Allouche, 1995). The expression of FGFRs is widely distributed on different hematopoietic progenitor cells and stromal cells, and FGFs play an important role in hematopoietic stem cell homeostasis. FGFs have been shown to sustain the proliferation of hematopoietic progenitor cells, maintaining their primitive phenotype. Basic FGF (bFGF, FGF-2) stimulates the formation of an adherent stromal cell layer in human LTBMCs, and promotes hematopoietic cell development. FGF-2 synergizes with other hematopoietic growth factors to enhance *in vitro* colony formation by several classes of hematopoietic progenitor cells (Kashiwakura and Takahashi, 2005).

6.2.5.17 *Other cytokines*

Some modulators of hematopoiesis including hormones and neurotransmitters are released from the nerves in the bone marrow (Rameshwar *et al.*, 1997; Maloof *et al.*, 2001). The neurotransmitter SP has a role in regulating the hematopoietic and immune systems. The hematopoietic effects of SP are mediated through interactions with its high affinity receptor, neurokinin-1 (NK-1) (Maggi and Giuliani, 1996). SP acts through indirect responses of mesenchymal cells and macrophages (Rameshwar *et al.*, 1997). SP is an undecapeptide that is the major peptide derived from the preprotachykinin-I (PPT-I) gene (Maggi and Giuliani, 1996). The NK-1 gene is expressed differently in bone marrow and neural cells (Rameshwar, Poddar, and Gascon, 1997). NK-1 is constitutively expressed at low levels in the brain (Yao *et al.*, 1999). The expression of NK-1 in the brain is further upregulated by inflammatory mediators (Abrahams *et al.*, 1999; Yao *et al.*, 1999). In contrast, in unstimulated stroma, the NK-1 receptor is low to undetectable and is detected when these cells are stimulated with cytokines and other factors associated with hematopoietic stimulation (Rameshwar, Poddar, and Gascon, 1997; Rameshwar *et al.*, 2001). Among the supporting hematopoietic cells, NK-1 and NK-2 expression has been detected in macrophages, fibroblasts, and endothelial cells (Rameshwar *et al.*, 2001). Non-stromal cells in the BM can also express the NK-1 receptor. These include T, B, and dendritic cells (Rameshwar *et al.*, 2002).

Adrenergic agents can also affect hematopoiesis. The α1-selective adrenergic agonist, methoxamine, and to a much lesser extent, the α2-agonist clonidine, proved to exert an inhibitory action when added to the GM-CFU assay. Presence of α1-adrenergic receptors (α1-ARs) were shown on bone marrow cells (Maestroni and Conti, 1994). Norepinephrine and dopamine were found in both short-term bone marrow cultures and LTBMCs (Maestroni *et al.*, 1998). Bone marrow catecholamines might derive, in part, from bone marrow lymphocytes or from their precursors. Catecholamines inhibit both lymphocyte function and hematopoiesis (Maestroni, 2006).

Other cytokines such as TNF, IGF-2, hepatocyte growth factor, IGF-1, macrophage inhibitory protein-1α, oncostatin M (OSM), TGF-β, IL-15, IL-21, PF4, and CD40, Wnts also controls proliferation, cellular differentiation, and other functions in most hematopoietic cells in certain degree.

6.2.6 Interactions

6.2.6.1- *Gap junctions*

Gap junctions (GJs) are membrane-spanning channels that facilitate intercellular communication by allowing small signaling molecules such as calcium ions, inositol phosphates, and cyclic nucleotides to pass from cell to cell. Gap junction-mediated intercellular communication allows groups of cells in contact to respond to environmental stimuli in a coordinated manner. Different tissue types express different connexins (Cxs), with Cx43 being the predominant GJ protein expressed by hematopoietic tissues (Cancelas *et al.*, 2000; Montecino-Rodriguez and Dorshkind, 2001). In the bone marrow, most Cx43 was found to be associated with endosteal and adventitial stromal cells, megakaryocytes, fat cell membranes, and osteoblasts and only rarely associated with all kinds of hematopoietic cells (Krenacs and Rosendaal, 1998). Cx43 expression in the stroma is critical during the early growth and regeneration of the hematopoietic system (Rosendaal, 1995; Hurtado *et al.*, 2004).

6.3 Leukemia and its microenvironment

6.3.1 Effect of leukemia on the bone marrow microenvironment

BCR-ABL-induced leukemia suppresses normal hematopoiesis. This is an active process involving secretion of the cell death-inducing factor, 24p3, by leukemia cells (Lin *et al.*, 2005). Leukemic cell growth was recently shown to disrupt normal HPC niches in the bone marrow and creates abnormal microenvironments that recruit and sequester transplanted human CD34$^+$ cells. Leukemia cells constituted the major source of SCF in the malignant microenvironment. SCF is an HPC growth factor and a chemoattractant believed to play a role in HSC localization to endosteal niches (Ashman, 1999). In long term transplantation, niches in leukemic mice failed to preserve the human CD34$^+$ cell pool size. Furthermore, the human CD34$^+$ cells failed to mobilize into the peripheral circulation in response to cytokine stimulation. Neutralization of SCF secreted by leukemic cells inhibited CD34$^+$ cell migration into malignant niches, normalized CD34$^+$ cell numbers, and restored CD34$^+$ cell mobilization in leukemic mice (Colmone *et al.*, 2008). The overexpression of c-myb often observed in leukemic cells leads to autocrine cytokine expression that, in turn, contributes to the growth advantage of leukemic cells (Szczylik *et al.*, 1993).

6.3.2 Leukemia and bone marrow angiogenesis

A number of studies report that various types of leukemia are associated with increased microvascular density in the bone marrow, which has been interpreted as an effect of leukemia on angiogenesis (Di Raimondo, 2003; Pule *et al.*, 2002). Others have suggested that this increased microvascular density may be related to anemia rather than leukemia because they both generate hypoxia in the bone marrow (Di Raimondo *et al.*, 2000). Veiga and colleagues showed that leukemia cells interact with endothelial cells and promote endothelial cells proliferation and migration (Veiga *et al.*, 2006). Litwin and colleagues demonstrated that angiogenesis in acute myeloid leukemia (AML) occurs in response to microenvironmental factors, rather than being directly due to the effect of leukemia cells (Litwin *et al.*, 2002). The origin of these microvascular endothelial cells is not consistently reported. A few investigators showed that some of the microvascular endothelial cells originate from leukemia cells (Gunsilius *et al.*, 2000). However, work in our laboratory using a xenograft leukemia model in non-obese diabetic/severe combined immunodeficient mice showed that human t(4;11) leukemia cells, reported to have a hemangioblastic phenotype, recruit murine blood vessels but do not directly contribute to the walls of the blood vessels in leukemia infiltrated areas (Hu *et al.*, 2009). These contradictory results may be because certain types of leukemia retain the ability to produce endothelium and others do not, or differences in the techniques used to test the ability of leukemia cells to make endothelium between different research groups. Regardless of whether the new vessels are produced by leukemia cells or existing blood vessels, the increase of bone marrow angiogenesis and marrow VEGF level is an adverse prognostic factor for some forms of leukemia (Giles *et al.*, 2003).

Exposure to leukemia cells can lead to changes in the endothelium that promote leukemia progression. For instance, BMECs exposed to leukemia cells can promote the survival of leukemia cells (Veiga *et al.*, 2006). Leukemia cells actively stimulate bone marrow endothelium, promote *de novo* angiogenesis, and induce neovascularization in the leukemic bone marrow. Soluble factors present in the leukemic bone marrow microenvironment promote the proliferation, migration, and morphogenesis of bone marrow endothelial cells, which are critical processes in tumor angiogenesis. Most acute lymphoblastic leukemia (ALL) patients have increased blood plasma and urine levels of basic fibroblast growth factor (bFGF; FGF-2). VEGF levels are increased in some patients with leukemia. Furthermore, leukemia cells produce connective tissue growth factor (CTGF; CCN2); and IL-8 (Veiga *et al.*, 2006). Treatment with chemotherapy reversed the increased bone marrow angiogenesis (Kvasnicka *et al.*, 2004). Antiangiogenic therapy is a potential choice for the leukemia management, though this therapy has not been extensively studied in clinical trials except for multiple myeloma. Some antileukemia chemotherapeutic agents have antiangiogenic function. Several clinical trials combine antiangiogenic drugs with chemotherapy.

6.3.3 Altered interactions between leukemia cells and the microenvironment

Contact with the ECM may anchor developing hematopoietic cells in the marrow cavity until they are ready for release into the bloodstream (Gordon, 1988). Hence, defective interactions with the ECM could account for the release of phenotypically immature leukemic blasts cells into the circulation. In LTMBCs, normal hematopoietic progenitor cells have a selective growth advantage (Coulombel, Eaves, and Eaves, 1983; Coulombel *et al.*, 1985). Leukemic cell growth disrupts normal HPC bone marrow niches and creates abnormal microenvironments that sequester transplanted human CD34+ (HPC-enriched) cells. The advantage of leukemic cells could result from defective adhesive interactions with ECM (Gordon *et al.*, 1987). Primitive CML progenitors cycled continuously regardless of the presence or absence of an adherent feeder layer, which shows an insensitivity to the regulation normally exercised by stromal cells (Eaves *et al.*, 1986). More precise information about the cause and consequences of defective adhesion is required for proper evaluation of its clinical and biological significance.

6.3.3.1 *Altered interactions between leukemia cells and stromal cells*

Adhesion receptors mediate cell-cell and cell-ECM interactions. These interactions are required for the residence of stem cells and progenitors in the marrow, as well as homing to the marrow. These interactions also play an important role in regulating cell behavior by directly activating signal pathways or by modulating responses to growth factors. Based on domain structure and function, adhesion molecules can be divided into integrins, cadherins, selectins, mucin-like family members, and members of the immunoglobulin family of receptors. Significant evidence shows that abnormal expression or function of cell adhesion molecules contributes to the aberrant behavior observed in a number of hematological diseases (Prosper and Verfaillie, 2001). For instance, the type of adhesion receptors expressed on leukemic cells often differs from that on non-leukemia cells at the same stage of differentiation.

6.3.3.2 *Adhesion defects*

Although Philadelphia chromosome positive $CD34^+$ cells in patients with CML express $\beta1$ integrins, the integrin function is defective. CML $CD34^+$ cells are less adherent to stromal feeders layers and to fibronectin than their normal counterparts. Engagement of $\alpha4\beta1$ and $\alpha5\beta1$ integrins on CML $CD34^+$ cells does not affect the growth of CML progenitors. Lack of integrin-mediated regulation of CML progenitor growth and increased ability of CML $CD34^+$ cells to survive in the absence of interactions with the bone marrow microenvironment may contribute to the massive expansion of the malignant stem and progenitor cell population in CML (Prosper and Verfaillie, 2001).

Adhesive defects have also been described in B-lineage ALL (Geijtenbeek *et al.*, 1999). In this study, $CD10^+$ ALL cells in some patients were found to lack any LFA-1-mediated adhesion caused by absent LFA-1 surface expression. $CD10^+$ ALL cells from other patients expressed LFA-1 that could not be activated by the phorbol ester phorbol 12-myristate 13-acetate (PMA), whereas the $CD10^-$ cells expressed a functional LFA-1. Still other patients contained $CD10^+$ cells that did not increase expression of VLA-4 like normal B-lymphocytes.

6.3.4 Microenvironment-mediated leukemia drug resistance

SDF-1-α and its cognate receptor, CXCR4, have recently emerged as critical mediators of stromal/leukemia cell interactions. Both surface and intracellular CXCR4 levels were found to be elevated in a subset of AML cases. CXCR4 was recently reported to be expressed at higher levels in cases of AML associated with an internal tandem duplication (ITD) type of mutation of the gene that encodes fetal liver tyrosine kinase-3 (FLT3). The specific synthetic peptides and CXCR4 inhibitors, such as AMD3100 and second generation inhibitor AMD3465, were shown to mobilize leukemia cells and stem cells into the circulation and made them more susceptible to chemotherapy-induced or FLT3 inhibitor induced cell death in a xenograft leukemia model (Burger and Peled, 2009).

In CML, the highly expressed chimeric fusion protein p210Bcr-Abl downregulated CXCR4 expression, and this was shown to be associated with cell migration defects in CML. The tyrosine kinase inhibitors, imatinib or bafetinib, may restore CXCR4 expression leading to the migration of CML cells to bone marrow microenvironment niches, which in turn results in acquisition of stroma-mediated chemoresistance of CML progenitor cells. In KBM5 and K562 cells, imatinib, INNO-406, and IFN-α all increased CXCR4 expression and migration. This increase in CXCR4 levels on CML progenitor cells was likewise found in samples from CML patients treated with imatinib or IFN-α. Imatinib induced G0-G1 cell cycle block in CML cells, which was further enhanced in a mesenchymal stem cell (MSC) coculture system. MSC coculture protected KBM-5 cells from imatinib-induced cell death. These antiapoptotic effects were abrogated by the CXCR4 antagonist AMD3465 or by inhibitor of integrin-linked kinase QLT0267. Altogether, these findings suggest that the upregulation of CXCR4 by imatinib promotes migration of CML cells to bone marrow stroma, causing the G0-G1 cell cycle arrest and hence ensuring the survival of quiescent CML progenitor cells (Jin *et al.*, 2008).

6.3.4.1 *Cell adhesion mediated-drug resistance*

Stromal cells have been shown to protect certain kinds of leukemia cells from apoptosis induced by chemotherapy. This phenomenon was called cell adhesion mediated-drug resistance (CAM-DR). Interactions between α4β1 and its ligand, VCAM, may play a role in chemotherapy resistance in ALL and chronic lymphocytic leukemia (CLL) (de la Fuente *et al.*, 2002; Li and Dalton, 2006). B-ALL cells arrested in G1 phase of the cell cycle and eventually underwent apoptosis when incubated with cytarabine or etoposide. However, contact with stroma prevented both cell cycle arrest and apoptosis of B-ALL and CLL cells. Interaction with fibronectin did not enhance survival of ALL cells during chemotherapy exposure. However, interaction with the other ligand of the VLA-4, VCAM, provided maximal protection from cytarabine- and etoposide-induced cell death (Li and Dalton, 2006).

The molecular mechanisms underlying this protection occurred through activation of the Bim (a Bcl-2 homology domain 3 (BH3) -containing protein) ERK1/2 (extracellular signal-related kinase) signaling pathway. Stimulation of this pathway promoted phosphorylation and proteasome-dependent degradation of Bim. β1 integrin-mediated reduction in Bim levels in K562 cells is independent of Erk1/2 pathway activation. In K562 cells, β1 integrin mediated adhesion reduced Bim levels in part by increasing proteasomal degradation. β1 integrin-mediated regulation of Bim in K562 CML cells was demonstrated to be partly a result of increased proteasomal-mediated degradation of Bim protein levels, and proteasome inhibitors prevented Bim degradation. Increased degradation of Bim was not related to activation of the mitogen-activated protein kinase pathway, as adhesion of K562 cells caused a reduction in phospho- ERK1/2 levels. In addition, pharmacological inhibition of mitogen-activated protein/ERK kinases with PD98059 did not increase Bim levels. Reducing Bim levels by short hairpin RNA targeting inhibited imatinib and mitoxantrone-induced cell death. These results showed that β1 integrin-mediated adhesion regulates Bim degradation and may contribute to the minimal residual disease associated with many hematopoietic malignancies, and particularly CML. Together this evidence indicates that disrupting β1 integrin-mediated regulation of Bim degradation may increase the efficacy of drugs, including imatinib, used to treat hematopoietic malignancies (Hazlehurst, Argilagos, and Dalton, 2007).

6.3.4.2 Soluble factor mediated-drug resistance (SM-DR)

A number of soluble factors promote cell proliferation and survival, and by providing survival signals promote the development of resistance. Some of the soluble factors that provide proliferative and survival signals include VEGF, bFGF, SDF-1, IL-6, nitric oxide, IL-3 and G-CSF, M-CSF, GM-CSF, and the TNF super family members BAFF (B cell-activating factor of the TNF family) and a proliferation-inducing ligand (APRIL) (Li and Dalton, 2006).

Growth factor receptor mutations occur during the process of transformation in a number of forms of leukemia. Mutations of the receptors for stem cell factor (c-kit) (Heinrich *et al.*, 2002), G-CSF (Gits *et al.*, 2006), PDGF receptor (Cools *et al.*, 2004), FGF-R (Macdonald, Reiter, and Cross, 2002), and the TPO receptor (c-mpl) (Kilpivaara and Levine, 2008) can occur as part of the transformation process. Receptors that are constitutively active lead to downstream signaling, leading to proliferation, cell survival and changes in cell-to-cell interaction. Cytokine effects are dependent on the expression of the normal or mutant receptor on malignant cells. Leukemic cells express a number of cytokine, chemokine and growth factor receptors. VEGFR represents a family of receptors that are involved in blood vessel development, but are also expressed by leukemia cells leading to survival, growth, and migration. Leukemic cell VEGFRs respond to autocrine as well as paracrine VEGFs. VEGF signaling increases the Bcl-2/Bax ratio and inhibits apoptosis. While these growth factors may play a role in the development of resistance, unless there is a receptor mutation, targeting the specific cell survival pathways promoted by these factors may not reverse resistance because resistant leukemia cells often develop other methods of eluding chemotherapy, including genetic mutations, cell cycle arrest, or through direct cell contact and adhesion.

6.4 Summary

The bone marrow microenvironment is an ideal place to support normal and malignant hematopoiesis. We are just beginning to understand the way that interactions with the microenvironment promote normal hematopoiesis and contribute to leukemia progression. New therapies targeting the interactions between leukemia cells and the microenvironment may prove to be effective treatments, either alone or in combination with other cancer therapies.

References

Abboud, C.N. and Lichtman, M.A. (2006) Structure of the marrow and the hematopoietic microenvironment, in *Williams Hematology*, 7th edn (eds M.A. Lichtman, E. Beutler, T.J. Kipps *et al.*), McGraw-Hill, New York, pp. 45–52.

Abrahams, L.G., Reutter, M.A., Mccarson, K.E., and Seybold, V.S. (1999) Cyclic AMP regulates the expression of neurokinin1 receptors by neonatal rat spinal neurons in culture. *Journal of Neurochemistry*, **73**, 50–58.

Allouche, M. (1995) Basic fibroblast growth factor and hematopoiesis. *Leukemia*, **9**, 937–942.

Arai, F., Hirao, A., Ohmura, M. *et al.* (2004) Tie2/angiopoietin-1 signaling regulates hematopoietic stem cell quiescence in the bone marrow niche. *Cell*, **118**, 149–161.

Arai, F. and Suda, T. (2007) Maintenance of quiescent hematopoietic stem cells in the osteoblastic niche. *Annals of the New York Academy of Sciences*, **1106**, 41–53.

Ashman, L.K. (1999) The biology of stem cell factor and its receptor C-kit. *International Journal of Biochemistry and Cell Biology*, **31**, 1037–1051.

Aumailley, M., Bruckner-Tuderman, L., Carter, W.G. *et al.* (2005) A simplified laminin nomenclature. *Matrix Biology*, **24**, 326–332.

Avecilla, S.T., Hattori, K., Heissig, B. *et al.* (2004) Chemokine-mediated interaction of hematopoietic progenitors with the bone marrow vascular niche is required for thrombopoiesis. *Nature Medicine*, **10**, 64–71.

Avraham, H., Scadden, D.T., Chi, S. *et al.* (1992) Interaction of human bone marrow fibroblasts with megakaryocytes: role of the c-kit ligand. *Blood*, **80**, 1679–1684.

Bainton, D.F., Maloney, M.A., Patt, H.M., and Stern, R. (1986) Characterization of rabbit stromal fibroblasts derived from red and yellow bone marrow. *Journal of Experimental Medicine*, **163**, 400–413.

Balkwill, F. and Mantovani, A. (2001) Inflammation and cancer: back to Virchow? *Lancet*, **357**, 539–545.

Barker, H.F., Hamilton, M.S., Ball, J. *et al.* (1992) Expression of adhesion molecules LFA-3 and N-CAM on normal and malignant human plasma cells. *British Journal of Haematology*, **81**, 331–335.

Becker, P.S., Li, Z., Potselueva, T. *et al.* (1996) Laminin promotes differentiation of NB4 promyelocytic leukemia cells with all-*trans* retinoic acid. *Blood*, **88**, 261–267.

Belaid-Choucair, Z., Lepelletier, Y., Poncin, G. *et al.* (2008) Human bone marrow adipocytes block granulopoiesis through neuropilin-1-induced granulocyte colony-stimulating factor inhibition. *Stem Cells*, **26**, 1556–1564.

Bellamy, W.T., Richter, L., Frutiger, Y., and Grogan, T.M. (1999) Expression of vascular endothelial growth factor and its receptors in hematopoietic malignancies. *Cancer Research*, **59**, 728–733.

Bianco, P., Robey, P.G., and Simmons, P.J. (2008) Mesenchymal stem cells: revisiting history, concepts, and assays. *Cell Stem Cell*, **2**, 313–319.

Blair, P. and Flaumenhaft, R. (2009) Platelet alpha-granules: basic biology and clinical correlates. *Blood Reviews*, **23**, 177–189.

Bord, S., Beavan, S., Ireland, D. *et al.* (2001) Mechanisms by which high-dose estrogen therapy produces anabolic skeletal effects in postmenopausal women: role of locally produced growth factors. *Bone*, **29**, 216–222.

Borojevic, R., Carvalho, M.A., Correa-Junior, J.D. *et al.* (2003) Stroma-mediated granulocytemacrophage colony-stimulating factor (GM-CSF) control of myelopoiesis: spatial organisation of intercellular interactions. *Cell Tissue Research*, **313**, 55–62.

Broxmeyer, H.E. (1999) The hematopoietic system: principles of therapy with hematopoietically active cytokines, in *Cytokines in the Treatment of Hematopoietic Failure* (eds A. Ganser and D. Hoelzer), Marcel Dekker, Inc., New York, Basel, Hong Kong.

Broxmeyer, H.E., Orschell, C.M., Clapp, D.W. *et al.* (2005) Rapid mobilization of murine and human hematopoietic stem and progenitor cells with AMD3100, a CXCR4 antagonist. *Journal of Experimental Medicine*, **201**, 1307–1318.

Burger, J.A. and Peled, A. (2009) CXCR4 antagonists: targeting the microenvironment in leukemia and other cancers. *Leukemia*, **23**, 43–52.

Campbell, A.D., Long, M.W., and Wicha, M.S. (1987) Haemonectin, a bone marrow adhesion protein specific for cells of granulocyte lineage. *Nature*, **329**, 744–746.

Cancelas, J.A., Koevoet, W.L., De Koning, A.E. *et al.* (2000) Connexin-43 gap junctions are involved in multiconnexin-expressing stromal support of hemopoietic progenitors and stem cells. *Blood*, **96**, 498–505.

Charbord, P. (2001) Mediators involved in the control of hematopoiesis by the microenvironment, in *Hematopoiesis: a Developmental Approach* (ed. L.I. Zon), Oxford University Press, Inc., New York, pp. 702–717.

Cohen, T., Nahari, D., Cerem, L.W. *et al.* (1996) Interleukin 6 induces the expression of vascular endothelial growth factor. *Journal of Biological Chemistry*, **271**, 736–741.

Collins, M.K., Marvel, J., Malde, P., and Lopez-Rivas, A. (1992) Interleukin 3 protects murine bone marrow cells from apoptosis induced by DNA damaging agents. *Journal of Experimental Medicine*, **176**, 1043–1051.

Colmone, A., Amorim, M., Pontier, A.L. *et al.* (2008) Leukemic cells create bone marrow niches that disrupt the behavior of normal hematopoietic progenitor cells. *Science*, **322**, 1861–1865.

Colognato, H. and Yurchenco, P.D. (2000) Form and function: the laminin family of heterotrimers. *Developmental Dynamics*, **218**, 213–234.

Cools, J., Stover, E.H., Wlodarska, I. *et al.* (2004) The FIP1L1-PDGFRalpha kinase in hypereosinophilic syndrome and chronic eosinophilic leukemia. *Current Opinion in Hematology*, **11**, 51–57.

Corre, J., Barreau, C., Cousin, B. *et al.* (2006) Human subcutaneous adipose cells support complete differentiation but not self-renewal of hematopoietic progenitors. *Journal of Cellular Physiology*, **208**, 282–288.

Coulombel, L., Eaves, A.C., and Eaves, C.J. (1983) Enzymatic treatment of long-term human marrow cultures reveals the preferential location of primitive hemopoietic progenitors in the adherent layer. *Blood*, **62**, 291–297.

Coulombel, L., Eaves, C., Kalousek, D. *et al.* (1985) Long-term marrow culture of cells from patients with acute myelogenous leukemia. Selection in favor of normal phenotypes in some but not all cases. *Journal of Clinical Investigation*, **75**, 961–969.

Dankbar, B., Padro, T., Leo, R. *et al.* (2000) Vascular endothelial growth factor and interleukin-6 in paracrine tumor-stromal cell interactions in multiple myeloma. *Blood*, **95**, 2630–2636.

Dar, A., Goichberg, P., Shinder, V. *et al.* (2005) Chemokine receptor CXCR4-dependent internalization and resecretion of functional chemokine SDF-1 by bone marrow endothelial and stromal cells. *Nature Immunology*, **6**, 1038–1046.

De La Fuente, M.T., Casanova, B., Moyano, J.V. *et al.* (2002) Engagement of alpha4beta1 integrin by fibronectin induces in vitro resistance of B chronic lymphocytic leukemia cells to fludarabine. *Journal of Leukocyte Biology*, **71**, 495–502.

Dennis, J.E. and Charbord, P. (2002) Origin and differentiation of human and murine stroma. *Stem Cells*, **20**, 205–214.

Devine, S.M., Bartholomew, A.M., Mahmud, N. *et al.* (2001) Mesenchymal stem cells are capable of homing to the bone marrow of non-human primates following systemic infusion. *Experimental Hematology*, **29**, 244–255.

Devine, S.M. and Hoffman, R. (2000) Role of mesenchymal stem cells in hematopoietic stem cell transplantation. *Current Opinion in Hematology*, **7**, 358–363.

Dimascio, L., Voermans, C., Uqoezwa, M. *et al.* (2007) Identification of adiponectin as a novel hemopoietic stem cell growth factor. *Journal of Immunology*, **178**, 3511–3520.

D'ippolito, G., Diabira, S., Howard, G.A. *et al.* (2004) Marrow-isolated adult multilineage inducible (MIAMI) cells, a unique population of postnatal young and old human cells with extensive expansion and differentiation potential. *Journal of Cell Science*, **117**, 2971–2981.

Di Raimondo, F. (2003) Angiogenesis in hematology: a field of active research. *Leukocyte Research*, **27**, 571–573.

Di Raimondo, F., Palumbo, G.A., Azzaro, M.P., and Giustolisi, R. (2000) Angiogenesis in acute myeloid leukemia. *Blood*, **96**, 3656–3657.

Eaves, A.C., Cashman, J.D., Gaboury, L.A. *et al.* (1986) Unregulated proliferation of primitive chronic myeloid leukemia progenitors in the presence of normal marrow adherent cells. *Proceedings of the National Academy of Sciences of the United States of America*, **83**, 5306–5310.

Ellis, S.J. and Tanentzapf, G. (2009) Integrin-mediated adhesion and stem-cell-niche interactions. *Cell Tissue Research*, **339**, 121–130.

Ferrara, N. (2004) Vascular endothelial growth factor: basic science and clinical progress. *Endocrine Reviews*, **25**, 581–611.

Fibbe, W.E. and Willemze, R. (1991) The role of interleukin-1 in hematopoiesis. *Acta Haematology*, **86**, 148–154.

Fibbe, W.E., Van Damme, J., Billiau, A. *et al.* (1988a) Human fibroblasts produce granulocyte-CSF, macrophage-CSF, and granulocytemacrophage-CSF following stimulation by interleukin-1 and poly(rI).poly(rC). *Blood*, **72**, 860–866.

Fibbe, W.E., Van Damme, J., Billiau, A. *et al.* (1988b) Interleukin 1 induces human marrow stromal cells in long-term culture to produce granulocyte colony-stimulating factor and macrophage colony-stimulating factor. *Blood*, **71**, 430–435.

Funk, P.E., Stephan, R.P., and Witte, P.L. (1995) Vascular cell adhesion molecule 1-positive reticular cells express interleukin-7 and stem cell factor in the bone marrow. *Blood*, **86**, 2661–2671.

Gallagher, J.T., Spooncer, E., and Dexter, T.M. (1983) Role of the cellular matrix in haemopoiesis. I. Synthesis of glycosaminoglycans by mouse bone marrow cell cultures. *Journal of Cell Science*, **63**, 155–171.

Gaynor, R.B. (2003) A role for extracellular matrix binding receptors in regulating hematopoietic growth factor signaling. *Proceedings of the National Academy of Sciences of the United States of America*, **100**, 13737–13738.

Geijtenbeek, T.B., Van Kooyk, Y., Van Vliet, S.J. *et al.* (1999) High frequency of adhesion defects in B-lineage acute lymphoblastic leukemia. *Blood*, **94**, 754–764.

Genever, P.G., Wilkinson, D.J., Patton, A.J. *et al.* (1999) Expression of a functional N-methyl-D-aspartate-type glutamate receptor by bone marrow megakaryocytes. *Blood*, **93**, 2876–2883.

Giannopoulos, K., Mertens, D., Buhler, A. *et al.* (2009) The candidate immunotherapeutical target, the receptor for hyaluronic acid-mediated motility, is associated with proliferation and shows prognostic value in B-cell chronic lymphocytic leukemia. *Leukemia*, **23**, 519–527.

Giles, F.J., Stopeck, A.T., Silverman, L.R. *et al.* (2003) SU5416, a small molecule tyrosine kinase receptor inhibitor, has biologic activity in patients with refractory acute myeloid leukemia or myelodysplastic syndromes. *Blood*, **102**, 795–801.

Gimble, J.M., Robinson, C.E., Wu, X., and Kelly, K.A. (1996) The function of adipocytes in the bone marrow stroma: an update. *Bone*, **19**, 421–428.

Gits, J., Van Leeuwen, D., Carroll, H.P. *et al.* (2006) Multiple pathways contribute to the hyperproliferative responses from truncated granulocyte colony-stimulating factor receptors. *Leukemia*, **20**, 2111–2118.

Gordon, M.Y. (1988) Extracellular matrix of the marrow microenvironment. *British Journal of Haematology*, **70**, 1–4.

Gordon, S. (2007) The macrophage: past, present and future. *European Journal of Immunology*, **37** (Suppl. 1), S9–17.

Gordon, M.Y., Riley, G.P., Watt, S.M. and Greaves, M.F. (1987) Compartmentalization of a haematopoietic growth factor (GM-CSF) by glycosaminoglycans in the bone marrow microenvironment. *Nature*, **326** (6111), 403–405.

Gu, Y.C., Kortesmaa, J., Tryggvason, K. *et al.* (2003) Laminin isoform-specific promotion of adhesion and migration of human bone marrow progenitor cells. *Blood*, **101**, 877–885.

Guba, S.C., Sartor, C.I., Gottschalk, L.R. *et al.* (1992) Bone marrow stromal fibroblasts secrete interleukin-6 and granulocytemacrophage colony-stimulating factor in the absence of inflammatory stimulation: demonstration by serum-free bioassay, enzyme-linked immunosorbent assay, and reverse transcriptase polymerase chain reaction. *Blood*, **80**, 1190–1198.

Guerriero, A., Worford, L., Holland, H.K. *et al.* (1997) Thrombopoietin is synthesized by bone marrow stromal cells. *Blood*, **90**, 3444–3455.

Gunsilius, E., Duba, H.C., Petzer, A.L. *et al.* (2000) Evidence from a leukaemia model for maintenance of vascular endothelium by bone-marrow-derived endothelial cells. *Lancet*, **355**, 1688–1691.

Gupta, P., Oegema, T.R. Jr., Brazil, J.J., *et al.* (1998) Structurally specific heparan sulfates support primitive human hematopoiesis by formation of a multimolecular stem cell niche. *Blood*, **92**, 4641–4651.

Gupta, D., Treon, S.P., Shima, Y. *et al.* (2001) Adherence of multiple myeloma cells to bone marrow stromal cells upregulates vascular endothelial growth factor secretion: therapeutic applications. *Leukemia*, **15**, 1950–1961.

Hamilton, J.A. and Anderson, G.P. (2004) GM-CSF Biology. *Growth Factors*, **22**, 225–231.

Hazlehurst, L.A., Argilagos, R.F., and Dalton, W.S. (2007) Beta1 integrin mediated adhesion increases Bim protein degradation and contributes to drug resistance in leukaemia cells. *British Journal of Haematology*, **136**, 269–275.

Heinrich, M.C., Blanke, C.D., Druker, B.J., and Corless, C.L. (2002) Inhibition of KIT tyrosine kinase activity: a novel molecular approach to the treatment of KIT-positive malignancies. *Journal of Clinical Oncology*, **20**, 1692–1703.

Heinrich, P.C., Behrmann, I., Haan, S. *et al.* (2003) Principles of interleukin (IL)-6-type cytokine signalling and its regulation. *Biochemical Journal*, **374**, 1–20.

Heissig, B., Hattori, K., Dias, S. *et al.* (2002) Recruitment of stem and progenitor cells from the bone marrow niche requires MMP-9 mediated release of kit-ligand. *Cell*, **109**, 625–637.

Hotamisligil, G.S., Shargill, N.S., and Spiegelman, B.M. (1993) Adipose expression of tumor necrosis factor-alpha: direct role in obesity-linked insulin resistance. *Science*, **259**, 87–91.

House, M.G., Kohlmeier, L., Chattopadhyay, N. *et al.* (1997) Expression of an extracellular calcium-sensing receptor in human and mouse bone marrow cells. *Journal of Bone and Mineral Research*, **12**, 1959–1970.

Hu, Z., Li, X.M., Jorgensen, M.L., and Slayton, W.B. (2009) MLL/AF-4 leukemic cells recruit new blood vessels but do not incorporate into capillaries in culture or in a NOD/SCID xenograft model. *Leukemia*, **23**, 990–993.

Huising, M.O., Stet, R.J., Savelkoul, H.F., and Verburg-Van Kemenade, B.M. (2004) The molecular evolution of the interleukin-1 family of cytokines; IL-18 in teleost fish. *Developmental and Comparative Immunology*, **28**, 395–413.

Hurtado, S.P., Balduino, A., Bodi, E.C. *et al.* (2004) Connexin expression and gap-junction-mediated cell interactions in an in vitro model of haemopoietic stroma. *Cell Tissue Research*, **316**, 65–76.

Hynes, R.O., Schwarzbauer, J.E., and Tamkun, J.W. (1987) Isolation and analysis of cDNA and genomic clones of fibronectin and its receptor. *Methods in Enzymology*, **144**, 447–463.

Jin, L., Tabe, Y., Konoplev, S. *et al.* (2008) CXCR4 up-regulation by imatinib induces chronic myelogenous leukemia (CML) cell migration to bone marrow stroma and promotes survival of quiescent CML cells. *Molecular Cancer Therapy*, **7**, 48–58.

Kacena, M.A. and Horowitz, M.C. (2006) The role of megakaryocytes in skeletal homeostasis and rheumatoid arthritis. *Current Opinion in Rheumatology*, **18**, 405–410.

Kacena, M.A., Gundberg, C.M., and Horowitz, M.C. (2006) A reciprocal regulatory interaction between megakaryocytes, bone cells, and hematopoietic stem cells. *Bone*, **39**, 978–984.

Kaplan, R.N. and Bussel, J.B. (2004) Differential diagnosis and management of thrombocytopenia in childhood. *Pediatric Clinics North America*, **51**, 1109–1140, xi.

Kartsogiannis, V., Zhou, H., Horwood, N.J. *et al.* (1999) Localization of RANKL (receptor activator of NF kappa B ligand) mRNA and protein in skeletal and extraskeletal tissues. *Bone*, **25**, 525–534.

Kashiwakura, I. and Takahashi, T.A. (2005) Fibroblast growth factor and ex vivo expansion of hematopoietic progenitor cells. *Leukemia Lymphoma*, **46**, 329–333.

Kaushansky, K., Lin, N., and Adamson, J.W. (1988) Interleukin 1 stimulates fibroblasts to synthesize granulocyte-macrophage and granulocyte colony-stimulating factors. Mechanism for the hematopoietic response to inflammation. *Journal of Clinical Investigation*, **81**, 92–97.

Kelm, R.J., Hair, G.A., Mann, K.G., and Grant, B.W. Jr. (1992) Characterization of human osteoblast and megakaryocyte-derived osteonectin (SPARC). *Blood*, **80**, 3112–3119.

Khazen, W., M'bika, J.P., Tomkiewicz, C. *et al.* (2005) Expression of macrophage-selective markers in human and rodent adipocytes. *FEBS Letters*, **579**, 5631–5634.

Kiel, M.J., Yilmaz, O.H., Iwashita, T. *et al.* (2005) SLAM family receptors distinguish hematopoietic stem and progenitor cells and reveal endothelial niches for stem cells. *Cell*, **121**, 1109–1121.

Kilpivaara, O. and Levine, R.L. (2008) JAK2 and MPL mutations in myeloproliferative neoplasms: discovery and science. *Leukemia*, **22**, 1813–1817.

Klein, G. (1995) The extracellular matrix of the hematopoietic microenvironment. *Experientia*, **51**, 914–926.

Kreis, T. and Vale, R. (1993) *Guidebook to the Extracellular Matrix and Adhesion Proteins*, Oxford University Press., New York

Kreis, T. and Vale, R. (1999) *Extracellular Matrix, Anchor, and Adhesion Proteins*, Oxford University Press, Inc., New York.

Krenacs, T. and Rosendaal, M. (1998) Connexin43 gap junctions in normal, regenerating, and cultured mouse bone marrow and in human leukemias: their possible involvement in blood formation. *American Journal of Pathology*, **152**, 993–1004.

Kuznetsov, S.A., Mankani, M.H., Gronthos, S. *et al.* (2001) Circulating skeletal stem cells. *Journal of Cell Biology*, **153**, 1133–1140.

Kvasnicka, H.M., Thiele, J., Staib, P. *et al.* (2004) Reversal of bone marrow angiogenesis in chronic myeloid leukemia following imatinib mesylate (STI571) therapy. *Blood*, **103**, 3549–3551.

Lamagna, C., Aurrand-Lions, M., and Imhof, B.A. (2006) Dual role of macrophages in tumor growth and angiogenesis. *Journal of Leukocyte Biology*, **80**, 705–713.

Lapidot, T. and Kollet, O. (2002) The essential roles of the chemokine SDF-1 and its receptor CXCR4 in human stem cell homing and repopulation of transplanted immune-deficient NOD/SCID and NOD/SCID/B2m(null) mice. *Leukemia*, **16**, 1992–2003.

Laterveer, L., Lindley, I.J., Hamilton, M.S. *et al.* (1995) Interleukin-8 induces rapid mobilization of hematopoietic stem cells with radioprotective capacity and long-term myelolymphoid repopulating ability. *Blood*, **85**, 2269–2275.

Leng, S.X. and Elias, J.A. (1997) Interleukin-11. *International Journal of Biochemistry and Cell Biology*, **29**, 1059–1062.

Levesque, J.P., Takamatsu, Y., Nilsson, S.K. *et al.* (2001) Vascular cell adhesion molecule-1 (CD106) is cleaved by neutrophil proteases in the bone marrow following hematopoietic progenitor cell mobilization by granulocyte colony-stimulating factor. *Blood*, **98**, 1289–1297.

Li, L. and Xie, T. (2005) Stem cell niche: structure and function. *Annual Review of Cell and Developmental Biology*, **21**, 605–631.

Li, X.M., Hu, Z., Jorgenson, M.L. *et al.* (2008) Bone marrow sinusoidal endothelial cells undergo nonapoptotic cell death and are replaced by proliferating sinusoidal cells in situ to maintain the vascular niche following lethal irradiation. *Experimental Hematology*, **36**, 1143–1156.

Li, Z.W. and Dalton, W.S. (2006) Tumor microenvironment and drug resistance in hematologic malignancies. *Blood Reviews*, **20**, 333–342.

Lin, H., Monaco, G., Sun, T. *et al.* (2005) Bcr-Abl-mediated suppression of normal hematopoiesis in leukemia. *Oncogene*, **24**, 3246–3256.

Link, D.C. (2000) Mechanisms of granulocyte colony-stimulating factor-induced hematopoietic progenitor-cell mobilization. *Seminars in Hematology*, **37**, 25–32.

Litwin, C., Leong, K.G., Zapf, R. *et al.* (2002) Role of the microenvironment in promoting angiogenesis in acute myeloid leukemia. *American Journal of Hematology*, **70**, 22–30.

Macdonald, D., Reiter, A., and Cross, N.C. (2002) The 8p11 myeloproliferative syndrome: a distinct clinical entity caused by constitutive activation of FGFR1. *Acta Haematology*, **107**, 101–107.

Mackie, E.J. (2003) Osteoblasts: novel roles in orchestration of skeletal architecture. *International Journal of Biochemistry and Cell Biology*, **35**, 1301–1305.

Maestroni, G.J. (2006) Sympathetic nervous system influence on the innate immune response. *Annals of the New York Academy of Sciences*, **1069**, 195–207.

Maestroni, G.J. and Conti, A. (1994) Modulation of hematopoiesis via alpha 1-adrenergic receptors on bone marrow cells. *Experimental Hematology*, **22**, 313–320.

Maestroni, G.J., Cosentino, M., Marino, F. *et al.* (1998) Neural and endogenous catecholamines in the bone marrow. Circadian association of norepinephrine with hematopoiesis? *Experimental Hematology*, **26**, 1172–1177.

Maggi, C.A. and Giuliani, S. (1996) Evidence that tachykinin NK2 receptors modulate resting tone in the rat isolated small intestine. *British Journal of Pharmacology*, **118**, 1262–1268.

Makgoba, M.W., Sanders, M.E., Ginther Luce, G.E. *et al.* (1988) ICAM-1 a ligand for LFA-1-dependent adhesion of B, T and myeloid cells. *Nature*, **331**, 86–88.

Maloof, P.B., Joshi, D.D., Qian, J. *et al.* (2001) Induction of preprotachykinin-I and neurokinin-1 by adrenocorticotropin and prolactin. Implication for neuroendocrineimmunehematopoietic axis. *Journal of Neuroimmunology*, **112**, 188–196.

Matsunaga, T., Takemoto, N., Sato, T. *et al.* (2003) Interaction between leukemic-cell VLA-4 and stromal fibronectin is a decisive factor for minimal residual disease of acute myelogenous leukemia. *Nature Medicine*, **9**, 1158–1165.

Metcalf, D. (1985) The granulocytemacrophage colony-stimulating factors. *Science*, **229**, 16–22.

Miharada, K., Hiroyama, T., Sudo, K. *et al.* (2008) Lipocalin 2-mediated growth suppression is evident in human erythroid and monocyte/macrophage lineage cells. *Journal of Cellular Physiology*, **215**, 526–537.

Montecino-Rodriguez, E. and Dorshkind, K. (2001) Regulation of hematopoiesis by gap junction-mediated intercellular communication. *Journal of Leukocyte Biology*, **70**, 341–347.

Montuori, N., Selleri, C., Risitano, A.M. *et al.* (1999) Expression of the 67-kDa laminin receptor in acute myeloid leukemia cells mediates adhesion to laminin and is frequently associated with monocytic differentiation. *Clinical Cancer Research*, **5**, 1465–1472.

Moore, K.A. and Lemischka, I.R. (2006) Stem cells and their niches. *Science*, **311**, 1880–1885.

Muruganandan, S., Roman, A.A., and Sinal, C.J. (2009) Adipocyte differentiation of bone marrow-derived mesenchymal stem cells: cross talk with the osteoblastogenic program. *Cellular and Molecular Life Science*, **66**, 236–253.

Nagasawa, T. (2007) The chemokine CXCL12 and regulation of HSC and B lymphocyte development in the bone marrow niche. *Advances in Experimental Medical Biology*, **602**, 69–75.

Naveiras, O., Nardi, N.V., Wenzel, P.L. *et al.* (2009) Bone-marrow adipocytes as negative regulators of the haematopoietic microenvironment. *Nature*, **460**, 259–263 .

Nishikawa, M., Ozawa, K., Tojo, A. *et al.* (1993) Changes in hematopoiesis-supporting ability of C3H10T1/2 mouse embryo fibroblasts during differentiation. *Blood*, **81**, 1184–1192.

Nuttall, M.E. and Gimble, J.M. (2004) Controlling the balance between osteoblastogenesis and adipogenesis and the consequent therapeutic implications. *Current Opinion in Pharmacology*, **4**, 290–294.

Pang, L., Weiss, M.J., and Poncz, M. (2005) Megakaryocyte biology and related disorders. *Journal of Clinical Investigation*, **115**, 3332–3338.

Parker, A., Clarke, J.B., Busby, W.H., and Clemmons, D.R. Jr. (1996) Identification of the extracellular matrix binding sites for insulin-like growth factor-binding protein 5. *The Journal of Biological Chemistry*, **271**, 13523–13529.

Paul, S.R., Bennett, F., Calvetti, J.A. *et al.* (1990) Molecular cloning of a cDNA encoding interleukin 11, a stromal cell-derived lymphopoietic and hematopoietic cytokine. *Proceedings of the National Academy of Sciences of the United States of America*, **87**, 7512–7516.

Pittenger, M.F., Mackay, A.M., Beck, S.C. *et al.* (1999) Multilineage potential of adult human mesenchymal stem cells. *Science*, **284**, 143–147.

Pluznik, D.H. and Mergenhagen, S.E. (1986) Synergistic action of lipopolysaccharide and tumor-promoting phorbol esters: two-signal requirement for colony-stimulating factor production by murine bone marrow cells. *Experimental Hematology*, **14**, 1029–1036.

Porter, R.L. and Calvi, L.M. (2008) Communications between bone cells and hematopoietic stem cells. *Archives of Biochemistry and Biophysics*, **473**, 193–200.

Poulsen, L.W., Melsen, F., and Bendix, K. (1998) A histomorphometric study of haematological disorders with respect to marrow fibrosis and osteosclerosis. *Apmis*, **106**, 495–499.

Prockop, D.J. (1997) Marrow stromal cells as stem cells for nonhematopoietic tissues. *Science*, **276**, 71–74.

Prosper, F. and Verfaillie, C.M. (2001) Regulation of hematopoiesis through adhesion receptors. *Journal of Leukocyte Biology*, **69**, 307–316.

Pule, M.A., Gullmann, C., Dennis, D. *et al.* (2002) Increased angiogenesis in bone marrow of children with acute lymphoblastic leukaemia has no prognostic significance. *British Journal of Haematology*, **118**, 991–998.

Rafii, S., Mohle, R., Shapiro, F. *et al.* (1997) Regulation of hematopoiesis by microvascular endothelium. *Leukemia and Lymphoma*, **27**, 375–386.

Rameshwar, P., Poddar, A., and Gascon, P. (1997) Hematopoietic regulation mediated by interactions among the neurokinins and cytokines. *Leukemia and Lymphoma*, **28**, 1–10.

Rameshwar, P., Poddar, A., Zhu, G., and Gascon, P. (1997) Receptor induction regulates the synergistic effects of substance P with IL-1 and platelet-derived growth factor on the proliferation of bone marrow fibroblasts. *Journal of Immunology*, **158**, 3417–3424.

Rameshwar, P., Joshi, D.D., Yadav, P. *et al.* (2001) Mimicry between neurokinin-1 and fibronectin may explain the transport and stability of increased substance P immunoreactivity in patients with bone marrow fibrosis. *Blood*, **97**, 3025–3031.

Rameshwar, P., Gascon, P., Bandari, P.S. *et al.* (2002) Structural similarity between the bone marrow extracellular matrix protein and neurokinin 1 could be the limiting factor in the hematopoietic effects of substance P. *Canadian Journal of Physiology and Pharmacology*, **80**, 475–481.

Reddy, S.V. and Roodman, G.D. (1998) Control of osteoclast differentiation. *Critical Reviews in Eukaryotic Gene Expression*, **8**, 1–17.

Reyes, M., Lund, T., Lenvik, T. *et al.* (2001) Purification and ex vivo expansion of postnatal human marrow mesodermal progenitor cells. *Blood*, **98**, 2615–2625.

Ribatti, D., Nico, B., Crivellato, E., and Vacca, A. (2007) Macrophages and tumor angiogenesis. *Leukemia*, **21**, 2085–2089.

Ritchie, A., Gotoh, A., Gaddy, J. *et al.* (1997) Thrombopoietin upregulates the promoter conformation of p53 in a proliferation-independent manner coincident with a decreased expression of Bax: potential mechanisms for survival enhancing effects. *Blood*, **90**, 4394–4402.

Roodman, G.D. (1999) Cell biology of the osteoclast. *Experimental Hematology*, **27**, 1229–1241.

Rosendaal, M. (1995) Gap junctions in blood forming tissues. *Microscopic Research and Technique*, **31**, 396–407.

Sacchetti, B., Funari, A., Michienzi, S. *et al.* (2007) Self-renewing osteoprogenitors in bone marrow sinusoids can organize a hematopoietic microenvironment. *Cell*, **131**, 324–336.

Schick, P.K., Wojenski, C.M., He, X. *et al.* (1998) Integrins involved in the adhesion of megakaryocytes to fibronectin and fibrinogen. *Blood*, **92**, 2650–2656.

Schmitt, M., Schmitt, A., Rojewski, M.T. *et al.* (2008) RHAMM-R3 peptide vaccination in patients with acute myeloid leukemia, myelodysplastic syndrome, and multiple myeloma elicits immunologic and clinical responses. *Blood*, **111**, 1357–1365.

Sherr, C.J. (1990) Regulation of mononuclear phagocyte proliferation by colony-stimulating factor-1. *International Journal of Cell Cloning*, **8** (Suppl. 1), 46–60; discussion 60-42.

Shih, I.M. (1999) The role of CD146 (Mel-CAM) in biology and pathology. *Journal of Pathology*, **189**, 4–11.

Shiozawa, Y., Havens, A.M., Pienta, K.J., and Taichman, R.S. (2008) The bone marrow niche: habitat to hematopoietic and mesenchymal stem cells, and unwitting host to molecular parasites. *Leukemia*, **22**, 941–950.

Siler, U., Seiffert, M., Puch, S. *et al.* (2000) Characterization and functional analysis of laminin isoforms in human bone marrow. *Blood*, **96**, 4194–4203.

Sipe, J.B., Zhang, J., Waits, C. *et al.* (2004) Localization of bone morphogenetic proteins (BMPs)-2, -4, and -6 within megakaryocytes and platelets. *Bone*, **35**, 1316–1322.

Slayton, W.B., Li, X.M., Butler, J. *et al.* (2007) The role of the donor in the repair of the marrow vascular niche following hematopoietic stem cell transplant. *Stem Cells*, **25**, 2945–2955.

Soini, Y., Kamel, D., Apaja-Sarkkinen, M. *et al.* (1993) Tenascin immunoreactivity in normal and pathological bone marrow. *Journal of Clinical Pathology*, **46**, 218–221.

Sorrell, J.M. (1988) Ultrastructural localization of fibronectin in bone marrow of the embryonic chick and its relationship to granulopoiesis. *Cell Tissue Research*, **252**, 565–571.

Souza, L.M., Boone, T.C., Gabrilove, J. *et al.* (1986) Recombinant human granulocyte colony-stimulating factor: effects on normal and leukemic myeloid cells. *Science*, **232**, 61–65.

Spessotto, P., Zucchetto, A., Degan, M. *et al.* (2007) Laminin-332 (Laminin-5) is the major motility ligand for B cell chronic lymphocytic leukemia. *Matrix Biology*, **26**, 473–484.

Spooncer, E., Heyworth, C.M., Dunn, A., and Dexter, T.M. (1986) Self-renewal and differentiation of interleukin-3-dependent multipotent stem cells are modulated by stromal cells and serum factors. *Differentiation*, **31**, 111–118.

Steurer, M., Kern, J., Zitt, M. *et al.* (2008) Quantification of circulating endothelial and progenitor cells: comparison of quantitative PCR and four-channel flow cytometry. *BMC Research Notes*, **1**, 71.

Sullenbarger, B.A., Petitt, M.S., Chong, P. *et al.* (1995) Murine granulocytic cell adhesion to bone marrow hemonectin is mediated by mannose and galactose. *Blood*, **86**, 135–140.

Szczylik, C., Skorski, T., Ku, D.H. *et al.* (1993) Regulation of proliferation and cytokine expression of bone marrow fibroblasts: role of c-myb. *Journal of Experimental Medicine*, **178**, 997–1005.

Taichman, R.S. and Emerson, S.G. (1998) The role of osteoblasts in the hematopoietic microenvironment. *Stem Cells*, **16**, 7–15.

Tan, J.T., Dudl, E., Leroy, E. *et al.* (2001) IL-7 is critical for homeostatic proliferation and survival of naive T cells. *Proceedings of the National Academy of Sciences of the United States of America*, **98**, 8732–8737.

Tanaka, R., Owaki, T., Kamiya, S. *et al.* (2009) VLA-5-mediated adhesion to fibronectin accelerates hemin-stimulated erythroid differentiation of K562 cells through induction of VLA-4 expression. *The Journal of Biological Chemistry*, **284**, 19817–19825.

Taylor, P.R., Brown, G.D., Geldhof, A.B. *et al.* (2003) Pattern recognition receptors and differentiation antigens define murine myeloid cell heterogeneity ex vivo. *European Journal of Immunology*, **33**, 2090–2097.

Thiede, M.A., Smock, S.L., Petersen, D.N. *et al.* (1994) Presence of messenger ribonucleic acid encoding osteocalcin, a marker of bone turnover, in bone marrow megakaryocytes and peripheral blood platelets. *Endocrinology*, **135**, 929–937.

Thiele, J. and Kvasnicka, H.M. (2007) Myelofibrosis what's in a name? Consensus on definition and EUMNET grading. *Pathobiology*, **74**, 89–96.

Timmermans, F., Plum, J., Yoder, M.C. *et al.* (2009) Endothelial progenitor cells: identity defined?. *Journal of Cellular and Molecular Medicine*, **13**, 87–102.

Untergasser, G., Koeck, R., Wolf, D. *et al.* (2006) CD34+/CD133- circulating endothelial precursor cells (CEP): characterization, senescence and in vivo application. *Experimental Gerontology*, **41**, 600–608.

Veiga, J.P., Costa, L.F., Sallan, S.E. *et al.* (2006) Leukemia-stimulated bone marrow endothelium promotes leukemia cell survival. *Experimental Hematology*, **34**, 610–621.

Vivien, L., Benoist, C., and Mathis, D. (2001) T lymphocytes need IL-7 but not IL-4 or IL-6 to survive in vivo. *International Immunology*, **13**, 763–768.

Volpi, N., Petrini, M., Conte, A. *et al.* (1994) Effects of glycosaminoglycans on U-937 leukemia cell proliferation and differentiation: structurefunction relationship. *Experimental Cell Research*, **215**, 119–130.

Von Freeden-Jeffry, U., Vieira, P., Lucian, L.A. *et al.* (1995) Lymphopenia in interleukin (IL)-7 gene-deleted mice identifies IL-7 as a nonredundant cytokine. *Journal of Experimental Medicine*, **181**, 1519–1526.

Wakikawa, T., Shioi, A., Hino, M. *et al.* (1997) Thrombopoietin inhibits in vitro osteoclastogenesis from murine bone marrow cells. *Endocrinology*, **138**, 4160–4166.

Wang, M., Crisostomo, P.R., Herring, C. *et al.* (2006) Human progenitor cells from bone marrow or adipose tissue produce VEGF, HGF, and IGF-I in response to TNF by a p38 MAPK-dependent mechanism. *American Journal of Physiology: Regulatory, Integrative and Comparative Physiology*, **291**, R880–R884.

Wang, M., Zhang, W., Crisostomo, P. *et al.* (2007) STAT3 mediates bone marrow mesenchymal stem cell VEGF production. *Journal of Molecular and Cellular Cardiology*, **42**, 1009–1015.

Weber, J.M. and Calvi, L.M. (2009) Notch signaling and the bone marrow hematopoietic stem cell niche. *Bone,* **46**, 281–285.

White, H., Totty, N., and Panayotou, G. (1993) Haemonectin, a granulocytic-cell-binding protein, is related to the plasma glycoprotein fetuin. *European Journal of Biochemistry*, **213**, 523–528.

Wight, T.N., Kinsella, M.G., Keating, A., and Singer, J.W. (1986) Proteoglycans in human long-term bone marrow cultures: biochemical and ultrastructural analyses. *Blood*, **67**, 1333–1343.

Wodnar-Filipowicz, A. (2003) Flt3 ligand: role in control of hematopoietic and immune functions of the bone marrow. *News in Physiological Sciences*, **18**, 247–251.

Yan, Q.W., Yang, Q., Mody, N. *et al.* (2007) The adipokine lipocalin 2 is regulated by obesity and promotes insulin resistance. *Diabetes*, **56**, 2533–2540.

Yao, R., Rameshwar, P., Donnelly, R.J., and Siegel, A. (1999) Neurokinin-1 expression and co-localization with glutamate and GABA in the hypothalamus of the cat. *Brain Research: Molecular Brain Research*, **71**, 149–158.

Yokota, T., Oritani, K., Takahashi, I. *et al.* (2000) Adiponectin, a new member of the family of soluble defense collagens, negatively regulates the growth of myelomonocytic progenitors and the functions of macrophages. *Blood*, **96**, 1723–1732.

Yong, K.T., Ding, H., Roy, I. *et al.* (2009) Imaging pancreatic cancer using bioconjugated InP quantum dots. *ACS Nano*, **3**, 502–510.

Zhang, Y., Harada, A., Bluethmann, H. *et al.* (1995) Tumor necrosis factor (TNF) is a physiologic regulator of hematopoietic progenitor cells: increase of early hematopoietic progenitor cells in TNF receptor p55-deficient mice in vivo and potent inhibition of progenitor cell proliferation by TNF alpha in vitro. *Blood*, **86**, 2930–2937.

Zheng, G., Rao, Q., Wu, K. *et al.* (2000) Membrane-bound macrophage colony-stimulating factor and its receptor play adhesion molecule-like roles in leukemic cells. *Leukemia Research*, **24**, 375–383.

7

Microenvironment Factors Influencing Skeletal Metastases

Alessandro Fatatis[1], Julia A. D'Ambrosio[2], Whitney L. Jamieson[2], Danielle L. Jernigan[2] and Mike R. Russell[2]

[1] *Department of Pathology and Laboratory Medicine, Drexel University College of Medicine, Philadelphia, PA, USA*

[2] *Department of Pharmacology and Physiology, Drexel University College of Medicine, Philadelphia, PA, USA*

7.1 Introduction

> Cancer is no more than a disease of cells than a traffic jam is a disease of cars. A lifetime study of the internal-combustion engine would not help anyone understand our traffic problems.
>
> D.W. Smithers, *The Lancet* (1962).

> What did my cells do? They invented a different Plan, and now they are proceeding on their own ... They invert, transpose, alternate, transform themselves into cells unheard of, new cells without meaning, or with meaning contrary to the right meaning. There must be a right meaning and a wrong meaning; otherwise you die. My cells joke without faith, blindly.
>
> Umberto Eco, *Foucault's Pendulum* (1990).

A hypothetical dispute on the cause of cancer between a radiation oncologist and a man of letters – even if extremely eloquent – would at first appear based on unfair ground. However, until recently the idea expressed by the character in Umberto Eco's book was the one vastly favored among cancer researchers. In fact, for quite some time the dependency of cancer on predisposing conditions

Tumor Microenvironment Edited by Dietmar W. Siemann

offered by the tissue in which the malignant phenotypes first arise has been harshly criticized or even ignored. Today the scientific contention on which is most crucial for the clinical manifestation of a neoplasia – genetic alterations of individual cells or the supportive role exerted by the normal tissues surrounding them – has mostly evaporated. It is widely recognized that malignant cells – almost invariably generated by causal genetic insults – very often behave differently depending on their specific microenvironment. Indeed, a large body of evidence shows that cancer cells need the cooperation of neighboring normal cells in order to fully execute their lethal programs. If true for local growth and invasion at the site of the primitive tumor, this concept is equally critical for the distant dissemination of malignant cells and their metastatic progression into the parenchyma of secondary organs.

The occurrence of metastases represents the most feared complication of a solid neoplasia. From a clinical standpoint, we recognize several tumors for which metastases represent the worsening of a prognosis that was already dire even in their absence. Pancreatic and lung tumors as well as glioblastomas can indeed be included in this group, which is characterized by infrequent early detection, limited choices of first-line therapy and substantial failure of the affected primitive organ. On the other hand, tumors of the prostate and mammary glands are typically detected at a localized stage, successfully treated by surgery, irradiation, or combined approaches and produce manageable organ compromission. Still, an unacceptably high number of patients affected by otherwise moderately aggressive tumors will inevitably present with metastatic disease and eventually succumb to it.

Cancer metastasizes most commonly to bone, liver, and brain. The American Cancer Society estimates that in 2009, 562 340 people will die of cancer and almost all will have metastasis to some part of the body (American Cancer Society, 2009). Tumors with propensity for skeletal metastases include breast, prostate, and lung (Ashford, 2006) and it is anticipated that 350 000 of those diagnosed with cancer each year will die with bone metastasis (American Cancer Society, 2009; Mundy, 2002). The survival rate after diagnosis of skeletal metastasis is 3–5 years and the treatments currently available are almost exclusively palliative.

7.2 The bone microenvironment as a target for cancer cell dissemination

The skeleton is responsible for two major physiological functions, support for the musculature and storage of minerals. It is also the site of continuous metabolic activity, including maturation of hematopoietic precursors. The cortex or compact bone sustains the weight of the body and encloses cancellous bone, a porous, metabolically active tissue (Bussard, Gay, and Mastro, 2008). The trabeculae of cancellous bone are arranged in an organized network among which lies the marrow, composed of collagen, fatty tissue, and cells responsible for erythropoiesis and granulopoiesis among other blood-forming activities. To properly function, the bone undergoes remodeling and repair throughout the lifespan of the individual. Notably, these processes are modulated by a plethora of trophic molecules to which cancer cells might also be responsive. In fact, an important role for bone turnover in the preferential location of prostate cancer cells to the skeleton has been recently reported (Schneider *et al.*, 2005).

Long bones are anatomically divided into three regions: diaphysis, metaphysis, and epiphysis. The epiphysis is found at the end of long bones atop the growth plate whereas the diaphysis is the long, narrow, primarily cortical bone section. The metaphysis is located just below growth plate, near the end of long bones, and is made of primarily trabecular bone (Bussard, Gay, and Mastro, 2008). The metaphyseal region of long bones is a frequent site of metastases and highly vascularized. The supply of arterial blood to long bones is provided by a main nutrient artery, which subdivides into a fine network of progressively smaller arteries, which eventually empty into the marrow sinusoids. The blood flow in the sinusoids is rather slow in comparison with other tissues. Sinusoidal lumens are much larger than the average cell, thus providing an easy entry point for circulating cancer cells into the metaphyseal marrow (Bussard, Gay, and Mastro, 2008). This anatomical feature also argues against mechanical entrapment being solely responsible for a stochastic arrest of circulating cancer cells at the skeletal level (Glinskii *et al.*, 2005). In fact, specific adhesive interactions between circulating cancer cells and the marrow endothelial wall are believed to contribute to the selectivity that some tumors demonstrate in preferential dissemination to the skeleton. Notably, the endothelium of bone marrow sinusoids is unique in that it maintains constitutive expression of adhesion molecules on its luminal side. In contrast, virtually any other tissue in the body requires inflammatory conditions for the expression of the same adhesion molecules in the vascular lumen (Bussard, Gay, and Mastro, 2008).

7.3 Roles of the bone microenvironment in promoting the arrest of circulating cancer cells at the skeleton

The extreme propensity of certain tumors to colonize specific tissues in the body is defined as organ-tropism (Weiss, 1985). Cells that depart from primitive tumors use the hematogenous and lymphatic circulatory systems as two major avenues for dissemination. It is therefore plausible that, in addition to vascular patterns, the unique characteristics of certain endothelia could lead to the preferential arrival of cancer cells to selected organs (Weiss, 1985; Orr *et al.*, 2000b). There is evidence that circulating cancer cells express adhesion macromolecules that are compatible with those present on the surface of the endothelial cells lining the bone marrow sinusoids and use these interactions to end their journey at the skeletal level (Glinsky, 2007).

However, simply adhering to the marrow's endothelial wall will not suffice to invade the newly found tissue. Trafficking of cells in the bone marrow has been well established for hematopoietic progenitor cells after either clinical marrow transplantation or during fetal life, an event defined as 'homing.' Lymphocytes and leukocytes also respond to a large spectrum of chemoattractant molecules that regulate their extravasation into lymphoid organs and inflamed tissues, respectively (Butcher and Picker, 1996; Middleton *et al.*, 2002). Similarly to immune and hematopoietic stem cells, cancer cells can migrate from the luminal side of the sinusoids into the surrounding bone in response to molecules released by resident stromal cells. This event is ultimately responsible for the initial lodging of disseminated cancer cells and the formation of microscopic malignant foci in the skeleton.

7.3.1 Adhesion

After cancer cells have escaped from a primary tumor and gained access to blood circulation, one possible outcome is their arrest at a new organ site. However, even malignant cells could still be anchorage-dependent (a common feature of normal cells) and fail to survive when detached from the extracellular matrix (ECM), an event known as anoikis (Gilmore, 2005). Thus, establishing adhesive interactions with the endothelial wall is a very efficient way for cancer cells to avoid death. The arrest of cancer cells at the skeleton is most likely the result of specific contacts with endothelial cells of the marrow sinusoids. *In vitro* data have shown that both breast and prostate cancer cells adhere preferentially to human bone marrow cell lines when compared with endothelia from other tissue (Lehr and Pienta, 1998; Scott *et al.*, 2001). The events involved in the adhesion of cancer cells to the bone endothelium have been assimilated to those utilized by leukocytes (Kucia *et al.*, 2005), in which cells adhere to the vascular wall by undergoing a multistep process that includes rolling, adhesion, and then capture (Springer, 1994). This is not a simple event; a concerted effort from different types of adhesion molecules (described below) is required to develop a stable, firm arrest on the endothelial wall (Springer, 1995). Receptors capable of specifically binding these molecules have also been identified also on the surface of cells from different tumors. This implies that mechanisms in place to allow the trafficking of leukocyte and hematopoietic stem cells can be effectively usurped by circulating cancer cells during the initial phase of secondary organ colonization.

7.3.1.1 Selectins

The selectin family consists of three adhesion receptors, E-, L-, and P-selectins. Both E and P-selectins are expressed in response to inflammatory stimuli, whereas L-selectin is constitutively expressed in lymphocytes (Gout, Tremblay, and Huot, 2008). Although E-selectin can be induced in response to inflammatory stimuli, E-selectin is expressed constitutively in the vascular endothelia of skin and bones. E-selectin binds several different glycoprotein ligands, including P-selectin glycoprotein ligand-1 (PSGL-1), β2 integrins, hematopoietic cell E- and L-selectin ligand (HCELL), glycolipids, and the death receptor-3 (DR3).

Several lines of evidence point toward a role for selectins in cancer cell adhesion during metastasis and show a correlation between metastatic ability and PSGL-1 and HCELL expression (Barthel *et al.*, 2007). Selectin-mediated adhesion is thought to be an initial event, involved in the "rolling" step of cell adhesion but insufficient to strongly arrest leukocytes – as well as cancer cells – on the endothelium. Therefore, a successive adhesion event must occur to achieve firm capture, which may be mediated by the triggering of integrin activation (Gout, Tremblay, and Huot, 2008).

7.3.1.2 *Integrins*

There are 25 identified members of the integrin family, in which each member presents a dimeric structure assembled from a pool of 19 alpha subunits and 8 beta subunits. As transmembrane receptors for (ECM) proteins, integrins are involved in a number of different cellular processes and promote adhesion primarily through interaction with molecules such as intercellular adhesion molecule-1 (ICAM-1), vascular cell adhesion molecule-1 (VCAM-1) and platelet endothelial cell adhesion molecule-1 (PECAM-1), also known as CD31. The physiological role for integrins is to mediate the adhesion of leukocytes at sites of inflammation and the aggregation of platelets at sites of vascular injury. Although their role in metastasis is not as well defined as for selectins, several groups have shown that up-regulation of integrins – particularly $\alpha_V\beta_3$ – confers a metastatic advantage to breast, prostate, and colon cancer cells (Fornaro, Manes, and Languino, 2001). In addition, the use of $\alpha_V\beta_3$ mimetics in murine metastasis studies reduced breast cancer metastases to the bone (Konstantopoulos and Thomas, 2009). Integrins exert their role in the adhesion of circulating cells upon activation by intracellular signaling pathways recruited by cell surface receptors for soluble chemokines (Goel *et al.*, 2008).

7.3.1.3 *Chemokines*

Chemokines are a family of chemotactic cytokines involved in the trafficking of leukocytes to sites of inflammation (Balkwill, 2003). A role for small, soluble proteins in cell adhesion seems counterintuitive, yet chemokines have been shown to be involved in adhesion indirectly via successive activation of adhesion molecules. The most prominent example and widely studied mechanism involve the chemokine receptor CXCR4 and its ligand CXCL12. This chemokine/receptor pair has been shown to mediate the metastases of breast and prostate cancer cells, which were found to express CXCR4 on their surface. The binding of this receptor to its soluble chemokine ligand CXCL12 is able to upregulate several integrins, including $\alpha_V\beta_3$ (Müller *et al.*, 2001; Engl *et al.*, 2006; Mochizuki *et al.*, 2004; Lapteva *et al.*, 2004). Because CXCL12 is a soluble chemokine, its presentation to circulating cancer cells depends on its binding to cellular proteoglycans located on the surface of endothelial cells, which after producing the chemokine need to display it on their luminal surface (Imai *et al.*, 1999). This mechanism seems well established (Kuschert *et al.*, 1999; Netelenbos *et al.*, 2003), although the resistance to the shear stress of the blood flow offered by CXCL12 attached to the endothelium in this fashion indeed raises some concerns (Chapman *et al.*, 2000).

Interestingly, fractalkine (CX3CL1) and CXCL16 are unique chemokines that exist not only in a soluble form but also as transmembrane proteins. The latter, membrane-anchored form mediates the adhesion of circulating leukocytes at sites of inflammation *independently* of activation of additional adhesion molecules, unlike CXCL12/CXCR4 (Fong *et al.*, 1998; Umehara *et al.*, 2006). A role for fractalkine in cancer metastasis has recently been proposed, based on its constitutive expression by human bone marrow endothelial cells and the detection of its unique receptor

CX3CR1 on the surface of cancer cells from breast and prostate, among other tumors (Shulby *et al.*, 2004; Jamieson *et al.*, 2008; Nevo *et al.*, 2009). The rapid and firm adhesive interactions between fractalkine and CX3CR1 could provide selected cancer cells in transit through the marrow sinusoids with the advantage of arresting without the intervention of integrins. Thus, cancer cells expressing CX3CR1 may have an additional metastatic advantage over cells lacking this receptor on their surface.

7.3.2 Extravasation

Despite intravascular growth of cancer cells has been reported for lung metastases (Al-Mehdi *et al.*, 2000), the circulating cancer cells that succeed in adhering to the luminal side of bone marrow sinusoids will plausibly need to traverse the endothelial wall to invade the surrounding stroma. Similarly to adhesion, most of what is known about extravasation of cancer cells has been extrapolated from the diapedesis of leukocytes in inflamed tissues. Cells can cross the endothelial monolayer in two predominant ways: (i) migration between endothelial cells (paracellular); (ii) migration at non-junctional locations (transcellular) (Barthel *et al.*, 2007). In order for either of these events to occur, some rearrangement of the endothelial cell cytoskeleton is necessary (Miles *et al.*, 2008). A number of molecules responsible for these processes have been identified. Strikingly, both selectins and chemokines have been found to play a role also in extravasation. Rho-GTPases is a third group of molecules also involved in assisting the cytoskeletal rearrangement during migration (Miles *et al.*, 2008).

7.3.2.1 Selectins

Binding of E-selectin to counter-receptors can initiate intracellular signaling in both cancer cells and endothelial cells. Several studies point to the ability of E-selectin to create a more fluid endothelial layer. Most notably, the binding of E-selectin to DR3 has been shown to activate the promigratory mitogen-activated protein kinase (MAP kinase) (p38) pathway and the pro-survival extracellular signal-regulated kinase (ERK) pathway. Activation of both of these signaling pathways promotes the motility of both cancer cells and endothelial cells, as well as overall survival of the former in the secondary tissue (Gout, Tremblay, and Huot, 2008).

7.3.2.2 Chemokines

Chemokines promote the recruitment of leukocytes toward inflammatory sites and can act similarly to promote metastases by attracting disseminated cancer cells into a tissue after they have adhered to the endothelium. Although some soluble chemokines such as CXCL12 have been involved in mediating cancer cell adhesion, their involvement in the extravasation step might be more predominant (Gassmann *et al.*, 2009). CXCL12 is expressed in several tissues frequently targeted by secondary tumors (Müller *et al.*, 2001; Gassmann and Haier, 2008). This chemokine plays an important role in normal bone marrow physiology by regulating the homing of mesenchymal stem cells (MSCs). Based on this evidence, this chemokine has been studied extensively for its involvement in promoting cancer cell migration to the skeleton and was found strongly chemotactic for malignant prostate and breast cells, among others (Lee *et al.*, 2004; Sun *et al.*, 2003, 2005; Taichman *et al.*, 2002; Burger *et al.*, 2003). Another chemokine implicated in cell migration is fractalkine. Its membrane-anchored form, which effectively works as an adhesion molecule, can be also cleaved and shed from the cell surface. This soluble form chemoattracts CX3CR1-expressing cells and has been identified as a possible mediator of prostate cancer cells extravasation into the bone marrow (Shulby *et al.*, 2004; Jamieson *et al.*, 2008). In addition, recent studies have shown that neuroblastoma cells migrate across a monolayer of bone marrow endothelial cells toward a concentration gradient generated by this chemokine (Nevo *et al.*, 2009).

7.3.2.3 Rho GTPases

Rho GTPases, by switching from an active GTP-bound state to an inactive GTP-unbound state, regulate the stability of adherens junctions, the tight adhesive structures necessary for maintaining integrity of endothelial cell layers. During leukocyte transendothelial migration (TEM), endothelial adherens junctions are disrupted, thereby providing a physical space for leukocytes to cross the endothelium. For instance, the binding of leukocytes to endothelia causes integrin activation and consequent recruitment of RhoA. This event ultimately leads to rearrangement of actin fibers and provide the leukocytes with a port of entry into the inflamed tissues. Less is known about the role of Rho GTPases in cancer cell extravasation. However, it is known that RhoA and RhoC are indeed necessary for the TEM of breast cancer cells and that RhoA is activated in human brain microvascular endothelial (HBMEC) in response to cancer cell infiltration. Furthermore, it has been reported that RhoC inhibition decreased TEM of prostate cancer cell (Miles *et al.*, 2008).

7.3.3 The initial survival of isolated cancer cells and their progression into small malignant foci

During recent years, the survival and growth of cancer cells disseminated to the skeleton have been commonly investigated using animal models of metastatic disease. Bone lesions are induced by intravascular injection (Schneider *et al.*, 2005; Shimamura *et al.*, 2005; Gomes *et al.*, 2009) or direct intraosseous implantation of cancer cells (Corey *et al.*, 2002; Fisher *et al.*, 2002). The imaging of malignant lesions would then rely on bioluminescence and/or radiographic detection (van der Pluijm *et al.*, 2005). Although these approaches have led to significant progress and continue to provide necessary information, they are limited in the ability to study the very early stages of the metastatic process. For instance, bioluminescence or radiographic analyses detect bone lesions produced by large numbers of cancer cells, but are generally unable to identify single cancer cells or small foci (Wetterwald *et al.*, 2002; van der Pluijm *et al.*, 2005). Significant progress in generating enhanced bioluminescent probes that allow the imaging of only a few cells has been made, but only for subcutaneous detection (Rabinovich *et al.*, 2008). These technical limitations have significantly hampered our understanding of the initial stages of skeletal colonization and the preclinical evaluation of the efficacy of antimetastatic treatments.

However, a recently described approach is capable to successfully detect fluorescence-labeled cancer cells immediately after their arrest to the skeleton by employing bone cryosectioning and stereomicroscopy based imaging (Russell *et al.*, 2009).

This technique effectively circumvents the restrictions caused by studying macroscopic tumors and allows determining the histological features of different cells in the metastatic lesion and the spatial-relationships between isolated disseminated cancer cells and the surrounding bone microenvironment (Figure 7.1).

The systematic use of similar methods of investigation seems a promising strategy to identify malignant phenotypes with the highest propensity to immediately survive in the bone as well as the molecular factors and signaling events causally involved in the early stages of tissue colonization. By counteracting the transition of single cancer cells into relatively small malignant foci it could be possible to prevent not only the progression of recently extravasated cancer cells, but perhaps also avoid that cancer cells blocked in a dormant state resume their growth. It seems intuitive that preventing cancer cells to take control of the bone environment and

Figure 7.1 Arrival of cancer cells to the skeleton. Fluorescent human prostate cancer cells were inoculated in the left cardiac ventricle of immunocompromised mice and detected 24 hours later in the metaphysis of the tibia in close proximity of the growth plate. A full color version of this figure can be found in the color plate section.

thereby subverting the organ architecture to support the expansion of lesions, would guarantee superior chances to contain skeletal metastatic disease at a preclinical, indolent stage.

Significant evidence exists that cancer cells migrated into a foreign tissue need favorable conditions to survive and proliferate (Fidler, 2003; Cooper *et al.*, 2003). Paget assimilated the compatibility between migrating cancer cells and selected organs to the required affinity between seeds and a specific soil (Paget, 1989). It is now clear that a congenial bone microenvironment is paramount for the initial stages of secondary tissue colonization (Cooper *et al.*, 2003). Cancer cells failing to receive appropriate trophic support may either remain dormant or undergo cell death (Luzzi *et al.*, 1998; Aguirre-Ghiso, 2007; Weinberg, 2008) and cells disseminated to the skeleton make no exception (Alix-Panabières, Riethdorf, and Pantel, 2008). Indeed, support for the survival of isolated cancer cells can derive from the increased availability of growth factors produced by the degradation of the bone matrix. A self-sustaining mechanism involving cytokines secreted by cancer cells, the recruitment and activation of osteoclasts and consequent further release of survival factors by osteolytic events has been extensively described (Chung, 2003). Notably, the evidence accumulated at the cellular and molecular level for this 'vicious-cycle' (described in more depth below) currently dictates some of the treatment used in the clinical setting for advanced, bone-metastatic cancer patients (Loberg *et al.*, 2005).

For instance, when grown into a significant tumor mass, cancer cells can alter the surrounding marrow and compact bone in their favor, establishing plethoric interactions with bone-residing cells and ensuring their progression into a full-blown metastasis (Roodman, 2007). However, it seems reasonable that the production of secreted signals co-opting osteoclasts could be unattainable unless a critical tumor mass – and substantial secretion of cytokines – is achieved. This idea seems to be confirmed by a recent study showing that, in contrast to larger secondary bone tumors, smaller cancer foci failed to induce a significant recruitment of osteoclasts at the interface between cancer cells and the surrounding bone tissue (Russell *et al.*, 2009) (Figure 7.2). Thus, the initial survival and growth of cells disseminated to the skeleton appears to be independent of osteoclast-induced erosion of bone matrix and may rather rely on different mechanisms.

Immediately after extravasation cancer cells are few in number and particularly vulnerable to death (Karamanolakis *et al.*, 2002). A critical factor for the expansion of a tumor mass is the availability of blood supply. However, the extent of tissue vascularization and the need for blood vessels to provide oxygen and nutrients are crucial requirements for tumors larger than 1–2 mm and containing already thousands of cells (Naumov, Folkman, and Straume, 2009). Thus, at the very early stages of dissemination to the bone, cancer cells might not heavily depend on their angiogenic abilities but rather on pre-existing local factors, which are readily available as produced by the bone stroma to support autochthonous cells (Blair, Zaidi, and Schlesinger, 2002; Chaudhary, Hofmeister, and Hruska, 2004). In addition, cancer cells can also secrete molecules that either act in an autocrine fashion or induce bone cells to release growth factors and cytokines. These molecules will in turn stimulate the cancer cells in a paracrine fashion, thus promoting and fostering a

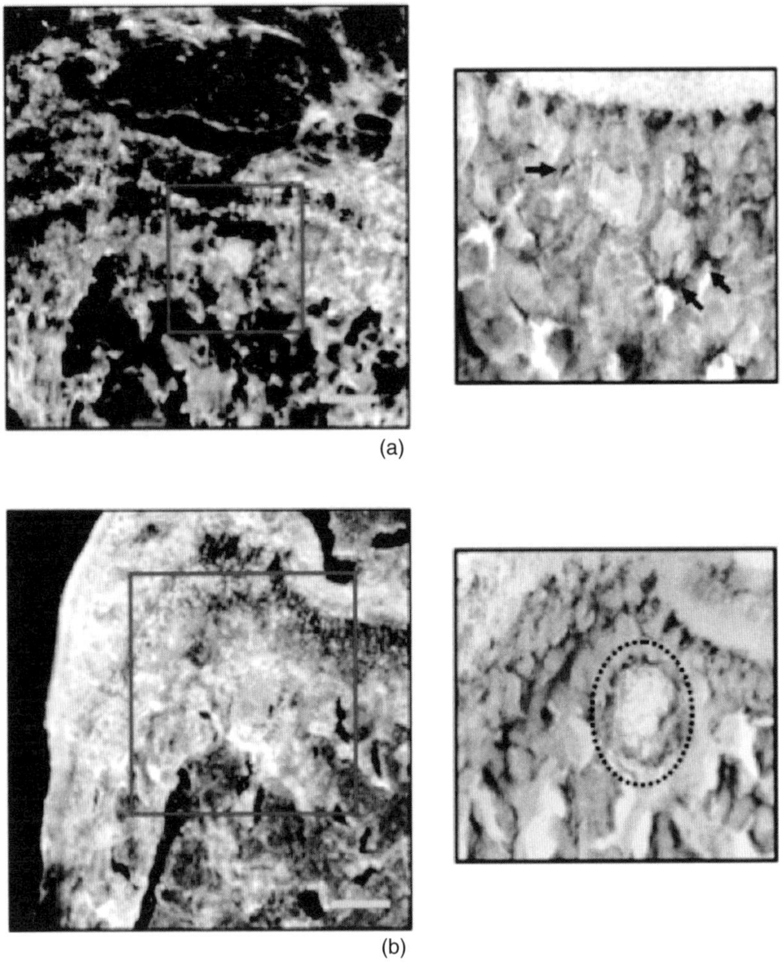

Figure 7.2 Spatial relationship of osteoclasts with bone metastatic tumor foci. Metastases with a cross-sectional area larger than $28 \times 10^3\,\mu m^2$ (b) were surrounded by a layer of active osteoclasts, identified by a red TRAcP staining. However, smaller metastases were spatially unrelated to osteoclasts (a), which appear sparsely distributed (arrows). Measurement bar is $100\,\mu m$. A full color version of this figure can be found in the color plate section.

give-and-take relationship within the initial bone lesion. The identification of these local factors and understanding their mechanism of action is therefore of paramount importance in the prevention and treatment of skeletal metastasis. Several promising candidates implicated in the establishment and progression of macroscopic skeletal metastases could be also contributing to the initial stages of bone colonization and are described in a number of excellent reviews (Deftos, 2000; Mundy, 2002; Yin, Pollock, and Kelly, 2005; Bonfil *et al.*, 2007; Kingsley *et al.*, 2007; Bussard, Gay, and Mastro, 2008; Buijs and van der Pluijm, 2009).

Here we discuss only some of these molecules and their receptors.

7.3.3.1 *Osteopntin*

Osteopontin (OPN) is a small phosphoprotein of the ECM that presents binding sites for CD44 (see below) and $\alpha_V\beta 3$ integrin and has been correlated with advanced stage and poor outcome of a wide variety of cancers (Wai and Kuo, 2008). Many clinical studies have demonstrated a high serum concentration of OPN in patients with bone metastatic disease, and for this reason OPN is now being considered as a biomarker with predictive value for skeletal dissemination. OPN has been shown to promote tumor progression through a number of different mechanisms. For example, it can prevent apoptosis and promote cell proliferation by signaling through the phosphoinositol-3 (PI3)-kinase/Akt pathway, particularly in osteoclasts in which it mediates functional activation (Hruska *et al.*, 1995). OPN has also been shown to aid in the degradation of the ECM, which may allow cancer cells to invade the surrounding stroma. This is accomplished through several mechanisms, whereby signaling in response to OPN can upregulate the expression of matrix metalloproteinases (MMPs) as well as stimulate urokinase plasminogen activator (uPA, see below), both of which have been shown to be elevated in patients with breast cancer metastases (Wai and Kuo, 2008). Lastly, OPN may also enhance tumor cell metastases by promoting cell adhesion and extravasation, evading host immune responses (particularly through the inhibition of nitric oxide synthase) and promoting neovascularization.

7.3.3.2 *Fibroblast growth factor*

The fibroblast growth factor (FGF) family consists at least 23 known genes, including FGF1/2, many of which have been shown to play an important role in both normal and malignant processes, including wound healing and neoplastic transformation. Many of the FGF ligands have been shown to participate in various mechanisms of tumor cell growth, motility, and invasiveness. For example, FGF2 has been shown to promote tumor angiogenesis as well as inhibit apoptosis (Kwabi-Addo, Ozen, and Ittmann, 2004). These molecules act in a typical growth factor fashion by binding to their receptors (designated FGFR1-4) and activating intracellular pathways supporting cell growth and motility, most notably the PI3K/Akt and mitogen-activated protein kinase (MAPK) pathways. Recent studies using tissue microarrays comprised of prostate cancer biopsies found that FGF ligand and receptors were overexpressed in a manner that correlated with grade of malignancy (Murphy *et al.*, 2010). Two of the most studied family members, acidic FGF (or FGF1) and basic FGF (or FGF2), have also been shown to be mediators of osteoblast proliferation in patients suffering from skeletal metastases, particularly in prostate cancer (Canalis *et al.*, 1987; Canalis, Centrella, and McCarthy, 1988). It has also been shown that a human cancer cell line that expresses FGF2 was able to activate osteoblasts to induce new bone formation *in vivo* (Izbicka *et al.*, 1996). These studies suggest that the FGF family of growth factors may be involved in promoting skeletal progression of metastases.

7.3.3.3 *Platelet-derived growth factor*

The platelet-derived growth factor (PDGF) family consists of soluble, disulfide-linked dimeric peptides that act as ligands at cognate platelet-derived growth factor receptors (PDGFRs). The PDGF isoforms themselves exist primarily as AA or BB homodimers or as an AB heterodimer. These ligands may then serve to activate receptor family members by binding to either α or β receptor homo- or heterodimers where they can stimulate intracellular growth and survival pathways (Heldin and Westermark, 1999). Both the α and β isoforms of the receptor have been implicated in cancer progression and metastasis, with additional evidence to support the role of PDGF-BB ligand. Many studies have shown that numerous tumor cell types may produce their own PDGF-BB, which acts in a paracrine fashion as a potent stimulator of both the osteoclastic and osteoblastic components of bone remodeling (Yi *et al.*, 2002). Other studies have focused on the role of the alpha-PDGFR in the growth and survival of disseminated prostate cancer and shown that the expression of PDGFRα correlates with advanced prostate cancer and skeletal metastases (Chott *et al.*, 1999). It has also been shown that cells expressing this receptor are able to activate downstream signaling pathways in response to human bone marrow, and that this mechanism may be responsible for skeletal lesions formed in an *in vivo* model of disseminated prostate cancer (Dolloff *et al.*, 2005, 2007; Russell *et al.*, 2009). It is likely that both PDGF ligands and their receptors are involved in the osteoblastic and osteoclastic mechanisms of bone remodeling, thereby promoting tumor cell expansion within the bone (Zhang *et al.*, 1991; Kubota *et al.*, 2002; Langley *et al.*, 2004; Yang, 2000).

7.3.3.4 *Epidermal growth factor*

The epidermal growth factor ligand (EGF) and its associated receptor (EGFR) are commonly expressed in many human cancers where they support a variety of tumor functions including proliferation, survival, and motility. EGF can regulate multiple aspects of metastasis, such as the induction of angiogenesis through upregulation of vascular endothelial growth factor (VEGF) mRNA expression (De Luca *et al.*, 2008). Recent studies have demonstrated that EGF/EGFR signaling may play a central role in skeletal metastases through tumor-mediated stimulation of osteoclasts function (De Luca *et al.*, 2008). The bone consists of a heterogeneous mixture of cells, including both epithelial and MSCs, the latter of which can differentiate into many of the bone resident cells such as adipocytes, osteoblasts, and chondrocytes. Bone MSCs have been shown to regulate osteoblastic and osteoclastic differentiation within the bone microenvironment, and numerous studies have shown that tumor cells may exploit this function to induce osteoclastogenesis. In this regard, EGF has been shown to act at its cognate receptor on bone MSCs to induce an increase in cell proliferation, as well as to act on osteoblast-like cells to inhibit their activity. The effects of EGFR inhibition of skeletal metastases have been demonstrated in the clinic, where several breast cancer patients enrolled in a study with gefitinib, a small-molecule inhibitor of EGFR, had significant relief of bone pain (von Minckwitz *et al.*, 2005). These findings were subsequently validated *in vitro*, where MSCs treated with gefitinib showed markedly decreased basal levels of EGFR signaling and reduced ability to induce osteoclast differentiation.

7.3.3.5 *Transforming growth factor*-β

The transforming growth factor-β (TGF-β) family consists of three closely related isoforms (TGF-β1, -β2, and -β3). These multifunctional cytokines elicit a diverse range of cellular responses that can both inhibit or promote tumor progression (Bierie and Moses, 2006). The mechanisms regulating this transition are not fully understood. It has been shown that metastatic cells not only loose responsiveness to the tumor suppressor functions of TGF-β but can actually release and respond to it by promoting tumor progression (Moses, Yang, and Pietenpol, 1990). TGF-β is normally stored in the bone matrix and is released during bone resorption. In addition, TGF-β can markedly increase the production of parathyroid hormone-related peptide (PTH-rP) by cancer cells during the metastatic process. Several studies have similarly implicated PTH-rP in the growth and metastasis of breast and prostate cancer cells (Iwamura *et al.*, 1994).

7.3.3.6 *Interleukins*

Interleukins (ILs) are proinflammatory cytokines that are generated by macrophages immediately after confronting an inflammatory stimulus. IL-1 family members consist of two agonist peptides, IL-1α and IL-1β, as well as an antagonist in the form of the soluble receptor, IL-1Rα (Apte *et al.*, 2006). Although the ILs are not typically secreted by normal cells (other than macrophages), malignant cells frequently secrete excess cytokines, including IL-1β. Many studies have shown that the autocrine release of IL-1β can act to increase the invasiveness of tumor cells, as well as stimulate the production and release of other growth factors and cytokines. There is direct evidence for the involvement of this ligand in the context of skeletal metastasis. For example, B16 melanoma cells injected via an intracardiac route are able to form skeletal metastases, however administration of IL-1Rα (the endogenous antagonist) reduced associated bone lesions (Anasagasti *et al.*, 1997). There are also several lines of study showing that IL-1β may be involved in the osteolytic phase of skeletal metastases. In addition, several groups have demonstrated that IL-1β is likely the active molecule responsible for the bone resorbing properties of supernatant derived from cultured myeloma cells. Finally, IL-1β can induce the expression of many cell adhesion molecules such as ICAM-1, VCAM-1, and endothelial-leukocyte adhesion molecule (ELAM) (Lust and Donovan, 1999).

7.3.3.7 *Endothelins*

The endothelin (ET) family consists of three members, which are small 21-amino-acid peptides with a relatively short half-life. Endothelin-1 is the most common circulating isoform and preferentially binds ET_A, one of two G-protein-coupled endothelin receptors. Activation of this receptor leads to a multitude of downstream effects including increase in intracellular Ca^{2+} levels, activation of protein kinase C, and the activation of numerous prosurvival pathways such as PI3-K and MAPK pathways (Carducci and Jimeno, 2006). Several different lines of research have shown that increased secretion of ET-1 and its subsequent signaling are involved in the progression of various cancers. ET-1 has also been shown to contribute to the osteoblastic nature of breast and prostate cancer bone metastasis by stimulating the growth of bone-resident osteoblasts (Nelson *et al.*, 1999). In addition, increased concentrations of ET-1 have been shown to reduce osteoclastic activity within the bone, with the net effect being increased bone deposition and a reduced bone turnover. Furthermore, this effect could be replicated by either administration of ET-1 to bone cells *in vitro* or by coculturing bone cells with prostate cancer cells. Because of the evidence for the involvement of ET-1 in osteoblastic lesions, clinical trials with antagonists of the ET_A receptor are underway in patients with hormone-refractory prostate cancer patients (Carducci and Jimeno, 2006).

7.3.3.8 CD44

CD44 is a glycoprotein receptor primarily for hyaluronic acid, although it may bind some other extracellular proteins as well. Although the link between CD44 and primary tumors is still uncertain, there is mounting evidence that this receptor is involved in skeletal metastases. Animal models have demonstrated that CD44 can promote breast cancer invasion, the expression of CD44 can differentiate between tumorigenic and non-tumorigenic breast cancer cells and also that sarcoma cells require expression of CD44 in order to metastasize (Weber *et al.*, 2002; Al-Hajj *et al.*, 2003). CD44 has been shown to aid in the metastatic process is by regulating the ability of cells to extravasate from the vasculature into the bone microenvironment, particularly by mediating adhesion of cancer cells to bone marrow endothelial cells (Draffin *et al.*, 2004). In addition, CD44 appears to be implicated in the reorganization of the bone matrix through its interaction with MMP-9 and cathepsin K. This association not only promotes the degradation of bone matrix collagen, but may also serve to activate TGF-β (a major factor in the *vicious cycle*) (Hill *et al.*, 2006). Lastly, CD44 has been shown to promote adhesion-dependent growth through its activation by hyaluronic acid and OPN and subsequent activation of prosurvival pathways, such as PI3K/Akt (Bates *et al.*, 2001; Kim *et al.*, 2005).

7.3.3.9 Urokinase

Urokinase (uPA) is a serine peptidase that is expressed in many normal cell types, where its traditional function is the generation of pericellular proteolysis during cell migration and tissue remodeling. It has been shown to be expressed at high levels in malignant tumors, and can promote malignant cell migration by activating proteases that are involved in the digestion of extracellular matrices (Danø *et al.*, 1985). In addition, there have been numerous studies showing that uPA is also able to activate TGF-β, which is involved in bone remodeling (Koeneman, Yeung, and Chung, 1999). Indeed, many highly aggressive human prostate cancer cell lines express greater amounts of uPA than normal tissue or well-differentiated prostatic tumors, and a loss of the uPA gene in human patients confers a better prognosis. Part of this action is likely caused by paracrine actions of secreted uPA from tumor cells that may act to promote proliferation of resident bone cells, resulting in an osteoblastic response (Achbarou *et al.*, 1994).

The list of bone tissue factors with a potential role in supporting the survival and growth of cancer cells during their progression into macroscopic and clinically relevant metastases is already extended and destined to expand. It is also evident that some factors exert a major role in the colonization of distant organs by some cancers more than others. This is the indication that, while growth in the bone stroma requires adaptation, not all tumors adapt by using the same molecular mechanisms and mediators. Therefore, the discovery and functional characterization of existing as well as additional supportive molecules and their relative significance for each type of tumor will greatly benefits from studies aimed to identify bone-metastatic gene signatures (Woelfle *et al.*, 2003; Kang *et al.*, 2003; Varambally *et al.*, 2005; Minn *et al.*, 2005; Vicent *et al.*, 2008). These approaches will reveal malignant phenotypes with the highest predisposition to colonize the skeleton. The potential gain is at least twofold: in addition to the ability to predict the site of secondary dissemination based on organ-specific metastatic signatures, these strategies will identify useful molecular targets for counteracting metastatic disease, either preventing initial homing and survival or impairing further progression (Horak and Steeg, 2005). This knowledge will be eventually translated into clinic and allow the adoption of therapeutic approaches tailored to metastatic disease associated to different tumors as well as distinct subtypes of the same malignancy (Nguyen, Bos, and Massagué, 2009).

7.3.4 The progression of small tumor foci into established macroscopic metastases

Skeletal secondary tumors that are small enough to be undetectable by imaging techniques such as radiography, positron emission tomography scan, or magentic resonance imaging, are also very often asymptomatic. The awareness of their presence would certainly make any patient (and their caring physicians) very anxious; nevertheless, if kept at this stage these lesions would still fail to create a substantial clinical problem. It is the progression of these initial small tumors into overt metastases that contributes to morbidity and mortality in the clinical setting. Despite significant efforts and some partial success (Tannock *et al.*, 2004), effective measures to either prevent the location of cancer cells to the skeleton or block the expansion of initial lesions within the bone tissue are still lacking. Quite understandably then, the prevalent line of treatment for bone metastatic patients is based on our acquired knowledge of macroscopic skeletal lesions (Bagi, 2005;

Coleman, 2007). The study of macroscopic metastases, using experimental animals as well as biopsied or postmortem specimens (Cawthorn *et al.*, 2009) is obviously much less problematic than investigating microscopic osseous lesions.

As mentioned above, a central event characterizing late metastatic progression in the bone is the recruitment of osteoclasts by cancer cells and the advantageous effects on the expansion of the tumor mass exerted by the resulting osteolytic activities (Chung, 2003; Guise *et al.*, 2006) (Figure 7.3). It has been reported that some malignant phenotypes produce molecules such as PTH-rP or IL-6, which stimulate osteoblasts to produce receptor activator of NF-kB ligand (RANKL). Both soluble and membrane-bound RANKL bind to RANK on the surface of osteoclast-precursor cells leading to osteoclastogenesis. Osteoclasts cause degradation of bone matrix, leading to release of trophic factors that sustain cancer cell proliferation (Roodman,

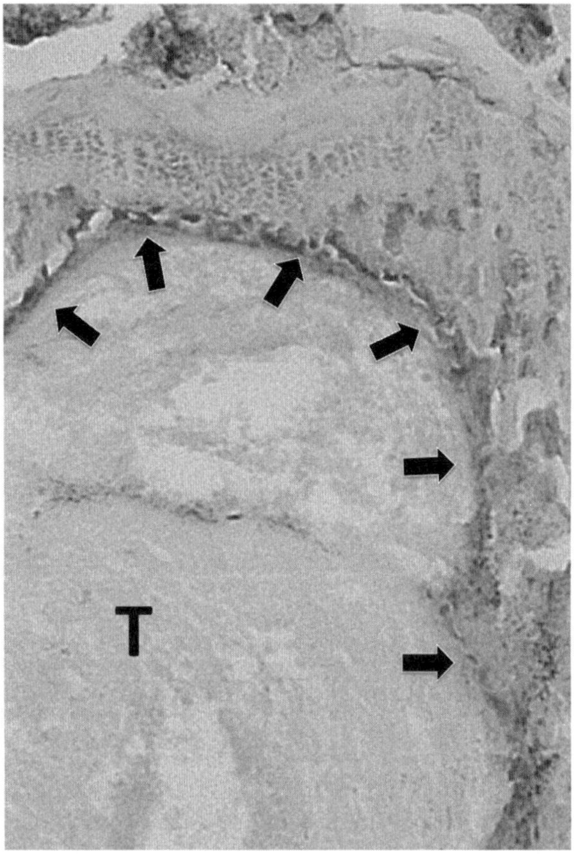

Figure 7.3 Bone resorption in established skeletal metastases. Active osteoclasts, identified by red TRAcP staining, appear as a continuous layer at the interface between tumor cells (T) and the surrounding bone tissue (arrows). This metastatic tumors was produced by human cancer cells growing for 4 weeks in the tibia of an immunocompromised mouse after being inoculated in the left cardiac ventricle.

2004). Thus, progressively more cancer cells will recruit and activate osteoclasts in a symbiotic vicious cycle that promotes the expansion of the tumor mass within the bone tissue (Kingsley *et al.*, 2007). Inhibition of molecules such as RANKL has been found effective in blocking skeletal progression and prolongs survival in animal models (Canon *et al.*, 2008). These osteolytic properties of bone tumors are characteristically attributed to some tumors, such as lung and breast carcinomas. However, it should be pointed out that, although prostate cancer among others is generally considered osteoblastic, this is a typical feature of established, late-stage metastases. In fact, areas of pronounced osteolysis are detected, alone or mixed with osteoblastic lesions, in at least 30% of patients with advanced prostate cancer (Cheville *et al.*, 2002). Thus, it seems plausible that the expansion of the majority of skeletal metastatic tumors is generally done at the expense of the bone matrix and thus requires its degradation.

Bone resorption processes can be effectively inhibited by bisphosphonates and osteoprotegerin. The first compounds, currently into their third generation, adhere on the surface of the bone lacunae in which osteoclasts reside, and are internalized by these cells, disrupting the biochemical processes responsible for resorption and inducing cell death (Russell *et al.*, 2007, 2008). Osteoprotegerin acts as a decoy receptor, impeding the association of RANKL with RANK, thus affecting osteoclast recruitment, functioning, and survival (Corey *et al.*, 2005). Administration of either bisphosphonates or osteoprotegerin has been shown to limit the growth of cancer cells in the skeleton of animal models (Lee *et al.*, 2002; Schneider *et al.*, 2005; Canon *et al.*, 2008). In the clinic, bisphosphonates are an integral part in the treatment regimen of patients with skeletal metastases (Body and Mancini, 2002) However, while these drugs are credited with a significant palliative role, the effect on overall survival presents discrepancies, probably based on the type of tumor investigated. For instance, a recent clinical trial in which zoledronic acid was compared to placebo in 422 advanced prostate cancer patients failed to show significant differences in disease progression, performance status and quality of life among the groups (Saad *et al.*, 2002). Similar results were provided by preclinical studies in which the progression of the bone metastatic disease from breast cancer cells was delayed but only transiently, and at later stages the total tumor burden per animal became equivalent to that in vehicle-treated animals (van der Pluijm *et al.*, 2005). On the other hand, trials conducted with lung cancer patients with evident bone metastases, showed significant increase in overall survival (Zarogoulidis *et al.*, 2009).

Advanced bone lesions are characterized by atypical tissue architecture, due to the altered balance of bone resorption and deposition by osteoclasts and osteoblasts, respectively. In addition, autochthonous cells like hematopoietic, mesenchymal and stromal cells, adipocytes, chondrocytes and endothelial cells coexist with immune cells migrated from the blood and lymphatic circulation (Figure 7.4). All of these cell populations are involved in spatial and functional interactions that differ from those occurring normally or would not take place at all (Orr *et al.*, 2000a). The degradation of the bone matrix will inherently generate harsh conditions due to the release of inorganic components, of which calcium is the main one. Physiological, extracellular ionized calcium levels are maintained bewteen a narrow range of 1.1–1.3 mmol/l. In active bone remodeling, extracellular calcium can reach levels

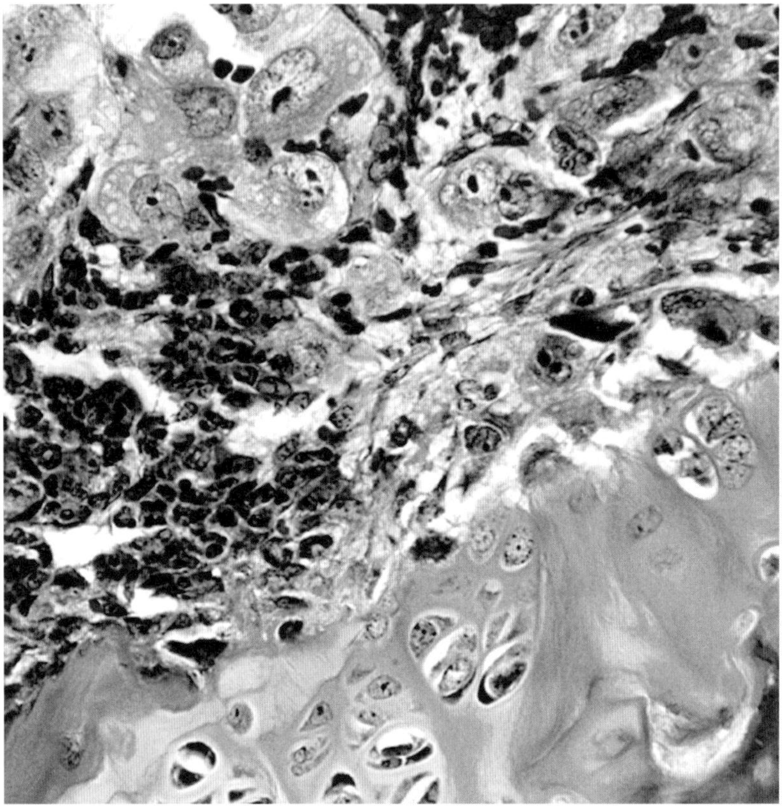

Figure 7.4 Multiple cell populations are present in metastatic bone lesions. Cancer cells (on top) are in close spatial relationship with autochthonous bone cells as well as immune cells migrated from the blood and lymphatic circulation. The collagen fibers of the bone tissue are stained in blue. A full color version of this figure can be found in the color plate section.

as high as 8–40 mmol/l and very possibly even more in the metastatic environment (Dvorak *et al.*, 2004). In addition, immune cells might exert direct as well as cytokine-mediated cytotoxic effects toward cancer cells (Pardoll, 2003; Dunn, Old, and Schreiber, 2004). The induction and completion of events leading to cell death, by either necrosis or apoptosis, almost invariably elicit alterations of intracellular calcium homeostasis (Orrenius, Zhivotovsky, and Nicotera, 2003; Roderick and Cook, 2008). This result can be caused by either decreasing the content of calcium in the intracellular stores or inducing its excessive influx from the extracellular space (Skryma *et al.*, 2000; Monteith *et al.*, 2007). There is evidence that cancer cells can acquire resistance to apoptosis by modifying the functioning of specific calcium channels (Prevarskaya, Skryma, and Shuba, 2004). In addition, changes in plasma membrane calcium channel expression and permeability can regulate metastatic potential (Prevarskaya *et al.*, 2007; Panner *et al.*, 2005; Panner and Wurster, 2006).

In this context, it is extremely interesting that normal human osteoblasts were deemed able to downregulate the influx of extracellular calcium only in bone-metastatic prostate and breast cancer cells, whereas cells from the same tumors but lacking the ability to target the bone were unaffected. These initial *in vitro* observations were also confirmed *ex vivo* with the same cells obtained from skeletal metastases produced in animals (D'Ambrosio and Fatatis, 2009). This ability to limit the increase in cytosolic calcium could be a determinant protective feature to evade from death or quiescence. The active role of osteoblasts in this mechanism further emphasizes the importance of the bone microenvironment in establishing favorable conditions to the progression of selected malignant phenotype from small foci to macroscopic metastases (Tapia-Vieyra and Mas-Oliva, 2001).

7.4 Concluding remarks

Skeletal metastases pose enormous problems for the prognosis and therapy of cancer patients. Whereas progress is constantly made in terms of therapeutic treatment for primitive solid tumors, the detection of cancer dissemination to the skeleton still represents a serious downturn for the majority of patients and many will eventually succumb to this complication.

It is therefore of utmost importance that molecular mediators and mechanisms supporting the homing and progression of cancer cells in the skeleton are identified. This goal will be achieved only when local microenvironment factors determining the bone tropism of cancer are identified and targeted by appropriate therapeutic strategies.

References

Achbarou, A. Kaiser, S. Tremblay, G. *et al.* (1994) Urokinase overproduction results in increased skeletal metastasis by prostate cancer cells in vivo. *Cancer Research*, **54**, 2372–2377.

Aguirre-Ghiso, J. (2007) Models, mechanisms and clinical evidence for cancer dormancy. *Nature Reviews Cancer*, **7**, 834–846.

Al-Hajj, Muhammad, Wicha, Max S., Benito-Hernandez, Adalberto, Morrison, Sean J., Clarke, Michael F. *et al.* (2003) Prospective identification of tumorigenic breast cancer cells. *Proceedings of the National Academy of Sciences of the United States of America*, **100**, 3983–3988.

Al-Mehdi, A.B., Tozawa, K., Fisher, A.B., Shientag, L., Lee, A., Muschel, R.J. *et al.* (2000) Intravascular origin of metastasis from the proliferation of endothelium-attached tumor cells: a new model for metastasis. *Nature Medicine*, **6**, 100–102.

Alix-Panabières, C., Riethdorf, S., and Pantel, K. (2008) Circulating tumor cells and bone marrow micrometastasis. *Clinical Cancer Research*, **14**, 5013–5021.

American Cancer Society (2009) Bone Metastasis Overview. American Cancer Society. www.cancer.org.

Anasagasti, M.J., Olaso, E., Calvo, F. *et al.* (1997) Interleukin 1-dependent and -independent mouse melanoma metastases. *Journal of the National Cancer Institute*, **89**, 645–651.

Apte, R.N., Dotan, S., Elkabets, M. *et al.* (2006) The involvement of IL-1 in tumorigenesis, tumor invasiveness, metastasis and tumor-host interactions. *Cancer and Metastasis Reviews*, **25**, 387–408.

Ashford, R.U., Pendlebury, S., Stalley, P.D., (2006) Management of metastatic disease of the appendicular skeleton. Current Orthopaedics, pp. 1–17.

Bagi, C.M. (2005) Targeting of therapeutic agents to bone to treat metastatic cancer. *Advanced Drug Delivery Reviews*, **57**, 995–1010.

Balkwill, F. (2003) Chemokine biology in cancer. *Seminars in Immunology*, **15**, 49–55.

Barthel, S.R., Gavino, J.D., Descheny, L., and Dimitroff, C.J. (2007) Targeting selectins and selectin ligands in inflammation and cancer. *Expert Opinion on Therapeutic Targets*, **11**, 1473–1491.

Bates, R.C., Edwards, N.S., Burns, G.F., and Fisher, D.E. (2001) A CD44 survival pathway triggers chemoresistance via lyn kinase and phosphoinositide 3-kinase/Akt in colon carcinoma cells. *Cancer Research*, **61**, 5275–5283.

Bierie, B. and Moses, H.L. (2006) Tumour microenvironment: TGFbeta: the molecular Jekyll and Hyde of cancer. *Nature Reviews Cancer*, **6**, 506–520.

Blair, H.C., Zaidi, M., and Schlesinger, P.H. (2002) Mechanisms balancing skeletal matrix synthesis and degradation. *Biochemical Engineering Journal*, **364**, 329–341.

Body, J.J. and Mancini, I. (2002) Bisphosphonates for cancer patients: why, how, and when? *Supportive Care in Cancer: Official Journal of the Multinational Association of Supportive Care in Cancer*, **10**, 399–407.

Bonfil, R.D., Chinni, S., Fridman, R., Kim, H.R., Cher, M. *et al.* (2007) Proteases, growth factors, chemokines, and the microenvironment in prostate cancer bone metastasis. *Urologic Oncology*, **25**, 407–411.

Buijs, J.T. and van der Pluijm, G. (2009) Osteotropic cancers: from primary tumor to bone. *Cancer Letters*, **273**, 177–193.

Burger, M. *et al.* (2003) Functional expression of CXCR4 (CD184) on small-cell lung cancer cells mediates migration, integrin activation, and adhesion to stromal cells. *Oncogene*, **22**, 8093–8101.

Bussard, K., Gay, C., and Mastro, A. (2008) The bone microenvironment in metastasis; what is special about bone? *Cancer and Metastasis Reviews*, **27**, 41–55.

Butcher, E.C. and Picker, L.J. (1996) Lymphocyte homing and homeostasis. *Science*, **272**, 60–66.

Canalis, E., Centrella, M., and McCarthy, T. (1988) Effects of basic fibroblast growth factor on bone formation in vitro. *The Journal of Clinical Investigation*, **81**, 1572–1577.

Canalis, E., Lorenzo, J., Burgess, W.H., and Maciag, T. (1987) Effects of endothelial cell growth factor on bone remodelling in vitro. *The Journal of Clinical Investigation*, **79**, 52–58.

Canon, J. *et al.* (2008) Inhibition of RANKL blocks skeletal tumor progression and improves survival in a mouse model of breast can cer bone metastasis. *Clinical and Experimental Metastasis*, **25**, 119–129.

Carducci, M.A. and Jimeno, A. (2006) Targeting bone metastasis in prostate cancer with endothelin receptor antagonists. *Clinical Cancer Research*, **12**, 6296s–6300s.

Cawthorn, T.R. *et al.* (2009) Mechanisms and pathways of bone metastasis: challenges and pitfalls of performing molecular research on patient samples. *Clinical and Experimental Metastasis*, **26**, 935–943.

Chapman, G.A. *et al.* (2000) The role of fractalkine in the recruitment of monocytes to the endothelium. *European Journal of Pharmacology*, **392**, 189–195.

Chaudhary, L.R., Hofmeister, A.M., and Hruska, K.A. (2004) Differential growth factor control of bone formation through osteoprogenitor differentiation. *Bone*, **34**, 402–411.

Cheville, J.C. *et al.* (2002) Metastatic prostate carcinoma to bone: clinical and pathologic features associated with cancer-specific survival. *Cancer*, **95**, 1028–1036.

Chott, A. *et al.* (1999) Tyrosine kinases expressed in vivo by human prostate cancer bone marrow metastases and loss of the type 1 insulin-like growth factor receptor. *American Journal of Pathology*, **155**, 1271–1279.

Chung, L.W. (2003) Prostate carcinoma bone-stroma interaction and its biologic and therapeutic implications. *Cancer*, **97**, 772–778.

Coleman, R. (2007) On the horizon: can bisphosphonates prevent bone metastases? *Breast (Edinburgh, Scotland)*, **16** (Suppl. 3), S21–S27.

Cooper, C. *et al.* (2003) Stromal factors involved in prostate carcinoma metastasis to bone. *Cancer*, **97**, 739–747.

Corey, E. *et al.* (2005) Osteoprotegerin in prostate cancer bone metastasis. *Cancer Research*, **65**, 1710–1718.

Corey, E. *et al.* (2002) Establishment and characterization of osseous prostate cancer models: intra-tibial injection of human prostate cancer cells. *The Prostate*, **52**, 20–33.

D'Ambrosio, J. and Fatatis, A. (2009) Osteoblasts modulate Ca(2+) signaling in bone-metastatic prostate and breast cancer cells. *Clinical and Experimental Metastasis*, **26**, 955–964.

Danø, K. *et al.* (1985) Plasminogen activators, tissue degradation, and cancer. *Advances in Cancer Research*, **44**, 139–266.

De Luca, A. *et al.* (2008) The role of the EGFR signaling in tumor microenvironment. *Journal of Cellular Physiology*, **214**, 559–567.

Deftos, L.J. (2000) Prostate carcinoma: production of bioactive factors. *Cancer*, **88**, 3002–3008.

Dolloff, N., Russell, M.R., Loizos, N., and Fatatis, A. (2007) Human bone marrow activates the Akt pathway in metastatic prostate cells through transactivation of the alpha-platelet-derived growth factor receptor. *Cancer Research*, **67**, 555–562.

Dolloff, N. *et al.* (2005) Bone-metastatic potential of human prostate cancer cells correlates with Akt/PKB activation by alpha platelet-derived growth factor receptor. *Oncogene*, **24**, 6848–6854.

Draffin, J.E. *et al.* (2004) CD44 potentiates the adherence of metastatic prostate and breast cancer cells to bone marrow endothelial cells. *Cancer Research*, **64**, 5702–5711.

Dunn, G.P., Old, L.J., and Schreiber, R.D. (2004) The immunobiology of cancer immunosurveillance and immunoediting. *Immunity*, **21**, 137–148.

Dvorak, M.M. *et al.* (2004) Physiological changes in extracellular calcium concentration directly control osteoblast function in the absence of calciotropic hormones. *Proceedings of the National Academy of Sciences of the United States of America*, **101**, 5140–5145.

Engl, T. *et al.* (2006) CXCR4 chemokine receptor mediates prostate tumor cell adhesion through alpha5 and beta3 integrins. *Neoplasia*, **8**, 290–301.

Fidler, I. (2003) The pathogenesis of cancer metastasis: the 'seed and soil' hypothesis revisited. *Nature Reviews Cancer*, **3**, 453–458.

Fisher, J.L. *et al.* (2002) An in vivo model of prostate carcinoma growth and invasion in bone. *Cell and Tissue Research*, **307**, 337–345.

Fong, A. *et al.* (1998) Fractalkine and CX3CR1 mediate a novel mechanism of leukocyte capture, firm adhesion, and activation under physiologic flow. *The Journal of Experimental Medicine*, **188**, 1413.

Fornaro, M., Manes, T., and Languino, L.R. (2001) Integrins and prostate cancer metastases. *Cancer and Metastasis Reviews*, **20**, 321–331.

Gassmann, P. and Haier, J. (2008) The tumor cell-host organ interface in the early onset of metastatic organ colonisation. *Clinical and Experimental Metastasis*, **25**, 171–181.

Gassmann, P. *et al.* (2009) CXCR4 regulates the early extravasation of metastatic tumor cells in vivo. *Neoplasia*, **11**, 651–661.

Gilmore, A.P. (2005) Anoikis. *Cell Death and Differentiation*, **12** (Suppl. 2), 1473–1477.

Glinskii, O. *et al.* (2005) Mechanical entrapment is insufficient and intercellular adhesion is essential for metastatic cell arrest in distant organs. *Neoplasia*, **7**, 522–527.

Glinsky, V. (2007) Intravascular cell-to-cell adhesive interactions and bone metastasis. *Cancer and Metastasis Reviews*, **25**, 531–540.

Goel, H.L., Li, J., Kogan, S., and Languino, L.R. (2008) Integrins in prostate cancer progression. *Endocrine-Related Cancer*, **15**, 657–664.

Gomes, R., Buttke, P., Paul, E., and Sikes, R. (2009) Osteosclerotic prostate cancer metastasis to murine bone are enhanced with increased bone formation. *Clinical and Experimental Metastasis*, **26**, 641–651.

Gout, S., Tremblay, P.L., and Huot, J. (2008) Selectins and selectin ligands in extravasation of cancer cells and organ selectivity of metastasis. *Clinical and Experimental Metastasis*, **25**, 335–344.

Guise, T. *et al.* (2006) Basic mechanisms responsible for osteolytic and osteoblastic bone metastases. *Clinical Cancer Research*, **12**, 6213s–6216s.

Heldin, C.H. and Westermark, B. (1999) Mechanism of action and in vivo role of platelet-derived growth factor. *Physiological Reviews*, **79**, 1283–1316.

Hill, A., McFarlane, S., Johnston, P.G., and Waugh, D.J. (2006) The emerging role of CD44 in regulating skeletal micrometastasis. *Cancer Letters*, **237**, 1–9.

Horak, C.E. and Steeg, P. (2005) Metastasis gets site specific. *Cancer Cell*, **8**, 93–95.

Hruska, K.A. Rolnick, F., Huskey, M. *et al.* (1995) Engagement of the osteoclast integrin alpha v beta 3 by osteopontin stimulates phosphatidylinositol 3-hydroxyl kinase activity. *Endocrinology*, **136**, 2984–2992.

Imai, K., Kobayahi, M., Wang, J. *et al.* (1999) Selective secretion of chemoattractants for haemopoietic progenitor cells by bone marrow endothelial cells: a possible role in homing of haemopoietic progenitor cells to bone marrow. *British Journal of Haematology*, **106**, 905–911.

Iwamura, M., Abrahamsson, P.A., Foss, K.A. *et al.* (1994) Parathyroid hormone-related protein: a potential autocrine growth regulator in human prostate cancer cell lines. *Urology*, **43**, 675–679.

Izbicka, E., Dunstan, C., Esparza, J. *et al.* (1996) Human amniotic tumor that induces new bone formation in vivo produces growth-regulatory activity in vitro for osteoblasts identified as an extended form of basic fibroblast growth factor. *Cancer Research*, **56**, 633–636.

Jamieson, W.L., Shimizu, S., D'Ambrosio, J.A. *et al.* (2008) CX3CR1 is expressed by prostate epithelial cells and androgens regulate the levels of CX3CL1/fractalkine in the bone marrow: potential role in prostate cancer bone tropism. *Cancer Research*, **68**, 1715–1722.

Kang, Y., Siegel, P.M., Shu, W. *et al.* (2003) A multigenic program mediating breast cancer metastasis to bone. *Cancer Cell*, **3**, 537–549.

Karamanolakis, D., Bogdanos, J., Sourla, A. *et al.* (2002) Molecular evidence-based use of bone resorption-targeted therapy in prostate cancer patients at high risk for bone involvement. *Molecular Medicine*, **8**, 667–675.

Kim, M.S., Park, M.J., Moon, E.J. *et al.* (2005) Hyaluronic acid induces osteopontin via the phosphatidylinositol 3-kinase/Akt pathway to enhance the motility of human glioma cells. *Cancer Research*, **65**, 686–691.

Kingsley, L.A., Fournier, P.G., Chirgwin, J.M., and Guise, T. (2007) Molecular biology of bone metastasis. *Molecular Cancer Therapeutics*, **6**, 2609–2617.

Koeneman, K.S., Yeung, F., and Chung, L.W. (1999) Osteomimetic properties of prostate cancer cells: a hypothesis supporting the predilection of prostate cancer metastasis and growth in the bone environment. *The Prostate*, **39**, 246–261.

Konstantopoulos, K. and Thomas, S.N. (2009) Cancer cells in transit: the vascular interactions of tumor cells. *Annual Review of Biomedical Engineering*, **11**, 177–202.

Kubota, K., Sakikawa, C., Katsumata, M. *et al.* (2002) Platelet-derived growth factor BB secreted from osteoclasts acts as an osteoblastogenesis inhibitory factor. *Journal of Bone and Mineral Research*, **17**, 257–265.

Kucia, M., Reca, R., Miekus, K. *et al.* (2005) Trafficking of normal stem cells and metastasis of cancer stem cells involve similar mechanisms: pivotal role of the SDF-1-CXCR4 axis. *Stem Cells*, **23**, 879–894.

Kuschert, G.S., Coulin, F., Power, C. A. *et al.* (1999) Glycosaminoglycans interact selectively with chemokines and modulate receptor binding and cellular responses. *Biochemistry*, **38**, 12959–12968.

Kwabi-Addo, B., Ozen, M., and Ittmann, M. (2004) The role of fibroblast growth factors and their receptors in prostate cancer. *Endocrine-Related Cancer*, **11**, 709–724.

Langley, R.R., Fan, D., Tsan, R.Z. *et al.* (2004) Activation of the platelet-derived growth factor-receptor enhances survival of murine bone endothelial cells. *Cancer Research*, **64**, 3727–3730.

Lapteva, N., Yang, A.G., Sanders, D.E., Strube, R.W., Chen, S.Y. (2004) CXCR4 knockdown by small interfering RNA abrogates breast tumor growth in vivo. *Cancer gene therapy*, **12**, 84–89.

Lee, B.C., Lee, T.H., Avraham, S., and Avraham, H.K. (2004) Involvement of the chemokine receptor CXCR4 and its ligand stromal cell-derived factor 1alpha in breast cancer cell migration through human brain microvascular endothelial cells. *Molecular Cancer Research*, **2**, 327–338.

Lee, Y.P., Schwarz, E. M., Davies, M. *et al.* (2002) Use of zoledronate to treat osteoblastic versus osteolytic lesions in a severe-combined-immunodeficient mouse model. *Cancer Research*, **62**, 5564–5570.

Lehr, J.E. and Pienta, K.J. (1998) Preferential adhesion of prostate cancer cells to a human bone marrow endothelial cell line. *Journal of the National Cancer Institute*, **90**, 118–123.

Loberg, R., Gayed, B., Olson, K., and Pienta, K.J. (2005) A paradigm for the treatment of prostate cancer bone metastases based on an understanding of tumor cell-microenvironment interactions. *Journal of Cellular Biochemistry*, **96**, 439–446.

Lust, J.A. and Donovan, K.A. (1999) The role of interleukin-1 beta in the pathogenesis of multiple myeloma. *Hematology/Oncology Clinics of North America*, **13**, 1117–1125.

Luzzi, K.J. *et al.* (1998) Multistep nature of metastatic inefficiency: dormancy of solitary cells after successful extravasation and limited survival of early micrometastases. *American Journal of Pathology*, **153**, 865–873.

Middleton, J., Neil, S., Wintle, J. *et al.* (2002) Leukocyte extravasation: chemokine transport and presentation by the endothelium. *Blood*, **100**, 3853–3860.

Miles, F.L., Pruitt, F.L., van Golen, K.L., and Cooper, C. (2008) Stepping out of the flow: capillary extravasation in cancer metastasis. *Clinical and Experimental Metastasis*, **25**, 305–324.

von Minckwitz, G., Jonat, W., Fasching, P. *et al.* (2005) A multicentre phase II study on gefitinib in taxane- and anthracycline-pretreated metastatic breast cancer. *Breast Cancer Research and Treatment*, **89**, 165–172.

Minn, A.J., Kang., Serganova, I. *et al.* (2005) Distinct organ-specific metastatic potential of individual breast cancer cells and primary tumors. *The Journal of Clinical Investigation*, **115**, 44–55.

Mochizuki, H., Matsubara, A., Teishima, J. *et al.* (2004) Interaction of ligand-receptor system between stromal-cell-derived factor-1 and CXC chemokine receptor 4 in human prostate cancer: a possible predictor of metastasis. *Biochemical and Biophysical Research Communications*, **320**, 656–663.

Monteith, G.R., Mcandrew, D., Faddy, H.M., and Roberts-Thomson, S.J. (2007) Calcium and cancer: targeting Ca2+ transport. *Nature Reviews Cancer*, **7**, 519–530.

Moses, H.L., Yang, E.Y., and Pietenpol, J.A. (1990) TGF-beta stimulation and inhibition of cell proliferation: new mechanistic insights. *Cell*, **63**, 245–247.

Müller, A., Homey, B., Soto, H. *et al.* (2001) Involvement of chemokine receptors in breast cancer metastasis. *Nature*, **410**, 50–56.

Mundy, G. (2002) Metastasis to bone: causes, consequences and therapeutic opportunities. *Nature Reviews Cancer*, **2**, 584–593.

Murphy, T., Darby, S., Mathers, M.E., and Gnanapragasam, V.J. (2010) Evidence for distinct alterations in the FGF axis in prostate cancer progression to an aggressive clinical phenotype. *Journal of Pathology*, **220**, 452–460.

Naumov, G.N., Folkman, J., and Straume, O. (2009) Tumor dormancy due to failure of angiogenesis: role of the microenvironment. *Clinical and Experimental Metastasis*, **26**, 51–60.

Nelson, J.B. *et al.* (1999) New bone formation in an osteoblastic tumor model is increased by endothelin-1 overexpression and decreased by endothelin A receptor blockade. *Urology*, **53**, 1063–1069.

Netelenbos, T. *et al.* (2003) Proteoglycans on bone marrow endothelial cells bind and present SDF-1 towards hematopoietic progenitor cells. *Leukemia*, **17**, 175–184.

Nevo, I. *et al.* (2009) The involvement of the fractalkine receptor in the transmigration of neuroblastoma cells through bone-marrow endothelial cells. *Cancer Letters*, **273**, 127–139.

Nguyen, D.X., Bos, P.D., and Massagué, J. (2009) Metastasis: from dissemination to organ-specific colonization. *Nature Reviews Cancer*, **9**, 274–284.

Orr, F.W., Lee, J., Duivenvoorden, W.C., and Singh, G. (2000a) Pathophysiologic interactions in skeletal metastasis. *Cancer*, **88**, 2912–2918.

Orr, F.W. *et al.* (2000b) Interactions between cancer cells and the endothelium in metastasis. *Journal of Pathology*, **190**, 310–329.

Orrenius, S., Zhivotovsky, B., and Nicotera, P. (2003) Regulation of cell death: the calcium-apoptosis link. *Nature Reviews Molecular Cell Biology*, **4**, 552–565.

Paget, S. (1989) The distribution of secondary growths in cancer of the breast. 1889. *Cancer and Metastasis Reviews*, **8**, 98–101.

Panner, A. *et al.* (2005) Variation of T-type calcium channel protein expression affects cell division of cultured tumor cells. *Cell Calcium*, **37**, 105–119.

Panner, A. and Wurster, R.D. (2006) T-type calcium channels and tumor proliferation. *Cell Calcium*, **40**, 253–259.

Pardoll, D. (2003) Does the immune system see tumors as foreign or self? *Annual Review of Immunology*, **21**, 807–839.

van der Pluijm, G. *et al.* (2005) Interference with the microenvironmental support impairs the de novo formation of bone metastases in vivo. *Cancer Research*, **65**, 7682–7690.

Prevarskaya, N., Skryma, R., Bidaux, G., Flourakis, M., Shuba, Y. *et al.* (2007) Ion channels in death and differentiation of prostate cancer cells. *Cell Death and Differentiation*, **14**, 1295–1304.

Prevarskaya, N., Skryma, R., and Shuba, Y. (2004) Ca^{2+} homeostasis in apoptotic resistance of prostate cancer cells. *Biochemical and Biophysical Research Communications*, **322**, 1326–1335.

Rabinovich, B.A., Ye, Y., Etto, T. *et al.* (2008) Visualizing fewer than 10 mouse T cells with an enhanced firefly luciferase in immunocompetent mouse models of cancer. *Proceedings of the National Academy of Sciences of the United States of America*, **105**, 14342–14346.

Roderick, H.L. and Cook, S. (2008) Ca^{2+} signalling checkpoints in cancer: remodelling Ca^{2+} for cancer cell proliferation and survival. *Nature Reviews Cancer*, **8**, 361–375.

Roodman, G.D. (2004) Mechanisms of bone metastasis. *The New England Journal of Medicine*, **350**, 1655–1664.

Roodman, G.D. (2007) Bone-breaking cancer treatment. *Nature Medicine*, **13**, 25–26.

Russell, M., Jamieson, W., Dolloff, N., and Fatatis, A. (2009) The alpha-receptor for platelet-derived growth factor as a target for antibody-mediated inhibition of skeletal metastases from prostate cancer cells. *Oncogene*, **28**, 412–421.

Russell, R.G., Watts, N.B., Ebetino, F.H., and Rogers, M.J. (2008) Mechanisms of action of bisphosphonates: similarities and differences and their potential influence on clinical efficacy. *Osteoporosis International*, **19**, 733–759.

Russell, R.G., Graham G., Xia, Z. *et al.* (2007) Bisphosphonates: an update on mechanisms of action and how these relate to clinical efficacy. *Annals of the New York Academy of Sciences*, **1117**, 209–257.

Saad, F., Gleason, D.M., Murray, R. *et al.* (2002) A randomized, placebo-controlled trial of zoledronic acid in patients with hormone-refractory metastatic prostate carcinoma. *Journal of the National Cancer Institute*, **94**, 1458–1468.

Schneider, A., Kalikin, L.M., Mattos, A.C. *et al.* (2005) Bone turnover mediates preferential localization of prostate cancer in the skeleton. *Endocrinology*, **146**, 1727–1736.

Scott, L.J., Clarke, N.W., George, N.J.R., Shanks, J.H., Testa, N.G., Lang, S.H. *et al.* (2001) Interactions of human prostatic epithelial cells with bone marrow endothelium: binding and invasion. *British Journal of Cancer*, **84**, 1417.

Shimamura, T., Amizuka, N., Li, M. *et al.* (2005) Histological observations on the microenvironment of osteolytic bone metastasis by breast carcinoma cell line. *Biomedical Research*, **26**, 159–172.

Shulby, S.A., Dolloff, N.G., Stearns, M.E., Meucci, O., Fatatis, A. *et al.* (2004) CX3CR1-fractalkine expression regulates cellular mechanisms involved in adhesion, migration, and survival of human prostate cancer cells. *Cancer Research*, **64**, 4693–4698.

Skryma, R., Mariot, P., Bourhis, X. L. *et al.* (2000) Store depletion and store-operated Ca2+ current in human prostate cancer LNCaP cells: involvement in apoptosis. *Journal of Physiology (London)*, **527** (Pt 1), 71–83.

Springer, T.A. (1994) Traffic signals for lymphocyte recirculation and leukocyte emigration: the multistep paradigm. *Cell*, **76**, 301–314.

Springer, T.A. (1995) Traffic signals on endothelium for lymphocyte recirculation and leukocyte emigration. *Annual Review of Physiology*, **57**, 827–872.

Sun, Y., Wang, J., Shelburne, C.E. *et al.* (2003) Expression of CXCR4 and CXCL12 (SDF-1) in human prostate cancers (PCa) in vivo. *Journal of Cellular Biochemistry*, **89**, 462–473.

Sun, Y.X., Schneider, A., Jung, Y. *et al.* (2005) Skeletal localization and neutralization of the SDF-1(CXCL12)/CXCR4 axis blocks prostate cancer metastasis and growth in osseous sites in vivo. *Journal of Bone and Mineral Research*, **20**, 318–329.

Taichman, R., Cooper, C., Keller, E.T. *et al.* (2002) Use of the stromal cell-derived factor-1/CXCR4 pathway in prostate cancer metastasis to bone. *Cancer Research*, **62**, 1832–1837.

Tannock, I.F., de Wit, R., Berry, W.R. *et al.* (2004) Docetaxel plus prednisone or mitoxantrone plus prednisone for advanced prostate cancer. *The New England Journal of Medicine*, **351**, 1502–1512.

Tapia-Vieyra, J.V. and Mas-Oliva, J. (2001) Apoptosis and cell death channels in prostate cancer. *Archives of Medical Research*, **32**, 175–185.

Umehara, H., Tanaka, M., Sawaki, T. *et al.* (2006) Fractalkine in rheumatoid arthritis and allied conditions. *Modern Rheumatology*, **16**, 124–130.

Varambally, S.Yu, J., Laxman, B. *et al.* (2005) Integrative genomic and proteomic analysis of prostate cancer reveals signatures of metastatic progression. *Cancer Cell*, **8**, 393–406.

Vicent, S., Luis-Ravelo, D., Antón, I. *et al.* (2008) A novel lung cancer signature mediates metastatic bone colonization by a dual mechanism. *Cancer Research*, **68**, 2275–2285.

Wai, P.Y. and Kuo, P.C. (2008) Osteopontin: regulation in tumor metastasis. *Cancer and Metastasis Reviews*, **27**, 103–118.

Weber, G.F., Bronson, R.T., Ilagan, J. *et al.* (2002) Absence of the CD44 gene prevents sarcoma metastasis. *Cancer Research*, **62**, 2281–2286.

Weinberg, R.A. (2008) The many faces of tumor dormancy. *APMIS*, **116**, 548–551.

Weiss, L. (1985) *Principles of Metastasis*, Academic Press, London.

Wetterwald, A., an der Pluijm, G., Que, I. *et al.* (2002) Optical imaging of cancer metastasis to bone marrow: a mouse model of minimal residual disease. *American Journal of Pathology*, **160**, 1143–1153.

Woelfle, U., Cloos, J., Sauter, G. *et al.* (2003) Molecular signature associated with bone marrow micrometastasis in human breast cancer. *Cancer Research*, **63**, 5679–5684.

Yang, D. (2000) Platelet-derived growth factor (PDGF)-AA: a self-imposed cytokine in the proliferation of human fetal osteoblasts. *Cytokine*, **12**, 1271–1274.

Yi, B., Williams, P.J., Niewolna, M., Wang, Y., Yoneda, T. *et al.* (2002) Tumor-derived platelet-derived growth factor-BB plays a critical role in osteosclerotic bone metastasis in an animal model of human breast cancer. *Cancer Research*, **62**, 917.

Yin, J., Pollock, C.B., and Kelly, K. (2005) Mechanisms of cancer metastasis to the bone. *Cell Research*, **15**, 57–62.

Zarogoulidis, K., Boutsikou, E., Zarogoulidis, P. *et al.* (2009) The impact of zoledronic acid therapy in survival of lung cancer patients with bone metastasis. *International Journal of Cancer*, **125**, 1705–1709.

Zhang, L., Leeman, E., Carnes, D.C., and Graves, D.T. (1991) Human osteoblasts synthesize and respond to platelet-derived growth factor. *American Journal of Physiology*, **261**, C348–C354.

8

Premetastatic Niches

Kevin L. Bennewith[1], Janine T. Erler[2] and Amato J. Giaccia[3]

[1]*Department of Integrative Oncology, British Columbia Cancer Research Centre, Vancouver, BC, Canada*
[2]*Section of Cell and Molecular Biology, The Institute of Cancer Research, London, UK*
[3]*Division of Radiation and Cancer Biology, Department of Radiation Oncology, Stanford University, Stanford, CA, USA*

8.1 Introduction

An estimated 90% of cancer-related deaths are attributed to the metastatic spread of cancer (Gupta and Massague, 2006; Steeg, 2006), highlighting the need for new and efficacious treatment strategies. A major problem for treating tumor metastases is our limited knowledge of the determinants of 'successful' metastatic growth in distant organs. However, the recent identification of bone marrow-derived cells (BMDCs) accumulating in 'premetastatic niches' and preparing the tissue within these metastatic target organs for the arrival of tumor cells (Kaplan *et al.*, 2005; Erler *et al.*, 2009) holds great promise for improving our understanding of how normal cells promote the survival and growth of metastatic tumors.

The premetastatic niche is a fundamentally important new concept emerging in the metastasis field, with implications for the treatment and prevention of metastatic disease. Premetastatic niches can be defined as localized microenvironments that form in metastatic target organs prior to the arrival of metastatic tumor cells, and consist of a collection of specific proteins and BMDCs. Premetastatic niches are thought to be fertile regions of tissue that facilitate the invasion, survival, and/or proliferation of metastatic tumor cells, providing a highly novel mechanism for the promotion of metastasis. Indeed, interest in the premetastatic niche is growing rapidly and the concept is compelling – a metastatic tumor stimulates an organism's bone marrow cells to mobilize and prepare distant sites for the future arrival of disseminated tumor cells. If premetastatic niches are found to be absolutely essential for metastatic tumor growth, disruption of these niches would eradicate metastatic disease.

Tumor Microenvironment Edited by Dietmar W. Siemann

However, our understanding of premetastatic niches is in a relatively early stage of development at this time, and the premetastatic niche idea is somewhat controversial. Critics of the premetastatic niche point to the large number of review articles relative to research articles on the subject, and correctly suggest that a great deal more data must be generated to demonstrate that specific BMDC subtypes are *required* for metastatic growth and therefore represent viable therapeutic targets. Unfortunately, premetastatic niche formation is a dynamic process that is inherently difficult to study in isolation. Concurrent (or subsequent) metastatic dissemination from primary tumors and the mere presence of metastatic tumor cells in an animal complicates the interpretation of data regarding the development, maintenance, and therapeutic impact of premetastatic niches before and after the arrival of metastatic tumor cells in the niche. Nevertheless, a number of important observations have started to shed light on the premetastatic niche, and recent evidence indicates that the solid tumor microenvironment within metastatic primary tumors plays an essential role in the recruitment of BMDCs to distant metastatic target organs. This chapter will summarize current knowledge about premetastatic niche formation, including the role of hypoxia-induced secreted proteins in the development of premetastatic niches and the possibility of therapeutically targeting premetastatic niches to treat (or prevent) metastatic disease.

8.2 'Seeds' influencing the 'Soil'

Metastasis is a remarkably inefficient, multistep process. The determinants of 'successful' metastatic growth in a given organ are poorly understood, but there is substantial evidence to suggest that tumor cells and host tissue both play important roles in metastasis. A great deal of work has been devoted to understanding the genetic and phenotypic characteristics of metastatic tumor cells, and tumor cells clearly must have (or develop) the ability to metastasize before dissemination from a primary tumor can occur. However, the majority of tumor cells that escape from a primary tumor mass either die in the circulation or fail to invade into distant organs (Gupta and Massague, 2006; Steeg, 2006). Furthermore, tumor cells that enter metastatic target organs can die or lie dormant for extended periods of time (Townson and Chambers, 2006); only a small proportion survive and proliferate to form micrometastatic tumor foci. The continued growth of micrometastatic tumor foci into macrometastatic tumors requires a switch to active angiogenesis (Holmgren *et al.*, 1995; Naumov *et al.*, 2006; Gao *et al.*, 2008), representing a collaboration between the metastatic tumor cells and surrounding host tissue.

The role of host tissues in allowing (or even promoting) metastatic tumor growth was somewhat controversial in the past. Work by Stephen Paget in the late 1800s led to his 'seed-and-soil' hypothesis (reproduced in Paget, 1989), suggesting that metastatic tumor cells (seeds) must localize in suitable host tissues (soil) in order to produce metastatic tumor foci. Thus, the propensity of a given tumor cell type to preferentially metastasize and grow in a distant target organ is related to the inherent (or acquired) properties of the tumor cells and the target organ itself. Some tissues are more receptive to a given metastasizing tumor cell type, which can explain the tendency of tumor cells to metastasize to some organs more often than other organs in a way that cannot be explained by differences in blood flow. In addition to target organ-specific growth of metastases, metastatic tumor foci seem to grow preferentially in specific areas of some tissues (Cameron *et al.*, 2000). These observations imply that differences in the regional tissue microenvironment may influence the survival and proliferation of metastatic tumor cells.

Recent evidence indicates that primary tumors actively modify the metastatic soil to promote subsequent metastatic tumor growth (Figure 8.1). An important step forward came when Kaplan *et al.* (2005) observed significant accumulation and aggregation of BMDCs in metastatic target organs after implantation of metastatic tumors. The BMDCs aggregated into specific microregions of tissues before the arrival of metastatic tumor cells, and the presence of BMDC aggregates correlated with subsequently increased metastatic growth in these tissues. It was in this paper that the term 'premetastatic niche' was first coined. The premetastatic niche is made up of both cellular and extracellular matrix (ECM) components, and is mediated by signals provided by the primary tumor (Figure 8.2).

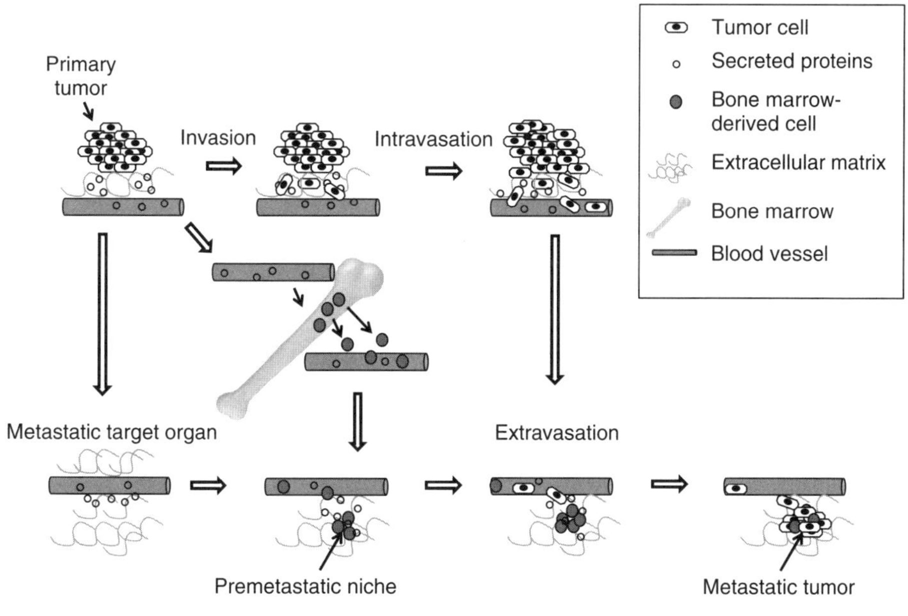

Figure 8.1 The multiple steps of metastasis including formation of the premetastatic niche.

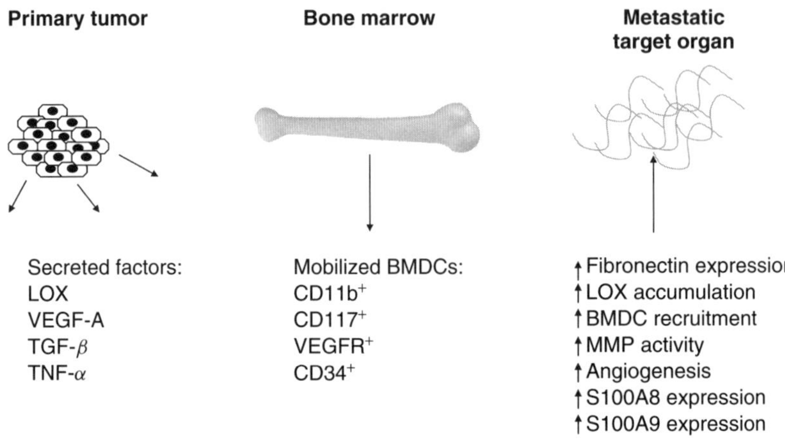

Figure 8.2 Pre-metastatic niche formation requires collaboration between the primary tumor, BMDC, and the ECM in metastatic target organs. LOX is the only tumor-secreted protein identified to date that is directly involved in formation of premetastatic niches.

8.3 Cellular components of premetastatic niches

8.3.1 BMDC Heterogeneity in premetastatic niches

Significant accumulation and aggregation of BMDCs have been observed in metastatic target organs after subcutaneous implantation of Lewis lung carcinoma (LLC) cells or B16 melanoma cells (Kaplan *et al.*, 2005). The accumulating BMDCs expressed cell surface proteins indicative of immature hematopoietic progenitor cells (HPCs) including vascular endothelial growth factor receptor-1 (VEGFR-1; Flt-1), with subsets of BMDCs also expressing CD133, CD34, CD117 (c-Kit), and/or CD11b (Mac-1). BMDCs in these premetastatic sites were therefore heterogeneous in terms of maturation, with many cells remaining in a more immature progenitor state.

The BMDC surface marker heterogeneity at premetastatic sites implies that formation of the premetastatic niche is a dynamic process, and an essential prerequisite to understanding the role of the premetastatic niche in tumor metastasis is to precisely define the identity of BMDCs that are directly involved in promoting metastatic growth. This is not a straightforward endeavor; while the number of total BMDCs in the lungs increases with time after primary tumor implantation (Kaplan *et al.*, 2005), the accumulating BMDCs are heterogeneous. Little is known about how individual populations of BMDCs increase or decrease over time in premetastatic niches, and the role that each cell type may play in enhancing metastatic growth. The number of a given BMDC-type observed in metastatic target organs may represent the summation of BMDC recruitment from the bloodstream, proliferation, turnover, differentiation within the tissue, and/or changes in expression of cell surface markers. It is certainly conceivable that any or all of these factors may change as a function of time after tumor implantation. Nevertheless, the identification of significant numbers of VEGFR-1$^+$ cells (Kaplan *et al.*, 2005) and CD11b$^+$ cells (Erler *et al.*, 2009) accumulating in premetastatic niches has led to studies investigating the role of these cells in promoting metastatic growth.

8.3.2 VEGFR-1$^+$ cells

An increasing amount of evidence supports the concept that BMDCs contribute to vasculogenesis that occurs over a range of disease states (Rafii *et al.*, 2008) including primary tumor growth (Asahara *et al.*, 1999; Davidoff *et al.*, 2001; Lyden *et al.*, 2001; Nolan *et al.*, 2007). While most studies have concentrated on the role of endothelial progenitor cells (EPCs) in remodeling of vasculature, evidence is accumulating that HPCs are important for postnatal vasculogenesis (Hattori *et al.*, 2001; Lyden *et al.*, 2001; Grant *et al.*, 2002; De Palma *et al.*, 2003). The precise role of HPCs in vasculogenesis is largely unknown, but HPCs expressing VEGFR-1 (Flt-1) are thought to be involved in the stabilization of new blood vessels (Lyden *et al.*, 2001; Grunewald *et al.*, 2006). VEGFR-1 is a cell surface receptor for vascular endothelial growth factor (VEGF) and placental growth factor (PlGF), and VEGFR-1$^+$ BMDCs are mobilized from the bone marrow in response to VEGF (Hattori *et al.*, 2001; Rafii *et al.*, 2003; Kopp *et al.*, 2006). Microregional VEGF levels in ischemic tissues are known to activate resident endothelial cells during angiogenesis, and VEGF may also promote tissue-specific homing of BMDCs close to newly forming blood vessels (Grunewald *et al.*, 2006).

Kaplan *et al.* (2005) found VEGFR-1$^+$ BMDCs accumulating in the lungs of mice bearing metastatic LLC or B16 tumors beginning 12 days after implantation. This timeframe was consistent with the VEGFR-1-dependent increase in pulmonary matrix metalloproteinase-9 (MMP-9) expression previously reported in LLC tumor-bearing mice (Hiratsuka *et al.*, 2002). Pulmonary BMDC aggregation was observed near terminal bronchioles and distal alveoli, both common areas of metastatic growth in lung tissue. Metastatic tumor cells were subsequently observed almost exclusively (>95%) associated with clusters of BMDCs. When taken with the known roles of VEGFR-1$^+$ BMDCs in angiogenesis and vasculogenesis, these data suggest that aggregating BMDCs in premetastatic niches may prepare fertile local lung microenvironments for the subsequent arrival and growth of metastatic tumor cells.

8.3.3 CD11b$^+$ cells

The most abundant BMDC-type observed in premetastatic niches formed by metastatic human or murine mammary tumors express CD11b (Erler *et al.*, 2009). CD11b (or Mac-1) is a cell surface integrin (α_M) expressed on a variety of myeloid lineage cells (granulocytes, neutrophils, monocytes, and macrophages), natural-killer (NK) cells, and a subset of B-cells. CD11b$^+$ myeloid cells are increased in some primary tumors (Du *et al.*, 2008; Shojaei and Ferrara, 2008; Yang *et al.*, 2008), and exhibit a range of functions that may enhance metastatic tumor growth (De Palma *et al.*, 2005; Serafini *et al.*, 2006; Ahn and Brown, 2008; Du *et al.*, 2008; Shojaei and Ferrara, 2008; Yang *et al.*, 2008). CD11b$^+$ cells have been implicated in enhancing tumor cell invasion (Du *et al.*, 2008; Yang *et al.*, 2008), angiogenesis (De Palma *et al.*, 2005; Shojaei and Ferrara, 2008), and vasculogenesis (Ahn and Brown, 2008) in various systems. CD11b^{+} cells can also express Gr-1, a cell surface protein expressed by murine myeloid lineage cells that is often used as an indication of myeloid cell differentiation. CD11b^{+} Gr-1^{+} cells represent a heterogeneous mixture of cells, but an important subset of CD11b^{+} Gr-1^{+} cells are myeloid-derived suppressor cells (MDSCs). MDSCs are part of the normal immune response that prevents damage to host tissues caused by prolonged inflammation (Berzofsky *et al.*, 2004). In naïve BALB/c mice, CD11b^{+} Gr-1^{+} cells are found primarily in the bone marrow and can also be detected in smaller numbers ($<4\%$) in the spleen and bloodstream. However, changes in cytokine homeostasis induced by infection, immune system stress, or tumor growth can lead to accumulation of CD11b^{+} Gr-1^{+} cells in lymphoid organs and the bloodstream (Serafini *et al.*, 2004). MDSCs are potent inactivators of T-cell and NK-cell-mediated immune responses against tumor cells (Serafini *et al.*, 2006; Liu *et al.*, 2007), and monocytes protect cardiomyocytes from hypoxia-induced cell death through CD11b-dependent cell–cell adhesion (DaCosta *et al.*, 2003). Emerging evidence also suggests that MDSCs induce a tumor-promoting phenotype in macrophages (Sinha *et al.*, 2007) and MDSCs are thought to increase primary tumor growth (Du *et al.*, 2008; Shojaei and Ferrara, 2008; Yang *et al.*, 2008), in part by manipulating and suppressing the innate immune response against tumor cells. However, our data demonstrated CD11b^{+} cell accumulation and enhanced growth of metastatic human breast tumor cells grown in immunocompromised athymic Nu/Nu mice (Erler *et al.*, 2009), suggesting that metastatic growth would not be affected in this system by CD11b-mediated suppression of T-cell activity. Interestingly, CD11b^{+} cells can also co-express VEGFR-1 (Kusmartsev *et al.*, 2008), and VEGFR-1^{+} cells at premetastatic sites are thought to enhance metastatic growth by VEGF-dependent angiogenesis (Kaplan *et al.*, 2005). While anti-VEGF therapy could be used to target VEGFR-1-mediated blood vessel formation in the premetastatic niche, CD11b^{+} Gr-1^{+} myeloid cells render tumor vasculature resistant to anti-VEGF antibody inhibition (Shojaei and Ferrara, 2008). Taken together, the range of functions attributed to CD11b^{+} cells and the inherent heterogeneity of BMDCs expressing CD11b highlight the complexity and importance of studying CD11b^{+} cells in metastatic target organs.

8.4 ECM components of premetastatic niches

8.4.1 Fibronectin

The ECM connects cells together in tissues and assists in cell migration during wound healing. One essential component of the ECM is fibronectin (FN), which mediates cellular interactions with the ECM including cell adhesion and migration. A compact, soluble form of FN is a major component of blood plasma (300 μg/ml) and is made mostly by hepatocytes in the liver. Less soluble, fibrillar forms of FN are made by fibroblasts and a variety of other cell types through alternative mRNA splicing from the FN gene, generating a large number of variants with different cell-adhesion, solubility, and ligand binding properties (Pankov and Yamada, 2002). Fibrillar FN acts as a protein scaffold, binding cells to the ECM by interacting with a number of cell surface integrins (Plow *et al.*, 2000) including $\alpha 4\beta 1$ (very late antigen-4; VLA-4). FN also binds to important biological molecules including collagen, fibrin, and lysyl oxidase (LOX). FN is important for cell adhesion, migration, differentiation, and for processes of embryogenesis and wound healing.

FN also plays an important role in formation of the premetastatic niche. Three days after implantation of LLC or B16 tumor cells, increased expression of FN can be observed in pulmonary fibroblasts and fibroblast-like cells (Kaplan *et al.*, 2005) in metastatic target organs. Deposition of FN secreted from orthotopic mammary tumors has also been observed within a few days of tumor implantation (Erler *et al.*, 2009), indicating that at least some of the FN that accumulates in the lungs after tumor implantation is derived from the primary tumor. FN levels increased up to 14 days post-tumor implantation (Kaplan *et al.*, 2005; Erler *et al.*, 2009), and pulmonary FN accumulation was observed adjacent to bronchioles, which are common sites of metastatic growth. Aggregates of BMDCs are observed in lung tissue adjacent to regions of increased FN staining (Kaplan *et al.*, 2005; Erler *et al.*, 2009), and VEGFR-1^+ cells in the premetastatic niche express the FN receptor VLA-4 ($\alpha 4\beta 1$) providing one potential explanation for the aggregation of VEGFR-1^+ VLA-4^+ cells in premetastatic niches (Kaplan *et al.*, 2005). However, the mechanism of BMDC invasion into premetastatic niches was unknown at the time.

The clustering of BMDCs into premetastatic niches in different metastatic target organs is tumor-type specific. LLC tumors induced premetastatic niche formation only in the lungs and liver, both of which are primary metastatic target organs for this tumor type. Similarly, B16 tumors induced BMDC accumulation and clustering in B16 metastatic target organs including the lungs, liver, testis, spleen, and kidney (Kaplan *et al.*, 2005). These intriguing observations suggest that different primary tumor-types can modify selected organs, and tumor-specific chemokines and/or cytokines are thought to be involved in site selection in metastasis (Balkwill, 2004).

8.4.2 LOX

LOX is an amine oxidase that cross-links collagen and elastin molecules in the ECM (Csiszar, 2001; Kagan and Li, 2003). LOX is important for the morphogenesis and repair of connective tissues, particularly in aging and diseased tissues (Maki *et al.*, 2005; Warburton and Shi, 2005). The critical role of LOX in normal development is underscored by the perinatal lethality of $LOX^{-/-}$ mice because of structural and functional defects in the cardiovascular system (Maki *et al.*, 2002; Hornstra *et al.*, 2003), with additional defects observed in the lungs and skin (Maki *et al.*, 2005). LOX is localized in the extracellular compartment of skin, aorta, heart, lung, liver, and cartilage (Kagan *et al.*, 1986; Trackman *et al.*, 1990; Hayashi *et al.*, 2004), and is also found in skeletal muscle, kidney, and pancreas in humans (Kagan and Li, 2003). The expression and activity of LOX in cultured vascular smooth muscle cells are increased by transforming growth factor-β (TGF-β) (Gacheru *et al.*, 1997; Shanley *et al.*, 1997), platelet-derived growth factor (PDGF) (Green *et al.*, 1995), and granulocyte–macrophage colony-stimulating factor (GM-CSF) (Weissen-Plenz *et al.*, 2008). Aberrant LOX expression and/or activity has been linked to a number of disease states including fibrosis of the lung, liver, or kidney (Kagan, 1994; Desmouliere *et al.*, 1997; Di Donato *et al.*, 1997; Tzortzaki *et al.*, 2003), neurodegeneration (Gilad *et al.*, 2001), Alzheimer's disease (Gilad *et al.*, 2005), and cardiovascular disease (Maki *et al.*, 2002; Rodriguez *et al.*, 2002; Hornstra *et al.*, 2003). LOX also plays an important role in cancer progression and metastasis, with increased LOX expression associated with metastasis and poor prognosis in patients with head and neck cancers (Erler *et al.*, 2006; Le *et al.*, 2009; Sakai *et al.*, 2009), oral cancers (Albinger-Hegyi *et al.*, 2010), and breast cancer (Erler *et al.*, 2006).

LOX is translated as a preproprotein, and is subjected to cleavage of its N-terminal signal peptide and glycosylation prior to secretion. LOX is secreted as a catalytically inactive 50 kDa proprotein (Trackman *et al.*, 1992) that is cleaved by extracellular proteases (Cronshaw *et al.*, 1995; Panchenko *et al.*, 1996; Uzel *et al.*, 2001) to release the 32 kDa mature LOX enzyme and a propeptide. LOX enzymatic activity requires copper bound in the active site of LOX (Gacheru *et al.*, 1990) and the presence of lysyl tyrosylquinone as a covalently integrated organic cofactor (Wang *et al.*, 1996). The enzymatic activity of LOX produces NH_4 and H_2O_2 as byproducts, and a fluorescence-based assay to measure LOX activity based on H_2O_2 production has been developed (Trackman *et al.*, 1981; Palamakumbura and Trackman, 2002). LOX displays chemoattractant properties for human peripheral blood mononuclear cells (Lazarus *et al.*, 1995) and vascular smooth muscle cells (Li *et al.*, 2000); LOX also increases breast cancer cell migration and invasion (Kirschmann *et al.*, 2002; Payne *et al.*, 2005; Erler *et al.*, 2006). At least four lysyl oxidase-like proteins (LOXL-1 to LOXL-4) have been identified (Csiszar, 2001), with LOXL-1 being the most similar to LOX in terms of sequence homology (Kagan and Li, 2003), post-translational processing, and inhibition by β-aminoproprionitrile (BAPN) (Borel *et al.*, 2001; Csiszar, 2001).

Despite the extracellular production of mature LOX, it has been identified in the cytosol and nuclei of cultured vascular smooth muscle cells and fibroblasts (Li *et al.*, 1997). Rapid entry of mature LOX into NIH3T3 cells with subsequent increasing concentration in the nucleus has also been observed (Nellaiappan *et al.*, 2000; Kagan and Li, 2003). The nuclear localization of LOX suggests intranuclear activity of the enzyme, and indeed intracellular LOX interacts with histone H1 (Kagan *et al.*, 1983; Giampuzzi *et al.*, 2003) and has been implicated in modifying gene expression (Giampuzzi *et al.*, 2000; Oleggini *et al.*, 2007). LOX can oxidize basic globular proteins, but not acidic globular proteins (Kagan *et al.*, 1984), likely altering protein–protein and/or protein–nucleic acid interactions (Lucero and Kagan, 2006). The intracellular activities of mature LOX and of the non-enzymatic LOX propeptide (Palamakumbura *et al.*, 2004) are areas of active investigation.

We recently demonstrated that LOX secreted by hypoxic tumor cells is critical for BMDC accumulation and premetastatic niche formation in metastatic target organs (Erler *et al.*, 2009). To date, LOX is the only tumor-secreted factor that has been shown to be directly involved in premetastatic niche formation. LOX expression is increased in hypoxic tumor cells (Denko *et al.*, 2003; Erler *et al.*, 2006), providing an important link between the primary tumor microenvironment and formation of premetastatic niches in distant organs. Enzymatically active LOX secreted by hypoxic tumor cells is found virtually exclusively colocalized with FN in the lungs of tumor-bearing mice by seven days after tumor implant. LOX can bind to FN, and this interaction enhances the enzymatic activity of LOX (Fogelgren *et al.*, 2005). Once in the lungs, LOX cross-links collagen IV in the basement membrane, and both CD11b$^+$ myeloid cells and c-Kit$^+$ HPCs readily adhere to collagen IV that has been cross-linked (Erler *et al.*, 2009). It is worth noting that while LLC and B16 tumors induce pulmonary accumulation of HPCs (Kaplan *et al.*, 2005), we have found that metastatic human and murine mammary tumors induce remarkably elevated numbers of CD11b$^+$ myeloid lineage cells in addition to lower levels of c-Kit$^+$ HPCs (Erler *et al.*, 2009). BMDCs in contact with cross-linked collagen IV exhibit increased expression of MMP-2 (Erler *et al.*, 2009), which degrades collagen in the ECM. Regions of degraded collagen are more readily invaded by CD11b$^+$ cells, c-Kit$^+$ HPCs, and tumor cells (Erler *et al.*, 2009). Thus, the enzymatic activity of LOX in premetastatic niches creates localized areas of ECM that are primed for invasion by BMDCs and tumor cells.

8.4.3 MMPs

MMPs are a large family of enzymes involved in ECM turnover (Lochter *et al.*, 1998; Mott and Werb, 2004). Before the description of premetastatic niches in metastatic target organs, Hiratsuka *et al.* (2002) observed increased expression of enzymatically active MMP-9 in the lungs of mice after subcutaneous implantation of metastatic LL3-Lewis lung carcinoma (LL3-LLC) cells or B16 melanoma cells. MMP activity can release sequestered VEGF from the ECM to promote angiogenesis (Bergers *et al.*, 2000; Rundhaug, 2005; Du *et al.*, 2008), and we have found that MMP-2 expressed by invading CD11b$^+$ BMDCs helps to produce areas of basement membrane that are more permissive for tumor cell invasion (Erler *et al.*, 2009). MMPs therefore contribute to a lung microenvironment that promotes invasion and subsequent growth of metastatic tumor cells.

Increased MMP-9 expression in premetastatic lung tissue is dependent on VEGF-A secreted from primary tumors and on the tyrosine kinase activity of VEGFR-1 (Hiratsuka *et al.*, 2002). MMP-9 expression is observed in pulmonary endothelial cells and Mac-1$^+$ (CD11b$^+$) monocytes in the lungs of tumor-bearing mice. MMP-9 was elevated specifically in the lungs, which was attributed to a higher density of VEGFR-1 on pulmonary endothelial cells at the time (Hiratsuka *et al.*, 2002), but is likely also related to the accumulation of VEGFR-1$^+$ BMDCs during premetastatic niche formation (Kaplan *et al.*, 2005). Indeed, VLA-4$^+$ cells demonstrate increased MMP expression when bound to FN (Huhtala *et al.*, 1995; Yakubenko *et al.*, 2000), and VEGFR-1$^+$ VLA-4$^+$ BMDCs have been identified in premetastatic niches in regions of FN accumulation (Kaplan *et al.*, 2005). Furthermore, increased pulmonary MMP-9 expression was not observed until 12 days after tumor implant (Hiratsuka *et al.*, 2002), which is entirely consistent with the arrival of BMDCs in premetastatic lung tissue (Kaplan *et al.*, 2005).

8.4.4 S100A8 and S100A9

S100A8 and S100A9 are secreted, inflammatory-related proteins that can act as chemoattractants for monocytes and neutrophils (Lackmann *et al.*, 1993; Devery *et al.*, 1994; Hobbs *et al.*, 2003). While neither protein is detectable in LLC or B16 tumor cells, increased expression of S100A8 and S100A9 can be observed in the lungs of mice bearing LLC or B16 tumors (Hiratsuka *et al.*, 2006). Increased expression was found primarily in VEGFR-2^+ endothelial cells of premetastatic lung tissue compared to lungs from naïve mice. S100A8 and A9 upregulation *in vitro* and *in vivo* could be reduced by inhibition of VEGF-A, tumor necrosis factor-α (TNF-α), or TGF-β, but the precise mechanism is unknown. S100A8 and A9 were found to be chemoattractants for alveolar macrophages, LLC cells, and B16 cells (Hiratsuka *et al.*, 2006). S100A8 and A9 were subsequently found to induce expression of serum amyloid A3 (SAA3) in premetastatic lungs (Hiratsuka *et al.*, 2008), although the precise molecular mechanism is unknown. SAA3 has previously been shown to play a role in chemoattraction of phagocytes (Liang *et al.*, 2000; He *et al.*, 2003; Larson *et al.*, 2003). SAA3 is thought to activate nuclear factor-kappaB (NF-κB) by signaling through Toll-like receptor-4 (TLR4) in premetastatic lung tissue, and NF-κB is known to stimulate production of proinflammatory cytokines by myeloid and lymphoid cells (Naugler and Karin, 2008). Taken together, these data suggest that increased expression of S100A8, S100A9, and SAA3 by lung endothelial cells after tumor implant may help to attract CD11b$^+$ cells and tumor cells to metastatic target organs.

8.5 Premetastatic niche formation precedes metastatic growth

The presence of metastatic tumor cells is not required for formation of the premetastatic niche. Indeed, accumulation of FN, LOX, and aggregation of BMDCs in premetastatic niches occurs prior to the arrival of detectable numbers of disseminated tumor cells (Kaplan *et al.*, 2005; Erler *et al.*, 2009). Furthermore, formation of premetastatic niches can be induced in tumor-free mice after intraperitoneal injection of serum-free conditioned media preincubated with metastatic tumor cells *in vitro*. We have found that conditioned media derived from human and murine mammary tumor cells incubated in hypoxia induces BMDC accumulation in metastatic target organs that can be disrupted when LOX is inhibited with a small molecule inhibitor (BAPN) or inhibitory antibody (Erler *et al.*, 2009). In addition, BMDC accumulation did not occur after daily injections of conditioned media derived from tumor cells stably expressing LOX shRNA. Interestingly, conditioned media obtained from hypoxic tumor cells that contained high levels of enzymatically active LOX was a more potent inducer of the premetastatic niche than conditioned media derived from normoxic tumor cells that contained significantly lower levels of enzymatically active LOX (Bennewith, Erler, and Giaccia, unpublished data). Thus, premetastatic niche formation was dependent on the availability of enzymatically active LOX secreted from hypoxic metastatic tumor cells.

BMDCs may affect the local cytokine milieu within metastatic target organs, independent of the presence or absence of metastatic tumor cells. It is therefore possible that BMDC accumulation induced by metastatic tumors may 'precondition' target organs to promote subsequent metastatic growth. Indeed, the induction of pulmonary premetastatic niches by injecting LOX-containing conditioned media stimulated the metastatic growth of largely non-metastatic mammary tumors expressing LOX shRNA (Erler *et al.*, 2009). Enhanced $CD11b^+$ cell accumulation and pulmonary metastasis of LOX shRNA tumors occurred when exogenous LOX was injected daily during primary tumor growth, and even when LOX was provided only for 2 weeks prior to tumor implant (Erler *et al.*, 2009). Thus, lung tissue can be preconditioned by LOX, resulting in BMDC accumulation, premetastatic niche formation, and metastatic growth of typically nonmetastatic tumors.

Evidence that the site-specificity of metastasis is at least partly mediated by proteins secreted by tumor cells comes from a set of experiments in which conditioned media derived from B16 cells was used to modify the metastatic site-specificity of LLC cells (Kaplan *et al.*, 2005). Mice were injected with conditioned media derived from B16 cells *in vitro* to initiate BMDC accumulation in B16 metastatic target organs. LLC cells were then implanted subcutaneously, and injection of B16 conditioned media was continued during tumor growth. Interestingly, metastatic foci of LLC cells were observed in B16 metastatic target organs, indicating that the proteins secreted from B16 tumor cells influenced the subsequent sites of metastatic growth of LLC tumor cells. Whether or not the B16 media was also affecting the LLC cells in the primary tumor is an open question, and the identity of the secreted protein(s) responsible for site-specific metastasis is unknown. Nevertheless, these data underscore the importance of tumor-derived secreted proteins in formation of premetastatic niches and enhancing metastatic growth independent from the presence of metastatic tumor cells in the organ.

Some tumor cells are capable of growing into tumor cell foci in murine lungs after intravenous (i.v.) injection. Tumor cells injected intravenously go primarily to the lungs, allowing the assessment of survival and growth of tumor cells in a foreign environment without the need for dissemination from a primary tumor. Evidently, the development of tumor cell foci after injection of large numbers (typically $>50\,000$) of tumor cells in this model does not require significant numbers of BMDCs in the lungs *prior* to tumor cell injection. Whether or not BMDCs accumulate after i.v. tumor cell injection and promote growth of tumor cell foci in the lungs is unknown, but it is certainly feasible given the relatively rapid increase in BMDC accumulation observed after tumor implant (Kaplan *et al.*, 2005; Erler *et al.*, 2009).

8.5.1 BMDCs and hypoxia

Cellular oxygenation status plays a central role in solid tumor progression (Graeber *et al.*, 1996), angiogenesis (Fong, 2008), and metastasis (Cairns *et al.*, 2003; Sullivan and Graham, 2007; Bristow and Hill, 2008) through increased expression of a variety of genes (Denko *et al.*, 2003). Preferential accumulation of $CD11b^+$

cells is observed in hypoxic regions of primary tumors (Bennewith, unpublished data), and recruitment of CD45$^+$ myeloid cells (including CD11b$^+$ cells) to primary tumors is dependent on the transcriptional activity of hypoxia-inducible factor-1 (HIF-1) (Du *et al.*, 2008). BMDCs are recruited to injured or ischemic normal tissue (Ceradini and Gurtner, 2005), and primitive hematopoietic cells are found preferentially in hypoxic 'niches' in the bone marrow (Kaplan *et al.*, 2007; Parmar *et al.*, 2007). Thus BMDCs accumulate in hypoxic normal and neoplastic tissue, suggesting that oxygenation levels in premetastatic sites may influence BMDC accumulation. However, despite evidence of severe hypoxia in micrometastases (Li *et al.*, 2007; Li and O'Donoghue, 2008), the role of regional oxygenation status within the premetastatic niche itself is unknown.

It is conceivable that site-specific binding of FN, BMDCs, and/or the expression of CD11b and VEGFR-1 on invading BMDCs may be related to regional oxygenation status in metastatic target organs. The FN receptor VLA-5 is hypoxia-inducible (Indovina *et al.*, 2006; Spangenberg *et al.*, 2006), and VEGFR-1 expression is increased in CD11b$^+$ cells during oxidative stress (Kusmartsev *et al.*, 2008). The cell surface proteins CD11b (Scannell *et al.*, 1995; Simms and D'Amico, 1995; Harmon *et al.*, 2004; Jiang *et al.*, 2005) and VEGFR-1 (Gerber *et al.*, 1997) are also hypoxia-inducible. VEGF is a well-known hypoxia-induced protein essential for angiogenesis (Liao and Johnson, 2007), and is also involved in recruitment of VEGFR-1$^+$ cells to sites of vasculogenesis (Grunewald *et al.*, 2006). The chemokine stromal-derived factor-1 (SDF-1) is also induced by hypoxia (Hitchon *et al.*, 2002; Ceradini *et al.*, 2004), and SDF-1 interaction with the chemokine receptor CXCR4 is involved in site-directed recruitment of BMDCs (Bonig *et al.*, 2004 Kucia *et al.*, 2004; Papayannopoulou, 2004) and CXCR4$^+$ tumor cells (Kaifi *et al.*, 2005; Kucia *et al.*, 2005; Burger and Kipps, 2006). Interestingly, hypoxic lung tissue has been observed in areas of localized ECM remodeling (as occurs with LOX activity), fibrosis, and inflammation (Polosukhin *et al.*, 2007; Steinke *et al.*, 2008) due to decreased microregional perfusion. Inflammation (including influx of an array of leukocytes) is recognized as a promoter of tumor progression (Coussens and Werb, 2002; Lu *et al.*, 2006; Rollins, 2006), and the range of chemokines and growth factors produced by inflammatory leukocytes create an environment that promotes tumor progression, proliferation, and survival (Coussens and Werb, 2002; Lu *et al.*, 2006; Rollins, 2006). Thus while hypoxia in the primary tumor microenvironment induces premetastatic niche formation at distant sites via secretion of LOX, it is possible that reduced perfusion (or even ischemia) within premetastatic niches may further enhance niche formation and subsequent metastatic growth.

8.6 Therapeutic targeting of the premetastatic niche

By improving our understanding of how premetastatic niches develop and the role of premetastatic niches in enhancing metastatic tumor growth, a variety of therapeutic strategies to therapeutically target premetastatic niches can be developed. Immunological inhibition of VEGFR-1$^+$ cells or of the FN receptor VLA-4 has been shown to reduce BMDC accumulation in metastatic target organs and inhibit metastatic growth (Kaplan *et al.*, 2005). Similarly, inhibition of enzymatically active LOX with the small molecule inhibitor BAPN or with a LOX antibody prevented premetastatic niche formation and metastatic growth of human and murine mammary tumors (Erler *et al.*, 2009). LOX secreted by hypoxic tumor cells is an excellent therapeutic target to decrease premetastatic niche formation and metastatic growth. LOX is essential for premetastatic niche formation, and LOX also influences metastatic dissemination of cells from primary tumors (Erler *et al.*, 2006). Furthermore, inhibition of LOX is less likely to induce pleiotropic side effects compared to targeting BMDCs based on immunological inhibition of cell surface markers.

While the mechanisms underlying the formation of premetastatic niches are being uncovered, very little is known about the *maintenance* of premetastatic niches after they have formed. Proteins secreted by hypoxic tumor cells such as LOX are required to initiate premetastatic niches, but we do not know if continual secretion of LOX by primary tumors or continued activity of LOX within premetastatic niches are required to maintain premetastatic niches over time. Presumably, after metastatic tumor cells have arrived in premetastatic niches, paracrine secretion of LOX and other factors could act locally even after removal of the primary tumor to maintain the niche. However, what if premetastatic niches have formed in metastatic target organs, but metastatic tumor cells have not yet arrived? The stability of the premetastatic niche in the absence of metastasizing tumor cells is unknown.

If the premetastatic niche, once formed, is a chronic condition that develops in metastatic target organs of tumor-bearing hosts, then removal of a primary tumor after niche formation (before metastatic tumor cell dissemination) should not affect niches in metastatic target organs. Since premetastatic niches can improve the metastatic growth of minimally metastatic tumors (Erler *et al.*, 2009), then subsequent tumors that develop in niche-bearing hosts may have increased metastatic potential and growth.

If the premetastatic niche is a transient phenomenon that is reversible after removal of the niche-forming stimuli, a host of therapeutic strategies become available. Surgical resection of the primary tumor is one possibility, and we do not know the influence of primary tumor removal on maintenance of the premetastatic niche or the fate of BMDCs in metastatic target organs. Therapeutically targeting hypoxic cells in the primary tumor using hypoxia-activated cytotoxins (e.g., tirapazamine (Zeman *et al.*, 1986) or PR-104 (Patterson *et al.*, 2007)) or small molecule inhibitors of the HIFs may also disrupt premetastatic niches. Combination of HIF inhibitors with therapeutic inhibition of tumor-secreted proteins such as LOX may represent a viable and effective strategy to target the premetastatic niche while also limiting the development of additional premetastatic niches.

The timing of therapeutic intervention may be essential when targeting the premetastatic niche. Because premetastatic niches develop prior to the arrival of metastatic tumor cells, the value of targeting premetastatic niches after metastatic disease is present must be considered. It is conceivable that maximal therapeutic benefit may be realized by targeting formation of the premetastatic niche in patients with localized primary tumors at risk of developing metastatic disease in order to *prevent* metastatic growth before gross metastatic disease is detectable. However, in order to select patients that have premetastatic niches in metastatic target organs and are therefore at risk for developing metastatic disease, we must develop reliable methods to detect components of premetastatic niches in metastatic target organs and/or the bloodstream.

8.7 Evidence for premetastatic niches in the clinic

Clinical data is emerging to indicate that BMDC accumulation and premetastatic niche formation may occur in cancer patients. VEGFR-1$^+$ cells have been identified in some clinical tissues, both associated with metastatic tumor cells and in regions of tissue that do not contain detectable tumor cells (Kaplan *et al.*, 2005). MMP-9 levels are elevated in the lungs of patients with a range of solid primary tumors (Hiratsuka *et al.*, 2002). LOX and CD11b$^+$ cells have also been observed in a variety of metastatic tissues by staining of tissue microarrays (Erler *et al.*, 2009). Thus, many of the important players involved in premetastatic niches have been observed in clinical biopsies of metastatic tissue. However, metastatic tissues are seldom biopsied on a routine basis, particularly before gross evidence of disseminated metastatic disease in the tissue. Thus, alternative methods of detection are required to identify patients that have developed premetastatic niches with undetectable metastatic disease, who are at high risk of growing metastatic tumors.

Many components of the premetastatic niche are transported to metastatic target organs through the bloodstream, providing an easily accessible source for detecting circulating proteins and/or cells. Enzymatically active LOX can be detected in the plasma, and we have found that circulating levels of LOX are related to the presence of metastatic disease in patients with prostate cancer or head and neck cancer (Bennewith, Erler, and Giaccia unpublished data). Elevated levels of CD11b$^+$ cells are found in the peripheral blood of patients with cancer of the head and neck, lung, or breast (Young and Lathers, 1999; Almand *et al.*, 2001), and in patients with renal cell carcinoma (Zea *et al.*, 2005; Kusmartsev *et al.*, 2008). Moreover, increased numbers of circulating CD11b$^+$ cells have recently been correlated with clinical stage and metastatic tumor burden in a mixed group of patients with a variety of tumor types (Diaz-Montero *et al.*, 2009). These findings illustrate the feasibility of measuring premetastatic niche components in the circulation of cancer patients, although further correlation of these data with identification of patients that have developed premetastatic niches (and with therapeutic outcome) is required before blood samples can be used to direct therapeutic targeting of the premetastatic niche.

8.8 Concluding remarks

Emerging evidence supports the idea that the primary tumor microenvironment can influence the microenvironment within distant metastatic target organs. The identification of BMDCs in premetastatic niches has opened new avenues of research into the role that host tissues may (inadvertently) play in promoting the survival and growth of metastatic tumor cells. While a number of BMDC types are found in varying amounts in metastatic target organs, the common thread seems to be the relationship of BMDC accumulation with enhanced metastatic growth. The involvement of BMDCs in promoting metastatic tumor growth has enormous potential to improve our understanding of the metastatic process and to design novel therapies to target metastatic disease. By studying the determinants of 'successful' metastatic growth, we will be better equipped to design more effective strategies to treat or prevent metastatic cancer.

References

Ahn, G.O. and Brown, J.M. (2008) Matrix metalloproteinase-9 is required for tumor vasculogenesis but not for angiogenesis: role of bone marrow-derived myelomonocytic cells. *Cancer Cell*, **13** (3), 193–205.

Albinger-Hegyi, A., Stoeckli, S.J., Schmid, S., *et al.* (2010) Lysyl oxidase expression is an independent marker of prognosis and a predictor of lymph node metastasis in oral and oropharyngeal squamous cell carcinoma (OSCC). *International Journal of Cancer*, **126** (11), 2653–2662.

Almand, B., Clark, J.I., Nikitina, E., *et al.* (2001) Increased production of immature myeloid cells in cancer patients: a mechanism of immunosuppression in cancer. *Journal of Immunology*, **166** (1), 678–689.

Asahara, T., Masuda, H., Takahashi, T., *et al.* (1999) Bone marrow origin of endothelial progenitor cells responsible for postnatal vasculogenesis in physiological and pathological neovascularization. *Circulation Research*, **85** (3), 221–228.

Balkwill, F. (2004) Cancer and the chemokine network. *Nature Reviews Cancer*, **4** (7), 540–550.

Bergers, G., Brekken, R., McMahon, G., *et al.* (2000) Matrix metalloproteinase-9 triggers the angiogenic switch during carcinogenesis. *Nature Cell Biology*, **2** (10), 737–744.

Berzofsky, J.A., Terabe, M., Oh, S., *et al.* (2004) Progress on new vaccine strategies for the immunotherapy and prevention of cancer. *Journal of Clinical Investigation*, **113** (11), 1515–1525.

Bonig, H., Priestley, G.V., Nilsson, L.M., Jiang, Y. and Papayannopoulou, T. (2004) PTX-sensitive signals in bone marrow homing of fetal and adult hematopoietic progenitor cells. *Blood*, **104** (8), 2299–2306.

Borel, A., Eichenberger, D., Farjanel, J., *et al.* (2001) Lysyl oxidase-like protein from bovine aorta. Isolation and maturation to an active form by bone morphogenetic protein-1. *Journal of Biological Chemistry*, **276** (52), 48944–48949.

Bristow, R.G. and Hill, R.P. (2008) Hypoxia and metabolism. Hypoxia, DNA repair and genetic instability. *Nature Reviews Cancer*, **8** (3), 180–192.

Burger, J.A. and Kipps, T.J. (2006) CXCR4: a key receptor in the crosstalk between tumor cells and their microenvironment. *Blood*, **107** (5), 1761–1767.

Cairns, R.A., Khokha, R. and Hill, R.P. (2003) Molecular mechanisms of tumor invasion and metastasis: an integrated view. *Current Molecular Medicine*, **3** (7), 659–671.

Cameron, M.D., Schmidt, E.E., Kerkvliet, N., *et al.* (2000) Temporal progression of metastasis in lung: cell survival, dormancy, and location dependence of metastatic inefficiency. *Cancer Research*, **60** (9), 2541–2546.

Ceradini, D.J. and Gurtner, G.C. (2005) Homing to hypoxia: HIF-1 as a mediator of progenitor cell recruitment to injured tissue. *Trends in Cardiovascular Medicine*, **15** (2), 57–63.

Ceradini, D.J., Kulkarni, A.R., Callaghan, M.J., *et al.* (2004) Progenitor cell trafficking is regulated by hypoxic gradients through HIF-1 induction of SDF-1. *Nature Medicine*, **10** (8), 858–864.

Coussens, L.M. and Werb, Z. (2002) Inflammation and cancer. *Nature*, **420** (6917), 860–867.

Cronshaw, A.D., Fothergill-Gilmore, L.A. and Hulmes, D.J. (1995) The proteolytic processing site of the precursor of lysyl oxidase. *Biochemical Journal*, **306** (Pt. 1), 279–284.

Csiszar, K. (2001) Lysyl oxidases: a novel multifunctional amine oxidase family. *Progress in Nucleic Acid Research and Molecular Biology*, **70**, 1–32.

DaCosta, M.L., Yao, Z., MacPherson, B.C., Jayakar, D.V. and Jeevanandam, V. (2003) Brief hypoxia conditions monocytes to protect reperfused cardiocytes against cell death via the CD11b receptor. *Journal of Heart and Lung Transplantation*, **22** (9), 979–985.

Davidoff, A.M., Ng, C.Y., Brown, P., *et al.* (2001) Bone marrow-derived cells contribute to tumor neovasculature and, when modified to express an angiogenesis inhibitor, can restrict tumor growth in mice. *Clinical Cancer Research*, **7** (9), 2870–2879.

Denko, N.C., Fontana, L.A., Hudson, K.M., *et al.* (2003) Investigating hypoxic tumor physiology through gene expression patterns. *Oncogene*, **22** (37), 5907–5914.

De Palma, M., Venneri, M.A., Roca, C. and Naldini, L. (2003) Targeting exogenous genes to tumor angiogenesis by transplantation of genetically modified hematopoietic stem cells. *Nature Medicine*, **9** (6), 789–795.

De Palma, M., Venneri, M.A., Galli, R., *et al.* (2005) Tie2 identifies a hematopoietic lineage of proangiogenic monocytes required for tumor vessel formation and a mesenchymal population of pericyte progenitors. *Cancer Cell*, **8** (3), 211–226.

Desmouliere, A., Darby, I., Costa, A.M., *et al.* (1997) Extracellular matrix deposition, lysyl oxidase expression, and myofibroblastic differentiation during the initial stages of cholestatic fibrosis in the rat. *Laboratory Investigation*, **76** (6), 765–778.

Devery, J.M., King, N.J. and Geczy, C.L. (1994) Acute inflammatory activity of the S100 protein CP-10. Activation of neutrophils in vivo and in vitro. *Journal of Immunology*, **152** (4), 1888–1897.

Diaz-Montero, C.M., Salem, M.L., Nishimura, M.I., Garrett-Mayer, E., Cole, D.J. and Montero, A.J. (2009) Increased circulating myeloid-derived suppressor cells correlate with clinical cancer stage, metastatic tumor burden, and doxorubicin-cyclophosphamide chemotherapy. *Cancer Immunology and Immunotherapy*, **58** (1), 49–59.

Di Donato, A., Ghiggeri, G.M., Di Duca, M., *et al.* (1997) Lysyl oxidase expression and collagen cross-linking during chronic adriamycin nephropathy. *Nephron*, **76** (2), 192–200.

Du, R., Lu, K.V., Petritsch, C., *et al.* (2008) HIF1alpha induces the recruitment of bone marrow-derived vascular modulatory cells to regulate tumor angiogenesis and invasion. *Cancer Cell*, **13** (3), 206–220.

Erler, J.T., Bennewith, K.L., Nicolau, M., *et al.* (2006) Lysyl oxidase is essential for hypoxia-induced metastasis. *Nature*, **440** (7088), 1222–1226.

Erler, J.T., Bennewith, K.L., Cox, T.R., *et al.* (2009) Hypoxia-induced lysyl oxidase is a critical mediator of bone marrow cell recruitment to form the pre-metastatic niche. *Cancer Cell*, **6** (15), 35–44.

Fogelgren, B., Polgar, N., Szauter, K.M., *et al.* (2005) Cellular fibronectin binds to lysyl oxidase with high affinity and is critical for its proteolytic activation. *Journal of Biological Chemistry*, **280** (26), 24690–24697.

Fong, G.H. (2008) Mechanisms of adaptive angiogenesis to tissue hypoxia. *Angiogenesis*, **11** (2), 121–140.

Gacheru, S.N., Thomas, K.M., Murray, S.A., *et al.* (1997) Transcriptional and post-transcriptional control of lysyl oxidase expression in vascular smooth muscle cells: effects of TGF-beta 1 and serum deprivation. *Journal of Cellular Biochemistry*, **65** (3), 395–407.

Gacheru, S.N., Trackman, P.C., Shah, M.A., *et al.* (1990) Structural and catalytic properties of copper in lysyl oxidase. *Journal of Biological Chemistry*, **265** (31), 19022–19027.

Gao, D., Nolan, D.J., Mellick, A.S., Bambino, K., McDonnell, K. and Mittal, V. (2008) Endothelial progenitor cells control the angiogenic switch in mouse lung metastasis. *Science*, **319** (5860), 195–198.

Gerber, H.P., Condorelli, F., Park, J. and Ferrara, N. (1997) Differential transcriptional regulation of the two vascular endothelial growth factor receptor genes. Flt-1, but not Flk-1/KDR, is up-regulated by hypoxia. *Journal of Biological Chemistry*, **272** (38), 23659–23667.

Giampuzzi, M., Botti, G., Di Duca, M., *et al.* (2000) Lysyl oxidase activates the transcription activity of human collagene III promoter. Possible involvement of Ku antigen. *Journal of Biological Chemistry*, **275** (46), 36341–36349.

Giampuzzi, M., Oleggini, R. and Di Donato, A. (2003) Demonstration of in vitro interaction between tumor suppressor lysyl oxidase and histones H1 and H2: definition of the regions involved. *Biochimica et Biophysica Acta*, **1647** (1-2), 245–251.

Gilad, G.M., Kagan, H.M. and Gilad, V.H. (2001) Lysyl oxidase, the extracellular matrix-forming enzyme, in rat brain injury sites. *Neuroscience Letters*, **310** (1), 45–48.

Gilad, G.M., Kagan, H.M. and Gilad, V.H. (2005) Evidence for increased lysyl oxidase, the extracellular matrix-forming enzyme, in Alzheimer's disease brain. *Neuroscience Letters*, **376** (3), 210–214.

Graeber, T.G., Osmanian, C., Jacks, T., *et al.* (1996) Hypoxia-mediated selection of cells with diminished apoptotic potential in solid tumors. *Nature*, **379** (6560), 88–91.

Grant, M.B., May, W.S., Caballero, S., *et al.* (2002) Adult hematopoietic stem cells provide functional hemangioblast activity during retinal neovascularization. *Nature Medicine*, **8** (6), 607–612.

Green, R.S., Lieb, M.E., Weintraub, A.S., *et al.* (1995) Identification of lysyl oxidase and other platelet-derived growth factor-inducible genes in vascular smooth muscle cells by differential screening. *Laboratory Investigation*, **73** (4), 476–482.

Grunewald, M., Avraham, I., Dor, Y., *et al.* (2006) VEGF-induced adult neovascularization: recruitment, retention, and role of accessory cells. *Cell*, **124** (1), 175–189.

Gupta, G.P. and Massague, J. (2006) Cancer metastasis: building a framework. *Cell*, **127** (4), 679–695.

Harmon, D., Lan, W. and Shorten, G. (2004) The effect of aprotinin on hypoxia-reoxygenation-induced changes in neutrophil and endothelial function. *European Journal of Anaesthesiology*, **21** (12), 973–979.

Hattori, K., Dias, S., Heissig, B., *et al.* (2001) Vascular endothelial growth factor and angiopoietin-1 stimulate postnatal hematopoiesis by recruitment of vasculogenic and hematopoietic stem cells. *Journal of Experimental Medicine*, **193** (9), 1005–1014.

Hayashi, K., Fong, K.S., Mercier, F., Boyd, C.D., Csiszar, K. and Hayashi, M. (2004) Comparative immunocytochemical localization of lysyl oxidase (LOX) and the lysyl oxidase-like (LOXL) proteins: changes in the expression of LOXL during development and growth of mouse tissues. *Journal of Molecular Histology*, **35** (8–9), 845–855.

He, R., Sang, H. and Ye, R.D. (2003) Serum amyloid A induces IL-8 secretion through a G protein-coupled receptor, FPRL1/LXA4R. *Blood*, **101** (4), 1572–1581.

Hiratsuka, S., Nakamura, K., Iwai, S., *et al.* (2002) MMP9 induction by vascular endothelial growth factor receptor-1 is involved in lung-specific metastasis. *Cancer Cell*, **2** (4), 289–300.

Hiratsuka, S., Watanabe, A., Aburatani, H. and Maru, Y. (2006) Tumor-mediated upregulation of chemoattractants and recruitment of myeloid cells predetermines lung metastasis. *Nature Cell Biology*, **8** (12), 1369–1375.

Hiratsuka, S., Watanabe, A., Sakurai, Y., *et al.* (2008) The S100A8-serum amyloid A3-TLR4 paracrine cascade establishes a pre-metastatic phase. *Nature Cell Biology*, **10** (11), 1349–1355.

Hitchon, C., Wong, K., Ma, G., Reed, J., Lyttle, D. and El-Gabalawy, H. (2002) Hypoxia-induced production of stromal cell-derived factor 1 (CXCL12) and vascular endothelial growth factor by synovial fibroblasts. *Arthritis and Rheumatism*, **46** (10), 2587–2597.

Hobbs, J.A., May, R., Tanousis, K., *et al.* (2003) Myeloid cell function in MRP-14 (S100A9) null mice. *Molecular Cell Biology*, **23** (7), 2564–2576.

Holmgren, L., O'Reilly, M.S. and Folkman, J. (1995) Dormancy of micrometastases: balanced proliferation and apoptosis in the presence of angiogenesis suppression. *Nature Medicine*, **1** (2), 149–153.

Hornstra, I.K., Birge, S., Starcher, B., Bailey, A.J., Mecham, R.P. and Shapiro, S.D. (2003) Lysyl oxidase is required for vascular and diaphragmatic development in mice. *Journal of Biological Chemistry*, **278** (16), 14387–14393.

Huhtala, P., Humphries, M.J., McCarthy, J.B., Tremble, P.M., Werb, Z. and Damsky, C.H. (1995) Cooperative signaling by alpha 5 beta 1 and alpha 4 beta 1 integrins regulates metalloproteinase gene expression in fibroblasts adhering to fibronectin. *Journal of Cell Biology*, **129** (3), 867–879.

Indovina, P., Ferrante, A., Rainaldi, G. and Santini, M.T. (2006) Hypoxia and ionizing radiation: changes in adhesive properties and cell adhesion molecule expression in MG-63 three-dimensional tumor spheroids. *Cell Communication and Adhesion*, **13** (3), 185–198.

Jiang, Y., Xue, Z.H., Shen, W.Z., *et al.* (2005) Desferrioxamine induces leukemic cell differentiation potentially by hypoxia-inducible factor-1[alpha] that augments transcriptional activity of CCAAT//enhancer-binding protein-[alpha]. *Leukemia*, **19** (7), 1239–1247.

Kagan, H.M. (1994) Lysyl oxidase: mechanism, regulation and relationship to liver fibrosis. *Pathology Research and Practice*, **190** (9–10), 910–919.

Kagan, H.M. and Li, W. (2003) Lysyl oxidase: properties, specificity, and biological roles inside and outside of the cell. *Journal of Cellular Biochemistry*, **88** (4), 660–672.

Kagan, H.M., Vaccaro, C.A., Bronson, R.E., Tang, S.S. and Brody, J.S. (1986) Ultrastructural immunolocalization of lysyl oxidase in vascular connective tissue. *Journal of Cell Biology*, **103** (3), 1121–1128.

Kagan, H.M., Williams, M.A., Calaman, S.D. and Berkowitz, E.M. (1983) Histone H1 is a substrate for lysyl oxidase and contains endogenous sodium borotritide-reducible residues. *Biochemical and Biophysical Research Communication*, **115** (1), 186–192.

Kagan, H.M., Williams, M.A., Williamson, P.R. and Anderson, J.M. (1984) Influence of sequence and charge on the specificity of lysyl oxidase toward protein and synthetic peptide substrates. *Journal of Biological Chemistry*, **259** (18), 11203–11207.

Kaifi, J.T., Yekebas, E.F., Schurr, P., *et al.* (2005) Tumor-cell homing to lymph nodes and bone marrow and CXCR4 expression in esophageal cancer. *Journal of National Cancer Institute*, **97** (24), 1840–1847.

Kaplan, R.N., Psaila, B. and Lyden, D. (2007) Niche-to-niche migration of bone-marrow-derived cells. *Trends in Molecular Medicine*, **13** (2), 72–81.

Kaplan, R.N., Riba, R.D., Zacharoulis, S., *et al.* (2005) VEGFR1-positive haematopoietic bone marrow progenitors initiate the pre-metastatic niche. *Nature*, **438** (7069), 820–827.

Kirschmann, D.A., Seftor, E.A., Fong, S.F., *et al.* (2002) A molecular role for lysyl oxidase in breast cancer invasion. *Cancer Research*, **62** (15), 4478–4483.

Kopp, H.G., Ramos, C.A. and Rafii, S. (2006) Contribution of endothelial progenitors and proangiogenic hematopoietic cells to vascularization of tumor and ischemic tissue. *Current Opinion in Hematology*, **13** (3), 175–181.

Kucia, M., Jankowski, K., Reca, R., *et al.* (2004) CXCR4-SDF-1 signalling, locomotion, chemotaxis and adhesion. *Journal of Molecular Histology*, **35** (3), 233–245.

Kucia, M., Reca, R., Miekus, K., *et al.* (2005) Trafficking of normal stem cells and metastasis of cancer stem cells involve similar mechanisms: pivotal role of the SDF-1-CXCR4 axis. *Stem Cells*, **23** (7), 879–894.

Kusmartsev, S., Eruslanov, E., Kubler, H., *et al.* (2008) Oxidative stress regulates expression of VEGFR1 in myeloid cells: link to tumor-induced immune suppression in renal cell carcinoma. *Journal of Immunology*, **181** (1), 346–353.

Kusmartsev, S., Su, Z., Heiser, A., *et al.* (2008) Reversal of myeloid cell-mediated immunosuppression in patients with metastatic renal cell carcinoma. *Clinical Cancer Research*, **14** (24), 8270–8278.

Lackmann, M., Rajasekariah, P., Iismaa, S.E., *et al.* (1993) Identification of a chemotactic domain of the pro-inflammatory S100 protein CP-10. *Journal of Immunology*, **150** (7), 2981–2991.

Larson, M.A., Wei, S.H., Weber, A., Weber, A.T. and McDonald, T.L. (2003) Induction of human mammary-associated serum amyloid A3 expression by prolactin or lipopolysaccharide. *Biochemical and Biophysical Research Communications*, **301** (4), 1030–1037.

Lazarus, H.M., Cruikshank, W.W., Narasimhan, N., Kagan, H.M. and Center, D.M. (1995) Induction of human monocyte motility by lysyl oxidase. *Matrix Biology*, **14** (9), 727–731.

Le, Q.T., Harris, J., Magliocco, A.M., *et al.* (2009) Validation of lysyl oxidase as a prognostic marker for metastasis and survival in head and neck squamous cell carcinoma: Radiation Therapy Oncology Group trial 90-03. *Journal of Clinical Oncology*, **27** (26), 4281–4286.

Li, X.-F., Carlin, S., Urano, M., Russell, J., Ling, C.C. and O'Donoghue, J.A. (2007) Visualization of hypoxia in microscopic tumors by immunofluorescent microscopy. *Cancer Research*, **67** (16), 7646–7653.

Li, W., Liu, G., Chou, I.N. and Kagan, H.M. (2000) Hydrogen peroxide-mediated, lysyl oxidase-dependent chemotaxis of vascular smooth muscle cells. *Journal of Cellular Biochemistry*, **78** (4), 550–557.

Li, W., Nellaiappan, K., Strassmaier, T., Graham, L., Thomas, K.M. and Kagan, H.M. (1997) Localization and activity of lysyl oxidase within nuclei of fibrogenic cells. *Proceedings of the National Academy of Sciences of the United States of Amarica*, **94** (24), 12817–12822.

Li, X.F. and O'Donoghue, J.A. (2008) Hypoxia in microscopic tumors. *Cancer Letters*, **264** (2), 172–180.

Liang, T.S., Wang, J.M., Murphy, P.M. and Gao, J.L. (2000) Serum amyloid A is a chemotactic agonist at FPR2, a low-affinity *N*-formylpeptide receptor on mouse neutrophils. *Biochemical and Biophysical Research Communications*, **270** (2), 331–335.

Liao, D. and Johnson, R.S. (2007) Hypoxia: a key regulator of angiogenesis in cancer. *Cancer Metastasis Reviews*, **26** (2), 281–290.

Liu, C., Yu, S., Kappes, J., *et al.* (2007) Expansion of spleen myeloid suppressor cells represses NK cell cytotoxicity in tumor-bearing host. *Blood*, **109** (10), 4336–4342.

Lochter, A., Sternlicht, M.D., Werb, Z. and Bissell, M.J. (1998) The significance of matrix metalloproteinases during early stages of tumor progression. *Annals of the New York Academy Sciences*, **857**, 180–193.

Lu, H., Ouyang, W. and Huang, C. (2006) Inflammation, a key event in cancer development. *Molecular Cancer Research*, **4** (4), 221–233.

Lucero, H.A. and Kagan, H.M. (2006) Lysyl oxidase: an oxidative enzyme and effector of cell function. *Cellular and Molecular Life Sciences*, **63** (19-20), 2304–2316.

Lyden, D., Hattori, K., Dias, S., *et al.* (2001) Impaired recruitment of bone-marrow-derived endothelial and hematopoietic precursor cells blocks tumor angiogenesis and growth. *Natural Medicine*, **7** (11), 1194–1201.

Maki, J.M., Rasanen, J., Tikkanen, H., *et al.* (2002) Inactivation of the lysyl oxidase gene Lox leads to aortic aneurysms, cardiovascular dysfunction, and perinatal death in mice. *Circulation*, **106** (19), 2503–2509.

Maki, J.M., Sormunen, R., Lippo, S., Kaarteenaho-Wiik, R., Soininen, R. and Myllyharju, J. (2005) Lysyl oxidase is essential for normal development and function of the respiratory system and for the integrity of elastic and collagen fibers in various tissues. *American Journal of Pathology*, **167** (4), 927–936.

Mott, J.D. and Werb, Z. (2004) Regulation of matrix biology by matrix metalloproteinases. *Current Opinion in Cellular Biology*, **16** (5), 558–564.

Naugler, W.E. and Karin, M. (2008) NF-kappaB and cancer-identifying targets and mechanisms. *Current Opinion in Genetics and Development*, **18** (1), 19–26.

Naumov, G.N., Akslen, L.A. and Folkman, J. (2006) Role of angiogenesis in human tumor dormancy: animal models of the angiogenic switch. *Cell Cycle*, **5** (16), 1779–1787.

Nellaiappan, K., Risitano, A., Liu, G., Nicklas, G. and Kagan, H.M. (2000) Fully processed lysyl oxidase catalyst translocates from the extracellular space into nuclei of aortic smooth-muscle cells. *Journal of Cellular Biochemistry*, **79** (4), 576–582.

Nolan, D.J., Ciarrocchi, A., Mellick, A.S., *et al.* (2007) Bone marrow-derived endothelial progenitor cells are a major determinant of nascent tumor neovascularization. *Genes and Development*, **21** (12), 1546–1558.

Oleggini, R., Gastaldo, N. and Di Donato, A. (2007) Regulation of elastin promoter by lysyl oxidase and growth factors: cross control of lysyl oxidase on TGF-beta1 effects. *Matrix Biology*, **26** (6), 494–505.

Paget, S. (1989) The distribution of secondary growths in cancer of the breast. 1889. *Cancer Metastasis Reviews*, **8** (2), 98–101.

Palamakumbura, A.H., Jeay, S., Guo, Y., *et al.* (2004) The propeptide domain of lysyl oxidase induces phenotypic reversion of ras-transformed cells. *Journal of Biological Chemistry*, **279** (39), 40593–40600.

Palamakumbura, A.H. and Trackman, P.C. (2002) A fluorometric assay for detection of lysyl oxidase enzyme activity in biological samples. *Analytical Biochemistry*, **300** (2), 245–251.

Panchenko, M.V., Stetler-Stevenson, W.G., Trubetskoy, O.V., Gacheru, S.N. and Kagan, H.M. (1996) Metalloproteinase activity secreted by fibrogenic cells in the processing of prolysyl oxidase. Potential role of procollagen C-proteinase. *Journal of Biological Chemistry*, **271** (12), 7113–7119.

Pankov, R. and Yamada, K.M. (2002) Fibronectin at a glance. *Journal of Cell Science*, **115** (Pt. 20), 3861–3863.

Papayannopoulou, T. (2004) Current mechanistic scenarios in hematopoietic stem/progenitor cell mobilization. *Blood*, **103** (5), 1580–1585.

Parmar, K., Mauch, P., Vergilio, J.A., Sackstein, R. and Down, J.D. (2007) Distribution of hematopoietic stem cells in the bone marrow according to regional hypoxia. *Proceedings of the National Academy of Sciences of the United States of America*, **104** (13), 5431–5436.

Patterson, A.V., Ferry, D.M., Edmunds, S.J., *et al.* (2007) Mechanism of action and preclinical antitumor activity of the novel hypoxia-activated DNA cross-linking agent PR-104. *Clinical Cancer Research*, **13** (13), 3922–3932.

Payne, S.L., Fogelgren, B., Hess, A.R., *et al.* (2005) Lysyl oxidase regulates breast cancer cell migration and adhesion through a hydrogen peroxide-mediated mechanism. *Cancer Research*, **65** (24), 11429–11436.

Plow, E.F., Haas, T.A., Zhang, L., Loftus, J. and Smith, J.W. (2000) Ligand binding to integrins. *Journal of Biological Chemistry*, **275** (29), 21785–21788.

Polosukhin, V.V., Lawson, W.E., Milstone, A.P., *et al.* (2007) Association of progressive structural changes in the bronchial epithelium with subepithelial fibrous remodeling: a potential role for hypoxia. *Virchows Archiv*, **451** (4), 793–803.

Rafii, S., Avecilla, S., Shmelkov, S., *et al.* (2003) Angiogenic factors reconstitute hematopoiesis by recruiting stem cells from bone marrow microenvironment. *Annals of the New York Academy Science*, **996**, 49–60.

Rafii, D.C., Psaila, B., Butler, J., Jin, D.K. and Lyden, D. (2008) Regulation of vasculogenesis by platelet-mediated recruitment of bone marrow-derived cells. *Arteriosclerosis, Thrombosis, and Vascular Biology*, **28** (2), 217–222.

Rodriguez, C., Raposo, B., Martinez-Gonzalez, J., Casani, L. and Badimon, L. (2002) Low density lipoproteins downregulate lysyl oxidase in vascular endothelial cells and the arterial wall. *Arteriosclerosis, Thrombosis, and Vascular Biology*, **22** (9), 1409–1414.

Rollins, B.J. (2006) Inflammatory chemokines in cancer growth and progression. *European Journal of Cancer*, **42** (6), 760–767.

Rundhaug, J.E. (2005) Matrix metalloproteinases and angiogenesis. *Journal of Cellular and Molecular Medicine*, **9** (2), 267–285.

Sakai, M., Kato, H., Sano, A., *et al.* (2009) Expression of lysyl oxidase is correlated with lymph node metastasis and poor prognosis in esophageal squamous cell carcinoma. *Annals of Surgical Oncology*, **16** (9), 2494–2501.

Scannell, G., Waxman, K., Vaziri, N.D., *et al.* (1995) Hypoxia-induced alterations of neutrophil membrane receptors. *Journal of Surgical Research*, **59** (1), 141–145.

Serafini, P., De Santo, C., Marigo, I., *et al.* (2004) Derangement of immune responses by myeloid suppressor cells. *Cancer Immunology and Immunotherapy*, **53** (2), 64–72.

Serafini, P., Borrello, I. and Bronte, V. (2006) Myeloid suppressor cells in cancer: recruitment, phenotype, properties, and mechanisms of immune suppression. *Seminars in Cancer Biology*, **16** (1), 53–65.

Shanley, C.J., Gharaee-Kermani, M., Sarkar, R., *et al.* (1997) Transforming growth factor-beta 1 increases lysyl oxidase enzyme activity and mRNA in rat aortic smooth muscle cells. *Journal of Vascular Surgery*, **25** (3), 446–452.

Shojaei, F. and Ferrara, N. (2008) Refractoriness to antivascular endothelial growth factor treatment: role of myeloid cells. *Cancer Research*, **68** (14), 5501–5504.

Simms, H. and D'Amico, R. (1995) Regulation of polymorphonuclear neutrophil CD16 and CD11b/CD18 expression by matrix proteins during hypoxia is VLA-5, VLA-6 dependent. *Journal of Immunology*, **155** (10), 4979–4990.

Sinha, P., Clements, V.K., Bunt, S.K., Albelda, S.M. and Ostrand-Rosenberg, S. (2007) Cross-talk between myeloid-derived suppressor cells and macrophages subverts tumor immunity toward a type 2 response. *Journal of Immunology*, **179** (2), 977–983.

Spangenberg, C., Lausch, E.U., Trost, T.M., *et al.* (2006) ERBB2-mediated transcriptional up-regulation of the alpha5beta1 integrin fibronectin receptor promotes tumor cell survival under adverse conditions. *Cancer Research*, **66** (7), 3715–3725.

Steeg, P.S. (2006) Tumor metastasis: mechanistic insights and clinical challenges. *Nature Medicine*, **12** (8), 895–904.

Steinke, J.W., Woodard, C.R. and Borish, L. (2008) Role of hypoxia in inflammatory upper airway disease. *Current Opinion in Allergy and Clinical Immunology*, **8** (1), 16–20.

Sullivan, R. and Graham, C.H. (2007) Hypoxia-driven selection of the metastatic phenotype. *Cancer and Metastasis Reviews*, **26** (2), 319–331.

Townson, J.L. and Chambers, A.F. (2006) Dormancy of solitary metastatic cells. *Cell Cycle*, **5** (16), 1744–1750.

Trackman, P.C., Pratt, A.M., Wolanski, A., *et al.* (1990) Cloning of rat aorta lysyl oxidase cDNA: complete codons and predicted amino acid sequence. *Biochemistry*, **29** (20), 4863–4870.

Trackman, P.C., Bedell-Hogan, D., Tang, J. and Kagan, H.M. (1992) Post-translational glycosylation and proteolytic processing of a lysyl oxidase precursor. *Journal of Biological Chemistry*, **267** (12), 8666–8671.

Trackman, P.C., Zoski, C.G. and Kagan, H.M. (1981) Development of a peroxidase-coupled fluorometric assay for lysyl oxidase. *Analytical Biochemistry*, **113** (2), 336–342.

Tzortzaki, E.G., Tischfield, J.A., Sahota, A., Siafakas, N.M., Gordon, M.K. and Gerecke, D.R. (2003) Expression of FACIT collagens XII and XIV during bleomycin-induced pulmonary fibrosis in mice. *Anatomical Record Part A: Discoveries in Molecular, Cellular and Evolutionary Biology*, **275** (2), 1073–1080.

Uzel, M.I., Scott, I.C., Babakhanlou-Chase, H., *et al.* (2001) Multiple bone morphogenetic protein 1-related mammalian metalloproteinases process pro-lysyl oxidase at the correct physiological site and control lysyl oxidase activation in mouse embryo fibroblast cultures. *Journal of Biological Chemistry*, **276** (25), 22537–22543.

Wang, S.X., Mure, M., Medzihradszky, K.F., *et al.* (1996) A crosslinked cofactor in lysyl oxidase: redox function for amino acid side chains. *Science*, **273** (5278), 1078–1084.

Warburton, D. and Shi, W. (2005) Lo, and the niche is knit: lysyl oxidase activity and maintenance of lung, aorta, and skin integrity. *American Journal of Pathology*, **167** (4), 921–922.

Weissen-Plenz, G., Eschert, H., Volker, W., *et al.* (2008) Granulocyte macrophage colony-stimulating factor deficiency affects vascular elastin production and integrity of elastic lamellae. *Journal of Vascular Research*, **45** (2), 103–110.

Yakubenko, V.P., Lobb, R.R., Plow, E.F. and Ugarova, T.P. (2000) Differential induction of gelatinase B (MMP-9) and gelatinase A (MMP-2) in T lymphocytes upon alpha(4)beta(1)-mediated adhesion to VCAM-1 and the CS-1 peptide of fibronectin. *Experimental Cell Research*, **260** (1), 73–84.

Yang, L., Huang, J., Ren, X., *et al.* (2008) Abrogation of TGF beta signaling in mammary carcinomas recruits Gr-1+CD11b+ myeloid cells that promote metastasis. *Cancer Cell*, **13** (1), 23–35.

Young, M.R. and Lathers, D.M. (1999) Myeloid progenitor cells mediate immune suppression in patients with head and neck cancers. *International Journal of Immunopharmacology*, **21** (4), 241–252.

Zea, A.H., Rodriguez, P.C., Atkins, M.B., *et al.* (2005) Arginase-producing myeloid suppressor cells in renal cell carcinoma patients: a mechanism of tumor evasion. *Cancer Research*, **65** (8), 3044–3048.

Zeman, E.M., Brown, J.M., Lemmon, M.J., Hirst, V.K. and Lee, W.W. (1986) SR-4233: a new bioreductive agent with high selective toxicity for hypoxic mammalian cells. *International Journal of Radiation Oncology, Biology, Physics*, **12** (7), 1239–1242.

9

Hypoxia, Anerobic Metabolism, and Interstitial Hypertension

Michael F. Milosevic
Department of Radiation Oncology, University of Toronto, Princess Margaret Hospital, Toronto, Canada

9.1 Introduction

The biologic behavior of cancer cells is determined not only by the intrinsic characteristics of those cells but also by the local microenvironment in which they exist. The structure, organization, and function of the vasculature and interstitium is abnormal in most solid, malignant tumors compared to normal tissues and contributes to a hostile metabolic milieu characterized by hypoxia, anaerobic metabolism, low pH, and high interstitial fluid pressure (IFP). These features of the microenvironment have all been associated to varying degrees with more aggressive tumor behavior and impaired response to radiotherapy or chemotherapy (Bristow and Hill, 2008; Lunt, Chaudary, and Hill, 2009). New drugs that target the microenvironment directly, or aspects of tumor biology that indirectly contribute to microenvironmental dysfunction, are now being tested in the clinic with the aim of improving response to conventional treatments and overall patient outcome (Moeller, Richardson, and Dewhirst, 2007; Brown and Wilson, 2004).

In this chapter, the pathophysiology of the abnormal tumor microenvironment will be reviewed with particular attention to the complex, dynamic interplay among the cellular, vascular, and interstitial compartments. The focus throughout will be on aspects of the microenvironment that are relevant to the assessment and treatment of human tumors in patients. Ways of measuring hypoxia, anaerobic metabolism, and IFP will be reviewed, along with the impact of microenvironmental abnormalities on the biologic and clinical behavior of tumors, response to treatment, and patient outcomes.

Tumor Microenvironment Edited by Dietmar W. Siemann

9.2 Pathophysiology of the tumor microenvironment

9.2.1 Hypoxia

Hypoxia develops in tumors when the metabolic demand for oxygen exceeds availability. This is thought to begin early in tumor development as oxygen consumption by the growing tumor mass outstrips the delivery capacity of the vasculature. Tumor hypoxia can broadly be classified as either acute or chronic according to the temporal characteristics and underlying pathophysiology. Acute and chronic hypoxia coexist in most tumors, although the relative balance between the two and the contributions of each to the overall hypoxic state is not known in most circumstances. There is mounting evidence to indicate that acute and chronic hypoxia may influence tumor behavior and response to treatment in different ways (Bristow and Hill, 2008; Dewhirst, Cao, and Moeller, 2008).

Oxygen supply and consumption are tightly controlled and closely balanced in most normal tissues. However, in tumors, supply and consumption are often decoupled due to loss of normal physiologic regulation and changes in molecular signaling that provide selective growth and survival advantages. Oxygen levels below 10–15 mmHg lead to activation of the hypoxia-inducible factors 1 and 2 (HIF1 and HIF2), which influence the expression of over 100 genes involved in angiogenesis, metabolism, pH regulation, proliferation, metastasis formation, and a range of other molecular and cellular processes (Semenza, 2009). Vascular endothelial growth factor (VEGF) produced by tumor and stromal cells stimulates new vessel formation via endothelial cell proliferation, migration, and survival. Platelet derived growth factor (PDGF) promotes maturation and stability of new vessels by enhancing pericyte recruitment and interaction with endothelial cells. Although angiogenesis is upregulated in most tumors, the controls that regulate this process under normal physiologic conditions are lost, resulting in a vascular network that is structurally and functionally abnormal and inefficient at delivering oxygen and other nutrients (Jain, 1988, 2005).

Paradoxically, the limited oxygen delivery capacity of the inefficient vasculature that forms from unregulated angiogenesis is probably the most important factor contributing to the persistence and progression of hypoxia as tumors grow. Microregional variability in vessel location, size, structure, orientation, and interconnectivity contribute to differences in geometric blood flow resistance, vessel density, and intervessel distance (Jain, 1988; Dewhirst, Cao, and Moeller, 2008). Hemoconcentration, stiffening of red cells (in hypoxia or acidic environments), and intravascular coagulation may focally increase viscous flow resistance. As a result, perfusion rates often vary dramatically from one region to another within the same tumor and among tumors of the same type. There may also be excessively long intravascular path lengths between the arterial and venous ends of the circulation: blood is depleted of oxygen early during passage through the tumor and becomes ineffective as a means of supplying downstream tissues. Oxygen concentration in the tumor interstitium diminishes with increasing distance from vessels in a manner commensurate with the consumption rate of cells, and typically approaches zero beyond distances of

150–200 μm. Collectively, these factors contribute to poor oxygen delivery efficiency and the development of a basal level of chronic hypoxia.

The spatial distribution of oxygen is not stable in tumors, but rather varies over time driven by periodic microregional fluctuations in blood flow that lead to acute or cycling hypoxia (Dewhirst, Cao, and Moeller, 2008). There are least two, often superimposed, cyclical blood flow patterns in tumors: one very slow with a period on the order of days, and the second faster with a period of 20–60 minutes. Small regions within tumors appear to cycle together and global tumor blood flow often does not change substantially, suggesting that cycling hypoxia is the result of blood flow redistribution (Brurberg, Graff, and Rofstad, 2003). The slow changes are thought to result from vascular remodeling. The faster fluctuations reflect inherently unstable blood flow patterns because of the chaotic vascular architecture in tumors. Focal, random perturbations in blood flow caused by residual arterial vasomotion, transient vessel obstruction, or vessel intussusception are amplified by this instability to produce major redistributions of blood flow from one region of a tumor to another.

Oxygen consumption by cancer cells is also an important determinant of tumor hypoxia. Oxygen consumption rates in tumors typically are intermediate in range between normal tissues with low and high metabolic activities (Vaupel, Kallinowski, and Okunieff, 1989). However, there may be substantial spatial and temporal variability within individual tumors and from one tumor to the next. Biomechanical models have suggested that an increase in oxygen consumption may have a much more profound effect on the development of hypoxia than a similar reduction in oxygen delivery under some anatomic and physiologic conditions (Secomb *et al.*, 1993).

The cumulative result of imbalances among the many factors influencing oxygen supply and consumption in tumors is temporal and spatial variability in oxygen concentration. At any point in time, there is a continuum of oxygen concentrations in most tumors that varies from anoxia at one extreme to very high levels typical of normal tissues at the other. The activation or suppression of cellular and metabolic processes, the induction of genes involved in adaptation to hypoxia, and tumor response to radiotherapy or chemotherapy depend on these oxygenation patterns in a dynamic and interactive manner.

9.2.2 Anerobic metabolism and extracellular acidosis

Energy production in well oxygenated normal tissues is mainly via aerobic oxidative phosphorylation, as summarized in Figure 9.1. At low tissue pO_2 levels, aerobic metabolism declines and alternate metabolic pathways are activated to maintain the energy supply necessary to support crucial cellular processes. It has been known for many years that solid tumors preferentially utilize anaerobic metabolism even in the presence of oxygen. despite markedly lower energy yields, a phenomenon known as the Warburg effect (Warburg, Posener, and Negelein, 1930). This partial shift to a less efficient mode of energy production is thought to convey a selective growth and survival advantage.

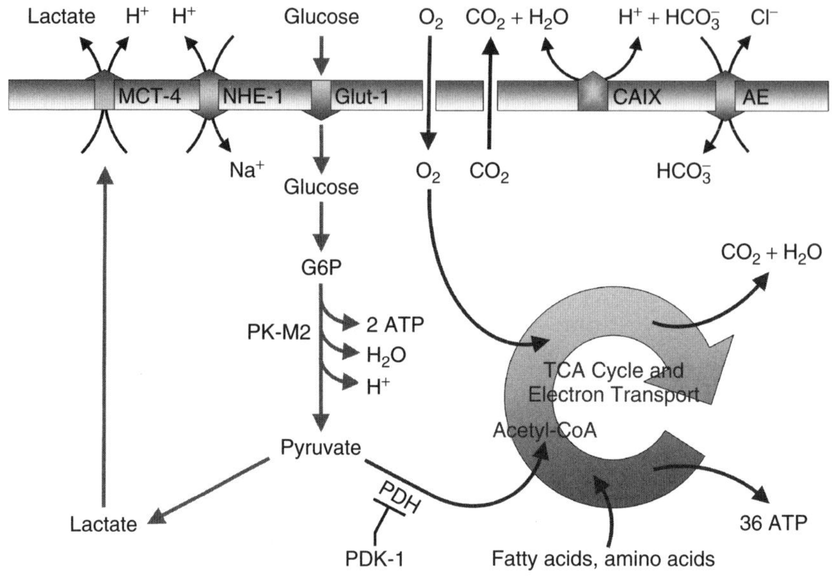

Figure 9.1 Aerobic and anerobic metabolism in tumors. Glucose uptake by cells is actively regulated by specific transporters like Glut-1. Once in the cytoplasm, glucose is converted to pyruvate with the liberation of 2 ATP molecules. Under normal conditions, pyruvate enters the mitochondria and is converted to acetyl-CoA, which fuels the tricarboxylic acid (TCA) cycle and electron transport chain to yield an additional 36 molecules of ATP. Aerobic metabolism is suppressed at pO_2 levels below 10 mmHg and in most solid tumors independent of O_2 concentration. This metabolic 'switch' to glycolysis in tumors is mediated by a variety of HIF dependent (Table 9.1) and independent mechanisms, including upregulation of glycolytic enzymes (Brahimi-Horn, Chiche, and Pouyssegur, 2007; Denko, 2008; Semenza, 2009), expression of the M2 isoform of PK (Christofk *et al.*, 2008), and inhibition of PDH activity by PDK-1 (Wigfield *et al.*, 2008) Under these conditions, pyruvate is converted to lactate and actively exported from cells by the monocarboxyl transporter MCT-4. Intracellular pH is tightly regulated by ion exchangers (MCT-4, NSE-1, AE) and the bicarbonate and other buffer systems. Increased production of H^+ coupled with decreased vascular efficiency leads to acidification of the extracellular tumor microenvironment. AE: anion exchanger, ATP: adenosine triphosphate, CAIX: carbonic anhydrase IX, G6P: glucose-6-phosphate, Glut-1: glucose transporter 1, LDH-A: lactate dehydrogenase A, MCT-4: monocarboxyl transporter 4, NHE-1: sodium hydrogen exchanger 1, PDH: pyruvate dehydrogenase, PDK-1: pyruvate dehydrogenase kinase 1, PK: pyruvate kinase, TCA: tricarboxylic acid.

The molecular mechanisms responsible for this switch to an anaerobic phenotype in tumors are not completely understood, but are thought to involve aspects of the tumor microenvironment including the availability of oxygen and nutrients, and altered expression of oncogenes and tumor suppressors that influence signaling via the HIF1, phosphoinositide 3-kinase (PI3K), and mammalian target of rapamycin (mTOR) pathways. HIF1 upregulation in tumors is thought to play a central role in this metabolic switch, driven by numerous factors including cycling hypoxia early in tumor development, acidosis, reactive oxygen species (ROS), and mutations of the Ras, von Hippel–Lindau (VHL), and PTEN

Table 9.1 HIF1 mediate regulation of tumor cell metabolism.

HIF1 mediated effect	Biochemical effect	Metabolic effect
↑ Glucose transporters	↑ Glucose uptake by cells	↑ Anerobic glycolysis
↑ Glycolytic enzymes	↑ Conversion of glucose to pyruvate and lactate	↑ Anerobic glycolysis
↑ PDK-1	↓ PDH and conversion of pyruvate to acetyl-CoA	↓ Aerobic metabolism
MYC dependent ↓ in mitochondrial biogenesis	↓ Mitochondrial density	↓ Aerobic metabolism
↑ BNIP3	Mitochondrial autophagy	↓ Aerobic metabolism
↑ CAIX, MCT-4, NHE-1	pH regulation	Normal intracellular pH, acidic extracellular pH

CAIX: carbonic anhydrase IX, HIF1: hypoxia inducible factor 1, MCT-4: monocarboxyl transporter 4, NHE-1: sodium hydrogen exchanger 1, PDH: pyruvate dehydrogenase, PDK-1: pyruvate dehydrogenase kinase 1.

genes, among others (Brahimi-Horn, Chiche, and Pouyssegur, 2007; Denko, 2008; Semenza, 2009; Wouters and Koritzinsky, 2008; Dewhirst, Cao, and Moeller, 2008; Gillies, Robey, and Gatenby, 2008). HIF1 in turn upregulates all of the enzymes involved in glycolysis, inhibits the conversion of pyruvate to acetyl-CoA (which fuels the tricarboxylic acid (TCA) cycle), inhibits mitochondrial biogenesis, and promotes mitochondrial autophagy, as summarized in Table 9.1. HIF1 interacts with mTOR in a complex manner to influence protein synthesis, tumor proliferation, and oxygen consumption (Wouters and Koritzinsky, 2008). Activation of the MYC oncogene results in increased expression of glycolytic enzymes through both an interaction with HIF and a direct effect on gene transcription (Gillies, Robey, and Gatenby, 2008; Brahimi-Horn, Chiche, and Pouyssegur, 2007).

The upregulation of anaerobic glucose metabolism in tumors has several important ramifications. More glucose is required to meet the cellular metabolic demand because of the lower energy efficacy of anerobic metabolism, and glucose uptake in tumors is often substantially higher than in normal tissues (Vaupel, Kallinowski, and Okunieff, 1989). High rates of lactic acid production, together with the limitations of the abnormal tumor vasculature to remove metabolic waste, lead to the development of a pH gradient between the intracellular and extracellular spaces (Gillies, Robey, and Gatenby, 2008). Intracellular pH is tightly regulated in the normal or slightly alkaline range by a variety of active transport and buffering mechanisms (Figure 9.1) even in the setting of increased acid production. Extracellular pH is typically much lower in the range of 6.8–7.2, and varies both spatially and temporally. In general, there is poor correlation between oxygen levels and pH at the microregional level (Helmlinger *et al.*, 1997). High lactate levels in tumors may directly modulate molecular processes involved in malignant progression (Walenta and Mueller-Klieser, 2004). Under some physiologic conditions, lactate may be utilized rather than glucose as a substrate for oxidative metabolism (Sonveaux *et al.*, 2008).

9.2.3 High interstitial fluid pressure (IFP) and reduced convection

IFP is highly regulated in most normal, non-diseased tissues, and remains close to atmospheric levels (−3 to +3 mmHg). In contrast, IFP is elevated in virtually all solid malignant tumors and typically is in the range of 10–40 mmHg (Lunt *et al.*, 2008). The lowest IFP values have been reported in lymphomas (often <10 mmHg) and the highest in melanomas (up to 100 mmHg in patients with progressive disease unresponsive to treatment).

Elevated IFP arises in tumors as a consequence of vascular abnormalities, including altered blood flow resistance and high permeability to small molecules and water, coupled to structural and functional changes in the interstitium, as summarized in Figure 9.2. Lymphatic vessels, while present within and immediately adjacent to tumors, are generally non-functional and have limited capacity to drain away excess interstitial fluid (Leu *et al.*, 2000). There may also be higher concentrations

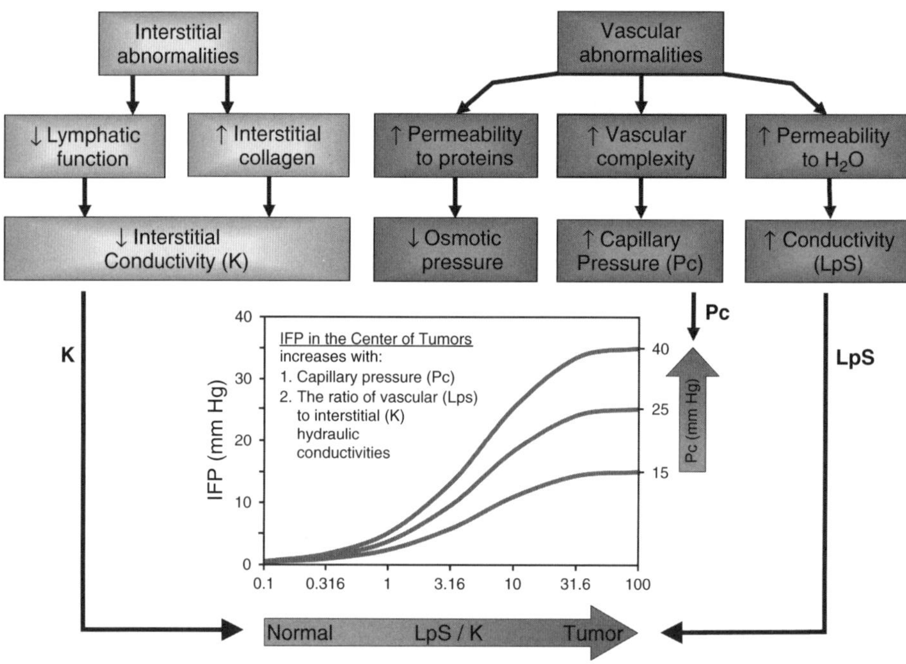

Figure 9.2 Pathophysiology of high IFP in tumors. Structural and functional abnormalities of the vasculature, lymphatics, and interstitium contribute to high vascular flow resistance and capillary pressure (Pc), high vessel wall hydraulic conductivity (LpS), and low interstitial hydraulic conductivity (K). Fluid leaks from vessels into the interstitium but is trapped there and accumulates causing the IFP to rise. The steady state IFP in the center of tumors increases as a function of the LpS/K ratio and the capillary pressure, as shown in the graph. In many tumors, LpS/K is >30. Under these circumstances, IFP almost equals the average capillary pressure and is relatively uniform throughout the central tumor volume, dropping to near zero over a few millimeters at the tumor–normal tissue interface (Baxter and Jain, 1989; Boucher, Baxter, and Jain, 1990; Boucher and Jain, 1992).

of collagen, elastin, and other extracellular matrix elements in tumors that hinder interstitial fluid flow, or regions of necrosis where free fluid motion is enhanced (Heldin *et al.*, 2004; Jain, 1987). In addition, extrapolating from studies of normal subcutaneous tissue, interstitial function may be dynamically regulated and able to respond over short times to environmental stimuli or cytokines like PDGF (Wiig, Rubin, and Reed, 2003). As a consequence of these abnormalities, vascular hydraulic conductivity is usually much higher than interstitial conductivity. Fluid leaks from the vessels, accumulates in the interstitium and causes the interstitial pressure to rise until it nearly equals the average capillary pressure (Baxter and Jain, 1989; Boucher, Baxter, and Jain, 1990; Boucher and Jain, 1992). Capillary pressure may also be higher than normal in tumors because of high geometric and viscous flow resistance, which in turn explains the very high IFPs (near systemic blood pressure levels) that have been reported in some tumors.

Biomechanical models of vascular and interstitial fluid convection suggest that IFP should be uniformly elevated throughout much of the central volume of tumors and fall precipitously to near zero at the interface with the surrounding normal tissue (Baxter and Jain, 1989). This has been confirmed in small experimental tumors (Boucher, Baxter, and Jain, 1990), although reproducible region-to-region variation in IFP has been observed in human cervix cancers (Milosevic *et al.*, 2001). This may be explained by greater regional heterogeneity in capillary pressure, vascular hydraulic conductivity or interstitial conductivity in these larger tumors.

9.3 Evaluating the tumor microenvironment

There are a variety of approaches for measuring hypoxia, the consequences of anerobic metabolism, and IFP in tumors, each with advantages and disadvantages. The ideal measurement technique for a particular microenvironmental parameter would be noninvasive and repeatable over time without altering the underlying physiology, have high sensitive, specificity and spatial resolution, and be useable both in the laboratory and clinic. Many of the currently available techniques are invasive, requiring either insertion of needles or removal of all or part of the tumor. However, functional and metabolic imaging strategies based on computerized tomography (CT), magnetic resonance (MR) imaging, positron emission tomography (PET), or single photon emission computed tomography (SPECT) are evolving rapidly and will eventually overcome many of the current limitations.

An important consideration when evaluating the tumor microenvironment is spatial and temporal heterogeneity. Tumor hypoxia, for example, has been shown to vary as much within individual tumors as among different tumors of the same type (Wong *et al.*, 1997). Intratumoral heterogeneity in the expression of carbonic anydrase IX (CAIX), a frequently used hypoxia biomarker regulated by HIF, is illustrated in Figure 9.3 (Iakovlev *et al.*, 2007). Intratumoral heterogeneity implies the need for multiple measurements in individual tumors if representative and reproducible estimates are to be obtained, and has important implications for the design of experiments studies and clinical trials (Pintilie *et al.*, 2009).

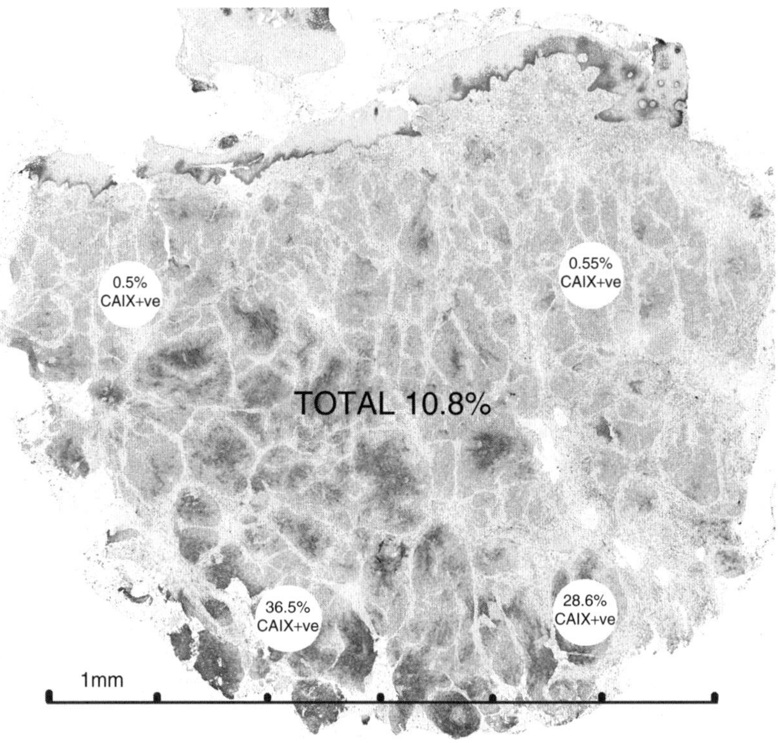

Figure 9.3 Expression of the endogenous hypoxia biomarker carbonic anhydrase IX (CAIX) in a biopsy from a patient with cervix cancer. The circles correspond to the sampling area typical of a tissue microarray (TMA) core. CAIX expression in the sampling areas ranges from a low of 0.55% to a high of 36.5%. This highlights the intratumoral heterogeneity that is typical of many microenvironmental parameters, and the need for multiple samples in individual tumors to assure reproducible results. In general, for immunohistochemical studies of tumor hypoxia, thorough histologic sectioning of at least two biopsies, each 4 mm or greater in diameter and containing at least 70% tumor, is required (Iakovlev *et al.*, 2007). Reprinted by permission from Macmillan Publishers Ltd, *Laboratory Investigation*, **87** (12), 1206–1217, 2007.

9.3.1 Hypoxia

Table 9.2 summarizes the commonly used approaches for measuring hypoxia in tumors. Most of these techniques measure a combination of acute and chronic hypoxia. Research in this area was revitalized in the early 1990s with the introduction of polarographic electrodes coupled to automated linear actuators that facilitated direct measurement of tumor oxygen concentration in patients. Polarographic measurements demonstrated conclusively that human tumors are hypoxic, and provided compelling evidence linking hypoxia with poor patient outcome following treatment. However, electrodes have several disadvantages that limited widespread use, including a high degree of operator dependence and difficulties with evaluating deep seated tumors.

Table 9.2 Techniques for measuring hypoxia in tumors.

Technique	Advantages	Disadvantages
Oxygen electrodes Polarographic Optical fluorescence	Multiple measurement leading to thorough spatial sampling of tumor; correlated with clinical outcomes	Invasive; accessible tumors only; operator dependent; large measurement volume at electrode tip; no information about microregional distribution of hypoxia
Bioreductive drugs Pimonidazole EF5	Direct measure of hypoxia; suitable for use with biopsies or surgical specimens; microregional distribution of hypoxia	Invasive; drug administration in advance of specimen collection; careful tumor sampling because of heterogeneity
Endogenous biomarkers CAIX, GLUT-1, HIF1, HIF2, LOX, Osteopontin, VEGF	Suitable for use with archival biopsies or surgical specimens without the need for prior drug administration; microregional distribution of hypoxia	Invasive; indirect measure of hypoxia; careful tumor sampling because of heterogeneity
Comet assay	Measures hypoxia related DNA damage relevant to radiation resistance	Invasive; indirect measure of hypoxia; requires prior irradiation to induce DNA damage; no information about microregional distribution of hypoxia
PET and SPECT ^{18}F-MISO, ^{18}F-FMISO ^{18}F-EF5, ^{18}F-FAZA ^{64}Cu-ATSM, ^{123}I-IAZA	Non-invasive; suitable for spatial mapping of hypoxia and serial assessment over time	No consensus about preferred hypoxia imaging isotope; false results from unbound isotope in tissue; limited spatial resolution
MR imaging BOLD	Non-invasive; suitable for spatial mapping of hypoxia and serial assessment over time	Evolving technique; indirect measure of hypoxia (hemoglobin saturation); strong dependence on blood flow

BOLD: blood oxygen level dependent, CAIX: carbonic anhydrase IX, Cu-ATSM: copper(II)-diacetyl-bis(N(4)-methylthiosemicarbazone, EF5: 2-(2-nitro-1*H*-imidazol-1-yl)-*N*-(2,2,3,3,3-pentafluoropropyl) acetamide, FAZA: fluoroazomycin, FMISO: fluoromisonidazole, GLUT: glucose transporter, HIF: hypoxia-inducible factor, IAZA: iodoazomycin, LOX: lysyl oxidase, MISO: misonidazole, VEGF: vascular endothelial growth factor.

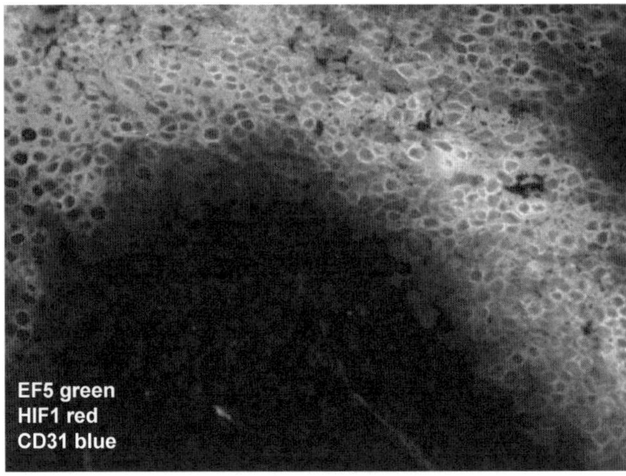

Figure 9.4 Micro-regional distribution of hypoxia in a SiHa cervix cancer xenograft illustrating co-localization between the bioreductive nitroimidazole drug EF5 (green) and increased expression of the endogenous hypoxia marker HIF1 (red) at a distance from blood vessels (CD31, blue). Image courtesy of D. Hedley. A full color version of this figure can be found in the color plate section.

Tissue based approaches to measuring tumor hypoxia in biopsies or excised tissue have the advantage of providing information about the microregional distribution of oxygen concentration in relation to cancer cells, vessels, stroma, and necrosis. Bioreductive drugs like the nitroimidazoles are selectively metabolized and bound in hypoxic cells. Immunohistochemistry and image analysis allow semiquantitative or quantitative evaluation of bound drug in hypoxic regions, as shown in Figure 9.4, and flow cytometry can provide an estimate of the proportion of hypoxic cells although with loss of information about the underlying tissue architecture. The binding of these drugs in tissue is determined in a complex manner by the oxygen concentration between the time of drug administration and tumor removal (usually 24 hours or longer), drug concentration in tissue over that time, and the availability in tissue of the enzymes required for drug metabolism. Therefore, binding reflects a combination of acute and chronic hypoxia, although an indication of the relative contribution of each to the overall hypoxic state of the tumor can be obtained by the sequential administration of different nitroimidazoles separated in time by several hours.

Bioreductive drugs, while capable of providing a robust measure of hypoxia at the microregional level, have the disadvantage of needing to be administered in advance of tumor biopsy or excision. This has limited the usefulness of these agents as markers of hypoxia in large multi-center clinical trials. Consequently, considerable effort has been expended to identify reliable endogenous biomarkers of hypoxia that can be applied to simple biopsies or archival tissue. HIF1 and HIF2, as well as related genes and proteins involved in angiogenesis, metabolism, pH regulation, and other cellular and metabolic processes (Table 9.2), have been explored as biomarkers of hypoxia and shown to correlate with biologic and clinical endpoints. However, there can be considerable variability in the usefulness of these markers from one tumor type to the next, in part because expression is not regulated exclusively by hypoxia but rather as a complex function of numerous hypoxia dependent and independent pathways. This suggests that it may be necessary to use multiple complementary biomarkers of hypoxia to obtain optimal results. Hypoxic gene signatures have recently been identified that appear to provide clinically relevant prognostic information independent of tumor type (Buffa *et al.*, 2010).

Molecular imaging approaches for measuring tumor hypoxia using PET or SPECT are based on hypoxia sensitive imaging tracers, usually nitroimidazoles, labeled with positron or gamma emitters respectively. The tracer binds in hypoxic regions and undergoes radioactive decay, which is detected and spatially localized to form a three-dimensional image of the distribution of hypoxia. Among the available radionuclide hypoxia imaging tracers, ^{18}F-fluoromisonidazole (FMISO) has been used most extensively although other tracers offer distinct advantages that may drive their widespread adoption in the future (Krohn, Link, and Mason, 2008). PET and SPECT imaging of hypoxia offer several advantages over other approaches: they are minimally invasive, repeatable over time, and allow evaluation of the entire tumor, thereby overcoming concerns about intratumoral heterogeneity and sampling error. However, unbound tracer in tumor at the time of imaging can confound the interpretation. This problem can be minimized by optimizing the timing of the imaging in relation to the radioactive and pharmacokinetic half lives of the tracer, or by continuous dynamic imaging and kinetic modeling of tracer uptake and metabolism (Thorwarth *et al.*, 2005).

9.3.2 Anerobic metabolism and extracellular acidosis

Hypoxia driven or constitutive upregulation of anerobic metabolism in tumors results in increased consumption of glucose, which is the basis for PET imaging of metabolic activity using the radiolabeled glucose analog ^{18}F-fluorodeoxyglucose (FDG). FDG is transported into cells and phosphorylated in the same way as glucose (Figure 9.1), thereby trapping it in the intracellular space where it undergoes positron decay. Many experimental and human tumors show increased FDG PET activity as an indication of elevated glucose consumption, and PET imaging is increasingly being utilized as a means of staging tumors, planning radiotherapy, and monitoring response to treatment. FDG uptake prior to treatment may convey important prognostic information for some tumors (Kidd *et al.*, 2007; Machtay *et al.*, 2009). While associations between hypoxia and PET FDG activity have been reported (Rajendran *et al.*, 2004), they are generally weak possibly because of variability in the extent to which individual tumors express hypoxia dependent versus constitutive (hypoxia independent) upregulation of anerobic metabolism (Busk *et al.*, 2008).

The metabolic products of increased anerobic glycolysis in tumors can be measured using electrodes, magnetic resonance spectroscopy (MRS) or optical techniques. pH electrodes provided the first evidence that tumors are more acidic than normal tissues (Tannock and Rotin, 1989). However, the utility of pH electrodes is limited because they measure a combination of intracellular and extracellular pH and have the same general disadvantages as oxygen electrodes. Intra- and extracellular pH can be measured using ^{31}P MRS, although obtaining results with adequate signal-to-noise and spatial resolution within a reasonable acquisition time can be challenging because of the low sensitivity. Nevertheless, much of the currently available data indicating the presence of a pH gradient between the intracellular and extracellular spaces were derived using this approach (Gillies *et al.*, 2002). ^{1}H MRS is more sensitive than ^{31}P MRS and allows metabolic changes in tumors to be monitored over time using endogenous biochemical markers such as lactate. In the future, nuclear spin polarization transfer and hyperpolarization techniques will dramatically improve the ability of MRS to interrogate metabolic pathways in much greater detail using low sensitivity probes like ^{13}C (Brindle, 2008).

Ex vivo bioluminescence imaging (BLI) of tumor metabolism is a technique that allows high resolution mapping of the distribution of tumor metabolites, including glucose, lactate, and adenosine triphosphate (ATP) (Walenta, Schroeder, and Mueller-Kliesser, 2002). Cryostat sections from flash-frozen tumor biopsies are immersed in a solution containing enzymes that link the metabolite of interest to a luciferase. As the tumor warms, light is emitted with an intensity that is proportional to metabolite concentration and collected with a high resolution camera. This technique requires careful attention to specimen acquisition to assure metabolite preservation. Multiple tissue sections need to be analyzed to obtain a reproducible estimate of average metabolite concentration. However, BLI offers several important advantages including ease of translation to the clinic, microscopic spatial localization of metabolites, quantitation of metabolite concentration with the use of appropriate standards, and spatial coregistration of metabolic maps with tumor histology or immunohistochemical biomarkers.

9.3.3 IFP

IFP in human tumors most often has been measured using a fluid filled wick-in-needle system connected to a pressure transducer (Lunt *et al.*, 2008). The pressure in the system rises after insertion of the needle into the tumor until it equilibrates with the tumor pressure. Recently, semiconductor interstitial pressure transducer systems have been developed (Ozerdem and Hargens, 2005) and implantable transducers with miniaturized radiotelemetry systems have been used to continuously monitor IFP over time and in response to treatment (Schnell *et al.*, 2008). The possibility of evaluating IFP non-invasively using MR imaging has also been explored, although the results have not correlated well with direct needle based measurements (Haider *et al.*, 2007; Gulliksrud, Brurberg, and Rofstad, 2009; Hassid *et al.*, 2006). In theory, IFP should be relatively uniform throughout the central volume of tumors (Baxter and Jain, 1989), which should reduce the need for extensive sampling with needle-based approaches. Nevertheless, in practice, multiple measurements are required in individual tumors (Milosevic *et al.*, 2001).

9.4 Biologic and therapeutic implications

The microenvironment can have profound effects on the biologic and clinical behavior of tumors and response to treatment. A comprehensive discussion is beyond the scope of this chapter, and many of these issues will be addressed in greater detail elsewhere. Here, aspects of the tumor microenvironment that have the greatest potential to influence clinical outcomes will be highlighted. These are summarized in Table 9.3.

9.4.1 Hypoxia

Hypoxia has long been known to inhibit response to radiation treatment through a direct biophysical effect on repair of radiation induced DNA damage (Brown and

Table 9.3 Biologic and therapeutic implications of the abnormal tumor microenvironment.

Hypoxia	Anerobic metabolism and acidosis	High IFP and reduced convection
Activation of HIF-dependent genes	↑ Glucose consumption	Biomarker of angiogenesis and impaired drug delivery
Altered regulation of HIF-independent pathways, including mTOR and UPR	Cellular adaptation – conservation of oxygen and glucose, ↓ ROS	Enhanced permeability and retention effect – ↓ macromolecule delivery
Genomic instability	Genomic instability	Vascular instability and redistribution of tumor blood flow
Selection of apoptotic-resistant cells	Selection of apoptotic-resistant cells	Peritumoral transport of cytokines involved in tumor progression and lymph node metastases
Malignant progression, invasion, and metastases formation	Malignant progression, invasion, and metastases formation	↑ Cell proliferation
Treatment resistance – RT and chemotherapy	Treatment resistance – RT and chemotherapy	Expression of genes involved in angiogenesis and lymphangiogenesis
↓ Clinical outcomes in head and neck, cervix and prostate cancer patients treated with RT or surgery	↓ Clinical outcome in head and neck or cervix cancer patients treated with RT	↓ Clinical outcomes in cervix cancer patients treated with RT

HIF: hypoxia inducible factor, mTOR: mammalian target of rapamycin, ROS: reactive oxygen species, RT: radiotherapy, UPR: unfolded protein response.

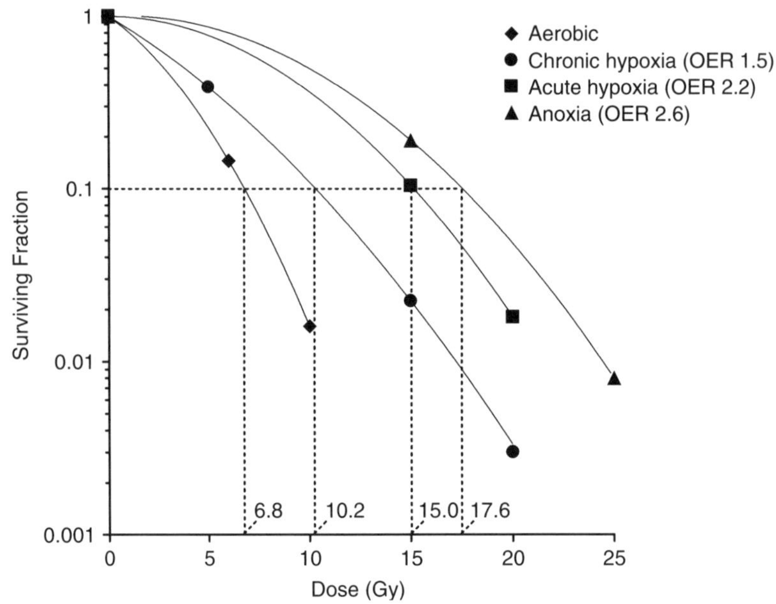

Figure 9.5 Clonogenic cell survival curves for H1299 human lung carcinoma cells irradiated under different oxygen condition. Aerobic: 21% O_2 for 6 hours prior to irradiation (IR). Anoxia: 0% O_2 for 6 hours prior to IR. Acute hypoxia: 0.2% O_2 for 6 hours prior to IR. Chronic hypoxia: 0.2% O_2 for 72 hours prior to IR. The oxygen enhancement ratio (OER) for chronically hypoxic cells was significantly reduced compared to anoxic or acutely hypoxic cells (1.5 versus 2.2 and 2.6, respectively) because of decreased expression of proteins involved in the homologous recombination DNA repair pathway. Image courtesy of N. Chan and R. Bristow.

Wilson, 2004). This is reflected in the oxygen enhancement ratio (OER), which is defined as the quotient of the doses required to achieve a particular biologic effect under hypoxic and oxic conditions. As illustrated in Figure 9.5, the OER for cells irradiated following anoxia is typically in the range of 2.5–3 (Brown and Wilson, 2004). While the response to large single fractions of radiation is determined mainly by the profoundly hypoxic fraction of cells in a tumor, the response to fractionated radiotherapy may be more strongly influenced by cells at intermediate oxygen levels in the range of 0.5–20 mmHg (Wouters and Brown, 1997).

There is increasing evidence to indicate that acute and chronic hypoxia may influence radiation response in different ways. Acute hypoxia appears to have a greater effect on HIF activation than chronic hypoxia (Dewhirst, Cao, and Moeller, 2008). Tumor reoxygenation during radiotherapy may also activate HIF, mediated by ROS and recovery of protein translation, which often is suppressed in hypoxic and nutrient deprived environments. HIF influences the radiation response of tumor cells in a complex manner relating to regulation of metabolism, proliferation, and apoptosis, and HIF dependent proangiogenic cytokines such as VEGF are thought to have a radioprotective effect on vascular endothelial cells (Moeller, Richardson, and Dewhirst, 2007). On the other hand, chronically low oxygen levels in the range of

1–2 mmHg (0.2%) lead to reduced translation of proteins like RAD51 and BRCA2 that are involved in the homologous recombination DNA repair pathway (Bristow and Hill, 2008). As a result, the OER for chronically hypoxic cells adapted to survive in low oxygen conditions may be lower than for acute hypoxia, as illustrated in Figure 9.5. These interrelated effects of acute and chronic hypoxia may combine to either enhance or inhibit radiation response depending on microenvironmental conditions. Strategies to modulate HIF expression or DNA repair are being explored as ways of tailoring radiation treatment in individual patients.

It is now widely recognized that hypoxia, in addition to modulating radiation response, has a much broader impact on the biologic and clinical behavior of tumors that translates to other therapeutic modalities, including surgery and chemotherapy. Reduced DNA repair contributes to genetic instability and accumulation of DNA damage in cells. Altered regulation of HIF and other pathways, including mTOR and the unfolded protein response (UPR), influence malignant progression, and metastasis formation while also allowing cells to adapt and survive in a nutrient-deprived environment (Wouters and Koritzinsky, 2008). In general, hypoxic tumors behave more aggressively and are associated with worse clinical outcomes even in surgically treated patients, in part because of metastases that are present at diagnosis (Brizel *et al.*, 1996; Hockel *et al.*, 1996). Hypoxia can reduce the effectiveness of cytotoxic chemotherapy by selecting cells that are resistant to apoptosis, upregulated genes involved in the development of drug resistance or modulating repair of drug induced DNA damage (Bristow and Hill, 2008; Brown and Wilson, 2004). However, hypoxia can also be exploited to therapeutic advantage as a means of converting a non-toxic prodrug to its toxic form (Brown and Wilson, 2004). This is the basis for hypoxic cytotoxins such as tirapazamine, which are selectively activated in hypoxic regions of tumors and produce cooperative killing of cells that may be relatively resistant to radiotherapy and other treatments.

9.4.2 Anerobic metabolism and extracellular acidosis

The switch to anerobic glycolysis in tumors, with acidification of the extracellular microenvironment and accumulation of other metabolic waste products, has wide reaching implications for tumor behavior and response to treatment, as summarized in Table 9.3 (Denko, 2008; Gatenby *et al.*, 2006; Gillies, Robey, and Gatenby, 2008; Gillies *et al.*, 2002). Anerobic metabolism is thought to provide a survival advantage for cancer cells by conserving oxygen and other metabolic substrates, shunting of glucose metabolites from energy production to anabolic processes or lower the concentration of toxic ROS. In addition, low extracellular pH has been implicated in the selection of aggressive tumor phenotypes that are resistant to apoptosis, more likely to invade locally and more likely to metastasize. One hypothesis to explain the increased propensity for invasion and metastasis formation is based on the premise that cancerous cells are selected over time for their ability to survive in an acidic environment, whereas the normal cells that surround the tumor remain more dependent on maintenance of normal extracellular conditions (Gatenby *et al.*, 2006). Hydrogen ion flux from the periphery of the tumor into the surrounding tissue causes death of these normal cells and degradation of the extracellular matrix, which is manifested as more aggressive biologic and clinical behavior. Acidosis, like hypoxia, may also contribute to genetic instability and the emergence of phenotypes with greater metastatic capacity (Gillies *et al.*, 2002). Low extracellular pH and high lactate concentration have both been associated with impaired response to radiotherapy (Gillies *et al.*, 2002; Quennet *et al.*, 2006; Tannock and Rotin, 1989). The pH difference between the intra- and extracellular spaces inhibits the cellular uptake of weakly basic chemotherapy drugs like doxorubicin and enhances the uptake of weak acids like cisplatin (Tannock and Rotin, 1989; Gerweck, Vijayappa, and Kozin, 2006). In general, hypoxia, acidic pH, and other manifestations of anerobic metabolism coexist in tumors and appear to exert similar effects on biologic behavior and treatment response, although how they interact to influence overall tumor behavior is poorly understood.

9.4.3 High IFP and reduced convection

The physiologic association between abnormal vasculature function and interstitial hypertension in tumors implies that IFP may be a useful biomarker of angiogenesis and response to vascular targeted treatments (Lunt *et al.*, 2008). Drugs that either disrupt the existing vasculature or inhibit new vessel formation (small and large molecular inhibitors of angiogenesis) have been associated with reductions in IFP and improved therapeutic outcome when combined with radiotherapy or conventional cytotoxic chemotherapy. This probably reflects restoration of the balance between pro- and antiangiogenic cytokines that exists under normal conditions, leading to improved vascular efficiency and drugs delivery (Jain, 2005).

An important consequence of high IFP in the center of tumors is a reduction in the hydrostatic pressure gradient for fluid flow across capillary walls from the intravascular space to the interstitial space, and therefore also a reduction in the convective transport of monoclonal antibodies, liposomes, and other macromolecules. In tumors where trans-capillary hydraulic conductivity is much higher than interstitial conductivity (Figure 9.2), IFP almost equals the average capillary pressure and transcapillary convection is largely abolished. In addition, high IFP (which is relatively uniform throughout the central volume of tumors) reduces the hydrostatic gradient that drives fluid flow and convection through the interstitium. Reduced transcapillary and interstitial pressure gradients and convection are thought to contribute to vascular instability (Mollica, Jain, and Netti, 2003; Rofstad *et al.*, 2009) and redistribution of blood flow (Baish, Netti, and Jain, 1997), imaging contrast agents (Zhao, Salmon, and Sarntinoranont, 2007), and cytokines involved in the development of lymph node metastases (Jain, Tong, and Munn, 2007). Reduced interstitial convection also contributes to the enhanced permeability and retention (EPR) effect (Iyer *et al.*, 2006), a phenomenon whereby drugs and other therapeutic agents with high molecular weights are trapped in the tumor interstitium.

9.5 Clinical implications

There are many studies indicting that the microenvironment is abnormal in human tumors, regardless of anatomic site of origin. However, most of the clinical evidence linking hypoxia, altered glucose metabolism, or high IFP to poor patient outcome comes from studies of head and neck cancer or cervix cancer.

9.5.1 Head and neck cancer

Hypoxia is a prominent feature of most head and neck squamous cancers. A meta-analysis of almost 400 patients who had electrode measurements of hypoxia performed prior to radiotherapy showed that those with hypoxic tumors had significantly lower overall survival, independent of clinical prognostic factors, and treatment technique (Nordsmark *et al.*, 2005). The anatomic sites of failure were not evaluated but many individual studies have reported reduced local control with radiotherapy in patients with hypoxic head and neck tumors. The relationship between the endogenous hypoxia biomarkers HIF2 and CAIX and patient outcome was analyzed retrospectively in archival tissue from 198 patients who participated in the Continuous Hyperfractionated Accelerated Radiotherapy (CHART) randomized trial (Koukourakis *et al.*, 2006). This trial failed to demonstrate a benefit overall of altered treatment fractionation, but increased expression of HIF2 or CAIX was associated with reduced local control and survival after correction for clinical variables. Furthermore, the two biomarkers together provided greater predictive information than either individually presumably because they are associated with different signaling pathways. There also was a trend toward a greater benefit of hyperfractionated and accelerated radiotherapy in well-oxygenated tumors. This study identifies several important points, including the usefulness of biomarkers to help interpret the results of clinical trials and the need to assess multiple, complementary biomarkers given the complexity of the genetic and molecular changes in most human tumors. The latter was also highlighted in a large single institutional study in which a panel of five hypoxic biomarkers was used to stratify patients according to outcome. The 5-year overall survival rates ranging from less than 20% in patients who had increased expression of all five markers to greater than 80% when none was expressed (Le *et al.*, 2007). A recently reported hypoxic gene signature, derived in part from patients with head and neck cancer, appears to be highly prognostic and translatable to other cancers (Buffa *et al.*, 2010).

Head and neck cancers have high levels of anaerobic metabolism and high glucose requirements, as evidenced by avid FDG uptake and high lactate levels. FDG PET scans are increasingly being used to plan and deliver radiotherapy to patients with head and neck tumors (Troost *et al.*, 2010), and an association between FDG uptake and patient outcome has been reported (Machtay *et al.*, 2009). In two small series, high tumor lactate levels evaluated using BLI were found to be associated with an elevated risk of metastases at diagnosis, an increased likelihood of recurrence in lymph nodes and at distant sites following radiotherapy, and lower overall survival (Brizel *et al.*, 2001; Walenta *et al.*, 1997). In a larger study of 140 patients, pyruvate dehydrogenase kinase 1 (PDK-1), a HIF regulated enzyme that inhibits pyruvate dehydrogenase (PDH) and plays an important role in maintaining anerobic metabolism (Table 9.1 and Figure 9.1), was differentially expressed in head and cancer compared to normal tissues and independently predictive of poor survival (Wigfield *et al.*, 2008).

There is little doubt that microenvironmental factors influence head and neck cancer behavior and response to treatment. An important question that follows directly from this conclusion is whether modification of the microenvironment can improve the outlook for patients, or rather is this predetermined prior to diagnosis and irreversible because of genetic changes that occur early in cancer development. Many small clinical studies of hypoxia modification in head and neck cancer or other tumors have failed to demonstrate a benefit. However, a meta-analysis of approximately 10 000 patients from over 60 randomized clinical trials of radiotherapy with or without hypoxia modification showed significant improvement in both local tumor control and patient survival, particularly among those with head and neck cancer (Overgaard, 2007). This is an important result because it provides proof-of-concept that the adverse consequences of hypoxia can be overcome at least partially with appropriate intervention prior to standard treatments like radiotherapy. There is also evidence from head and neck cancer studies to suggest that imaging and tissue-based hypoxia biomarkers may be useful as a means of selecting appropriate patients for hypoxia modification, thereby increasing the likelihood of therapeutic benefit (Overgaard *et al.*, 2005; Rischin *et al.*, 2006).

9.5.2 Cervix cancer

Several studies have established a link between tumor hypoxia and survival in cervix cancer patients treated with radiotherapy. In addition, there is clinical evidence to indicate that hypoxia is associated with more aggressive cervical tumors having a greater proportion of apoptosis resistant cells and a higher risk of lymph node metastases at diagnosis (Fyles *et al.*, 2002; Hockel *et al.*, 1999). While hypoxia is a strong determinate of local control in head and neck cancer patients who receive radiotherapy, the evidence is less convincing in patients with cervix cancer. Instead, patients with hypoxic cervical tumors, regardless of whether they undergo surgery or receive radiotherapy, are more likely to develop recurrent disease in untreated lymph nodes or at distant metastatic sites (Fyles *et al.*, 2002).

IFP is elevated in most solid malignant tumors in patients. However, only two studies, both in patients with cervix cancer, have explored the relationship between IFP and outcome following radiotherapy (Milosevic *et al.*, 2001; Yeo *et al.*, 2009). In the larger of the two studies, which was conducted at Princess Margaret Hospital, patients underwent pretreatment assessment of both tumor hypoxia using a needle probe and IFP prior to radiotherapy (Fyles *et al.*, 2006; Milosevic *et al.*, 2001). There was no correlation between hypoxia and IFP. Both biomarkers were found to be strong, independent predictive of patient outcome but were associated with different patterns of disease recurrence. Hypoxia was important only in patients without lymph node metastases at diagnosis, and mainly predicted for distant metastatic failure. In contrast, high IFP was associated with both local and distant recurrence. The predictive effect of IFP has now been confirmed in a second, independent study (Yeo *et al.*, 2009).

FDG PET activity is high in cervix cancer and has been associated with lymph node metastases at diagnosis and lower patient survival following radiotherapy, independent of clinical prognostic factors (Kidd *et al.*, 2007). In addition, BLI of tumor biopsies from 34 patients with cervix cancer showed a correlation between lactate concentration and the presence of lymph node metastases at diagnosis, as well as between lactate concentration and survival following radiotherapy (Walenta *et al.*, 2000).

9.6 Summary

The microenvironment of most solid malignant tumors, regardless of anatomic site of origin, is abnormal as a result of the complex and dynamic interplay among the cellular, vascular, and interstitial compartments. Hypoxia, anaerobic metabolism, and high IFP arise because of vascular and cellular dysfunction and, in turn, strongly influence vascular and cellular/molecular behavior. As part of this interaction, tumor cells evolve adaptive mechanisms that provide a selective survival advantage. Treatments that target aspects of the abnormal microenvironment directly, or the vasculature or molecular machinery that contribute to microenvironmental dysfunction, have the potential to improve treatment outcomes for patients whether used alone or in combination with currently available treatments like radiotherapy or chemotherapy.

References

Baish, J.W., Netti, P.A., and Jain, R.K. (1997) Transmural coupling of fluid flow in microcirculatory network and interstitium in tumors. *Microvascular Research*, **53**, 128–141.

Baxter, L. and Jain, R. (1989) Transport of fluid and macromolecules in tumors I. Role of interstitial pressure and convection. *Microvascular Research*, **37**, 77–104.

Boucher, Y., Baxter, L.T., and Jain, R.K. (1990) Interstitial pressure gradients in tissue isolated and subcutaneous tissues: implications for therapy. *Cancer Research*, **50**, 4478–4484.

Boucher, Y. and Jain, R.K. (1992) Microvascular pressure is the principal driving force for interstitial hypertension in solid tumors: implications for vascular collapse. *Cancer Research*, **52**, 5110–5114.

Brahimi-Horn, M.C., Chiche, J., and Pouyssegur, J. (2007) Hypoxia signalling controls metabolic demand. *Current Opinion in Cell Biology*, **19**, 223–229.

Brindle, K. (2008) New approaches for imaging tumour responses to treatment. *Nature Reviews Cancer*, **8**, 94–107.

Bristow, R.G. and Hill, R.P. (2008) Hypoxia and metabolism. Hypoxia, DNA repair and genetic instability. *Nature Reviews Cancer*, **8**, 180–192.

Brizel, D.M., Schroeder, T., Scher, R.L. *et al.* (2001) Elevated tumor lactate concentrations predict for an increased risk of metastases in head-and-neck cancer. *International Journal of Radiation Oncology, Biology, Physics*, **51**, 349–353.

Brizel, D.M., Scully, S.P., Harrelson, J.M. *et al.* (1996) Tumor oxygenation predicts for likelihood of distant metastases in human soft tissue sarcoma. *Cancer Research*, **56**, 941–943.

Brown, J.M. and Wilson, W.R. (2004) Exploiting tumour hypoxia in cancer treatment. *Nature Reviews Cancer*, **4**, 437–447.

Brurberg, K.G., Graff, B.A., and Rofstad, E.K. (2003) Temporal heterogeneity in oxygen tension in human melanoma xenografts. *British Journal of Cancer*, **89**, 350–356.

Buffa, F.M., Harris, A.L., West, C.M., and Miller, C.J. (2010) Large meta-analysis of multiple cancers reveals a common, compact and highly prognostic hypoxia metagene. *British Journal of Cancer*, **102**, 428–435.

Busk, M., Horsman, M.R., Kristjansen, P.E. *et al.* (2008) Aerobic glycolysis in cancers: implications for the usability of oxygen-responsive genes and fluorodeoxyglucose-PET as markers of tissue hypoxia. *International Journal of Cancer*, **122**, 2726–2734.

Christofk, H.R., Vander Heiden, M.G., Harris, M.H. *et al.* (2008) The M2 splice isoform of pyruvate kinase is important for cancer metabolism and tumour growth. *Nature*, **452**, 230–233.

Denko, N.C. (2008) Hypoxia, HIF1 and glucose metabolism in the solid tumour. *Nature Reviews Cancer*, **8**, 705–713.

Dewhirst, M.W., Cao, Y., and Moeller, B. (2008) Cycling hypoxia and free radicals regulate angiogenesis and radiotherapy response. *Nature Reviews Cancer*, **8**, 425–437.

Fyles, A., Milosevic, M., Hedley, D. *et al.* (2002) Tumor hypoxia has independent predictor impact only in patients with node-negative cervix cancer. *Journal of Clinical Oncology*, **20**, 680–687.

Fyles, A., Milosevic, M., Pintilie, M. *et al.* (2006) Long-term performance of interstitial fluid pressure and hypoxia as prognostic factors in cervix cancer. *Radiotherapy and Oncology*, **80**, 132–137.

Gatenby, R.A., Gawlinski, E.T., Gmitro, A.F. *et al.* (2006) Acid-mediated tumor invasion: a multidisciplinary study. *Cancer Research*, **66**, 5216–5223.

Gerweck, L.E., Vijayappa, S., and Kozin, S. (2006) Tumor pH controls the in vivo efficacy of weak acid and base chemotherapeutics. *Molecular Cancer Therapeutics*, **5**, 1275–1279.

Gillies, R.J., Raghunand, N., Karczmar, G.S., and Bhujwalla, Z.M. (2002) MRI of the tumor microenvironment. *Journal of Magnetic Resonance Imaging*, **16**, 430–450.

Gillies, R.J., Robey, I., and Gatenby, R.A. (2008) Causes and consequences of increased glucose metabolism of cancers. *Journal of Nuclear Medicine*, **49** (Suppl. 2), 24S–42S.

Gulliksrud, K., Brurberg, K.G., and Rofstad, E.K. (2009) Dynamic contrast-enhanced magnetic resonance imaging of tumor interstitial fluid pressure. *Radiotherapy and Oncology*, **91**, 107–113.

Haider, M., Sitartchouk, I., Roberts, T.P.L. *et al.* (2007) Correlations between dynamic contrast enhanced magnetic resonance imaging derived measures of tumor microvasculature and interstitial fluid pressure in patients with cervix cancer. *Journal of Magnetic Resonance Imaging*, **25**, 153–159.

Hassid, Y., Furman-Haran, E., Margalit, R. *et al.* (2006) Noninvasive magnetic resonance imaging of transport and interstitial fluid pressure in ectopic human lung tumors. *Cancer Research*, **66**, 4159–4166.

Heldin, C.H., Rubin, K., Pietras, K., and Ostman, A. (2004) High interstitial fluid pressure - an obstacle in cancer therapy. *Nature Reviews Cancer*, **4**, 806–813.

Helmlinger, G., Yuan, F., Dellian, M., and Jain, R.K. (1997) Interstitial pH and pO2 gradients in solid tumors in vivo: high-resolution measurements reveal a lack of correlation. *Nature Medicine*, **3**, 177–182.

Hockel, M., Schlenger, K., Aral, B. *et al.* (1996) Association between tumor hypoxia and malignant progression in advanced cancer of the uterine cervix. *Cancer Research*, **56**, 4509–4515.

Hockel, M., Schlenger, K., Hockel, S., and Vaupel, P. (1999) Hypoxic cervical cancers with low apoptotic index are highly aggressive. *Cancer Research*, **59**, 4525–4528.

Iakovlev, V.V., Pintilie, M., Morrison, A. *et al.* (2007) Effect of distributional heterogeneity on the analysis of tumor hypoxia based on carbonic anhydrase IX. *Laboratory Investigation*, **87**, 1206–1217.

Iyer, A.K., Khaled, G., Fang, J., and Maeda, H. (2006) Exploiting the enhanced permeability and retention effect for tumor targeting. *Drug Discovery Today*, **11**, 812–818.

Jain, R.K. (1987) Transport of molecules in the tumor interstitium: a review. *Cancer Research*, **47**, 3039–3051.

Jain, R.K. (1988) Determinants of tumor blood flow: a review. *Cancer Research*, **48**, 2641–2658.

Jain, R.K. (2005) Normalization of tumor vasculature: an emerging concept in antiangiogenic therapy. *Science*, **307**, 58–62.

Jain, R.K., Tong, R.T., and Munn, L.L. (2007) Effect of vascular normalization by antiangiogenic therapy on interstitial hypertension, peritumor edema, and lymphatic metastasis: insights from a mathematical model. *Cancer Research*, **67**, 2729–2735.

Kidd, E.A., Siegel, B.A., Dehdashti, F., and Grigsby, P.W. (2007) The standardized uptake value for F-18 fluorodeoxyglucose is a sensitive predictive biomarker for cervical cancer treatment response and survival. *Cancer*, **110**, 1738–1744.

Koukourakis, M.I., Bentzen, S.M., Giatromanolaki, A. *et al.* (2006) Endogenous markers of two separate hypoxia response pathways (hypoxia inducible factor 2 alpha and carbonic anhydrase 9) are associated with radiotherapy failure in head and neck cancer patients recruited in the CHART randomized trial. *Journal of Clinical Oncology*, **24**, 727–735.

Krohn, K.A., Link, J.M., and Mason, R.P. (2008) Molecular imaging of hypoxia. *Journal of Nuclear Medicine*, **49** (Suppl. 2), 129S–148S.

Le, Q.T., Kong, C., Lavori, P.W. *et al.* (2007) Expression and prognostic significance of a panel of tissue hypoxia markers in head-and-neck squamous cell carcinomas. *International Journal of Radiation Oncology, Biology, Physics*, **69**, 167–175.

Leu, A.J., Berk, D.A., Lymboussaki, A. *et al.* (2000) Absence of functional lymphatics within a murine sarcoma: a molecular and functional evaluation. *Cancer Research*, **60**, 4324–4327.

Lunt, S.J., Chaudary, N., and Hill, R.P. (2009) The tumor microenvironment and metastatic disease. *Clinical and Experimental Metastasis*, **26**, 19–34.

Lunt, S.J., Fyles, A., Hill, R.P., and Milosevic, M. (2008) Interstitial fluid pressure in tumors: therapeutic barrier and biomarker of angiogenesis. *Future Oncology*, **4**, 793–802.

Machtay, M., Natwa, M., Andrel, J. *et al.* (2009) Pretreatment FDG-PET standardized uptake value as a prognostic factor for outcome in head and neck cancer. *Head and Neck*, **31**, 195–201.

Milosevic, M., Fyles, A., Hedley, D. *et al.* (2001) Interstitial fluid pressure predicts survival in patients with cervix cancer independent of clinical prognostic factors and tumor oxygen measurements. *Cancer Research*, **61**, 6400–6405.

Moeller, B.J., Richardson, R.A., and Dewhirst, M.W. (2007) Hypoxia and radiotherapy: opportunities for improved outcomes in cancer treatment. *Cancer Metastasis Reviews*, **26**, 241–248.

Mollica, F., Jain, R.K., and Netti, P.A. (2003) A model for temporal heterogeneities of tumor blood flow. *Microvascular Research*, **65**, 56–60.

Nordsmark, M., Bentzen, S.M., Rudat, V. *et al.* (2005) Prognostic value of tumor oxygenation in 397 head and neck tumors after primary radiation therapy. An international multi-center study. *Radiotherapy and Oncology*, **77**, 18–24.

Overgaard, J. (2007) Hypoxic radiosensitization: adored and ignored. *Journal of Clinical Oncology*, **25**, 4066–4074.

Overgaard, J., Eriksen, J.G., Nordsmark, M. *et al.* (2005) Plasma osteopontin, hypoxia, and response to the hypoxia sensitiser nimorazole in radiotherapy of head and neck cancer: results from the DAHANCA 5 randomised double-blind placebo-controlled trial. *The Lancet Oncology*, **6**, 757–764.

Ozerdem, U. and Hargens, A.R. (2005) A simple method for measuring interstitial fluid pressure in cancer tissues. *Microvascular Research*, **70**, 116–120.

Pintilie, M., Iakovlev, V., Fyles, A. *et al.* (2009) Heterogeneity and power in clinical biomarker studies. *Journal of Clinical Oncology*, **27**, 1517–1521.

Quennet, V., Yaromina, A., Zips, D. *et al.* (2006) Tumor lactate content predicts for response to fractionated irradiation of human squamous cell carcinomas in nude mice. *Radiotherapy and Oncology*, **81**, 130–135.

Rajendran, J.G., Mankoff, D.A., O'Sullivan, F. *et al.* (2004) Hypoxia and glucose metabolism in malignant tumors: evaluation by [18F]fluoromisonidazole and [18F]fluorodeoxyglucose positron emission tomography imaging. *Clinical Cancer Research*, **10**, 2245–2252.

Rischin, D., Hicks, R.J., Fisher, R. *et al.* (2006) Prognostic significance of [18F]-misonidazole positron emission tomography-detected tumor hypoxia in patients with advanced head and neck cancer randomly assigned to chemoradiation with or without tirapazamine: a substudy of Trans-Tasman Radiation Oncology Group Study 98.02. *Journal of Clinical Oncology*, **24**, 2098–2104.

Rofstad, E.K., Gaustad, J.V., Brurberg, K.G. *et al.* (2009) Radiocurability is associated with interstitial fluid pressure in human tumor xenografts. *Neoplasia*, **11**, 1243–1251.

Schnell, C.R., Stauffer, F., Allegrini, P.R. *et al.* (2008) Effects of the dual phosphatidylinositol 3-kinase/mammalian target of rapamycin inhibitor NVP-BEZ235 on the tumor vasculature: implications for clinical imaging. *Cancer Research*, **68**, 6598–6607.

Secomb, T.W., Hsu, R., Dewhirst, M.W. *et al.* (1993) Analysis of oxygen transport to tumor tissue by microvascular networks. *International Journal of Radiation Oncology, Biology, Physics*, **25**, 481–489.

Semenza, G.L. (2009) Regulation of cancer cell metabolism by hypoxia-inducible factor 1. *Seminars in Cancer Biology*, **19**, 12–16.

Sonveaux, P., Vegran, F., Schroeder, T. *et al.* (2008) Targeting lactate-fueled respiration selectively kills hypoxic tumor cells in mice. *Journal of Clinical Investigation*, **118**, 3930–3942.

Tannock, I.F. and Rotin, D. (1989) Acid pH in tumors and its potential for therapeutic exploitation. *Cancer Research*, **49**, 4373–4384.

Thorwarth, D., Eschmann, S.M., Paulsen, F., and Alber, M. (2005) A kinetic model for dynamic [18F]-Fmiso PET data to analyse tumour hypoxia. *Physics in Medicine and Biology*, **50**, 2209–2224.

Troost, E.G., Schinagl, D.A., Bussink, J. *et al.* (2010) Innovations in radiotherapy planning of head and neck cancers: role of PET. *Journal of Nuclear Medicine*, **51**, 66–76.

Vaupel, P., Kallinowski, F., and Okunieff, P. (1989) Blood flow, oxygen and nutrient supply, and metabolic microenvironment of human tumors: a review. *Cancer Research*, **49**, 6449–6465.

Walenta, S. and Mueller-Klieser, W.F. (2004) Lactate: mirror and motor of tumor malignancy. *Seminars in Radiation Oncology*, **14**, 267–274.

Walenta, S., Salameh, A., Lyng, H. *et al.* (1997) Correlation of high lactate levels in head and neck tumors with incidence of metastasis. *American Journal of Pathology*, **150**, 409–415.

Walenta, S., Schroeder, T., and Mueller-Klieser, W. (2002) Metabolic mapping with bioluminescence: basic and clinical relevance. *Biomolecular Engineering*, **18**, 249–262.

Walenta, S., Wetterling, M., Lehrke, M. *et al.* (2000) High lactate levels predict likelihood of metastases, tumor recurrence, and restricted patient survival in human cervical cancers. *Cancer Research*, **60**, 916–921.

Warburg, O., Posener, K., and Negelein, E. (1930) Ueber den stoffwechsel der tumoren, in *On Metabolism of Tumors* (ed. O. Warburg), Constable, London.

Wigfield, S.M., Winter, S.C., Giatromanolaki, A. *et al.* (2008) PDK-1 regulates lactate production in hypoxia and is associated with poor prognosis in head and neck squamous cancer. *British Journal of Cancer*, **98**, 1975–1984.

Wiig, H., Rubin, K., and Reed, R.K. (2003) New and active role of the interstitium in control of interstitial fluid pressure: potential therapeutic consequences. *Acta Anaesthesiologica Scandinavica*, **47**, 111–121.

Wong, R.K.W., Fyles, A., Milosevic, M. *et al.* (1997) Heterogeneity of polarographic oxygen tension measurements in cervix cancer: an evaluation of within and between tumor variability, probe position, and track depth. *International Journal of Radiation Oncology, Biology, Physics*, **39**, 405–412.

Wouters, B.G. and Brown, J.M. (1997) Cells at intermediate oxygen levels can be more important than the "hypoxic fraction" in determining tumor response to fractionated radiotherapy. *Radiation Research*, **147**, 541–550.

Wouters, B.G. and Koritzinsky, M. (2008) Hypoxia signalling through mTOR and the unfolded protein response in cancer. *Nature Reviews Cancer*, **8**, 851–864.

Yeo, S.G., Kim, J.S., Cho, M.J., and Kim, K.H. (2009) Interstitial fluid pressure as a prognostic factor in cervical cancer following radiation therapy. *Clinical Cancer Research*, **15**, 6201–6207.

Zhao, J., Salmon, H., and Sarntinoranont, M. (2007) Effect of heterogeneous vasculature on interstitial transport within a solid tumor. *Microvascular Research*, **73**, 224–236.

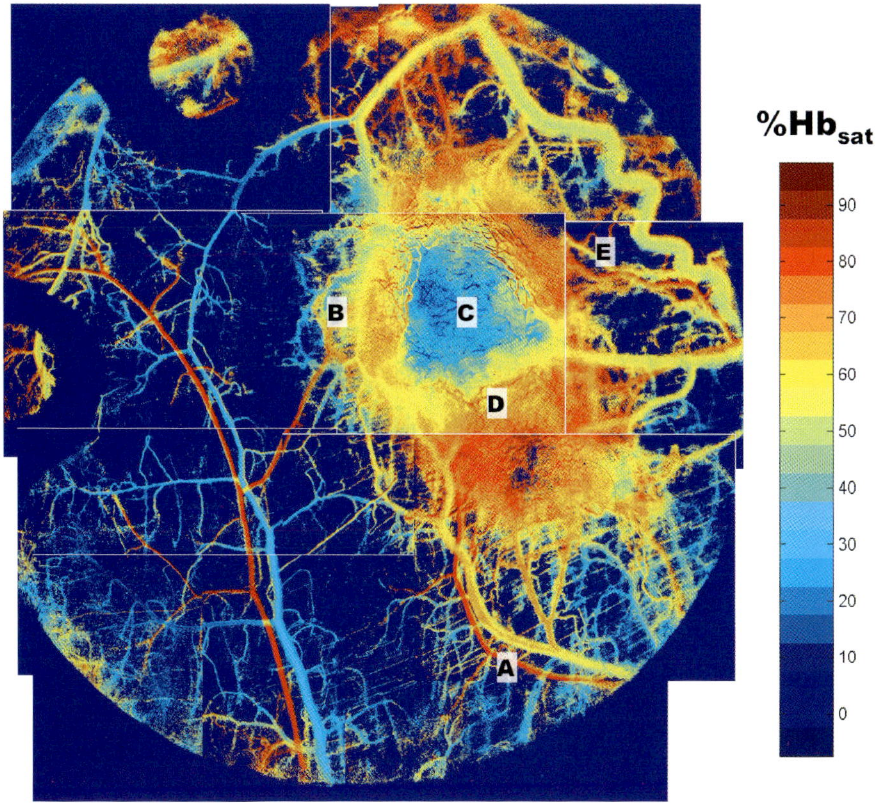

Figure 2.2 This mouse mammary tumor grown in a dorsal skin-fold window chamber was analyzed for regions of different hemoglobin saturations. This figure was reproduced with permission from Hardee, M.E., Dewhirst, M.W., Agarwal, N. and Sorg, B.S. (2009) Novel imaging provides new insights into mechanisms of oxygen transport in tumors. *Curr Mol Med*, 9, 435–441.

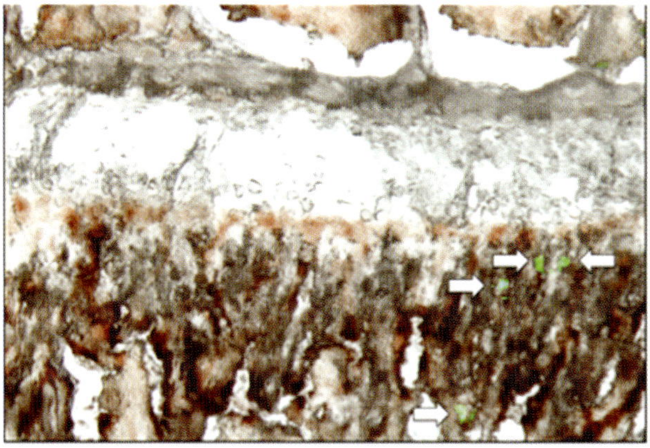

Figure 7.1 Arrival of cancer cells to the skeleton.

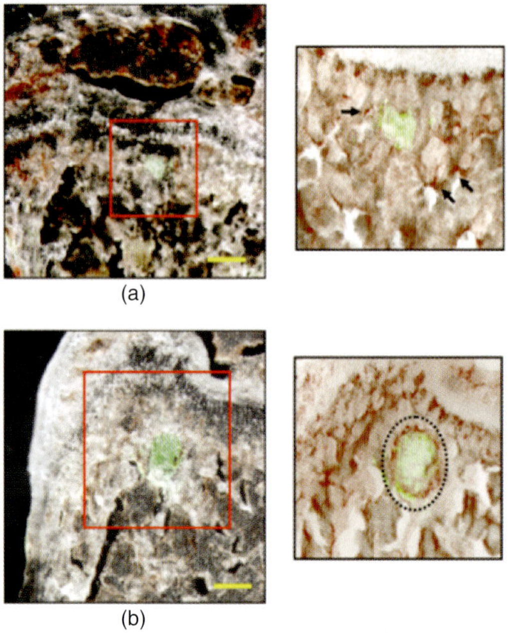

Figure 7.2 Spatial relationship of osteoclasts with bone metastatic tumor foci. Metastases with a cross-sectional area larger than $28 \times 10^3\,\mu m^2$ (b) were surrounded by a layer of active osteoclasts, identified by a red TRAcP staining. However, smaller metastases were spatially unrelated to osteoclasts (a), which appear sparsely distributed (arrows). Measurement bar is $100\,\mu m$.

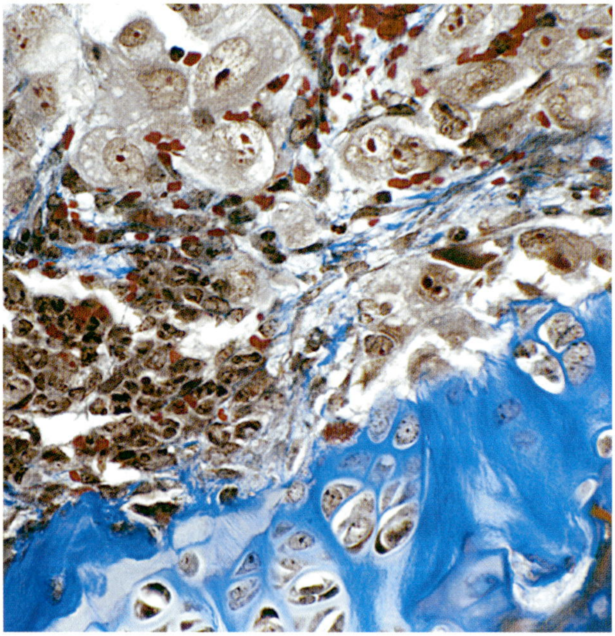

Figure 7.4 Multiple cell populations are present in metastatic bone lesions.

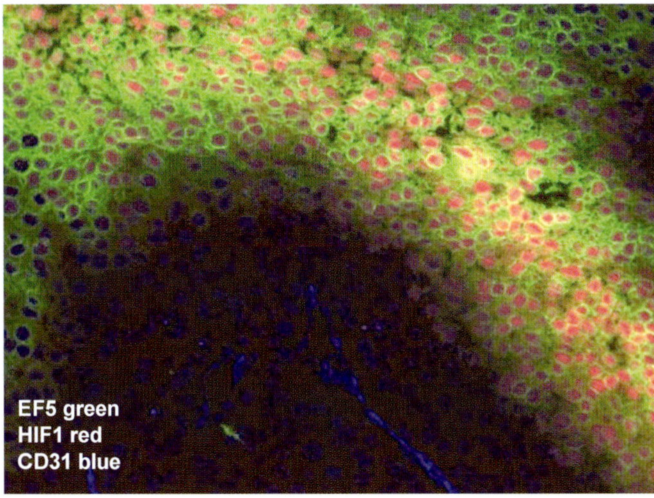

Figure 9.4 Micro-regional distribution of hypoxia in a SiHa cervix cancer xenograft illustrating co-localization between the bioreductive nitroimidazole drug EF5 (green) and increased expression of the endogenous hypoxia marker HIF1 (red) at a distance from blood vessels (CD31, blue). Image courtesy of D. Hedley.

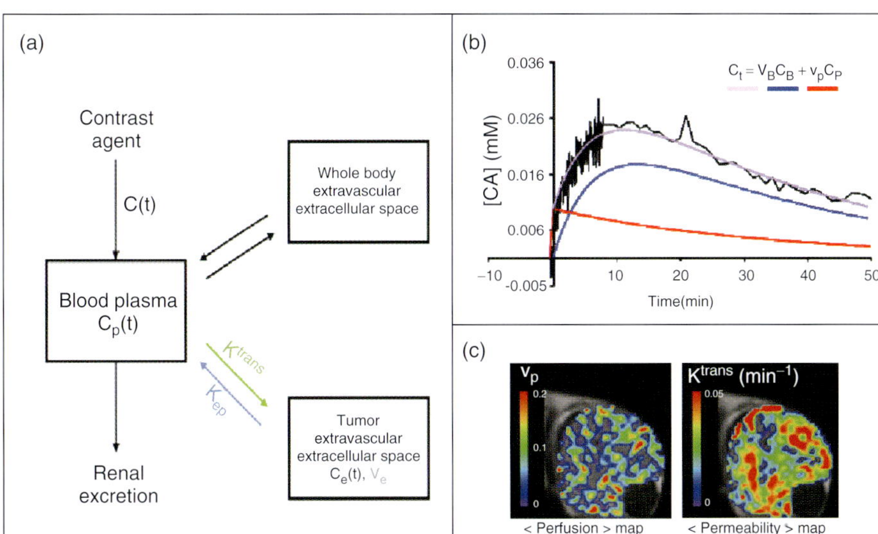

Figure 11.1 (a) Illustration of two-compartment model demonstrates the exchange of contrast between plasma and extravascular extracellular space. (b) Modelization of typical enhancement curve in the tumor (purple) after bolus injection of CA, dissociation into ESS curve (blue) and plasmic curve (red). (c) Typical perfusion (V_p) and permeability (K_{trans}) experimental tumor maps.

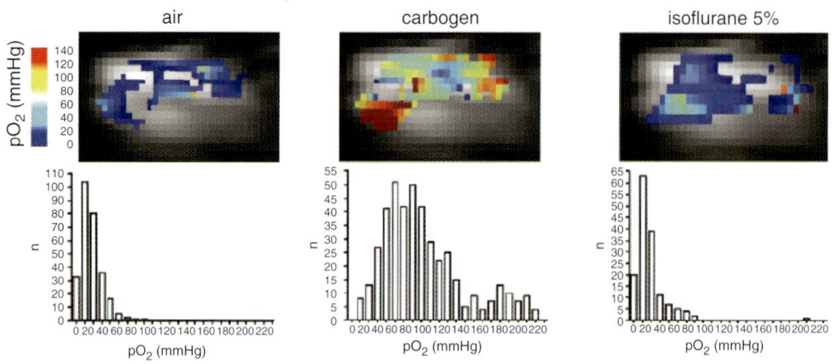

Figure 11.2 Typical oxygen maps of a typical tumour under air, carbogen, and isoflurane 5% (sacrifice) breathing conditions and their corresponding histograms. Adapted from Jordan, Cron, and Gallez (2009).

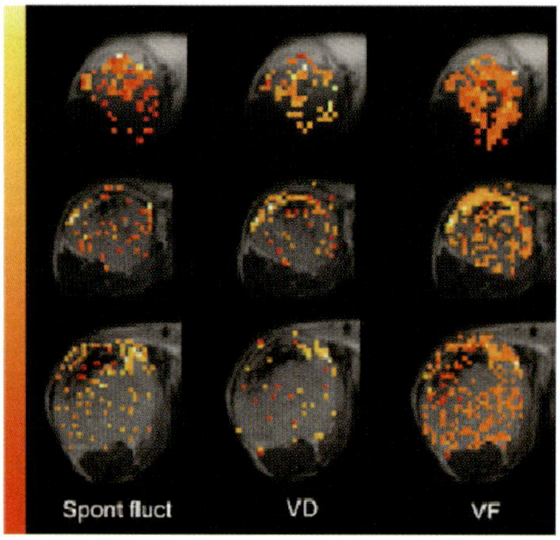

Figure 11.3 Co-registration of spontaneous fluctuations, vascular maturation, and vascular function maps for three seperate intramuscular FSa II fibrosarcoma tumors in mice. Adapted from Baudelet, Cron, and Gallez (2006).

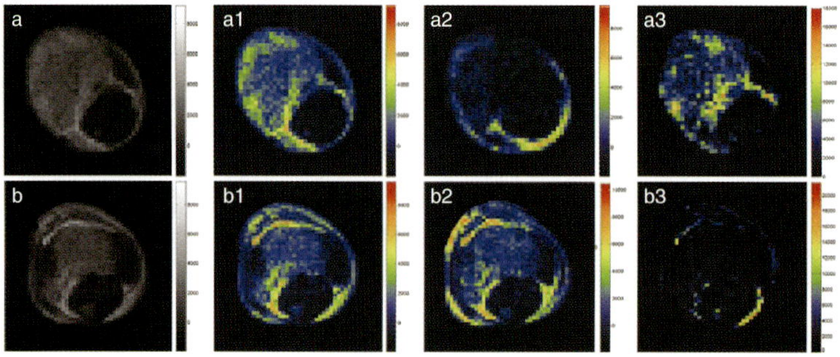

Figure 11.5 Axial slices of irradiated tumors obtained by T_2-weighted MRI. (Adapted from Radermacher *et al* (in press).)

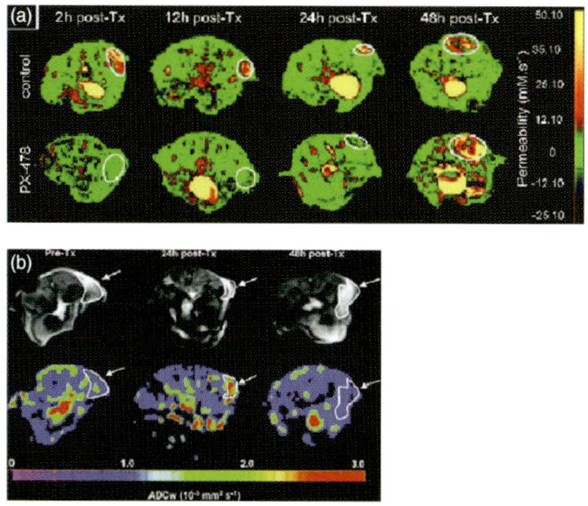

Figure 11.6 (a) Permeability maps of tumors at 2, 12, 24, and 48 hours following injection of vehicle (control) or PX-478 (drug) injection. Each image represents an axial slice of the mouse with the tumor area encircled. A substantial decrease in tumor permeability was observed as early as 2 hours following treatment and continuing until 24 hours in comparison to controls. (b) DW images at a b-value (top row) and corresponding diffusion maps (bottom row) of an HT-29 tumor-bearing mouse at 0, 24, and 48 hours following PX-478 injection. Decrease in tumor cellularity was noted at 24 and 36 hours following treatment as indicated with an increase in ADCw values. Each image represents an axial slice of the mouse with the tumor area encircled and indicated by an arrow. Adapted from Jordan *et al.* (2005).

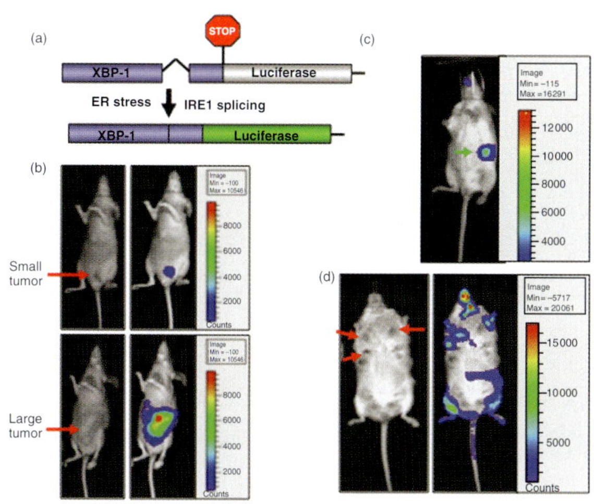

Figure 13.2 Activation of Ire1/XBP-1 pathway in implanted tumor xenografts and transgenic mouse model. (a) Schematic representation of XBP-luciferase reporter construct. (b) HT1 080 human fibrosarcoma cells stably expressing the XBP-luciferase reporter were implanted subcutaneously into nude mice. Luciferase activity was detected in both small tumor and large tumor (red arrow). (c) Transgenic mouse harboring the XBP–luciferase reporter gene. Highest basal luciferase activity was detected in the pancreas (green arrow). (d) Representative of double transgenic mouse harboring both the XBP–luciferase reporter gene and the MMTV-PyVT gene. Luciferase activity was detected in multiple spontaneous breast tumors.

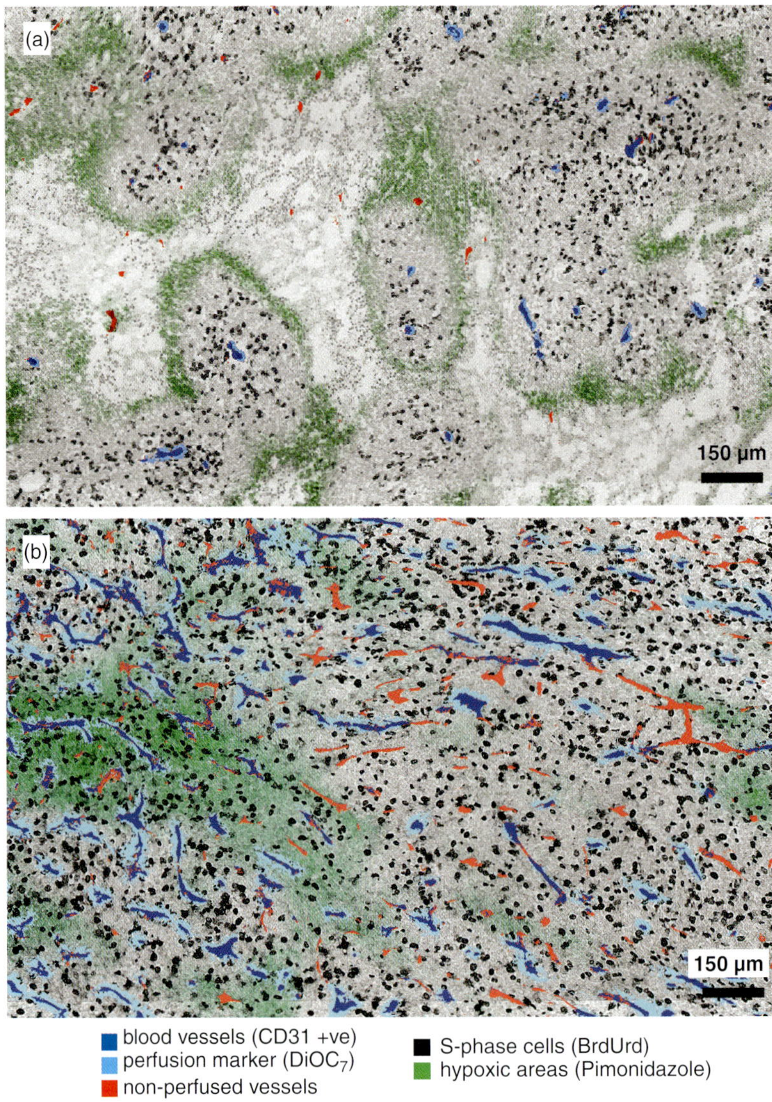

Figure 15.1 Examples of two tumor architectures commonly observed in experimental tumors grown subcutaneously in mice. (a) A human HCT-116 tumor xenograft, which exhibits a sparse vasculature where cords of tumor cells are often seen to surround individual vessels. (b) An SCCVII mouse tumor, which exhibits a dense vasculature and significant intermittent flow (Minchinton lab, unpublished).

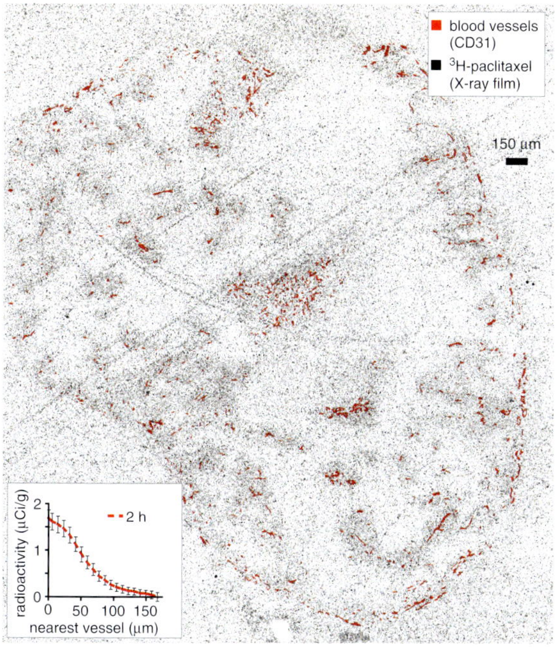

Figure 15.2 Composite image of ^{3}H-paclitaxel distribution (black) and CD31 positive vasculature (red) in an HCT-116 tumor xenograft 2 hour post i.v. injection of 10 mg/kg ^{3}H-paclitaxel. Adapted from Kyle (2007).

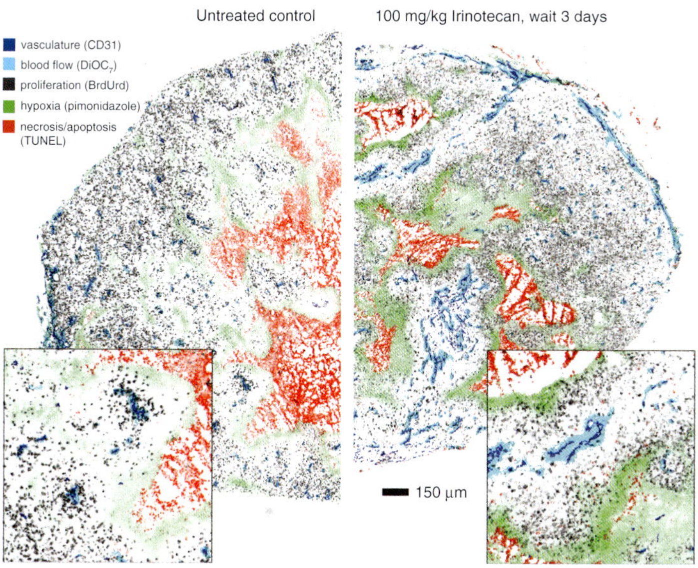

Figure 15.3 Microregional activity of 100 mg/kg Irinotecan i.v. 3 days following treatment. (Minchinton lab, unpublished).

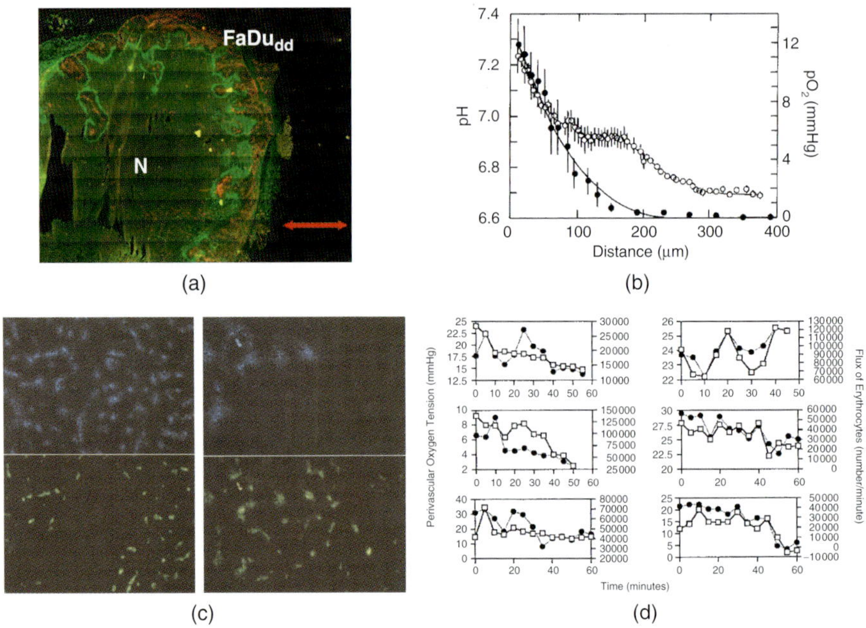

Figure 16.1 Illustrations of hypoxia in tumors. (a) Pseudocolor gray value composite image reconstructed after scanning histological sections from a head and neck xenograft (FaDu$_{dd}$) grown in nude mice. Images show hypoxia (green) and vessels (red).N denotes necrosis; red scale bar = $200\,\mu m$. (b) Combined measurements of pH and pO_2 profiles from human colon adenocarcinoma xenogafts (LS174T) grown in a window chamber in severe combined immunodeficient mice. Results show mean values of fluorescent ratio imaging for pH (○) and phosphorence quenching for pO_2 (●) for 24 profiles from seven tumors. (c) Histological sections from SCCVII carcinomas showing 'closing' (non-perfusion) of previously perfused microvessels (left) or 'opening' (perfusion) of previously non-perfused microvessels (right) during a 20-minute interval between intravenous injection of the fluorescent markers H.33342 (blue) and DiOC7 (green). (d) Representative figures showing the relationship between microvascular erythrocyte flux (●) and perivascular oxygen tensions (□) measured simultaneously in dorsal flap window chambers of Fisher-344 rats with implanted R3230Ac tumors. Erythrocytes were fluorescently labeled and observed with intravital video microscopy, while oxygenation was monitored using microelectrodes. Figures and data from Chaplin and Trotter (1991), Kimura *et al.* (1996), Helmlinger *et al.* (1997), Busk *et al.* (2008).

10

Hypoxia and the DNA Damage Response

Isabel M. Pires, Rachel Poole and Ester M. Hammond
Cancer Research UK/MRC Gray Institute for Radiation Oncology and Biology, University of Oxford, Oxford UK

10.1 Introduction

Regions of hypoxia are present in all solid tumors and can occur at early or late stages of tumor development. Levels of hypoxia range from near 0% pO_2 (anoxia) to 8% (Brown, 2007). Elegant studies using direct oxygen tension measures in numerous tumor types demonstrated a correlation between the level of hypoxia and prognosis, with lower oxygen levels associated with poorer prognosis (Brizel *et al.*, 1997; Hockel *et al.*, 1996; Nordsmark *et al.*, 1996a; Nordsmark, Overgaard, and Overgaard, 1996b). We have shown previously that severe hypoxia induces a robust DNA damage response (DDR). Although, interestingly this seems to occur in the absence of DNA damage detectable by either comet assay or the formation of p53 binding protein 1 (53BP1) foci (Hammond *et al.*, 2002; Hammond, Green, and Giaccia, 2003b; Bencokova *et al.*, 2009). In contrast, reoxygenation events, which occur as a consequence of irregular perfusion of the tumor, induce significant levels of DNA damage in a reactive oxygen species (ROS)-dependent manner (Brown and Giaccia, 1998; Hammond, Dorie, and Giaccia, 2003a; Brown, 1979). Failure to repair these lesions due to the loss of either repair pathways and/or p53 can then lead to increased genomic instability and tumor progression (Graeber *et al.*, 1996; Cairns, Kalliomaki, and Hill, 2001; Weinmann *et al.*, 2004). Because of their intrinsic connection, hypoxia and reoxygenation can be considered as two facets of the same stress. The focus of this chapter is the DDR induced by the tumor microenvironment and specifically conditions of low oxygen, hypoxia. First, we will discuss how hypoxia and reoxygenation can promote DDR induction and signaling. Then we will explore the role of hypoxia in deregulating DNA repair and how this can potentially be exploited for novel therapeutic strategies. We will conclude by

Tumor Microenvironment Edited by Dietmar W. Siemann

highlighting the impact of low oxygen on the proposed role of the DDR as a barrier to tumorigenesis (Bartkova *et al.*, 2005, 2006).

10.2 The DNA damage response

The human cell is constantly barraged by DNA damaging agents from both endogenous (e.g., metabolites) and exogenous (e.g., UV) sources. It has been estimated that cellular DNA is damaged over 10 000 times a day, each of which could be potentially lethal (Lindahl and Barnes, 2000). Throughout evolution a series of molecular pathways designed to detect, and ultimately repair DNA lesions have developed. These pathways are designated as the DDR. The DDR is generally considered to correspond to the concept of a classical signaling transduction pathway including signal, sensors, transducers, and effectors (Jackson and Bartek, 2009). In this case the signal is a lesion or lesions which includes: single strand breaks (SSBs) or double strand breaks (DSBs), oxidative base damage, bulky adducts, and mismatches.

10.2.1 The p53 tumor suppressor is induced by hypoxia

In 1996 Giaccia and colleagues described the hypoxia-induction of p53 and p53-dependent apoptosis (Graeber *et al.*, 1996). From their study it was concluded that a selection pressure to lose p53 activity occurred as a result of the hypoxic tumor microenvironment. The induction of p53 in hypoxic conditions was, in part, attributed to the observation that the levels of human homolog of mouse double minute 2 (hMDM2) decrease during exposure to hypoxia (Alarcon *et al.*, 1999). One of the principal roles of hMDM2 is to keep p53 in check by targeting it for proteosomal degradation (Haupt *et al.*, 1997). In the absence of hMDM2, p53 accumulates or stabilizes. This finding raised a pertinent question and gave perhaps the first hint that hypoxia-induced p53 was not behaving as might be expected. *Mdm2* is a target of p53, clearly containing p53 response elements and responding to increased levels of p53 and yet in the presence of hypoxia-induced p53, levels of hMDM2 fall (Koumenis *et al.*, 2001). In fact, further work showed that this was a widespread phenomenon; genes expected to be induced in response to hypoxia as a result of being characterized p53-targets were not induced (Gottifredi *et al.*, 2001). This led to the hypothesis and subsequent evidence to support it that, in response to hypoxia, p53 with transrepressive rather than transactivating capabilities is induced (Koumenis *et al.*, 2001). To clarify this point, it has been proposed that hypoxia-induced p53 transrepresses a unique set of responsive genes in order to drive the cell into apoptosis as opposed to repressing genes it would normally activate. Although p53 had been shown to transrepress individual genes before and since this finding, the hypoxia response remains unusual in that p53 seems to act exclusively as a repressor (Hammond and Giaccia, 2005; St. Clair *et al.*, 2004; Hammond *et al.*, 2006). The mechanism behind p53-mediated gene repression in hypoxic conditions is still somewhat unclear. The available data indicates that repression is through DNA binding and requires both transactivation domains of p53 to be intact (Hammond *et al.*, 2006; Johnson *et al.*, 2005). What is perhaps more intriguing is why the p53-stress response to hypoxia utilizes repression in contrast to other factors that induce a p53 response. One of the reasons behind this might be the repressive nature of hypoxia itself. In response to hypoxia both translation and transcription are decreased making it plausible that it is more efficient to repress targets rather than overcoming the cellular environment to induce apoptosis-promoting factors (Hammond and Giaccia, 2005). The prediction is that the p53-repressed genes will be survival factors which, when decreased during exposure to hypoxia, will drive the cell into apoptosis. Indeed, one of the p53-repressed genes identified in hypoxia was *Survivin*, although the biological relevance of this remains to be seen. This raises the interesting possibility of targeting this pathway for therapeutic gain. As the majority of tumor cells have deficient p53 signaling it may be possible to pharmacologically inhibit one or more of the p53-repressed genes to achieve the same goal, that is, apoptosis.

The finding that p53 could be stabilized and activated in response to hypoxia both *in vitro* and *in vivo* provided some of the first evidence that a DDR was being initiated in response to the tumor microenvironment. This led to the obvious question regarding the nature of the p53-inducing signal present in conditions of low oxygen. The p53 protein is heavily post-translationally modified and in particular, by kinases at the amino terminus. To our knowledge, in response to hypoxia p53 is phosphorylated at residues 15, 20, 37, and 46 (Poole and Hammond, unpublished data; Hammond *et al.*, 2002). Phosphorylation of serine 15 was of particular interest as this has a critical role in the stabilization of p53 through the masking of the hMDM2 interaction domain (Shieh *et al.*, 1997). There is also evidence that phosphorylation of serine 20 has a similar role, with some data suggesting that this is in fact the critical event in stabilization (Unger *et al.*, 1999; Dumaz *et al.*, 2001). The oxygen dependency of hypoxia-induced p53 phosphorylation proved informative as it was only detected in severe hypoxia ($<0.1\%$ O_2). These same levels of hypoxia were found to induce an S phase arrest or more accurately a complete cessation of DNA synthesis (Probst *et al.*, 1988, 1999; Probst, Gekeler, and Helftenbein, 1984). It had been hypothesized that this was likely due to a failure of ribonucleotide reductase to generate sufficient nucleotides (deoxyribonucleotide triphosphates, dNTPs) under hypoxic conditions as the activity of the enzyme is oxygen dependent (Reichard and Ehrenberg, 1983; Probst *et al.*, 1989). Although the activity of ribonucleotide reductase has not been measured in hypoxic conditions we recently demonstrated that dNTP levels drop rapidly therefore supporting this hypothesis (Pires *et al.*, 2010). A lack of dNTPs would be predicted to induce a replication block in both the initiation and elongation phases of DNA replication. This has been confirmed using DNA combing technology to investigate individual replication structures in cells exposed to hypoxia. A strong correlation was found between hypoxic conditions which induced both replication arrest and p53 stabilization/phosphorylation suggesting that one might be the signal for the other (Hammond *et al.*, 2002).

When replication forks stall, due for example, to a lack of nucleotides, regions of single stranded DNA (ssDNA) accumulate. These forks are subsequently coated by the replication factor replication protein A (RPA), which can be seen as discreet nuclear foci in S phase cells (Robison *et al.*, 2004; Hammond, Dorie, and Giaccia, 2004; Pires *et al.*, 2010; Wold and Kelly, 1988). These foci are known to be the signal for the recruitment and activation of the ataxia telangiectasia mutated and Rad3-related kinase (ATR) (Zou and Elledge, 2003; Cimprich *et al.*, 1996). ATR is one of the phosphatidyl inositol-3-kinase like kinases (PIKKs), which are generally accepted to be the core transducers of the DDR. ATR also requires the participation of its binding partner ATR interacting protein, the 9-1-1 sensor complex, Rad17, and mediators such as mediator of DNA-damage checkpoint 1 (MDC1), Claspin, and topoisomerase (DNA) II binding protein1 (TopBP1) (Jackson and Bartek, 2009; Cimprich and Cortez, 2008). Further work demonstrated that, in response to hypoxia-induced stalled replication, ATR signaling was initiated, leading to the phosphorylation of, amongst other targets, p53. It should also be noted that there has been some controversy as to whether hypoxia or hypoxia-associated acidosis induces p53 (Hammond, Dorie, and Giaccia, 2003a; Pan *et al.*, 2004).

The checkpoint kinase, Chk1, has also been identified as a target of the ATR kinase in response to replication stresses, including hypoxia (Hammond *et al.*, 2002; Hammond, Dorie, and Giaccia, 2004; Hammond, Freiberg, and Giaccia, 2005). Robust phosphorylation at residues serine 317 and serine 345 of Chk1 was observed in the early response to hypoxia. Both ATR and Chk1 have been shown to have an important role in preserving replication fork stability during exposure to replication stresses (Brown and Baltimore, 2003; Casper *et al.*, 2002; Lomonosov *et al.*, 2003). Loss or inhibition of either ATR or Chk1 leads to an increased sensitivity to hypoxia/reoxygenation. Most significantly, loss of ATR during hypoxia leads to the accumulation of comet-detectable DNA damage suggesting the collapse and breakage of stalled replication forks. These studies have not been extended to investigate whether this would also be the case when Chk1 is inhibited. However, it has been shown that if Chk1 is pharmacologically inhibited during reoxygenation, replication-restart rates are increased suggesting that the loss of Chk1 allows replication to proceed before reoxygenation-induced damage is repaired.

The ATM and ATR kinases are similar in that there is much redundancy in their downstream targets (Chaturvedi *et al.*, 1999; Liu *et al.*, 2000). However, the two kinases, both members of the PIKK family, are not redundant and this is thought to be primarily due to the different signals they respond to. Whereas, ATR responds to ssDNA generated at stalled replication forks as discussed previously ATM has traditionally been described as responding to DSBs. ATM is recruited to the sites of DSBs through the activity of the sensor complex Mre11-Rad50-Nbs1 (MRN) and promoted by mediators such as MDC1, 53BP1, ring finger protein (RNF)8, RNF168, and BRCA1 (Lee and Paull, 2007). An increasing number of reports indicate that ATM responds to more than DSBs, for example, studies of hypothermia, chloroquine exposure, heat, or insulin signaling (Pandita, Pandita, and Bhaumik, 2009; Loehberg *et al.*, 2007; Hunt *et al.*, 2007; Jeong *et al.*, 2009). Recently, ATM has been described as being active in hypoxic conditions (Bencokova *et al.*, 2009; Gibson, Bindra, and Glazer, 2005; Freiberg *et al.*, 2006a). The involvement of ATM in the hypoxic response was unexpected as both comet and 53BP1 foci assays indicate a lack of DSB accumulation during hypoxia. In response to severe hypoxia, ATM is rapidly phosphorylated and this at least in part, is a result of an autophosphorylation which has been described (Bakkenist and Kastan, 2003; Goodarzi *et al.*, 2004; Pellegrini *et al.*, 2006). There are reports indicating that as well as ATM becoming active as a result of autophosphorylation ATR can also carry out this role, that is, phosphorylate ATM (Stiff *et al.*, 2006). This has not been fully investigated in the hypoxia context as in order to do so, ATR would have to be inhibited in hypoxic conditions which would in turn induce ATM through the formation of DSBs. In support of the data indicating an absence of DSBs in hypoxia the MRN complex is not required to activate ATM in hypoxia and the ATM does not subsequently form nuclear foci (Bencokova *et al.*, 2009). This was also observed with other non-DNA damaging agents, for example, exposure to chroloquine (Bakkenist and Kastan, 2003). It is therefore unclear what the hypoxia-induced signal is that activates ATM in response to hypoxia, although it seems likely that it is aberrant chromatin structures/modifications arising from the presence of stalled replication forks. This is supported by the finding that whilst 53BP1 fails to form foci in hypoxic conditions, γH2AX an alternative marker of the DDR, does (Rappold *et al.*, 2001). The suggestion is that the γH2AX signal is at sites of chromosome abnormalities present at stalled replication forks whilst the lack of 53BP1 foci represent the lack of actual damage and also a possible failure of the mediator MDC1 to recruit RNF8 (Minter-Dykhouse *et al.*, 2008; Huen *et al.*, 2007; Kolas *et al.*, 2007; Mailand *et al.*, 2007; Wang and Elledge, 2007; Bencokova *et al.*, 2009). The role of γH2AX in hypoxia will be discussed further in a later section.

10.2.2 Reoxygenation induces DNA damage

Reoxygenation events are both frequent and rapid in tumors and have been effectively demonstrated using vessel mismatch experiments. In these studies tumor bearing animals are injected with a non-vasoactive drug, which travels through the bloodstream and diffuses into surrounding tissues, therefore highlighting them as functional. A short time after the drug is administered a second drug is added. The tumor is then sectioned and examined for the presence of both drugs. These reoxygenation events are highly relevant because they lead to an accumulation of DNA damage. This has been demonstrated through the accumulation of comet detectable damage, the appearance of reoxygenation-induced 53BP1 foci and the sustained phosphorylation of p53 in an ATM-dependent manner. ROS have been shown to be at least partially responsible for accumulation of damage after reoxygenation since ROS scavengers such as *N*-acetyl-L-cysteine attenuate this (Freiberg *et al.*, 2006a). The presence of bona fide DNA damage in reoxygenation suggested a further involvement of the DDR.

The cell cycle checkpoints ensure cell cycle arrest in response to DNA damage. The serine/threonine protein kinase Chk2 plays an important role in the cell cycle checkpoints. Upon activation by phosphorylation it can, in turn, phosphorylate a number of downstream targets including cell division cycle 25 homolog C (CDC25C),

BRCA1, and p53 and this contributes to cell cycle arrest, apoptosis, and/or repair. ATM dependent Chk2 phosphorylation at threonine 68 occurs in hypoxia in the absence of DNA damage and is maintained upon reoxygenation. Reoxygenation of hypoxic cells leads to a substantial G_2 arrest, which is dependent on Chk2 (Freiberg *et al.*, 2006a, 2006b; Gibson, Bindra, and Glazer, 2005). Chk2 activation in this situation phosphorylates the protein phosphatase CDC25C leading to its sequestration by the 14-3-3 protein. This prevents the removal of an inhibitory phosphate group on Cdc2, preventing its interaction with cyclin B. Cells lacking Chk2 do not arrest in G_2 during reoxygenation and exhibit higher levels of apoptosis, reduced clonogenic survival, and deficient phosphorylation of downstream targets, including BRCA1 (Kim *et al.*, 2007; Gibson, Bindra, and Glazer, 2006). Phosphorylated BRCA1 is required for homologous recombination (HR) after DNA damage and cells deficient in BRCA1 undergo apoptosis following DNA damage. It is possible therefore, that the increased sensitivity of Chk2-deficient cells to hypoxia/reoxygenation could be due, at least in part, to a failure to activate BRCA1 resulting in compromised repair of reoxygenation-associated damage. Some of the key differences in the DDR to hypoxia as opposed to hypoxia/reoxygenation are highlighted in Figure 10.1.

The biological consequences of the induction of a DDR in cells exposed to severe hypoxia unclear. At these oxygen tensions the most likely fate of these cells is necrosis and therefore little or no contribution to tumorigenesis. However, as discussed, rapid changes in the tumor vasculature and/or therapeutic reoxygenation events can and do occur raising the possibility that these previously arrested cells might be returned to replication-permissive conditions (Cardenas-Navia *et al.*, 2008; Brown, 2002; Jain, 2005). We have recently demonstrated that after acute periods (up to 8 hours) of hypoxia-induced replication arrest cells can indeed resume DNA synthesis and complete the cell cycle (Pires *et al.*, 2010). It has been suggested that one of the benefits to activating the DDR during hypoxia is to prepare the cell for the possibility for reoxygenation-induced replication restart and/or DNA damage. Loss or inhibition of key players in the response has been demonstrated to sensitize cells to hypoxia/reoxygenation (Hammond, Dorie, and Giaccia, 2004; Freiberg *et al.*, 2006a). An interesting aspect of the hypoxia-induced DDR is that it is dependent on the length of the hypoxic exposure. We have shown that the DDR is only induced during acute exposures (shorter than 12 hours). Longer or chronic exposures (12 hours or longer) lead to a decrease in the earlier induced DDR signaling, both at the level of phosphorylation events as well as of the protein levels, including those of Chk1 and TopBP1 (Pires *et al.*, 2010). Chronic exposures to severe hypoxia also lead to the dissociation of the replication machinery (Pires *et al.*, 2010). This abrogation affects the minichromosome maintenance complex helicase complex through a transcriptionally-mediated decrease in protein level. This is accompanied by the dissociation of the complex from the chromatin, alongside that of other replication factors such as hyperphosphorylated RPA. Importantly, this effect on the replication machinery is observable *in vivo* (Pires *et al.*, 2010). As a result chronically arrested cells are incapable of replication restart in response to reoxygenation and therefore do not contribute to continued tumorigenesis. In contrast, our studies showed that after shorter periods of hypoxia-induced replication arrest cells are able to resume replication in response to reoxygenation. Importantly, this occurs in the presence of

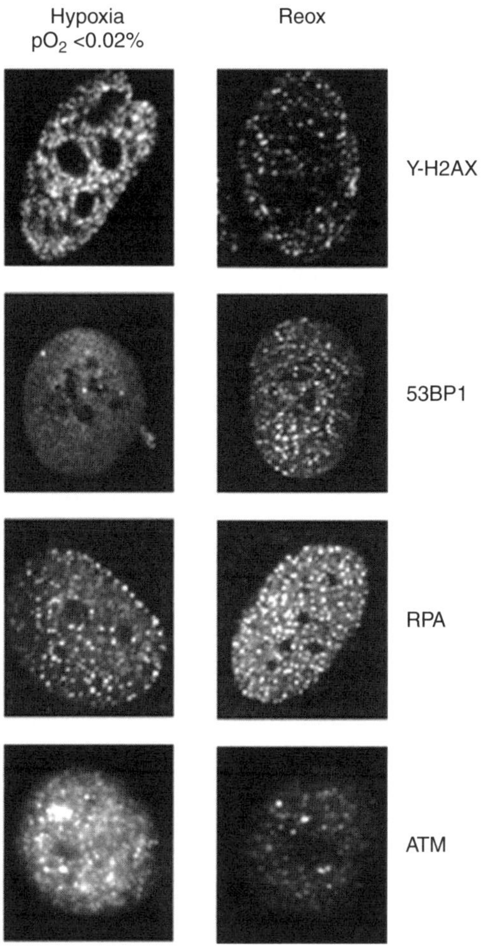

Figure 10.1 Severe hypoxia ($pO_2 < 0.02\%$) results in activation of the DNA damage response in the absence of detectable DNA damage. This DNA damage response differs to the classical response as seen by the distribution of various DDR proteins in the cell. In hypoxia γ-H2AX foci are formed, but differ from those formed in response to double strand breaks. RPA foci are seen due to ssDNA formed in the replication arrest. 53BP1 does not form foci, which supports the observation that DNA damage does not occur in hypoxia. ATM is phosphorylated and active, but not chromatin bound, and so fails to form foci. Upon reoxygenation DNA damage occurs leading to the formation of 53BP1 and ATM foci. The more classical double strand break-induced γ-H2AX foci are also present following reoxygenation.

ROS-induced DNA damage and when essential DNA repair pathways are somewhat inhibited (as discussed later in this chapter). Together, this indicates that cells that resume DNA synthesis after a reoxygenating event do so in the presence of DNA damage and with impaired DNA repair capabilities, which has the potential to lead to genomic instability, particularly in cells in tumors that have lost p53 signaling to apoptosis.

10.2.3 A novel role for H2AX in angiogenesis and tumor growth

A critical event during DDR signaling is the phosphorylation of histone H2AX (a variant of the H2A histone), to form γH2AX (Bonner *et al.*, 2008). This event initially occurs in chromatin sites flanking the lesion, progressing later to increasingly distant sites, and leading to a formation of large and defined γH2AX foci. This phosphorylation event has been proposed to function as a signal amplification and recruitment stage for other DDR signaling and DNA repair molecules (Stucki *et al.*, 2005; Ward and Chen, 2001; Kinner *et al.*, 2008). There is clear evidence that DSB-inducing stresses (such as ionizing radiation) induce γH2AX foci formation (Burma *et al.*, 2001). This has led to the widespread use of γH2AX as a marker of DNA damage. However, more recently, stress types which are not associated with DSBs, such as UVC and hypoxia have been found to promote γH2AX recruitment and foci formation (Bencokova *et al.*, 2009; Hammond, Dorie, and Giaccia, 2003a; Marti *et al.*, 2006). Perhaps not surprizingly, given the lack of actual breaks in these conditions these foci differ in terms of nuclear distribution from the traditional DSB induced ones, often presenting a more punctuate pan-nuclear pattern. A growing number of studies have demonstrated a role for H2AX as a tumor suppressor. Including evidence that a variety of cancers harbor mutated or deleted H2AX (Bonner *et al.*, 2008). It has also been observed that loss of H2AX makes p53-deficient mice more prone to tumor development (Bassing *et al.*, 2003).

Until recently hypoxia-induced DNA damage signaling was thought to only occur at extreme, almost anoxic levels that would be found surrounding the necrotic areas in tumors. Studies using higher oxygen concentrations (pO_2 0.2%) however, indicate that DDR signaling can be observed at these oxygen tensions (Chan *et al.*, 2008). Recently, it has been shown that ATR-dependent signaling to the histone H2AX occurs at oxygen levels as high as 1% in non-transformed endothelial cells and that this plays an important role in angiogenesis. In this study, genetic inactivation of H2AX was found to suppress tumor angiogenesis and tumor growth in xenograft models. H2AX was shown to be phosphorylated and form foci in endothelial cells during exposure to hypoxia. This occurred in an ATR-dependent manner and γH2AX foci colocalize with RPA, in accordance with published reports described previously (Bencokova *et al.*, 2009; Hammond, Dorie, and Giaccia, 2003a). As discussed, RPA foci have been shown to form as a result of ssDNA generated at stalled replication forks. However, in their study Economopoulou *et al.* demonstrate that hypoxic endothelial cells accumulate γH2AX and RPA foci whilst still proliferating and most interestingly that γH2AX is required to maintain proliferation in these conditions. This requirement for H2AX is suggested to be dependent on its role in DNA repair, which would in turn predict that repair factors downstream of H2AX may also have a role to play in tumor angiogenesis. This study has raised numerous questions, for example, what is the hypoxia-induced signal in these cells leading to the activation of ATR and subsequently H2AX? Economopoulou and colleagues propose that the hypoxia-induced repression of DNA repair could, in the endothelial cell model, be sufficient to induce the activation of ATR and γH2AX phosphorylation (Economopoulou *et al.*, 2009). Equally intriguing and perhaps relevant to this question is why, despite the accumulation of γH2AX foci, the endothelial cells exposed to hypoxia (1% O_2) do not show detectable ATM phosphorylation. Understanding this novel role for H2AX opens up new avenues for targeted therapies based on angiogenesis inhibition.

10.3 Hypoxia regulation of DNA repair

The hypoxic microenvironment increases genomic instability (Bindra, Crosby, and Glazer *et al.*, 2007; Bristow and Hill, 2008; Huang *et al.*, 2007). This has been demonstrated by various studies, including observations that either hypoxic exposure of cells or the tumor microenvironment in xenograft models increased the number of point mutations in various reporter gene loci assays (Li *et al.*, 2001; Papp-Szabo, Josephy, and Coomber, 2005; Reynolds, Rockwell, and Glazer, 1996). In addition, hypoxia–reoxygenation cycles are known to be able to induce DNA lesions, including oxidative damage (such as 8-oxoguanine and thymine glycols) and DNA strand breaks (Hammond, Dorie, and Giaccia, 2003a; Yuan *et al.*, 2000). Hypoxia exposure is also associated with other genetic aberrations, such as the occurrence of fragile sites, gene amplification, and over-replication (Coquelle *et al.*, 1997; Smith, Huang, and Wang, 1998; Young, Marshall, and Hill, 1988; Rice, Spiro, and Ling, 1985). Deficiencies in DNA repair pathways have been associated with cancer predisposition since ineffectively repaired DNA is one of the key factors for increased genomic instability and, consequently, promotion of tumorigenesis (Loeb, Loeb, and Anderson, 2003). DNA repair pathways are one of the key response pathways of the DDR. There are several of these mechanisms that are determined by the nature of the damage signal. However, they are not simply linear pathways, since many of them overlap. This is relevant in particular, in the case where one of them is down regulated or absent. For example, when functional base excision repair (BER) is absent a SSB will transform into a DSB, which will be then processed by HR (Tutt *et al.*, 2005). There are five main DNA damage repair pathways: BER, nucleotide excision repair (NER), mismatch repair (MMR), non-homologous end-joining (NHEJ), and HR. Figure 10.2 shows a schematic representation of the DNA repair pathways and examples of how they are deregulated by hypoxia.

10.3.1 The DNA repair pathways

The BER pathway has a critical role to play in preserving genomic integrity. Its importance is reiterated by the attempts to generate knockout mice of BER genes. These are either embryonic lethal, indicating an absolute requirement for the gene product, or show either slight or no phenotype, indicating a high degree of redundancy (Hakem, 2008; Robertson *et al.*, 2009). BER deals with lesions to bases and nucleotides which are associated to metabolic processes, for example, ROS, which can cause alkylation or oxidative damage to DNA. Examples of these lesions are 8-oxo-7, 8-dihydroguanine (8-oxo-G), depurination of DNA, and spontaneous cytosine deamination to uracil. The BER pathway consists of a recognition step by a glycosylase, followed by base removal, processing to A-P sites by endonucleases and finally filling in by polymerases. Key molecules in this pathway include XRCC1

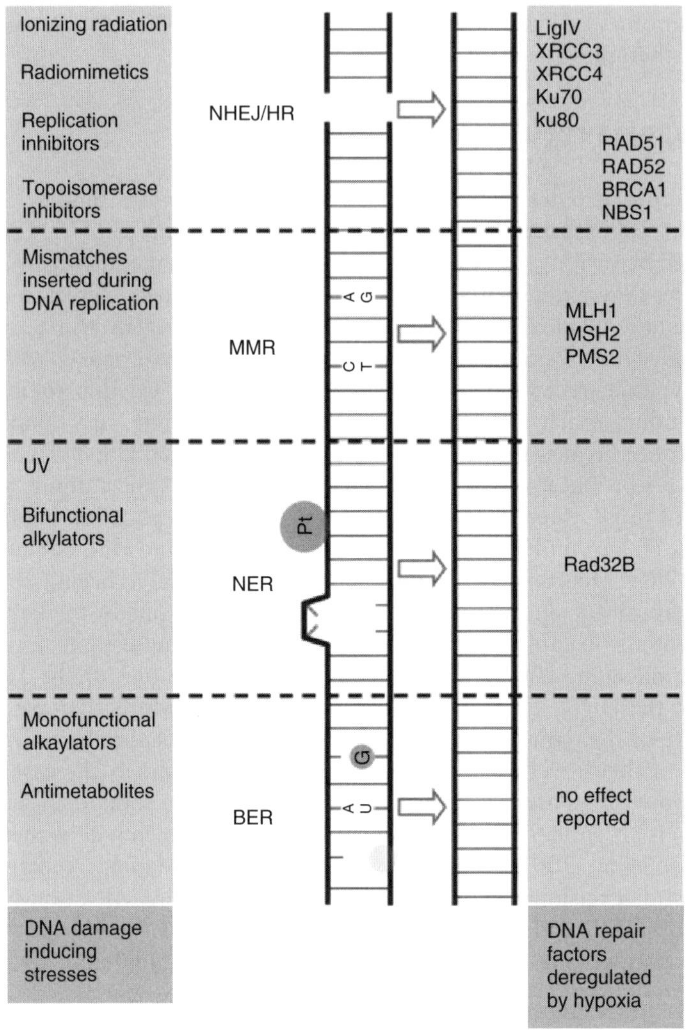

Figure 10.2 DNA damage is induced over 10 000 times a day and is a consequence of both endogenous and exogenous stress (shown on the left hand side of the diagram). A number of different lesions can be formed and these are repaired by distinct cellular pathways. In hypoxia components of these pathways are deregulated (some examples are shown on the right hand side). This deregulation can compromise repair of DNA damage upon reoxygenation and potentially lead to increased genomic instability.

X-ray repair complementing defective repair in Chinese hamster cells), flap structure-specific endonuclease 1, polymerase β, APE1. During BER repair a SSB is formed as an intermediate, this is effectively repaired with some additional enzymatic activities, for example, poly(ADP-ribose) polymerase (PARP). At this time hypoxia has not been formally demonstrated to affect the BER pathway, although it would seem unlikely that this is the case. PARP has been shown to be active in hypoxic conditions and in particular, to have a role in delaying replication restart after reoxygenation in order to prevent replication of damaged DNA (Hammond, unpublished data).

NER is the main repair pathway for the removal of bulky adducts induced by UV radiation and other environmental carcinogenic agents (Setlow and Carrier, 1964). Detection of damage by the NER pathway does not occur with the same degree of specificity as is the case for BER. NER-recognized lesions include bulky adducts (as a result of intrastrand crosslinks), cyclobutane pyrimidine dimers and 6-4 photo products. Inability to remove these lesions can lead to DNA replication or transcriptional blocks. NER consists of two pathways, global genome repair (GG-NER or GGR) and transcription coupled repair (TC-NER or TCR), which act at a genome-wide level or specifically in actively transcribed areas, respectively (van Hoffen *et al.*, 2003). Mechanistically, these two pathways differ only on the recognition of the lesion, which is followed by an incision of the DNA near the damaged area, repair of the area using the complementary strand as a template and a final ligation step. There is relatively little information available on the role of hypoxia in regulating NER. A substantial amount of evidence now exists demonstrating a significant role for microRNAs (miRs) in the hypoxia response. The most characterized is miR210 which has been shown to be hypoxia inducible and regulate processes relevant to tumor biology, including angiogenesis, cell proliferation, and apoptosis (Crosby *et al.*, 2009a). A recent report demonstrates that the hypoxia-induced expression of miR-373 represses the levels of the NER factor Rad23B and the HR factor Rad52 via interaction with the 3′ untranslated regions (UTRs) of the genes (Crosby *et al.*, 2009b). Like miR-210, miR-373 was found to be hypoxia-inducible through a hypoxia-inducible factor (HIF)-1-dependent mechanism.

The MMR pathway is important for the repair of potentially mutagenic incorrect nucleotide insertions and base modification for example, those that occur during normal replication and recombination. The mechanism of MMR is initiated by the identification of the mismatched base (the signal) by sensors such as the MutS orthologs mutS homolog (MSH)-2MSH6 and MSH2-MSH3 (Kolodner and Marsischky, 1999). This is followed by signal transduction for the recruitment of MMR factors (MutL orthologs MLH1-postmeiotic segregation increased (PMS)-2, MLH1-PMS1, and MLH1-MLH3) and the mismatch-containing strand is excised by factors such as the exonuclease EXO1. The excised strand is then finally resynthesized. Mutations of MLH1 and other MMR factors such as MSH2 and MSH6 are associated to both sporadic and hereditary cancers (Buermeyer *et al.* 1999). The first report of an effect of hypoxia on MMR regulation focuses on the effect on the *MLH1* gene (Mihaylova *et al.*, 2003). This gene was observed to be down-regulated at the mRNA level in various cell line models in hypoxic conditions. This was accompanied by the decrease of the protein levels of its heterodimer partner, PMS2. The decrease of both MLH1 and PMS2 was accompanied by a functional MMR defect, demonstrated by an increased mutagenesis in reporter genes, in both transient and stable assays. Proposed mechanisms for this deregulation of *MLH1* expression include hypoxia-induced chromatin modifications such as increased levels of dimethylation of histone H3 at lysine 9 at the proximity of the *MLH1* promoter, which correspond to a gene repressive chromatin signature (Chen *et al.*, 2006). *MSH2* and *MLH1* regulation has also been reported to involve signaling via the Myc/Max network (Bindra and Glazer, 2007a). The hypoxia-mediated regulation of *MSH2*, involves a coordinated response by the Myc/Max network via differential promoter occupancy. Specifically in response to hypoxia, an increase in promoter binding by repressive Mad1/Mnt-Max heterodimers as opposed to activating c-Myc/Max complexes occurs. This has been shown to occur independently of both HIF-1α and p53.

DSBs can be generated by a variety of agents, including ionizing radiation, chemotherapy agents, the collapse of replications forks, bulky adducts, and ROS. The importance of the accurate repair of DSBs is clear considering the observation that a single one of these lesions, when unrepaired, can lead to cell death (Helleday *et al.*, 2007; Valerie and Povirk, 2003). DSB repair is therefore considered essential for the preservation of the integrity of the genome and, when impaired can lead to increased tumorigenesis. Furthermore, there are various human syndromes that are associated with mutations in DSB repair pathways, such as ataxia telangiectasia (AT) for mutations of *ATM*, Nijmegen breakage syndrome (NBS) for mutations in *NBS1*, or familial breast cancer for mutations in *BRCA1/2* (Lavin, 2008; Antoccia *et al.*, 2006; Honrado, Benitez, and Palacios, 2005). DSB repair can be further divided into two pathways, NHEJ and HR.

NHEJ is considered a more error prone pathway, because it does not rely on a DNA template strand for repair accuracy and can lead to nucleotide gain or loss. It can also occur in all the stages of the cell cycle, but most predominantly in G_1. The pathway consists of the recognition of the DSB by the Ku70/80 heterodimer, which recruits the catalytic subunit of DNA-protein kinase (PK) (DNA-PKcs) as well as other factors. The gap is ligated by the XRCC4-Ligase IV complex (Mahaney, Meek, and Lees-Miller, 2009). NHEJ factors are affected by hypoxia; for example, Meng and colleagues reported a decrease in mRNA levels of *XRCC4*, *Ku70*, and *Ligase IV* (Meng *et al.*, 2005). Recently, a decrease in Ku70/ 80 protein levels in cervical carcinomas was found to be present in more hypoxic tumors and associated with altered p53 expression and increased angiogenesis (Lara *et al.*, 2008). Unlike NHEJ, HR occurs predominantly in the S and G_2 phases of the cell cycle because of its use of the second copy of the damaged DNA strand as a template allowing for error-free repair. During HR, DSBs are recognized by sensor complexes such as the MRN complex and its component Mre11 promotes a $5'-3'$ resection producing a ssDNA overhang. This structure is then covered by RPA before the recruitment of other HR factors (such as Rad51, Rad52, and BRCA1/2). This recruitment promotes strand invasion of the homologous strand and the pathway is completed by the resolution of the subsequent structures and ligation (Mimitou and Symington, 2009).

A number of reports have demonstrated that hypoxia induces the repression of both HR and NHEJ genes at the mRNA level, including, for example, *RAD51*, *BRCA1/2*, *XRCC3*, *XRCC4*, *Ku70*, and *Ligase IV* (Bindra *et al.*, 2004, 2005a, b; Meng *et al.*, 2005). In an earlier study, downregulation at the mRNA level of some of these genes in both normal fibroblasts and prostate cancer cell lines by hypoxia (pO_2 0.2%) did not always correlate with a subsequent decrease in protein expression (Meng *et al.*, 2005). These data highlight the importance of determining whether the hypoxia-mediated effect of DNA repair proteins has a functional consequence, that is, does it make a difference to repair efficiency. The downregulation of HR related genes such as *RAD51* has been further corroborated by microarray analysis of cancer cells exposed to hypoxia and has been shown to be functionally significant, that is, effecting the recombinogenic potential of the cells (Bindra *et al.*, 2004). Glazer and colleagues demonstrated that the hypoxia-dependent decrease of *RAD51* mRNA was independent of both the cell cycle and the HIF. Instead, they identified that the transcriptional repression of *RAD51* occurred via differential occupancy of the gene promoter by E2F factors (Bindra and Glazer, 2007b). Specifically, E2F1, an activating member of the family, is replaced by E2F4, associated to p130, forming a repressive complex. The same group elucidated that *BRCA1* hypoxia-dependent repression also occurred at the transcriptional level in a similar mechanism, after the identification of a common E2F biding site in both gene promoters (Bindra *et al.*, 2005a). Importantly, in either case this was accompanied by a decrease of the recombinogenic capacity of the cells. As previously mentioned, it was shown that miRs can regulate HR DNA repair genes; specifically it was demonstrated that mir-210 and mir-373 mediated the repression of *RAD52* (Crosby *et al.*, 2009b).

10.3.2 HIF1-Myc axis

HIF is a heterodimeric complex that can regulate gene expression in very specific conditions, primarily low oxygen tensions. This transcription factor specifically binds promoter DNA sequences known as hypoxia responsive elements (HREs). The complex is composed of alpha and beta subunits, of which the better characterized ones are HIF-1α and HIF-1β, respectively. The hypoxia specificity of HIF-1 function is related to the oxygen dependent regulation of HIF-1α stability. HIF biology will be addressed elsewhere in this book and has been extensively reviewed (Brahimi-Horn and Pouyssegur, 2009; Kaelin and Ratcliffe, 2008; Mole *et al.*, 2009; Semenza, 2003).

HIF-1α has been described in association with an induction of a cell cycle arrest in a DDR-independent manner. This HIF mediated regulation of the cell cycle was found to be dependent on $p21^{Cip1}$ and $p27^{Kip}$, both of which are cyclin-dependent kinase (CDK) inhibitors (Carmeliet *et al.*, 1998; Gardner *et al.*, 2001; Goda *et al.*, 2003). This was later shown to be at least partially dependent on the ability of HIF-1α to counteract the repressing activity of the Myc transcription factor at the promoter, in the so-called HIF1α-Myc pathway (Koshiji *et al.*, 2004; Mack *et al.*, 2003). The HIF1α–c-Myc pathway or axis has been recently reviewed (Yoo *et al.*, 2009). In brief, this mechanism has also been shown to be behind hypoxia-dependent transcriptional down regulation of several DNA repair genes, namely, the MMR components *MSH2*, *MSH6* (part of the MutSα complex), and *NBS1* (part of the MRN complex) (Koshiji *et al.*, 2005; To *et al.*, 2006). This was shown to be dependent on the competitive binding of the PAS-B domain of HIF1α to Sp1 sites and subsequent displacement of activating c-Myc (Gartel *et al.*, 2001). Regulation of gene expression through the HIF1α-c-Myc axis is independent of the DNA-binding and transactivation properties of HIF-1. Hypoxia has been shown to inhibit the expression of CDC25A leading to a cell cycle arrest. This was demonstrated to occur as a result of HIF-1α displacing c-Myc from the CDC25A gene promoter. Interestingly, although HIF-1α was necessary for this regulation it was not sufficient, indicating that other factors are also required (Hammer *et al.*, 2007). MSH2 has also been shown to be repressed in a similar manner; hypoxia-dependent loss of expression is associated to HIF-1α overexpression and has been correlated to the loss of p53 (Koshiji *et al.*, 2005). However, these observations are controversial; as previously mentioned, MLH1 regulation has been shown to occur via differential promoter occupancy by repressive rather than activating Myc/Max family members in a HIF-independent manner (Bindra and Glazer, 2007a). HIF1α has also been recently implicated in the regulation of the expression of NER factors for xeroderma pigmentosum complementation groups C and D in keratinocytes (Rezvani *et al.*, 2009). Together, these studies demonstrate that HIF-1 expression plays a significant role in repressing DNA repair in hypoxic conditions and so highlights another role for HIF-1 in driving tumor progression.

10.4 Context synthetic lethality: exploiting hypoxic deregulation of DNA repair

In the previous sections we discussed how hypoxia actively leads to repression of gene expression of various key DNA repair factors. It has emerged as a recent concept that this fact could be exploited to sensitize cells in hypoxic regions to radio and chemotherapy by the use of a phenomenon known as synthetic lethality (Bristow and Hill, 2008; Kaelin, 2005). Synthetic lethality has been exploited effectively for many years by yeast geneticists. It is based on the concept that inhibition or deregulation of one survival signaling pathway (let us call it A) or a second (B) alone does not affect cellular integrity but when both are inactivated simultaneously the effect is cellular toxicity. More recently a mammalian example of this was described using BRCA1- or BRCA2-deficient cell lines, which have inadequate HR signaling and were found to be exquisitely sensitive to inhibition of the BER repair pathway. This observation suggested the use of PARP inhibitors in combination with radio or chemotherapy, to selectively target, sensitize and kill tumors in patients carrying *BRCA1* or *BRCA2* mutations (Farmer *et al.*, 2005; Bryant *et al.*, 2005; Rodon, Iniesta, and Papadopoulos, 2009). Moreover, Ashworth and his group furthered this concept by showing sensitization with PARP1 inhibitors of cell lines in which other HR and DDR signaling factors were depleted, including RAD51, RPA, NBS1, ATR, ATM, CHK1, CHK2, and Fanconi anemia complementation group D2 (FANCD2) (McCabe *et al.*, 2006). As a consequence of these studies a number of orally available PARP inhibitors are currently in clinical trials for the treatment of BRCA-associated malignancies.

Hypoxia downregulates many DNA repair factors, including the HR factors mentioned above, leading to a new model of synthetic lethality: context synthetic lethality. This concept involves the exploitation of the partially compromised repair capacity of hypoxic regions in combination with the use of small molecule inhibitors of components of other repair pathways (such as PARP). The desired outcome of this treatment would be increased toxicity to the hypoxic tumor cells whilst sparing tissue at normal oxygen concentrations. One of the reasons this is such an attractive approach is that hypoxia is not specific to a cancer type; most if not all solid tumors have some degree of hypoxia. Initial studies have already shown promising results indicating radiosensitization of tumor cell lines exposed to hypoxia by PARP inhibitors, such as ABT-888 (Liu *et al.*, 2008). It was recently shown that the hypoxia-induced HR deregulation was associated with sensitization to ionizing radiation and chemotherapy agents inducing DNA crosslinks (Chan *et al.*, 2008). Furthermore, this radiosensitivity phenotype could not be observed in BRCA2-deficient cell lines reinforcing the HR dependency of this effect. More extensive work is required to explore which other DNA repair and DDR signaling pathways are 'druggable' as potential targets for inhibition and subsequent sensitization of hypoxic tumor cells. For example, signaling pathways already shown to increase tumor radio and chemo sensitization in the presence of defective DNA repair signaling include ATM and Chk1/2 signaling, NHEJ, and the proteosome, reviewed extensively elsewhere (Bristow *et al.*, 2007; Choudhury, Cuddihy, and Bristow, 2006; Helleday *et al.*, 2007; Martin, Lord, and Ashworth, 2008).

10.5 Conclusions

The recent model for the role of the DDR in tumorigenesis indicates that the damage response pathways are activated in early stages of tumor development (before induction of genomic instability), as a result of oncogene-induced replicative stress (Bartkova *et al.*, 2005; Gorgoulis *et al.*, 2005). This activated DDR acts as a tumorigenic barrier to induce cell cycle arrest, senescence, or apoptosis; however, these are dependent on the presence and functionality of p53, and other such tumor suppressors (Halazonetis, Gorgoulis, and Bartek, 2008). Loss of this barrier will allow tumor development. How does hypoxia fit in this model? In the proposed model hypoxia is considered to occur after the replicative stress is induced, occurring as a consequence of the increased hyperplasia and tumor growth and as a stimulus for the DDR tumorigenic barrier. Our own work demonstrates that hypoxia induces replicative stress but not until levels of severe hypoxia, almost anoxia, are reached. It is possible that this could occur reasonably early during tumor development, for example, before the effect of increased angiogenesis. However, more recent findings have shown that the DDR can also be induced by hypoxia in some cell types at much higher levels. It is important to also consider that HIF-1 is induced in response to much milder hypoxic conditions than are required to induce the DDR. Given the recently described role for HIF in deregulating DNA repair this means that early neoplastic lesions will effectively be turning down their repair capacity and therefore promoting tumorigenesis. It is well known that hypoxia is a major driver of tumor progression and it is equally clear that we have yet to understand fully the multiple and complex mechanisms involved.

References

Alarcon, R., Koumenis, C., Geyer, R.K. *et al.* (1999) Hypoxia induces p53 accumulation through MDM2 down-regulation and inhibition of E6-mediated degradation. *Cancer Research*, 59, 6046–6051.

Antoccia, A., Kobayashi, J., Tauchi, H. *et al.* (2006) Nijmegen breakage syndrome and functions of the responsible protein, NBS1. *Genome Dynamics*, **1**, 191–205.

Bakkenist, C.J. and Kastan, M.B. (2003) DNA damage activates ATM through intermolecular autophosphorylation and dimer dissociation. *Nature*, **421**, 499–506.

Bartkova, J., Horejsi, Z., Koed, K. *et al.* (2005) DNA damage response as a candidate anti-cancer barrier in early human tumorigenesis. *Nature*, **434**, 864–870.

Bartkova, J., Rezaei, N., Liontos, M. *et al.* (2006) Oncogene-induced senescence is part of the tumorigenesis barrier imposed by DNA damage checkpoints. *Nature*, **444**, 633–637.

Bassing, C.H., Suh, H., Ferguson, D.O. *et al.* (2003) Histone H2AX: a dosage-dependent suppressor of oncogenic translocations and tumors. *Cell*, **114**, 359–370.

Bencokova, Z., Kaufmann, M.R., Pires, I.M. *et al.* (2009) ATM activation and signaling under hypoxic conditions. *Molecular and Cellular Biology*, **29**, 526–537.

Bindra, R.S., Crosby, M.E., and Glazer, P.M. (2007) Regulation of DNA repair in hypoxic cancer cells. *Cancer Metastasis Reviews*, **26**, 249–260.

Bindra, R.S., Gibson, S.L., Meng, A. *et al.* (2005a) Hypoxia-induced down-regulation of BRCA1 expression by E2Fs. *Cancer Research*, **65**, 11597–11604.

Bindra, R.S., Schaffer, P.J., Meng, A. *et al.* (2005b) Alterations in DNA repair gene expression under hypoxia: elucidating the mechanisms of hypoxia-induced genetic instability. *Annals of the New York Academy of Sciences*, **1059**, 184–195.

Bindra, R.S. and Glazer, P.M. (2007a) Co-repression of mismatch repair gene expression by hypoxia in cancer cells: role of the Myc/Max network. *Cancer Letters*, **252** (1), 93–103.

Bindra, R.S. and Glazer, P.M. (2007b) Repression of RAD51 gene expression by E2F4/p130 complexes in hypoxia. *Oncogene*, **26**, 2048–2057.

Bindra, R.S., Schaffer, P.J., Meng, A. *et al.* (2004) Down-regulation of Rad51 and decreased homologous recombination in hypoxic cancer cells. *Molecular and Cellular Biology*, **24**, 8504–8518.

Bonner, W.M., Redon, C.E., Dickey, J.S. *et al.* (2008) GammaH2AX and cancer. *Nature Reviews Cancer*, **8**, 957–967.

Brahimi-Horn, M.C. and Pouyssegur, J. (2009) HIF at a glance. *Journal of Cell Science*, **122**, 1055–1057.

Bristow, R.G. and Hill, R.P. (2008) Hypoxia and metabolism. Hypoxia, DNA repair and genetic instability. *Nature Reviews Cancer*, **8**, 180–192.

Bristow, R.G., Ozcelik, H., Jalali, F. *et al.* (2007) Homologous recombination and prostate cancer: a model for novel DNA repair targets and therapies. *Radiotherapy and Oncology*, **83**, 220–230.

Brizel, D.M., Sibley, G.S., Prosnitz, L.R. *et al.* (1997) Tumor hypoxia adversely affects the prognosis of carcinoma of the head and neck. *International Journal of Radiation Oncology, Biology, Physics*, **38**, 285–289.

Brown, J.M. (1979) Evidence for acutely hypoxic cells in mouse tumours, and a possible mechanism of reoxygenation. *The British Journal of Radiology*, **52**, 650–656.

Brown, J.M. (2002) Tumor microenvironment and the response to anticancer therapy. *Cancer Biology and Therapy*, **1**, 453–458.

Brown, J.M. (2007) Tumor hypoxia in cancer therapy. *Methods in Enzymology*, **435**, 297–321.

Brown, E.J. and Baltimore, D. (2003) Essential and dispensable roles of ATR in cell cycle arrest and genome maintenance. *Genes and Development*, **17**, 615–628.

Brown, J.M. and Giaccia, A.J. (1998) The unique physiology of solid tumors: opportunities (and problems) for cancer therapy. *Cancer Research*, **58**, 1408–1416.

Bryant, H.E., Schultz, N., Thomas, H.D. *et al.* (2005) Specific killing of BRCA2-deficient tumours with inhibitors of poly(ADP-ribose) polymerase. *Nature*, **434**, 913–917.

Buermeyer, A.B., Deschênes, S.M., Baker S.M. and Liskay, R.M. (1999) Mammalian DNA mismatch repair. *Annual Review of Genetics*, **33**, 533–564.

Burma, S., Chen, B.P., Murphy, M. *et al.* (2001) ATM Phosphorylates Histone H2AX in Response to DNA Double-strand Breaks. *The Journal of Biological Chemistry*, **276**, 42462–42467.

Cairns, R.A., Kalliomaki, T., and Hill, R.P. (2001) Acute (cyclic) hypoxia enhances spontaneous metastasis of KHT murine tumors. *Cancer Research*, **61**, 8903–8908.

Cardenas-Navia, L.I., Mace, D., Richardson, R.A. *et al.* (2008) The pervasive presence of fluctuating oxygenation in tumors. *Cancer Research*, **68**, 5812–5819.

Carmeliet, P., Dor, Y., Herbert, J.M. *et al.* (1998) Role of HIF-1alpha in hypoxia-mediated apoptosis, cell proliferation and tumour angiogenesis. *Nature*, **394**, 485–490.

Casper, A.M., Nghiem, P., Arlt, M.F., and Glover, T.W. (2002) ATR Regulates Fragile Site Stability. *Cell*, **111**, 779–789.

Chan, N., Koritzinsky, M., Zhao, H. *et al.* (2008) Chronic hypoxia decreases synthesis of homologous recombination proteins to offset chemoresistance and radioresistance. *Cancer Research*, **68**, 605–614.

Chaturvedi, P., Eng, W.K., Zhu, Y. *et al.* (1999) Mammalian Chk2 is a downstream effector of the ATM-dependent DNA damage checkpoint pathway. *Oncogene*, **18**, 4047–4054.

Chen, H., Yan, Y., Davidson, T.L. *et al.* (2006) Hypoxic stress induces dimethylated histone H3 lysine 9 through histone methyltransferase G9a in mammalian cells. *Cancer Research*, **66**, 9009–9016.

Choudhury, A., Cuddihy, A., and Bristow, R.G. (2006) Radiation and new molecular agents part I: targeting ATM-ATR checkpoints, DNA repair, and the proteasome. *Seminars in Radiation Oncology*, **16**, 51–58.

Cimprich, K.A. and Cortez, D. (2008) ATR: an essential regulator of genome integrity. *Nature Reviews Molecular Cell Biology*, **9**, 616–627.

Cimprich, K.A., Shin, T.B., Keith, C.T., and Schreiber, S.L. (1996) cDNA cloning and gene mapping of a candidate human cell cycle checkpoint protein. *Proceedings of the National Academy of Sciences of the United States of America*, **93**, 2850–2855.

Coquelle, A., Pipiras, E., Toledo, F. *et al.* (1997) Expression of fragile sites triggers intrachromosomal mammalian gene amplification and sets boundaries to early amplicons. *Cell*, **89**, 215–225.

Crosby, M.E., Devlin, C.M., Glazer, P.M. *et al.* (2009a) Emerging roles of microRNAs in the molecular responses to hypoxia. *Current Pharmaceutical Design*, **15**, 3861–3866.

Crosby, M.E., Kulshreshtha, R., Ivan, M., and Glazer, P.M. (2009b) MicroRNA regulation of DNA repair gene expression in hypoxic stress. *Cancer Research*, **69**, 1221–1229.

Dumaz, N., Milne, D.M., Jardine, L.J., and Meek, D.W. (2001) Critical roles for the serine 20, but not the serine 15, phosphorylation site and for the polyproline domain in regulating p53 turnover. *The Biochemical Journal*, **359**, 459–464.

Economopoulou, M., Langer, H.F., Celeste, A. *et al.* (2009) Histone H2AX is integral to hypoxia-driven neovascularization. *Nature Medicine*, **15**, 553–558.

Farmer, H., McCabe, N., Lord, C.J. *et al.* (2005) Targeting the DNA repair defect in BRCA mutant cells as a therapeutic strategy. *Nature*, **434**, 917–921.

Freiberg, R.A., Hammond, E.M., Dorie, M.J. *et al.* (2006a) DNA damage during reoxygenation elicits a Chk2-dependent checkpoint response. *Molecular and Cellular Biology*, **26**, 1598–1609.

Freiberg, R.A., Krieg, A.J., Giaccia, A.J., and Hammond, E.M. (2006b) Checking in on hypoxia/reoxygenation. *Cell Cycle*, **5**, 1304–1307.

Gardner, L.B., Li, Q., Park, M.S. *et al.* (2001) Hypoxia inhibits G1/S transition through regulation of p27 expression. *Journal of Biological Chemistry*, **276**, 7919–7926.

Gartel, A.L., Ye, X., Goufman, E. *et al.* (2001) Myc represses the p21(WAF1/CIP1) promoter and interacts with Sp1/Sp3. *Proceedings of the National Academy of Sciences of the United States of America*, **98**, 4510–4515.

Gibson, S.L., Bindra, R.S., and Glazer, P.M. (2005) Hypoxia-induced phosphorylation of Chk2 in an ataxia telangiectasia mutated-dependent manner. *Cancer Research*, **65**, 10734–10741.

Gibson, S.L., Bindra, R.S., and Glazer, P.M. (2006) CHK2-dependent phosphorylation of BRCA1 in hypoxia. *Radiation Research*, **166**, 646–651.

Goda, N., Ryan, H.E., Khadivi, B. *et al.* (2003) Hypoxia-inducible factor 1alpha is essential for cell cycle arrest during hypoxia. *Molecular and Cellular Biology*, **23**, 359–369.

Goodarzi, A.A., Jonnalagadda, J.C., Douglas, P. *et al.* (2004) Autophosphorylation of ataxia-telangiectasia mutated is regulated by protein phosphatase 2A. *EMBO Journal*, **23**, 4451–4461.

Gorgoulis, V.G., Vassiliou, L.V., Karakaidos, P. *et al.* (2005) Activation of the DNA damage checkpoint and genomic instability in human precancerous lesions. *Nature*, **434**, 907–913.

Gottifredi, V., Shieh, S., Taya, Y., and Prives, C. (2001) p53 accumulates but is functionally impaired when DNA synthesis is blocked. *Proceedings of the National Academy of Sciences of the United States of America*, **98**, 1036–1041.

Graeber, T.G., Osmanian, C., Jacks, T. *et al.* (1996) Hypoxia-mediated selection of cells with diminished apoptotic potential in solid tumours. *Nature*, **379**, 88–91.

Hakem, R. (2008) DNA-damage repair; the good, the bad, and the ugly. *EMBO Journal*, **27**, 589–605.

Halazonetis, T.D., Gorgoulis, V.G., and Bartek, J. (2008) An oncogene-induced DNA damage model for cancer development. *Science*, **319**, 1352–1355.

Hammer, S., To, K.K., Yoo, Y.G. *et al.* (2007) Hypoxic suppression of the cell cycle gene CDC25A in tumor cells. *Cell Cycle*, **6**, 1919–1926.

Hammond, E.M., Denko, N.C., Dorie, M.J. *et al.* (2002) Hypoxia links ATR and p53 through replication arrest. *Molecular and Cellular Biology*, **22**, 1834–1843.

Hammond, E.M., Dorie, M.J., and Giaccia, A.J. (2003a) ATR/ATM targets are phosphorylated by ATR in response to hypoxia and ATM in response to reoxygenation. *Journal of Biological Chemistry*, **278**, 12207–12213.

Hammond, E.M., Green, S.L., and Giaccia, A.J. (2003b) Comparison of hypoxia-induced replication arrest with hydroxyurea and aphidicolin-induced arrest. *Mutation Research*, **532**, 205–213.

Hammond, E.M., Dorie, M.J., and Giaccia, A.J. (2004) Inhibition of ATR leads to increased sensitivity to hypoxia/reoxygenation. *Cancer Research*, **64**, 6556–6562.

Hammond, E.M., Freiberg, R.A., and Giaccia, A.J. (2005) The roles of Chk 1 and Chk2 in hypoxia and reoxygenation. *Cancer Letters*, **238** (2), 161–167.

Hammond, E.M., and Giaccia, A.J. (2005) The role of p53 in hypoxia-induced apoptosis. *Biochemical and Biophysical Research Communications*, **331**, 718–725.

Hammond, E.M., Mandell, D.J., Salim, A. *et al.* (2006) Genome-wide analysis of p53 under hypoxic conditions. *Molecular and Cellular Biology*, **26**, 3492–3504.

Haupt, Y., Maya, R., Kazaz, A., and Oren, M. (1997) Mdm2 promotes the rapid degradation of p53. *Nature*, **387**, 296–299.

Helleday, T., Lo, J., van Gent, D.C., and Engelward, B.P. (2007) DNA double-strand break repair: from mechanistic understanding to cancer treatment. *DNA Repair*, **6**, 923–935.

Hockel, M., Schlenger, K., Aral, B. *et al.* (1996) Association between tumor hypoxia and malignant progression in advanced cancer of the uterine cervix. *Cancer Research*, **56**, 4509–4515.

van Hoffen, A., Balajee, A.S., van Zeeland, A.A., and Mullenders, L.H. (2003) Nucleotide excision repair and its interplay with transcription. *Toxicology*, **193**, 79–90.

Honrado, E., Benitez, J., and Palacios, J. (2005) The molecular pathology of hereditary breast cancer: genetic testing and therapeutic implications. *Modern Pathology*, **18**, 1305–1320.

Huang, L.E., Bindra, R.S., Glazer, P.M., and Harris, A.L. (2007) Hypoxia-induced genetic instability – a calculated mechanism underlying tumor progression. *Journal of Molecular Medicine*, **85**, 139–148.

Huen, M.S., Grant, R., Manke, I. *et al.* (2007) RNF8 transduces the DNA-damage signal via histone ubiquitylation and checkpoint protein assembly. *Cell*, **131**, 901–914.

Hunt, C.R., Pandita, R.K., Laszlo, A. *et al.* (2007) Hyperthermia activates a subset of ataxia-telangiectasia mutated effectors independent of DNA strand breaks and heat shock protein 70 status. *Cancer Research*, **67**, 3010–3017.

Jackson, S.P. and Bartek, J. (2009) The DNA-damage response in human biology and disease. *Nature*, **461**, 1071–1078.

Jain, R.K. (2005) Normalization of tumor vasculature: an emerging concept in antiangiogenic therapy. *Science*, **307**, 58–62.

Jeong, I., Patel, A.Y., Zhang, Z. *et al.* (2009) Role of ataxia telangiectasia mutated in insulin signalling of muscle-derived cell lines and mouse soleus. *Acta Physiologica*, **198** (4), 465–475.

Johnson, T.M., Hammond, E.M., Giaccia, A., and Attardi, L.D. (2005) The p53QS transactivation-deficient mutant shows stress-specific apoptotic activity and induces embryonic lethality. *Nature Genetics*, **37**, 145–152.

Kaelin, W.G. Jr. (2005) The concept of synthetic lethality in the context of anticancer therapy. *Nature Reviews Cancer*, **5**, 689–698.

Kaelin, W.G. and Ratcliffe, P.J. Jr. (2008) Oxygen sensing by metazoans: the central role of the HIF hydroxylase pathway. *Molecular Cell*, **30**, 393–402.

Kim, B.M., Choi, J.Y., Kim, Y.J. *et al.* (2007) Reoxygenation following hypoxia activates DNA-damage checkpoint signaling pathways that suppress cell-cycle progression in cultured human lymphocytes. *FEBS Letters*, **581**, 3005–3012.

Kinner, A., Wu, W., Staudt, C., and Iliakis, G. (2008) Gamma-H2AX in recognition and signaling of DNA double-strand breaks in the context of chromatin. *Nucleic Acids Research*, **36**, 5678–5694.

Kolas, N.K., Chapman, J.R., Nakada, S. *et al.* (2007) Orchestration of the DNA-damage response by the RNF8 ubiquitin ligase. *Science*, **318**, 1637–1640.

Kolodner, R.D. and Marsischky, G.T. (1999) Eukaryotic DNA mismatch repair. *Current Opinion in Genetics and Development*, **9**, 89–96.

Koshiji, M., Kageyama, Y., Pete, E.A. *et al.* (2004) HIF-1alpha induces cell cycle arrest by functionally counteracting Myc. *EMBO Journal*, **23**, 1949–1956.

Koshiji, M., To, K.K., Hammer, S. *et al.* (2005) HIF-1alpha induces genetic instability by transcriptionally downregulating MutSalpha expression. *Molecular Cell*, **17**, 793–803.

Koumenis, C., Alarcon, R., Hammond, E. *et al.* (2001) Regulation of p53 by hypoxia: dissociation of transcriptional repression and apoptosis from p53-dependent transactivation. *Molecular and Cellular Biology*, **21**, 1297–1310.

Lara, P.C., Lloret, M., Clavo, B. *et al.* (2008) Hypoxia downregulates Ku70/80 expression in cervical carcinoma tumors. *Radiotherapy and Oncology*, **89**, 222–226.

Lavin, M.F. (2008) Ataxia-telangiectasia: from a rare disorder to a paradigm for cell signalling and cancer. *Nature Reviews Molecular cell Biology*, **9**, 759–769.

Lee, J.H. and Paull, T.T. (2007) Activation and regulation of ATM kinase activity in response to DNA double-strand breaks. *Oncogene*, **26**, 7741–7748.

Li, C.Y., Little, J.B., Hu, K. *et al.* (2001) Persistent genetic instability in cancer cells induced by non-DNA-damaging stress exposures. *Cancer Research*, **61**, 428–432.

Lindahl, T. and Barnes, D.E. (2000) Repair of endogenous DNA damage. *Cold Spring Harbor Symposia on Quantitative Biology*, **65**, 127–133.

Liu, S.K., Coackley, C., Krause, M. *et al.* (2008) A novel poly(ADP-ribose) polymerase inhibitor, ABT-888, radiosensitizes malignant human cell lines under hypoxia. *Radiotherapy and Oncology*, **88**, 258–268.

Liu, Q., Guntuku, S., Cui, X.S. *et al.* (2000) Chk1 is an essential kinase that is regulated by Atr and required for the G(2)/M DNA damage checkpoint. *Genes and Development*, **14**, 1448–1459.

Loeb, L.A., Loeb, K.R., and Anderson, J.P. (2003) Multiple mutations and cancer. *Proceedings of the National Academy of Sciences of the United States of America*, **100**, 776–781.

Loehberg, C.R., Thompson, T., Kastan, M.B. *et al.* (2007) Ataxia telangiectasia-mutated and p53 are potential mediators of chloroquine-induced resistance to mammary carcinogenesis. *Cancer Research*, **67**, 12026–12033.

Lomonosov, M., Anand, S., Sangrithi, M. *et al.* (2003) Stabilization of stalled DNA replication forks by the BRCA2 breast cancer susceptibility protein. *Genes and Development*, **17**, 3017–3022.

Mack, F.A., Rathmell, W.K., Arsham, A.M. *et al.* (2003) Loss of pVHL is sufficient to cause HIF dysregulation in primary cells but does not promote tumor growth. *Cancer Cell*, **3**, 75–88.

Mahaney, B.L., Meek, K., and Lees-Miller, S.P. (2009) Repair of ionizing radiation-induced DNA double-strand breaks by non-homologous end-joining. *Biochemical Journal*, **417**, 639–650.

Mailand, N., Bekker-Jensen, S., Faustrup, H. *et al.* (2007) RNF8 ubiquitylates histones at DNA double-strand breaks and promotes assembly of repair proteins. *Cell*, **131**, 887–900.

Marti, T.M., Hefner, E., Feeney, L. *et al.* (2006) H2AX phosphorylation within the G1 phase after UV irradiation depends on nucleotide excision repair and not DNA double-strand breaks. *Proceedings of the National Academy of Sciences of the United States of America*, **103**, 9891–9896.

Martin, S.A., Lord, C.J., and Ashworth, A. (2008) DNA repair deficiency as a therapeutic target in cancer. *Current Opinion in Genetics and Development*, **18**, 80–86.

McCabe, N., Turner, N.C., Lord, C.J. *et al.* (2006) Deficiency in the repair of DNA damage by homologous recombination and sensitivity to poly(ADP-ribose) polymerase inhibition. *Cancer Research*, **66**, 8109–8115.

Meng, A.X., Jalali, F., Cuddihy, A. *et al.* (2005) Hypoxia down-regulates DNA double strand break repair gene expression in prostate cancer cells. *Radiotherapy and Oncology*, **76**, 168–176.

Mihaylova, V.T., Bindra, R.S., Yuan, J. *et al.* (2003) Decreased expression of the DNA mismatch repair gene Mlh1 under hypoxic stress in mammalian cells. *Molecular and Cellular Biology*, **23**, 3265–3273.

Mimitou, E.P. and Symington, L.S. (2009) Nucleases and helicases take center stage in homologous recombination. *Trends in Biochemical Sciences*, **34**, 264–272.

Minter-Dykhouse, K., Ward, I., Huen, M.S. *et al.* (2008) Distinct versus overlapping functions of MDC1 and 53BP1 in DNA damage response and tumorigenesis. *Journal of Cell Biology*, **181**, 727–735.

Mole, D.R., Blancher, C., Copley, R.R. *et al.* (2009) Genome-wide association of hypoxia-inducible factor (HIF)-1alpha and HIF-2alpha DNA binding with expression profiling of hypoxia-inducible transcripts. *Journal of Biological Chemistry*, **284**, 16767–16775.

Nordsmark, M., Hoyer, M., Keller, J. *et al.* (1996a) The relationship between tumor oxygenation and cell proliferation in human soft tissue sarcomas. *International Journal Of Radiation Oncology, Biology, Physics*, **35**, 701–708.

Nordsmark, M., Overgaard, M., and Overgaard, J. (1996b) Pretreatment oxygenation predicts radiation response in advanced squamous cell carcinoma of the head and neck. *Radiotherapy and Oncology*, **41**, 31–39.

Pan, Y., Oprysko, P.R., Asham, A.M. *et al.* (2004) p53 cannot be induced by hypoxia alone but responds to the hypoxic microenvironment. *Oncogene*, **23**, 4975–4983.

Pandita, T.K., Pandita, S., and Bhaumik, S.R. (2009) Molecular parameters of hyperthermia for radiosensitization. *Critical Reviews in Eukaryotic Gene Expression*, **19**, 235–251.

Papp-Szabo, E., Josephy, P.D., and Coomber, B.L. (2005) Microenvironmental influences on mutagenesis in mammary epithelial cells. *International Journal of Cancer*, **116**, 679–685.

Pellegrini, M., Celeste, A., Difilippantonio, S. *et al.* (2006) Autophosphorylation at serine 1987 is dispensable for murine Atm activation in vivo. *Nature*, **443**, 222–225.

Pires, I.M., Bencokova, Z., Milani, M. *et al.* (2010) Chronic hypoxia leads to replisome disassembly and inhibition of the DNA damage response whereas acute hypoxia contributes to increased genomic instability. *Cancer Research*, **70** (3), 925–935.

Probst, H., Gekeler, V., and Helftenbein, E. (1984) Oxygen dependence of nuclear DNA replication in Ehrlich ascites cells. *Experimental Cell Research*, **154**, 327–341.

Probst, G., Riedinger, H.J., Martin, P. *et al.* (1999) Fast control of DNA replication in response to hypoxia and to inhibited protein synthesis in CCRF-CEM and HeLa cells. *Biological Chemistry*, **380**, 1371–1382.

Probst, H., Schiffer, H., Gekeler, V. *et al.* (1988) Oxygen dependent regulation of DNA synthesis and growth of Ehrlich ascites tumor cells in vitro and in vivo. *Cancer Research*, **48**, 2053–2060.

Probst, H., Schiffer, H., Gekeler, V., and Scheffler, K. (1989) Oxygen dependent regulation of mammalian ribonucleotide reductase in vivo and possible significance for replicon initiation. *Biochemical and Biophysical Research Communications*, **163**, 334–340.

Rappold, I., Iwabuchi, K., Date, T., and Chen, J. (2001) Tumor suppressor p53 binding protein 1 (53BP1) is involved in DNA damage-signaling pathways. *The Journal of Cell Biology*, **153**, 613–620.

Reichard, P. and Ehrenberg, A. (1983) Ribonucleotide reductase--a radical enzyme. *Science*, **221**, 514–519.

Reynolds, T.Y., Rockwell, S., and Glazer, P.M. (1996) Genetic instability induced by the tumor microenvironment. *Cancer Research*, **56**, 5754–5757.

Rezvani, H.R., Mahfouf, W., Ali, N. *et al.* (2009) Hypoxia-inducible factor-1{alpha} regulates the expression of nucleotide excision repair proteins in keratinocytes. *Nucleic Acids Research*, **38** (3), 797–809.

Rice, G.C., Spiro, I.J., and Ling, C.C. (1985) Detection of S-phase overreplication following chronic hypoxia using a monoclonal anti-BrdUrd. *International Journal Of Radiation Oncology, Biology, Physics*, **11**, 1817–1822.

Robertson, A.B., Klungland, A., Rognes, T., and Leiros, I. (2009) DNA repair in mammalian cells: base excision repair: the long and short of it. *Cellular and Molecular Life Sciences*, **66**, 981–993.

Robison, J.G., Elliott, J., Dixon, K., and Oakley, G.G. (2004) Replication protein A and the Mre11.Rad50.Nbs1 complex co-localize and interact at sites of stalled replication forks. *Journal of Biological Chemistry*, **279**, 34802–34810.

Rodon, J., Iniesta, M.D., and Papadopoulos, K. (2009) Development of PARP inhibitors in oncology. *Expert Opinion on Investigational Drugs*, **18**, 31–43.

Semenza, G.L. (2003) Targeting HIF-1 for cancer therapy. *Nature Reviews Cancer*, **3**, 721–732.

Setlow, R.B. and Carrier, W.L. (1964) The disappearance of thymine dimers from DNA: an error-correcting mechanism. *Proceedings of the National Academy of Sciences of the United States of America*, **51**, 226–231.

Shieh, S.Y., Ikeda, M., Taya, Y., and Prives, C. (1997) DNA damage-induced phosphorylation of p53 alleviates inhibition by MDM2. *Cell*, **91**, 325–334.

Smith, D.I., Huang, H., and Wang, L. (1998) Common fragile sites and cancer (review). *International Journal of Oncology*, **12**, 187–196.

St. Clair, S., Giono, L., Varmeh-Ziaie, S. *et al.* (2004) DNA damage-induced downregulation of Cdc25C is mediated by p53 via two independent mechanisms: one involves direct binding to the cdc25C promoter. *Molecular Cell*, **16**, 725–736.

Stiff, T., Walker, S.A., Cerosaletti, K. *et al.* (2006) ATR-dependent phosphorylation and activation of ATM in response to UV treatment or replication fork stalling. *EMBO Journal*, **25**, 5775–5782.

Stucki, M., Clapperton, J.A., Mohammad, D. *et al.* (2005) MDC1 directly binds phosphorylated histone H2AX to regulate cellular responses to DNA double-strand breaks. *Cell*, **123**, 1213–1226.

To, K.K., Sedelnikova, O.A., Samons, M. *et al.* (2006) The phosphorylation status of PAS-B distinguishes HIF-1alpha from HIF-2alpha in NBS1 repression. *EMBO Journal*, **25**, 4784–4794.

Tutt, A.N., Lord, C.J., McCabe, N. *et al.* (2005) Exploiting the DNA repair defect in BRCA mutant cells in the design of new therapeutic strategies for cancer. *Cold Spring Harbor Symposia on Quantitative Biology*, **70**, 139–148.

Unger, T., Juven-Gershon, T., Moallem, E. *et al.* (1999) Critical role for Ser20 of human p53 in the negative regulation of p53 by Mdm2. *EMBO Journal*, **18**, 1805–1814.

Valerie, K. and Povirk, L.F. (2003) Regulation and mechanisms of mammalian double-strand break repair. *Oncogene*, **22**, 5792–5812.

Wang, B. and Elledge, S.J. (2007) Ubc13/Rnf8 ubiquitin ligases control foci formation of the Rap80/Abraxas/Brca1/Brcc36 complex in response to DNA damage. *Proceedings of the National Academy of Sciences of the United States of America*, **104**, 20759–20763.

Ward, I.M. and Chen, J. (2001) Histone H2AX is phosphorylated in an ATR-dependent manner in response to replicational stress. *Journal of Biological Chemistry*, **22**, 22.

Weinmann, M., Jendrossek, V., Guner, D. *et al.* (2004) Cyclic exposure to hypoxia and reoxygenation selects for tumor cells with defects in mitochondrial apoptotic pathways. *FASEB Journal*, **18** (15), 1906–1908.

Wold, M.S. and Kelly, T. (1988) Purification and characterization of replication protein A, a cellular protein required for in vitro replication of simian virus 40 DNA. *Proceedings of the National Academy of Sciences of the United States of America*, **85**, 2523–2527.

Yoo, Y.G., Hayashi, M., Christensen, J., and Huang, L.E. (2009) An essential role of the HIF-1alpha-c-Myc axis in malignant progression. *Annals of the New York Academy of Sciences*, **1177**, 198–204.

Young, S.D., Marshall, R.S., and Hill, R.P. (1988) Hypoxia induces DNA overreplication and enhances metastatic potential of murine tumor cells. *Proceedings of the National Academy of Sciences of the United States of America*, **85**, 9533–9537.

Yuan, J., Narayanan, L., Rockwell, S., and Glazer, P.M. (2000) Diminished DNA repair and elevated mutagenesis in mammalian cells exposed to hypoxia and low pH. *Cancer Research*, **60**, 4372–4376.

Zou, L. and Elledge, S.J. (2003) Sensing DNA damage through ATRIP recognition of RPA-ssDNA complexes. *Science*, **300**, 1542–1548.

11

Non-Invasive Imaging of the Tumor Microenvironment

Bénédicte F. Jordan and Bernard Gallez

Biomedical Magnetic Resonance Unit, Louvain Drug Research Institute, Université Catholique de Louvain, Brussels, Belgium

11.1 Introduction

In vivo molecular imaging is an emerging discipline in medical research (Margolis *et al.*, 2007). Computed tomography (CT), positron emission tomography (PET), single photon emission computed tomography (SPECT), magnetic resonance imaging (MRI), electron paramagnetic resonance imaging (EPRI), and optical imaging are being used to visualize patho/physiological conditions non-invasively (Torigian *et al.*, 2007). More specifically, imaging endpoints in the tumor field include: diagnosis, hemodynamic characterization, metabolic characterization, as well as tumor response to treatment. In the clinic, MRI and PET have allowed to better identification of primary and metastatic lesions, and the techniques also potentially constitute powerful tools for individualization of treatment in the clinical setting.

11.2 Imaging tumor vasculature, perfusion, and angiogenesis

The microenvironment within tumors is significantly different from that in normal tissues. A major difference is seen in the chaotic vasculature of tumors, which results in unbalanced blood supply and significant perfusion heterogeneities.

Tumor Microenvironment Edited by Dietmar W. Siemann

11.2.1 Vessel perfusion and permeability

Microvascular parameters such as permeability and perfusion are of particular interest in the context of the abnormal tumor microvascular network. Useful imaging systems have been developed to monitor angiogenesis and the microvasculature *in vivo*, including dynamic contrast enhanced magnetic resonance imaging (DCE-MRI) (Choyke, Dwyer, and Knopp, 2003), PET and SPECT, CT, Doppler ultrasound, and optical imaging methods (reviewed in Jennings, Raghunand, and Gillies, 2008). Perfusion is the process by which the circulatory system delivers blood to tissues. By contrast, vascular permeability is expressed as a transfer coefficient of a solute per unit time, per volume of tissue between the vascular compartment (plasma) and the extracellular extravascular space (EES). Permeability has the units $cm\,s^{-1}$ and is functionally equivalent to a diffusion coefficient ($cm^2\,s^{-1}$) corrected for the presence of an intervening semipermeable membrane (Jennings, Raghunand, and Gillies, 2008).

11.2.1.1 DCE-MRI

DCE-MRI is the acquisition of serial MR images before, during, and after the administration of an intravenous contrast agent (CA) that produces time series images that enable pixel-by-pixel analysis of contrast kinetics within a tumor. These 'time-signal' curves can be analyzed with descriptive 'heuristic' tools, such as initial slope, time to peak, or rate of washout. These methods are valuable and easy to apply; however, they provide no insight into the underlying physiology. Moreover, they are highly dependent on the imaging protocol and scanner. Pharmacokinetic models provide a means of summarizing contrast enhancement data in terms of parameters that relate to underlying vascular anatomy and physiology. As described by Tofts *et al.* (1999), the essential features of a variety of models are covered by the generalized kinetic model. In it, the contribution of intravascular CA to tumor CA concentration is assumed to be negligible. The concentration of CA within the tumor is determined by the blood plasma concentration curve and two parameters, K^{trans} (the transfer constant) and the EES fractional volume index, V_e. In relating the shape of the enhancement curve obtained from dynamic MRI studies, K^{trans} controls the height of the C_{tumor} curve (C(t) on Figure 11.1) and $k_{ep} = K^{trans}/V_e$ controls the shape of the curve (k_{ep} is the redistribution rate constant); the smaller k_{ep}, the more delayed the enhancement (Figure 11.1). K^{trans} is a function of flow (perfusion) and permeability. So, the revealed kinetics of contrast wash-in and wash-out are used to characterize masses as malignant or benign and determine whether a tumor is responding to therapy. Therefore, treatment-induced changes in these parameters reflect changes in blood flow and/or permeability related parameters.

There is no consensus among the imaging community regarding what measured permeability veritably represents. In truth, imaging offers only an estimation of permeability, or an 'effective' permeability based on tracer enhancement kinetics, and thus, on this macroscopic level, the rate of enhancement is related to microvascular permeability through complex kinetic models.

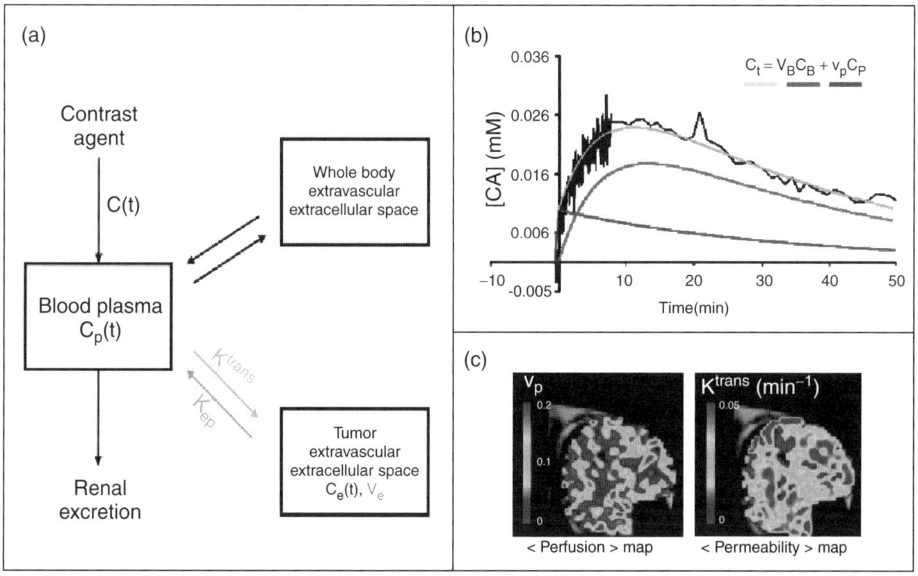

Figure 11.1 (a) Illustration of two-compartment model demonstrates the exchange of contrast between plasma and extravascular extracellular space. (b) Modelization of typical enhancement curve in the tumor (purple) after bolus injection of CA, dissociation into ESS curve (blue) and plasmic curve (red). (c) Typical perfusion (V_p) and permeability (K^{trans}) experimental tumor maps. A full color version of this figure can be found in the color plate section.

A distinction is made between diffusible and blood pool tracers because they behave differently and yield different information (Daldrup *et al.*, 1998; Padhani and Husband, 2001). Diffusible tracers range from freely diffusible tracers, such as ^2H_2O, and can often include small-molecular-weight CAs, such as chelates of gadolinium. Blood pool tracers are usually larger molecules and hence diffuse into the interstitium much more slowly, if at all. They are also cleared more slowly, yielding flatter vascular input functions. Slower kinetics allow for higher dynamic range in quantification of vascular leakage and acquisition of much higher-resolution images, which can be important in pathodiagnosis.

DCE-MRI has evolved from an experimental technique to a clinically feasible adjunct procedure that can be integrated into a standard morphologic imaging protocol. It does provide unique non-invasive functional information on the properties of tumors related to microcirculation (distribution volume, permeability, and perfusion). This information can improve diagnostic characterization, follow-up of therapy, and tumor staging; and it provides tools to facilitate advanced molecular imaging (Knopp *et al.*, 2001; Daldrup *et al.*, 1998; Padhani and Husband, 2001; Brasch *et al.*, 2000). Preclinical and clinical studies suggest that successful antivascular treatment results in a decrease in the rate of enhancement along with a decreased amplitude and a slower washout, and that poor response can result in persistent abnormal enhancement.

11.2.1.2 PET/SPECT

PET is based on the approximate collinearity of two photons simultaneously emitted by the annihilation of a positron and electron. The resolution of PET is limited by uncertainties arising from the approximation of collinearity and annihilation position caused by individual positron ranges (Cho, Jones, and Singh, 1993).

Generally, PET measures of tumor perfusion have used (^{15}O)-labeled radiotracers. The so-called steadystate method requires inhalation of $^{15}O\text{-}CO_2$ and the dynamic method requires an intravenous bolus injection of $^{15}O\text{-}H_2O$ (Jennings, Raghunand, and Gillies, 2008). A requirement for quantification of perfusion using dynamic methods is an accurate determination of an arterial input function, which can be obtained non-invasively in a purely arterial region of interest, such as the aorta.

11.2.1.3 Computed tomography (CT)

In X-ray CT, tissue contrast is based on variable attenuation coefficients of the object absorbing the X-rays. Hemodynamic parameters may be extracted from dynamic changes in X-ray attenuation caused by the intravenous injection of an iodinated CA. Perfusion CT data can deliver quantitative hemodynamic information, such as blood volume, blood flow, permeability surface-area product and mean transit time (MTT).

11.2.1.4 Doppler ultrasound

Medical ultrasonography is based on the pulse–echo and back-scattered echo waveforms. A transducer which converts electric power to acoustic power is used to transmit short and relatively broadband pulses through the tissue, which are attenuated at tissue interfaces due to absorption and scattering. The pressure from the backscattered signal is collected by the same or a second phase-sensitive transducer, and the output voltage is a radiofrequency trace, which is recorded as a function of the acoustical travel time in the tissue.

There are several different ultrasonic approaches designed specifically to measure blood flow including transit time, continuous-wave Doppler, pulsed and color Doppler, and power Doppler flowmeters, requiring the use of microbubbles (filled with air, perfluorocarbon, sulfur hexafluoride or nitrogen), which expand and contract because of pressure from the acoustical transmit pulse, and the primary mode of echogenicity is the impedance mismatch between the microbubble–blood interface, making them significantly more echogenic than normal tissue. Typical parameters that are estimated using Doppler ultrasound include: percent intratumor CA uptake, enhancement timing and pattern, percent blood volume fraction, RBC velocity, and perfusion; depending on the type of study and tracer used (Jennings, Raghunand, and Gillies, 2008).

11.2.2 Vessel size index

The vessel size index, which is a measure of the average blood vessel radius, can be estimated with MRI using the measured values of relaxivity rate ($\Delta R_2{}^*$) and ΔR_2 (where $R_2 = 1/T_2$) after administration of ultrasmall superparamagnetic iron oxide (USPIO) CAs (Dennie *et al.*, 1998; Tropres *et al.*, 2001). USPIO particles are CAs that consist of a superparamagnetic crystalline iron core with a biocompatible coating. When administered intravenously, USPIO particles remain within the intravascular space with a long half-life of up to several hours (Enochs *et al.*, 1999). The large susceptibility changes adjacent to the blood vessels created by the presence of USPIO particles produces an increase in the transverse MRI relaxation rates R_2 and $R_2{}^*$ of the surrounding tissue. The change in the $R_2{}^*$ ($\Delta R_2{}^*$) can be directly related to the tissue fractional blood volume (fBV). In contrast, ΔR_2 is strongly dependent on the microstructure of blood vessels. The fBV, as determined with USPIO CAs, has been shown to correlate with the tumor blood volume derived from the uptake of the perfusion marker Hoechst 33342 (Robinson *et al.*, 2005), and fBV was decreased in rodent tumors treated with vascular-targeted agents (Bentzen *et al.*, 2005; Robinson *et al.*, 2007; Ferretti *et al.*, 2005). The technique has also been successfully applied using vascular disrupting agents (Howe *et al.*, 2008). Because the Rv and fBV are independent of vascular permeability, they might provide more robust biomarkers that are complementary to K^{trans} as derived from DCE-MRI.

11.2.3 Vessel maturity and functionality

$T_2{}^*$-weighted ($T_2{}^*$w) gradient echo (GRE) MRI can be used to provide *in vivo* imaging of tumor vessel maturation and functionality (Abramovitch *et al.*, 1999; Neeman *et al.*, 2001; Gilead, Meir, and Neeman, 2004). This technique is sensitive to changes in deoxyhemoglobin content manifested as change in $T_2{}^*$, referred to as the blood oxygen level-dependent (BOLD) effect. Thereby a change in $T_2{}^*$w GRE signal intensity can result from changes in tumor blood hemoglobin oxygen saturation, hematocrit, blood volume or blood flow (red blood cells flux). GRE sequences are also sensitive to changes in plasma flow via changes in the apparent T_1 relaxation (Howe *et al.*, 2001; Baudelet and Gallez, 2002). Maturation of the tumor vasculature can be determined from changes in $T_2{}^*$w GRE images in response to hypercapnia (elevation of inhaled CO_2) (Neeman *et al.*, 2001; Gilead, Meir, and Neeman, 2004). Vascular reactivity to hypercapnia occurs in mature vessels covered with smooth muscle cells which respond to elevated levels of arterial CO_2. This method has been used to predict the vascular response to antiangiogenic and antivascular treatments (Abramovitch *et al.*, 1999; Gross *et al.*, 1999). On the other hand, functional status of the tumor vascular bed can be assessed from T2*-w GRE images during a carbogen (95% O_2, 5% CO_2) breathing challenge. High-oxygen gas breathing produces acute changes in blood oxygenation in the functional tumor vascular bed, resulting in the BOLD contrast effect. This technique has been used to assess the role of vessel maturation and functionality in tumor acute hypoxia (spontaneous fluctuations) (Baudelet, Cron, and Gallez, 2006).

11.3 Imaging tumor hypoxia: chronic and acute

Hypoxia, or low oxygenation, has emerged as an important factor in tumor biology and response to cancer treatment. It has been correlated with angiogenesis, tumor aggressiveness, local recurrence, and metastasis, and it appears to be a prognostic factor for several cancers, including those of the cervix, head and neck, prostate, pancreas, and brain. The relationship between tumor oxygenation and response to radiation therapy has been well established, but hypoxia also affects and is affected by some chemotherapeutic agents (Tatum *et al.*, 2007). Methods to measure absolute pO_2 encompass electron paramagnetic resonance (EPR) oximetry and ^{19}F relaxometry whereas indirect methods include BOLD MRI, PET tracers retained in hypoxic regions, and near-infrared spectroscopy (NIRS).

11.3.1 EPR oximetry

EPR is a MR method that detects only species containing unpaired electrons (Gallez and Swartz, 2004). One of the numerous applications of EPR is *in vivo* oximetry. Molecular oxygen is a triplet radical that possesses two unpaired electrons, which are responsible for its paramagnetism. However, no EPR spectra have been reported from oxygen dissolved in fluids near room temperature. There seems to be no possibility for the direct detection of oxygen in biological systems. The lines are so broadened as to be undetectable. However, indirect methods exist. Most of these methods rely on the paramagnetic properties of molecular oxygen, which acts as an efficient relaxer for other paramagnetic species (Gallez, Baudelet, and Jordan, 2004). The enhancement of relaxation rates scales linearly with the concentration of oxygen over a wide range of oxygen tensions. The lack of detectable levels of endogenous paramagnetic species makes it necessary to use exogenous paramagnetic materials. Two classes of stable paramagnetic materials are useful for oximetry purposes: soluble paramagnetic materials and insoluble particulate materials.

The use of spectroscopy offers the advantage of maximum sensitivity. For this purpose, the particulate materials can be introduced into the tissue. Thus, the method is only invasive the first time. After that, the measurements can be made non-invasively using a surface coil. This method is particularly well adapted for measurements of pO_2 from the same site over long periods of time, and offers the highest sensitivity in terms of pO_2. Variations in pO_2 of less than 1 mmHg can be detected using particulate materials.

While EPR spectroscopy provides local measurements, EPR imaging techniques provide spatially resolved measurements of these materials. The spatial distribution of free radicals can be performed utilizing magnetic field gradients in a manner similar to that of MRI. Spectral–spatial EPR imaging encodes both the spatial distribution of the spin probe and the spectral information, which allows mapping of molecular oxygen (Kuppusamy *et al.*, 2003). For this purpose, the use of soluble EPR materials such as trityl radicals is more convenient as they can diffuse in the whole tissue.

EPR oximetry provides a potentially very useful method to follow changes in the pO_2 under various physiological and pathophysiological conditions. In preliminary studies, measurements of tumor oxygenation using EPR have been carried out as a proof of principle in order to show that a specific method or a newly developed paramagnetic material was applicable for this purpose. In a second time, EPR oximetry was compared with other methods that provide direct or indirect measurements of tumor oxygenation: comparison with polarographic electrodes, the distribution of nitroimidazoles, the BOLD effect in MRI, and pO_2 recordings using OxyLite (reviewed in Gallez, Baudelet, and Jordan, 2004). More interestingly, the power of EPR oximetry to measure oxygen from the same site over long periods of time offers the possibility of developing novel approaches to modifying tumor oxygenation, and thus optimizing anticancer treatments. A transient increase in tumor oxygenation and perfusion may be beneficial if combined with radiotherapy or chemotherapy. The tumor oxygenation may be modified by changing the oxygen supply (change in hemoglobin saturation, change in perfusion) or by changing the local oxygen consumption. Several of these approaches have been successfully studied using *in vivo* EPR oximetry in experimental tumors (Gallez, Baudelet, and Jordan, 2004; Jordan *et al.*, 2000, 2002, 2007; Crokart *et al.*, 2005, 2007; Ansiaux *et al.*, 2006, 2009; Sonveaux *et al.*, 2004; Pogue *et al.*, 2003; Hou *et al.*, 2007; Elas *et al.*, 2008).

Two major challenges are now considered for moving this technology into the clinic: (i) assuring biocompatibility of the oxygen sensors in humans and (ii) modifying the instruments so that they can be used for humans instead of small animals (Swartz *et al.*, 2004).

11.3.2 ^{19}F relaxometry

^{19}F NMR spectroscopy and imaging of perfluorocarbon (PFC) emulsions, hydrocarbons whose protons have been replaced with fluorine nuclei, has been extensively exploited to measure the oxygen tension of biological systems in preclinical studies. The ^{19}F MR signal of the PFC is sensitive to the pO_2 of the surrounding tumor tissue, and acts as an oximeter. The principle behind ^{19}F MR oximetry relies on the linear increase of the NMR spin-lattice relaxation rate R_1 ($=1/T_1$) of PFC emulsions with increasing oxygen tension (Mason, Rodbumrung, and Antich, 1996). Molecular oxygen has a very high solubility in PFC emulsions, thus permitting oxygen tension measurements in locations where the PFC is sequestered from the PFC spin-lattice relaxation rates *in vivo*. ^{19}F MR oximetry provides a sensitive measure of tissue oxygen tension and is a powerful approach for monitoring tumor hypoxia. Several PFCs have been used for NMR oximetry, but hexafluorobenzene (HFB) is preferred (Mason, Rodbumrung, and Antich, 1996; Zhao, Jiang, and Mason, 2004), a PFC that has sixfold symmetry, with a single ^{19}F NMR resonance, and has a low sensitivity to temperature. Its spin lattice relaxation rate is highly sensitive to pO_2 and exhibits a linear relationship across the entire range of oxygenation.

Mason and colleagues have been successfully developing fluorocarbon relaxometry using echo planar imaging for dynamic oxygen mapping (FREDOM) MRI following direct intratumoral injection of the oxygen reporter molecule HFB. The technique was used to study tumor oxygenation after chemotherapy and radiotherapy and was compared with NIRS, fiber optic probes, oxygen needle electrodes, and the hypoxia marker pimonidazole (Hunjan *et al.*, 2001; Mason *et al.*, 2003; Zhao *et al.*, 2003; Zhao, Jiang, and Mason, 2004; Xia *et al.*, 2006).

An MRI fluorocarbon oximetry technique using snapshot inversion recovery was recently developed, with an effective in-plane spatial resolution (1.88 mm) similar to that of FREDOM (1.25 mm), and a temporal resolution of 1.5 minutes, which is shorter than that of FREDOM (6.5 minutes). The method was validated with simultaneous measurements from an MR-compatible fiber-optic probe in experimental tumors. The method provides a rapid way to map the tumor oxygenation and is particularly suitable to monitor acute changes of pO_2 in tumors (Jordan, Cron, and Gallez, 2009) (Figure 11.2).

It is important to note that the absolute pO_2 values determined from a range of different tumor models using FREDOM or the snapshot method usually show higher basal pO_2 values than standard invasive methods (eppendorf electrodes and fluorescence quenching fiber optic probes) (Robinson and Griffiths, 2004; Jordan, Cron, and Gallez, 2009), or EPR oximetry. Many factors can be involved in this discrepancy: difference in sampling volumes, *in vitro* calibration curves, slight sensitivity to changes in temperature, poor distribution of the PFC emulsion, higher solubility of oxygen in lipidic environment such as PFC than in aqueous environment. Nevertheless, the method remains highly valuable to detect relative changes in tumor oxygenation and provides an early indication of response to therapy.

The translation to the clinic is currently limited by the lack of development of coils in the clinical setting, and the lack of characterization of the PFCs in humans.

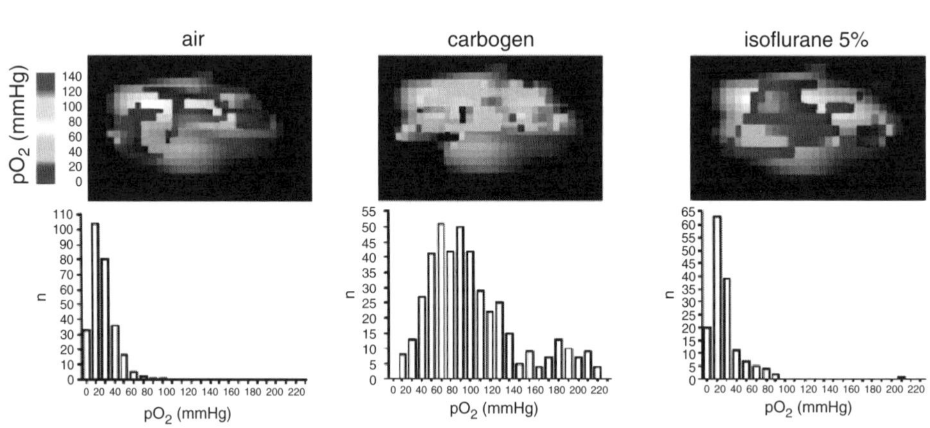

Figure 11.2 Typical oxygen maps of a typical tumour under air, carbogen, and isoflurane 5% (sacrifice) breathing conditions and their corresponding histograms. (Adapted from Jordan, Cron, and Gallez (2009).) A full color version of this figure can be found in the color plate section.

11.3.3 BOLD (R_2*)-MRI

The BOLD contrast mechanism in brain was first described by Ogawa *et al.* (1990) in rat studies using NMR at strong magnetic fields (7 and 8.4 T). Ogawa noticed that the contrast of very high resolution images acquired with a gradient-echo pulse sequence depicts anatomical details of the brain as numerous dark lines of varying thickness. By accentuating the susceptibility effect of dHb in the venous blood with gradient-echo techniques, he discovered that image contrast reflected the blood oxygen level. As it is now known, the phenomenon is indeed due to the field inhomogeneities induced by the endogenous MRI CA dHb (Howe *et al.*, 2001). In dHb, the iron (Fe^{2+}) is in a paramagnetic high spin state (d^4) as four out of six outer electrons are unpaired. The paramagnetic nature of dHb can modify the strength of the magnetic field passing through it. It enhances the R_2 ($= 1/T_2$) and $R_2^* = 1/T_2^*$ (transverse relaxation rates of water in blood and in the tissue surrounding the blood vessels). In the simplest model, these relaxation rates change linearly with deoxyhemoglobin concentration, which therefore acts as an endogenous CA for blood oxygenation.

MRI is expected to provide a sensitive index of the oxygenation status of tissue immediately adjacent to perfused microvessels. The BOLD MRI effect can be quantified by plotting the natural log of signal intensity against the echo time to derive tissue relaxivity (the power of the exponent is $-1/T_2^*$ or $-R_2^*$, the rate of relaxation). R_2^* maps reflect the structure and content of tissues, including the local concentration of deoxyhemoglobin in macrovessels and microvessels. R_2^* maps minimize, but do not completely eliminate, inflow effects that can contribute to contrast on single-echo gradient-echo images while retaining sensitivity to oxygenation changes that may be caused by alterations in blood flow (Tatum *et al.*, 2006).

Apart from its very large application in neuroscience, the use of BOLD contrast in tumor brings with it new challenges of understanding and interpretation. The underlying physiological changes are different from those of BOLD in the brain. In the brain, blood vessel size, density, and distribution are well characterized, in contrast to tumors. The assumptions used in models to describe BOLD contrast of fMRI in brain cannot therefore be applied to tumors. GRE imaging, which provides BOLD contrast, has been used to investigate parameters related to tumor vasculature, such as perfusion, blood oxygenation, blood vessel development, remodeling, and function, as well as to monitor the effects of therapy changing the tumor oxygenation. BOLD contrast changes were also more recently detected in rodent tumors with high spectral and spatial resolution (HiSS) MRI rather than with conventional GRE imaging.

BOLD MRI has become a useful tool for addressing important questions regarding the pathophysiology of tumors. However, as with any technique, it has both advantages and disadvantages. One advantage of BOLD MRI is that it is noninvasive and can be used to monitor real time changes of tumor oxygenation during pharmacological treatments. It does not require externally administered contrast medium or radioactive isotopes, it can be repeated as necessary, and flow dependence can be decoupled. BOLD MRI also has high spatial resolution, allowing it to address the issue of the spatial hetereogeneity of the tumor response. Carbogen induced changes in R_2^* or basal R_2^*, which reflect vascular development, may also

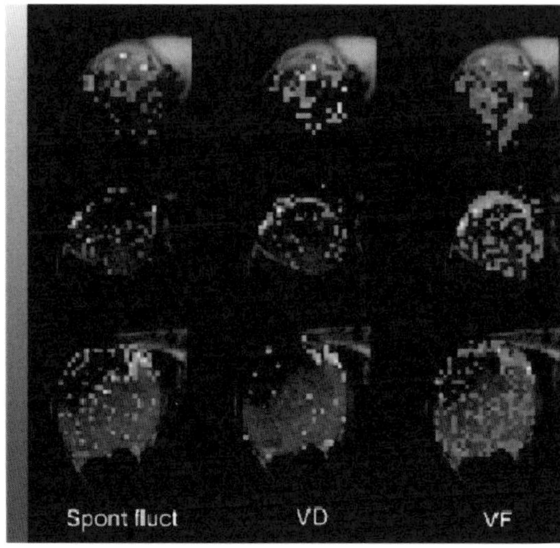

Figure 11.3 Co-registration of spontaneous fluctuations, vascular maturation, and vascular function maps for three seperate intramuscular FSa II fibrosarcoma tumors in mice. Left: spontaneous fluctuation maps determined by spectral analysis of $T_2{}^*$W GRE imaging during the 1 hour of air breathing. Middle: vascular maturation (VD) maps. Right: vascular function (VF) maps. The last two types of maps represent significant GRE MRI signal changes during air – 5%CO_2 breathing and carbogen breathing, respectively. The color scale is a function of the variance of the time series (for the spontaneous fluctuation maps) or yhe extent signal change (for the VD and VF maps). Parameters are in arbitrary units: from red (minimum) to yellow (maximum). Note that the proportion of mature and functional vessels are greater in tumor areas demonstrating spontaneous $T_2{}^*$W GRE signal intensity fluctuations than in non-fluctuation areas. (Adapted from Baudelet, Cron, and Gallez (2006).) A full color version of this figure can be found in the color plate section.

be monitored with BOLD MRI to predict radiotherapy sensitivity. BOLD MRI, in combination with hypercania and hyperoxia, is also an attractive method for assessing maturation and the functional state of the tumor (Baudelet, Cron, and Gallez, 2006) (Figure 11.3). Another appealing application is the study of changes in blood flow and oxygenation associated with the phenomenon of fluctuating hypoxia (Baudelet *et al.*, 2004).

As for disadvantages, BOLD MRI is unfortunately a non-quantitative method for monitoring tumor pO_2. This is the result of the extreme sensitivity of changes in $R_2{}^*$ to the basal state of tumor oxygenation and blood volume fraction. The intra- and intertumoral distribution of these parameters may be greatly heterogeneous, making it very difficult to compare estimated pO_2 changes between two regions or individuals. Even more problematic is the fact that the change in $R_2{}^*$ is not always indicative of the change in pO_2. Concomitant changes in blood volume, blood pH and metabolic status can lead to smaller-than-expected or even negative changes in $R_2{}^*$ (Baudelet and Gallez, 2002). Similarly, changes in oxygen consumption rate has been described to result in a lack of change in $R_2{}^*$ even though absolute pO_2 is increased (Jordan *et al.*, 2006).

11.3.4 PET

^{18}F-labeled fluoromisonidazole (^{18}F-FMISO) is probably the most widely used PET imaging agent for hypoxia. ^{18}F-FMISO is the prototype hypoxia imaging agent whose uptake is homogeneous in most normal tissues, reflecting its high partition coefficient that nears unity, and whose delivery to tumors is not limited by perfusion (Padhani *et al.*, 2007). ^{18}F-MISO accumulates in tissues by binding to intracellular macromolecules when $pO_2 < 10$ mmHg. Retention within tissues is dependent on nitroreductase activity (that is, on reduction status of a NO_2 group on the imidazole ring) and accumulation in hypoxic tissues over a range of blood flows has been observed. The rapid clearance that is characteristic of hydrophilic tracers can interfere with the ability to distinguish hypoxia from low blood flow. The background ^{18}F-FMISO signal can be corrected by subtracting or reporting a tumor–blood ratio. A tumor–blood ratio of 1.2 has been demonstrated as a reasonable cutoff between normoxia and hypoxia. The degree of hypoxia can be expressed as a quantity, defined by hypoxic fraction or volume, or by its severity, defined as the region with the lowest oxygen concentration and its relative level. Hypoxic volume can be determined by multiplying the number of pixels that exceed the threshold of 1.2 by the volume per pixel (Tatum *et al.*, 2006).

^{18}F-MISO is only sensitive to the presence of hypoxia in viable cells: ^{18}F-MISO is not retained in necrosis because the electron transport chain that reduces the nitroimidazole to a bioreductive alkylating agent is no longer active (Padhani *et al.*, 2007). Other limitations of 18F-MISO PET include the modest signal-to-noise ratio of raw ^{18}F-MISO PET images, the range of fractional hypoxic volume that is scattered across tumor types, which precludes a standard correlation of hypoxic volume with that of the tumor; and the difficulty to monitor acute temporal heterogeneity (Tatum *et al.*, 2006).

Nevertheless, ^{18}F-MISO PET is able to monitor the changing hypoxia status of lung tumors during radiotherapy (Koh *et al.*, 1995). Studies in sarcoma (Rajendran *et al.*, 2003) and head and neck cancer (Ng *et al.*, 2003; Rajendran *et al.*, 2004) have demonstrated a correlation of ^{18}F-FMISO uptake with poor outcome to radiation and chemotherapy. Other ^{18}F labeled nitroimidazoles are currently in development, including EF3 and fluoroazomycin arabinoside, for example (Tatum *et al.*, 2006).

11.3.5 NIRS

In recent years, NIRS has gained importance for non-invasive or minimally invasive diagnostic applications in cancer. This technology is based on differences of endogenous chromophores between cancer and normal tissues using either oxyhemoglobin or deoxyhemoglobin, lipid or water bands, or a combination of two or more of these as diagnostic markers. These marker bands provide a basis for the diagnosis and therapy monitoring of several cancers (Kondepati, Heise, and Backhaus, 2008).

The physiological relevance of NIRS for cancer studies, as manifested by the differences in absorption and tissue scattering parameters, can be traced back to the hemoglobin concentration (total, oxy, and deoxy forms), tissue hemoglobin oxygen saturation, and water and lipid content. NIRS data interpretation in the field of oncology relies on the following assumption: high total tissue hemoglobin concentration corresponds to angiogensis/elevated tissue blood volume; decreased tissue hemoglobin oxygenation saturation indicates tissue hypoxia driven by metabolically active tumor cells; and high water concentration suggests edema and increased cellularity.

Altered vasculature, oxygen dynamics, and oxy-hemoglobin and deoxy-hemoglobin concentrations in brain, muscle, mammary, and prostate cancers, or the response of these parameters to therapy, have been reported in animal models (Pham *et al.*, 2007; Kragh *et al.*, 2001; Gulsen *et al.*, 2002). Furthermore, hypoxia-related parameters have also been studied (Kim *et al.*, 2003; Gu *et al.*, 2003). Most of the investigations dealing with human tissues were on breast cancer. In those studies, total hemoglobin concentration and tissue hemoglobin oxygen saturation were calculated, and are expected to provide information on tumor angiogenesis and hypermetabolism. Other studies on human tissues include cervix, brain, skin, prostate, lung, head and neck, pancreas, and colorectal tissues (Kondepati, Heise, and Backhaus, 2008).

NIRS can offer options for optimizing the biological effect, accurate dosimetry, and monitoring treatment progression and efficacy. Thus, it has been applied to monitor changes in cancer tissues during chemotherapy, neoadjuvant therapy, photodynamic therapy (PDT), or radiation therapy of human head and neck tumors, breast cancer, skin tumors, cervical dysplasia, and rat ovarian and murine fibrosarcoma tumors. The effect of vascular-modifying or antiangiogenic agents on tumors has also been studied in mice (all reviewed in Kondepati, Heise, and Backhaus, 2008). Moreover, the development of fluorescent quantum dots and catheter-based imaging approaches will further advance clinical applications, even though they are still in their infancy.

11.4 Imaging tumor oxygen consumption

Tumor oxygenation depends on the balance between oxygen supply and consumption, and both factors are now considered in developing strategies to reduce tumor hypoxia in the field of radiosensitization. Theoretical simulation of oxygen handling in tumors suggests that decreasing the oxygen consumption rate of tumor tissue could be a particularly effective strategy to reduce the fraction of hypoxic tissue in solid tumors (Secomb, Hsu, and Dewhirst, 2004).

11.5 EPR oximetry

In view of its clinical importance, several techniques have been developed to measure oxygenation status in tissues and tumors in living animals and humans, but methods to measure rates of oxygen consumption are more limited. Three common *in vitro* methods are currently available to measure oxygen consumption: EPR oximetry, the Clark oxygen electrode, and the MitoXpress fluorescent assay. Among them,

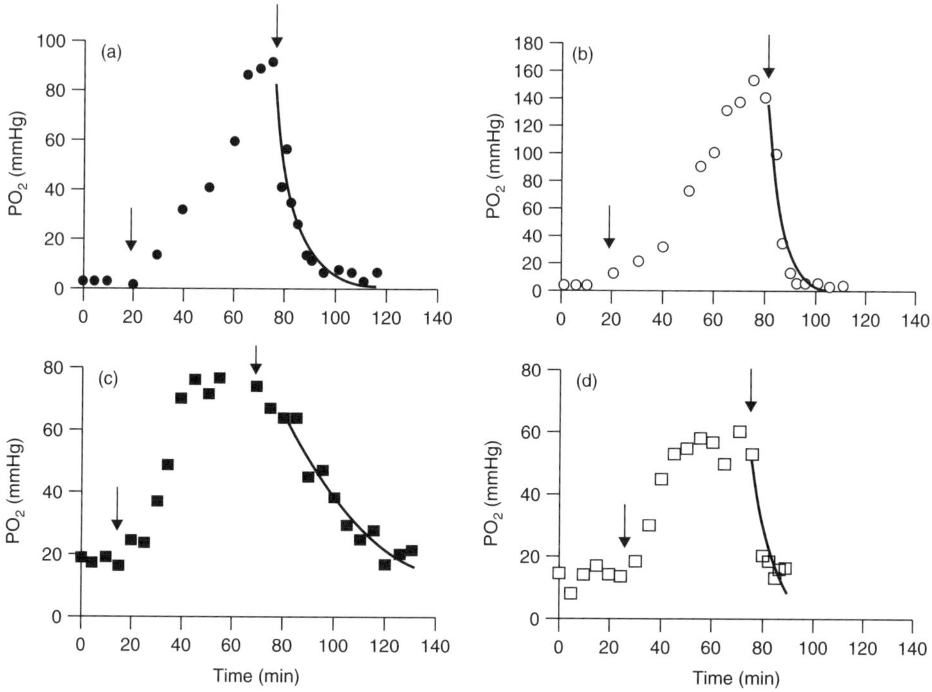

Figure 11.4 Breathing challenge in tumor and muscle of hyperthyriod and control mice, measured by EPR oximetry.First arrow : air to carbogen. Second arrow: carbogen to air. (a,b) Typical evolution of pO_2 during breathing challenge in tumor of control (●) and hyperthyriod mice (○). (c,d) Typical evolution of pO_2 during breathing challenge in muscle of control (■) and hyperthyroid mice (□). The graph indicates that tissues (including tumors) in hyperthyriod mice consume oxygen faster than in control mice, as compared while comparing kinetics constant. (Adapted from Diepart, Jordan and Gallez (2009).)

EPR has been shown to be the most sensitive and was a unique candidate for translation to *in vivo* measurements (Diepart *et al.*, 2010). An *in vivo* EPR protocol has recently been validated in a mouse tumor model, using a carbogen breathing challenge combined with drugs that are described to inhibit oxygen consumption by tumor cells (Diepart, Jordan and Gallez, 2009). Briefly, tumor-bearing mice are submitted to carbogen breathing until pO_2 reaches a maximum (plateau) value, then, the gas is switched back to air breathing and the kinetics of return to the basal level are estimated and reflect oxygen consumption by tumors *in vivo* (Figure 11.4). Further development with EPR imaging will give access to the mapping of oxygen consumption using this technique (data not shown).

11.5.1 ^{17}O MRI

The possibility of using ^{17}O NMR spectroscopy imaging techniques for monitoring labeled $H_2{}^{17}O$ as a tracer of oxygen utilization or tissue perfusion in the brain have been introduced a decade ago. These early studies can be divided into two distinct groups. The first group involved the simple ^{17}O NMR approach for detecting $H_2{}^{17}O$ directly. Non-localized ^{17}O NMR was used for most studies to estimate the cerebral metabolic rate of oxygen ($CMRO_2$) in the whole brain of experimental animals as well as in the human occipital lobe after inhalation of oxygen gas enriched with the ^{17}O isotope ($^{17}O_2$). All of these studies were conducted at relatively low fields, and demonstrated the dramatic limitations imposed by the low inherent sensitivity of ^{17}O NMR because of its low gyromagnetic ratio. These limitations had lead to the exploration of a second group of methods that attempt to circumvent the sensitivity limitations of ^{17}O NMR by using indirect detection through ^{17}O-coupled protons (Ronen *et al.*, 1998). However, the method presented difficulties for quantitatively correlating the proton signal changes to the concentrations of $H_2{}^{17}O$ because the water T_2 is sensitive to many physiologic parameters such as pH and temperature. These indirect ^{17}O detection approaches have therefore not been successful in imaging $CMRO_2$ quantitatively.

Recently, a direct ^{17}O MRI approach has been developed in order to measure cerebral oxygen consumption rate in animal models at high fields (9.4 T) during inhalation of $^{17}O_2$ (Zhu *et al.*, 2002, 2005). These method presents many advantages, including: specific detection of $H_2{}^{17}O$ without confounding signals from the $^{17}O_2$ bound to hemoglobin or $^{17}O_2$ dissolved in tissue space; stability of the ^{17}O isotope, which is non-radioactive; the experimental procedure of $^{17}O_2$ inhalation does not introduce any physiological perturbation in the animal; the natural abundance signal of $H_2{}^{17}O$, which can be accurately imaged, provides an internal reference for rigorously calibrating and calculating the absolute time-dependent $H_2{}^{17}O$ concentrations expressed as $[H_2{}^{17}O]$ in excess of the natural abundance $[H_2{}^{17}O]$ level in the arterial blood (Ca(t)) and brain tissue (Cb(t)) during $^{17}O_2$ inhalation; after the cessation of $^{17}O_2$ inhalation, the cerebral $H_2{}^{17}O$ concentration reaches a new steady state within a short time (6–10 minutes); this fast recovery should allow repeated CMRO2 measurements in the same subject and same experimental session; and, finally, at ultra-high fields the ^{17}O-NMR sensitivity is adequate to obtain relatively high-resolution images, with an 11-second acquisition time, which makes the ^{17}O approach possible for imaging CMRO2 within a short measurement duration (2 minutes). The application of the method to cerebral tumors still needs to be developed. However, the translation of this method to humans will be mostly hampered by the lack of ultra high field magnets in the clinical setting.

11.5.2 ^{15}O PET

The short half-life of ^{15}O led early observers to believe that it was unsuitable for use as a biological tracer. However, initial studies with this nuclide demonstrated its potential usefulness for *in vivo*, regional physiologic measurements. Subsequently, techniques were developed to measure cerebral blood flow (CBF), blood volume, and oxygen metabolism using intracarotid injection of ^{15}O-labeled radiopharmaceuticals and highly collimated scintillation probes to record the time course of radioactivity in the brain (Frackowiak *et al.*, 1980). The development of PET made possible the *in vivo*, non-invasive measurement of the absolute concentration of positron-emitting nuclides. Regional cerebral oxygen metabolism is measured using scan data obtained following the inhalation of ^{15}O-labeled oxygen; independent determinations of local blood flow and blood volume are also required for this measurement (Ter-Pogossian and Herscovitch, 1985). The method has been shown to be suitable for the study of human cerebral aerobic metabolism and blood flow under differing physiological conditions (Frackowiak *et al.*, 1988) and in a wide variety of neurological disorders (Sheppard *et al.*, 1983), as well as in brain tumors (Ito *et al.*, 1982).

11.5.3 Q-BOLD

As described earlier, the advent of BOLD MR imaging initiated by Ogawa *et al.* (1990) opened new opportunities to non-invasively study brain hemodynamics. BOLD approach capitalizes on the fact that deoxygenated blood has different magnetic susceptibility as compared to oxygenated blood (Ogawa *et al.*, 1990), which in turn has magnetic susceptibility similar to the tissue. Due to this effect, the deoxyhemoglobin containing part of the blood vessel network in the brain creates mesoscopic field inhomogeneities in the surrounding tissue leading to more rapid MRI signal decay than from standard T_2 decay alone. Because these field inhomogeneities are tissue specific, measuring the MRI signal decay rate may provide information on the tissue structure and functioning. Interestingly, a group has been developing a theoretical model of BOLD contrast that analytically connects the BOLD signal to hemodynamic parameters such as the deoxyhemoglobin containing blood volume (DBV), deoxyhemoglobin concentration, and oxygen extraction fraction (Yablonskiy and Haacke, 1994).

A subsequent work (Yablonskiy, 1998) quantitatively validated important features of the model in phantom studies and developed a theoretical background and experimental method (based on the gradient echo sampling of spin echo (GESSE) sequence) that allows the separation of mesoscopic field inhomogeneity effects from both macroscopic and microscopic inhomogeneities. Such separation allows one to take full advantage of the mesoscopic, tissue-specific magnetic field inhomogeneity effects to extract quantitative information about tissue hemodynamic properties. The first attempts to directly implement this method *in vivo* were encouraging but did not produce conclusive results. One reason was that the simplistic model used therein, describing brain tissue as a one compartment structure similar to water in phantom, was not sufficient to describe real brain tissue. The recent qBOLD MR signal model (He and Yablonskiy, 2007) of brain tissue incorporates prior knowledge about brain tissue composition and includes contributions from intravascular water in gray matter (GM) and white matter (WM), cerebrospinal fluid (CSF), or interstitial fluid (ISF) with a resonance frequency shift, and intravascular venous blood. Using MR images of human brain parenchyma obtained with a 2D GESSE sequence, the qBOLD model demonstrated an ability to measure brain hemodynamic parameters such as DBV, oxygen enhancement factor, ISF/CSF volume fraction, and frequency shift in the baseline activity state. Moreover, preliminary results (He and Yablonskiy, 2007) obtained from the qBOLD model were in a good agreement with previous PET studies. Finally, quantitative validation was performed using venous blood samples directly drawn from the rat superior sagittal sinus (He, Zhu, and Yablonskiy, 2008).

11.6 Imaging tumor interstitial fluid pressure (IFP)

Solid tumors have a raised interstitial fluid pressure (IFP) because of high vessel permeability, low lymphatic drainage, poor perfusion, and high cell density around the blood vessels. IFP forms a barrier to transcapillary transport which is an obstacle in tumor treatment, as it results in inefficient uptake of therapeutic agents (Heldin *et al.*, 2004).

In the past, assessment of tumor IFP has been done by invasive methods such as the wick-in-needle and the micropuncture technique (Heldin *et al.*, 2004; Jain, 1987). Lyng *et al.* (1997) attempted to use MRI for mapping IFP was based on correlating proton relaxation rates with IFP values obtained by the wick-in-needle technique in a study of melanoma xenografts. However, the results showed that both T_1 and T_2 relaxation rates did not correlate with the measured IFP. Subsequent DCE-MRI studies, using a bolus injection of gadolinium diethylene penta-acetic acid (GdDTPA), indicated the presence of disparities between the influx and outflux transcapillary transfer constants in breast tumors (Dadiani *et al.*, 2004). This disparity with the outflux exceeding the influx constant increased from the tumor rim to the center. This distribution is in accord with the profile of IFP distribution in tumors.

A non-invasive method which estimates IFP and its spatial distribution *in vivo* using DCE-MRI has been developed by the group of Degani in non-small-cell carcinoma xenografts and breast carcinoma xenografts (Hassid *et al.*, 2006). The MRI protocol consisted of slow infusion of the CA (GdDTPA) into the blood for about 2 hours, sequential acquisition of images before and during the infusion, and measurements of T_1 relaxation rates before infusion and after blood and tumor GdDTPA concentration reached a steady state. Analysis of the changes in T_1 relaxation rates yielded steady-state tissue GdDTPA concentration (mmol/tissue volume) maps of the tumors and their surrounding. These maps reflected inhibition of transfer due to elevated tumor IFP and transfer by convection in the tumor surrounding. Image analysis yield parametric images of steadystate tissue GdDTPA concentration with high values of this concentration outside the tumor boundaries, $\sim$1 mmol/l, declining in the tumor periphery to $\sim$0.5 mmol/l, and then steeply decreasing to low or null values. The distribution of steady-state tissue GdDTPA concentration reflects the distribution of IFP, showing an increase from the rim inward, with a high IFP plateau inside the tumor.

In the clinical setting, Haider *et al.* (2007) subjected patients with cervical cancer to DCE-MRI and measurement of tumor IFP, and found significant correlations between IFP and DCE-MRI-derived parameters. However, the correlations were weak, possibly because the DCE-MRI parameters were derived from a single region of interest encompassing the tumor rather than from voxel-by-voxel parametric images.

The group of Rofstad recently assessed tumor IFP via DCE-MRI using a different modelisation, a Kety analysis of DCE-MRI series. Significant correlations were found between the derived parameters and IFP in non necrotic tumors. However, no such correlation was found in necrotic tumors (Gulliksrud, Brurberg, and Rofstad, 2009).

11.7 Imaging tumor pH

A number of studies have shown that the extracellular pH (pHe) in cancers is typically lower than that in normal tissue and that an acidic pHe promotes invasive tumor growth in primary and metastatic cancers. Several techniques have been developed to measure and/or image tumor pH.

11.7.1 ^{31}P-MRS

Several techniques have been used for the measurement of tissue pH by magnetic resonance spectroscopy (MRS) and MRI (reviewed in Gillies *et al.*, 2004). Some of these techniques exploit endogenous MR resonances while others require the administration of exogenous agents. Tissue pH, particularly in tumors, can be estimated from the ^{31}P MR resonance of inorganic phosphate (Pi) (Stubbs *et al.*, 1992). Because intracellular Pi concentrations are 2–3 mM, compared to about 1.0 mM for extracellular, and because the intracellular volume fractions are generally greater than 50%, the chemical shift of the endogenous Pi resonance is generally assumed to reflect intracellular pH. The pH-sensitive ^{31}P MR resonance of 3-aminopropylphosphate (3-APP) has been used to measure extracellular pH of tumors in mice, with a chemical shift dependence of about 1 ppm per pH unit. This technique led to the observation that the external pH is lower than the internal pH in tumor xenografts, contrarily to normal tissues. Such reversal of pH gradients can negatively impact therapy, and has physiological and metabolic consequences to the tumor cells.

11.7.2 ^{19}F-MRS

Molecules with pH-sensitive ^{19}F resonances have also been developed (Gillies *et al.*, 2004). The spin 1/2 resonance of ^{19}F has advantages in that it has a high gyromagnetic ratio, relatively large chemical shift dispersion, and an almost total lack of endogenous resonances in normal tissues. Hence, resonances from exogenous agents are readily resolved. Mason and his coworkers have developed a fluorinated derivative of vitamin B6, 6-fluoropyridoxol, which readily enters cells. They have measured both the intracellular and extracellular pH in rodent tumors by resolving the pH-sensitive ^{19}F resonance arising from the two compartments (Mason, 1999). They have shown that the triflouromethylated derivative of pyridoxal is membrane impermeant and insensitive to temperature, with a pKa of 6.82. Griffith's group has investigated the application of the extracellular ^{19}F pH probe ZK-150471 (Ojugo *et al.*, 1999). Both 3-APP and ZK-150471 are cell-impermeant and report only the extracellular pH, although ^{31}P MRS of 3-APP offers the possibility of simultaneous measurement of intracellular pH from the Pi resonance.

11.7.3 ^{1}H-MRS

The ^{1}H nucleus offers the highest inherent sensitivity, and it is possible to image the spatial distribution of tissue pH *in vivo* by the use of probes with pH-sensitive ^{1}H resonances (Gillies *et al.*, 2004). For this purpose, an exogenously administered imidazole,2-imidazole-1-yl-3-ethoxycarbonylepropionic acid,, has been used. This compound has a pH-dependent chemical shift of the H-2 resonance in the 7–9 ppm range, is non-toxic, and membrane-impermeant. MRS and MRSI approaches are generally dependent on measuring the pH-dependent chemical shifts. Hence, such measurements are theoretically independent of the concentration of the pH probe. However, the spatial resolution is limited to 1–2 mm, which is lower spatially than other techniques based on imaging. A drawback to this approach is the relatively insensitivity of chemical shift to pH, since there is only 0.7 ppm shift over the entire titration range, at least at low magnetic field.

11.7.4 Magnetization transfer

An alternative approach to the measurement of tissue pH is to measure a pH-dependent chemical exchange-dependent magnetization (saturation) transfer (CEST) (Gillies *et al.*, 2004). Water in biological systems exists as bulk phase H_2O but also in association with the macromolecules and metabolites. Bound water generally has a shortened T_2 and is therefore not visible in a typical MR experiment. However, these two states of water are in dynamic equilibrium. Therefore, saturation of the magnetization of the bound water can be transferred to the bulk phase, which is the basis of the conventional magnetization transfer (MT) effect. Water interacts with macromolecules and solutes through either dipolar coupling, for example, through hydrogen bonds, or through chemical exchange, wherein the hydrogens of water exchange with hydrogens on ionizable groups. All such exchanges will be acid or base-catalyzed and thus have pH dependence. CEST approaches can be applied using exchange of hydrogens with ionizable groups that are either endogenous or exogenous. In all cases, CEST effects have to be separated from conventional MT effects, which are approximately symmetric. While endogenous signals have the advantage of being fully non-invasive and relatively straightforward to apply, they lack a full biophysical characterization and dynamic range that might be afforded by future CEST agents. A disadvantage of the exogenous agents, however, is the need to inject large amounts of the pH-sensitive agent in order to be able to image pH with acceptable sensitivity and spatial resolution. A challenge to the development of these agents is the optimization of the exchange rate. Too fast an exchange rate makes it more difficult to fully saturate the bound spins within energy deposition (specific absorption rate) limits. If spins are not fully saturated, the effects are nonsteady state and more difficult to quantify. However, too slow an exchange rate will significantly reduce the CEST effect. A strength of these MT approaches is the internal control of acquiring an image without saturation, yielding direct and robust quantification of effect. However, as for Gd based CAs, the quantification is critically dependent on knowing the concentration of contrast reagent.

11.7.5 Gd-Based pH sensitive contrast agents

The development of Gd-based CAs whose relaxivity is dependent on pH is quite new (Gillies *et al.*, 2004). Like CEST, these agents provide the possibility of imaging pH with spatial resolution comparable to that of standard MRI. The magnitude of the effect is dependent on the local concentration of CA, as well as the pH. Therefore, accurate methods must be developed with which to determine the spatial and time variant CA concentrations, in order to convert the observed relaxivity to a molar relaxivity and thus, pH. Gd(DOTA)-4AmP5^- contains four ionizable phosphonate groups that catalyze the pH-dependent hydrogen exchange to the Gd-bound water, as well as the pH-dependent molar relaxivity. This compound is non-toxic and membrane-impermeant and distributes with pharmacokinetics similar to that of the identically charged, but pH-independent, Gd(DOTP)5^-. Hence, one approach to uniquely compute a pH using such agents is by sequential administration of Gd(DOTP)5^- and Gd(DOTA)-4AmP5^-, calculating the time and spatially variant concentrations of Gd(DOTP)5^- from T_1-weighted images and then using these values to correct for the time- and spatially variant concentrations of Gd(DOTP)5^-. Despite these challenges, the dual injection protocol has been successfully applied to small brain tumors (Garcia-Martin *et al.*, 2006). Although the dual-injection protocol may be appropriate for imaging of pH in animal models, it will be difficult to perform a dual injection protocol in the clinic. Hence, there is a need to develop methods that can simultaneously measure relaxation and concentration in a single injection.

11.7.6 EPR

A number of pH sensitive nontoxic nitroxide probes have been developed in the past, which have been applied to various biological systems, including *in vivo* pH measurements in rodents (Khramtsov, 2005). A crucial advantage of EPR over NMR is more than three orders of magnitude in sensitivity. The presence of the ionizable group in the structure of paramagnetic molecule near the radical center results in the difference in the EPR spectra of its protonated and unprotonated forms (large pH effect on nitrogen hyperfine splitting). Strong pH effects on the EPR spectra of short-lived organic radicals with ionizable groups were reported in early EPR studies but were difficult to use for pH detection because of instability of the radicals. Therefore, the main limitation of the nitroxides application is their degradation in biological tissues to EPR-silent products which results in a narrow experimental time-window for detection. The studies of proton transport across lipid bilayers of liposomes caused by transmembrane pH gradient were one of the first applications of pH measurement by EPR (Khramtsov, Panteleev, and Weiner, 1989). Other *in vitro* applications included proteins and proteineous matrix (Khramtsov *et al.*, 1992), as well as fuctional polymers (Khramtsov *et al.*, 1992; Mäder *et al.*, 1998). The first *in vivo* applications involved measurement of pH values within rat and human skin to study influence of drug treatment on the microacidity (Kroll *et al.*, 2001). The first *in vivo* applications performed at low EPR frequency took advantage of a nitroxide probe to monitor the pH value inside the stomach of mice after administration of different antacids (Gallez *et al.*, 1996). Subsequently, EPR spectroscopy was used to monitor pH-induced degradation of implanted polymer in mice (Mäder *et al.*, 1996). More recently, spin pH probes were used to study the ischemic acidosis in isolated perfused rat heart (Khramtsov, 2005). Also, an attractive application is to use EPR/NMR co-imaging to monitor simultaneously pHe and pHi. Intra and extracellular targeting is obviously crucial in the current development of spin pH probes. Therefore, pH-sensitive nitroxides with ester groups were considered, with intracellular accumulation after hydrolysis, providing possibility of simultaneous detection of pHi and pHe, by combined administration of intra and extracellular probes. Also, efforts have been made to develop pH-sensitive tetraethyl substituted nitroxides of imidazol tyme with enhanced stability against reduction (Khramtsov, 2005). Finally, recent developments in dynamic nuclear polarization (DNP) (Gallagher *et al.*, 2008) and proton electron double resonance imaging (Potapenko *et al.*, 2006) allowed to work at low field in living animals with more sensitivity and even to image the probe distribution. Besides nitroxyl radicals, the recent development of triarylmethyls containing ionizable groups (triarylmethyl radicals such as Oxo63) has significantly expanded the potential for measuring pH thanks to their extreme stability toward tissue redox processes. Moreover, these probes are oxygen sensitive and therefore offer the possibility of dual pH and oxygen measurements (Bobko *et al.*, 2007).

11.8 Imaging tumor redox status

Redox imaging employing MR techniques (EPRI and MRI) with nitroxides as cell permeable redox sensitive CAs has been used for non-invasive monitoring of tissue redox status in animal models. The redox environment within the tumor cell is an important parameter that may determine the response of a tumor to certain chemotherapeutic agents, radiation, and bioreductive hypoxic cell cytotoxins (Hyodo *et al.*, 2008). Therefore, imaging of tissue redox status and monitoring the antioxidant level could be useful in tumor diagnosis and in the assessment of treatment response.

11.8.1 EPR

EPRI is an imaging modality, which detects unpaired electrons in such species as transition metal complexes and free radicals and, using magnetic field gradients, provides spatial distribution of free radicals (Kuppusamy *et al.*, 1994; Subramanian *et al.*, 2004). Free radicals are present at extremely low levels in tissues, below the detection limits of EPR. Whereas this was initially felt to be a limitation to the use of this technique, it was quickly discovered that agents containing unpaired electrons, or agents converted to such a compound *in vivo*, could be introduced into a living system and detected accurately (Hyodo *et al.*, 2008).

Nitroxides have been used as *in vivo* EPR spectroscopy/imaging probes to study redox mechanisms in disease models (Berliner and Fujii, 1985; Kuppusamy *et al.*, 2002; Yamada *et al.*, 2002; Hyodo *et al.*, 2008). Nitroxides were found to be in equilibrium between the nitroxide radical form *in vivo*, that can be detected by EPR, and the reduced form, the hydroxylamine, which is not detected by EPR because of its diamagnetic nature (Swartz, 1990). This equilibrium is dependent on the surrounding environment, specifically tissue oxygenation, and the levels of reducing equivalents of the tumor (Swartz, 1990). Cellular redox processes convert the compound between the two states. Therefore, the ratio of the two states is determined by the redox status within the cell. Since only the oxidized form of the nitroxide can be detected using EPR, signal intensity can be used as a surrogate marker for the relative amounts of the oxidized compound and, therefore, the relative redox activity. This property of nitroxides makes them ideal compounds for studying intracellular redox metabolism (Hyodo *et al.*, 2008). Although EPRI can detect nitroxide free radicals directly and obtain images of nitroxide free radical distribution as well as redox maps, the poor image resolution and lack of anatomic detail limit its use in the clinical setting presently (Hyodo *et al.*, 2008).

11.8.2 MRI

Nitroxide radicals have a single unpaired electron and can provide T_1-contrast similar to gadolinium complexes. Although nitroxides ($0.16{\sim}0.18\,mM^{-1}\,s^{-1}$) compare unfavorably to Gd^{3+}-containing agents in terms of relaxivity ($3.7\,mM^{-1}\,s^{-1}$), they are cell permeable and have a larger volume of distribution meaning they can provide useful T_1-contrast enhancement per unit volume (Hyodo *et al.*, 2008). The feasibility of using nitroxides as T_1-CAs was first examined by Brasch in 1983. However, at the time they were found not to be optimal MRI CAs since, *in vivo*, the paramagnetic nitroxide radicals were not stable. However, because of the EPRI experience with nitroxides, they were re-evaluated as functional redox sensitive probes using MRI.

There are many technical challenges using nitroxide CAs in MRI. Nitroxide relaxivity is 20 times less than Gd^{3+} complexes. Additionally, since nitroxides can react with reducing species in the body, they easily lose contrast ability. Therefore fast acquisition sequences and high field magnets are required (Hyodo *et al.*, 2008). The major advantages of using nitroxides in MRI as opposed to EPRI include the availability of high resolution MRI scanners for both human and small animal studies, multislice imaging capability, enhanced spatial and temporal resolution, and coregistration of images of tissue redox status with anatomical/functional (diffusion weighted magnetic resonance imaging (DW-MRI), blood flow, etc.) information that is available from MRI. The image enhancement in MRI is dependent on the original T_1 relaxation time of the tissue examined and its variation by free radicals, which is related to the accessibility of the radical to water protons. Therefore, image enhancement in tissue with a given T_1 should depend on the nitroxide concentration (Hyodo *et al.*, 2008).

Even though T_1-weighted MRI serves as an indirect detection modality of nitroxide CAs (contrarily to EPRI), the T_1-weighted spoiled GRE based dynamic MRI can give appropriate tumor redox status information with useful anatomic resolution. Functional nitroxide redox MRI could be used in the clinical setting to monitor redox changes in tumor and normal tissue in patients undergoing radiotherapy and other types of cancer treatment (Hyodo *et al.*, 2008).

11.9 Imaging tumor response

There is a crucial need to use non-invasive imaging to facilitate the evaluation of the responsiveness of experimental tumors in preclinical therapeutic studies. Because molecular and cellular changes precede macroscopic changes in tumor size, it would be ideal to have an assay that could quantify these changes in both clinical cancer therapeutics and preclinical drug trials. Early indicators of treatment response that could also provide information on the spatial heterogeneity of response would be of significant benefit for both experimental and clinical trials.

11.9.1 Imaging cell death

As the cell undergoes apoptosis, there are a number of potential steps in the process that could be imaged: increase in cellularity with DW-MRI, cytoplasmic lipid droplets associated with changes in the lipid structure during programmed cell death using ^{1}H-MRS, or changes in the cell membrane (such as externalization of phosphatidylserine (PS)) with targeted CAs or radiolabeled annexin V.

11.9.1.1 Radiolabeled annexin V

Radiolabeled annexin V, a human protein commonly with fluorescent markers for in vitro detection of apoptotic cells, can be used to image apoptosis *in vivo*. The method relies on its high affinity for the constitutive membrane aminophospholipid, PS. PS is normally restricted to the inner leaflet of the cell membrane. With the onset of apoptosis, PS is rapidly externalized on the cell surface, resulting in a 100 times increase of the number of annexin V binding sites during apoptosis (Blankenberg, 2003). This phenomenon is prior to DNA degradation in apoptotic cells, and follows caspase 3 activation. Annexin V is therefore a sensitive marker of the early to intermediate phases of apoptosis (called execution phase) (Blankenberg, 2003). Radiolabeled Annexin V will localize at sites of membrane bound PS expression *in vivo* following intravenous administration. Tait *et al.* (2008) described techniques for labeling annexin with ^{125}I, ^{123}I and ^{99m}Tc. Blankenberg *et al.* (2006) showed the first images with ^{99m}Tc-HYNIC-AnxV in anti-Fas treated mice. Subsequently,

Mandl *et al.* (2004) have demonstrated the feasibility of imaging apoptosis in tumors after chemotherapy using ^{99m}Tc-labeled AnxV (2004). The first human ^{99m}Tc-AnxV SPECT study was performed just prior to chemotherapy (day 1 or 2) and within 3 days after the first dose. However, the ^{99m}Tc radiotracer has some disadvantages and several groups have tried to overcome them. Fluorine-18 labeled Annexin V (18F-AnxV), a PET tracer, was designed as PET has higher resolution and is more quantitative than SPECT. Mice with cycloheximidine-induced hepatic apoptosis were successfully imaged with μPET 4 hours after ^{18}F-AnxV injection. The fast clearance of 18F-AnxV into the urine and the short half-life of ^{18}F make this tracer promising for early assessment of tumor response in cancer patients. Other attempts to label AnxV with a PET-isotope were made with iodine-124. μPET images showed specific ^{124}I-AnxV uptake in the liver of anti-Fas treated mice. Finally, because of the low tumor to background activity ratio the tested radiolabeled derivatives of AnxV are not optimal for response assessment of tumors. Therefore, the development of conjugates with potentially better biodistribution profile was initiated (De Saint-Hubert *et al.*, 2009).

11.9.1.2 MRI with targeted contrast agents

MR CAs, such as iron nanoparticulates (superparamagnetic iron oxide (SPIO) particles) can be conjugated to peptides that bind to characteristic changes in apoptotic cells. SPIOs act as small permanent magnets that diphase proton spins in the local environment thereby causing a local loss of MR signal in target tissues ('negative contrast') (Blankenberg, 2003). One such peptide recognizes the C2 domain of synaptotagmin I (a binding domain for anionic phospholipids such as PS). As stated earlier, PS are selectively expressed on the surface of apoptotic cells just after activation of the caspase cascade, and prior to morphologic changes. Iron-labeled synaptotagmin I (C2-SPIO) is therefore of interest for the MR detection of cell death. The method is probably not only specific to apoptosis since PS also becomes available to the CA during necrosis, where cell membranes are disrupted. The disadvantage of the method is the large quantities of material needed to see a significant decrease in the signal of apoptotic tissues (Blankenberg, 2003).

In a comparable way, cross-linked iron oxide (CLIO) nanoparticles were attached to Annexin V and tested in vitro (Schellenberger *et al.*, 2002). In the case of the C2-labeled SPIO particles, T_2-weighted MRI images of tumor-bearing C57/B16 mice treated with cyclophosphamide/ etoposide showed a significant and progressive decrease in tumor signal intensity compared with control animals. Similarly, T_2-weighted imaging sequences of camptothecin-treated Jurkat T cells (i.e., 65% apoptotic cells) incubated with increasing concentrations of Annexin V-CLIO particles indicated a stronger, dose-dependent decrease in signal intensity versus untreated control cells. Thus both described CAs seem to offer considerable potential for successful detection of apoptosis by MRI (Lahorte *et al.*, 2004).

Smaller PS-recognizing molecules have recently been suggested as alternatives to anxA5 (Laumonier *et al.*, 2006). *In vivo* imaging of cell death by MRI has been achieved by PS recognizing moieties coupled to paramagnetic gadolinium-based CAs

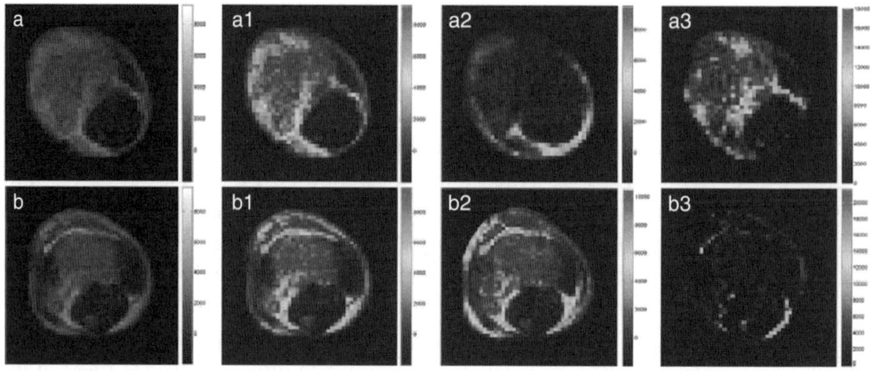

Figure 11.5 Axial slices of irradiated tumors obtained by T_2-weighted MRI. The first row of images represents the irradited tumor of a mouse that has been injected with USPIO-PEG750-E3 (a) and the second row of a mouse where USPIO-PEG750-E3Scramble was injected (b). The black and white images show anatomical images of irradiated tumors (a, b). Aside are the precontrast (A1, B1) and the the 3 hours post-contract images (A2, B2). Subtraction images (A3, B3) were obtained by subtracting the corresponding postcontrast images from the precontrast images. (Adapted from Radermacher *et al.* (in press).) A full color version of this figure can be found in the color plate section.

(Krishnan *et al.*, 2008) or to superparamagnetic iron oxide particles (Zhao *et al.*, 2001). Even though gadolinium complexes have potentially higher tumor accessibility due to their smaller size when compared with USPIO particles, USPIO particles have a higher relaxivity and they also avoid potential long-term toxic effects that can occur when using gadolinium derivatives (Geraldesa and Laurent, 2009). Recently, the PS-targeted hexapeptide, E3 (Laumonier *et al.*, 2006), was coupled to pegylated ultrasmall iron oxide (USPIO) nanoparticles that can be detected by multimodal imaging: MRI and by EPR. This marker was validated in experimental tumors with radiation induced high levels of apoptosis (Radermacher *et al.*, in press) (Figure 11.5).

11.9.1.3 ^{1}H-MRS

^{1}H MRS has the capacity to serially track a range of small molecules including cytoplasmic lipid droplets associated with changes in the lipid structure and fluidity of cellular membranes during the course of programmed cell death. Indeed, cells undergoing apoptosis have an associated increase in cytoplasmic neutral mobile lipid droplets composed of polyunsaturated fatty acids, cholesterol esters, and triglycerides (Blankenberg, 2003). It is therefore possible to directly quantify the fraction of apoptotic cells with water suppressed lipid ^{1}H MRS techniques, using existing research or clinical MR software and scanners. The advantage of ^{1}H MR spectroscopy, at least in the brain, is its ability to serially monitor lipid signals based on the innate characteristics of the tissue, without requiring localization of a CA. In tissues outside the brain, however, water suppressed lipid ^{1}H MR spectroscopy is challenging due to physiologic motion of lipids and non-specific signal from lipids from adipose tissue.

11.9.1.4 DW-MRI

DW imaging has become a major contrast for tissue assessment by MRI (Bammer, 2003; Norris, 2001). Demonstration of the remarkable sensitivity of water diffusion to ischemia in the brain initiated a new field within MRI, which prompted clinical interest in this novel MRI contrast. The potentials of diffusion MRI and MRS for oncology emerged more recently and are still under development.

Diffusion MRI pulse sequences incorporate two additional magnetic field gradients that make the intensity of the MR signal dependent on the mobility of the signal source, that is, water molecules (Bammer, 2003). The first of these two gradient pulses imparts a phase shift to each water molecule proportional to its initial location. The second gradient pulse will remove this phase shift if the water molecule remains at its original location. Any molecular movement between first and second pulses will lead to incomplete rephasing and signal attenuation. The amount of signal loss is a direct reflection of water mobility, that is, the greater signal loss implies greater molecular mobility. Practically, a series of DW images at different b-values (gradient values) can be used to calculate an apparent diffusion coefficient (ADC) for water by fitting the signal decay.

Enough data are in hand to indicate that ADC values are dominated by the presence of diffusion barriers (restriction effects). The physical diffusion coefficients of intra- and extracellular water are not known with certainty. Nevertheless, because water is 'trapped' (on the NMR timescale) inside of the cells, the apparent diffusion of water within the cells is lower than that in the extracellular space. Hence, the measured ADC is sensitive to cellularity.

Treatment of tumors may result in damage and/or killing of cells, thus altering the integrity of cell membranes or size of cells, thereby increasing the fractional volume of the interstitial space due. These changes have been shown to increase the diffusion of water in the damaged tumor tissue (Kauppinen, 2002; Moffat *et al.*, 2004; Ross *et al.*, 2003). Successful anticancer therapies are correlated to early increases in tumor ADC in both animals and humans (Jordan and Gillies, 2008). This is likely a consequence of reductions in cell volume, which are a general response to effective chemotherapy. Such predictions have been borne out by diffusion measurements in perfused cells and in *in vivo* systems wherein cell volume was accurately correlated with ADC. These findings have placed diffusion MRI in a new perspective for clinical oncology.

11.9.2 Imaging tumor metabolism

11.9.2.1 MRS(I)

MRS is a non-invasive technique for measuring biochemicals in tissue. It uses the same general principles and equipment as MRI. However, while MRI builds images using signals from ^{1}H nuclei in tissue water, present at high concentrations, MRS is used to measure signals from magnetic nuclei of tissue metabolites such as choline, creatine, and lactate that are present at much lower concentrations.

A number of NMR-active isotopes are available for interrogation *in vivo*. The most common nuclei used are ^{31}P and ^{1}H, which generate spectra that are dominated by resonances from endogenous metabolites (Gillies and Morse, 2005). Most advanced MRS techniques have been developed and applied first in the brain since it is possible to adjust the magnetic field to a high degree of homogeneity and the movement is also minimal compared to other tissues. In terms of cancer, however, there are compelling reasons to apply this approach to other tumor types because 95% of adult cancers are found outside the brain (Gillies and Morse, 2005). Beyond, Pi for pHi determination, the most relevant endogenous metabolites with regard to oncology are: choline (tCho), lactate (Lac), lipids; as well as *N*-acetyl aspartate (NAA) in brain tumors.

The interest in choline (tCho peak, detectable with both^{1}H and ^{31}P MRS) primarily comes from its utility as a biomarker for cancer staging and diagnosis. This was first documented by Negendank (1992) who provided strong evidence for elevated phosphomonoesters in lymphomas and head and neck cancers. Elevated choline (reflecting increased membrane synthesis and a higher cell turnover) has since been confirmed for breast, prostate, colon, cervical, and brain cancers and metastases, among others. Beyond diagnosis, choline has been suggested as a biomarker for response to anticancer therapies. Changes in choline are associated with positive responses in cancers in animal models and patients (Gillies and Morse, 2005).

The role of lactate in cancer prognosis and monitoring is also important. Lactate is elevated in late-stage brain cancers (Sijens *et al*., 1996) and is elevated in some intermediate-grade non-enhancing gliomas (Li *et al*., 2005), as well as in head and neck cancer metastases. With regard to therapy follow-up, lactate appears to increase early following induction of interferon/tumor necrosis factor-induced apoptosis in colon cancer cells (Lutz, Tome, and Cozzone, 2003). Lactate has also been reported to decrease *in vivo* with some chemotherapies, such as the antimetabolite, 5-fluorouracil (Aboagye *et al*., 1998), as well as with radiotherapy (Bhujwalla and Glickson, 1996).

The most advanced clinical applications in the field of MRS(I) include brain and prostate tumors. Indeed, MRSI studies have demonstrated dramatic spectral differences between normal brain tissue (low choline and high NAA) and prostate (low choline and high citrate) compared to brain (low NAA, high choline) and prostate (low citrate, high choline) tumors. MRSI seems to be able to discriminate necrosis (absence of all metabolites, except lipids, and lactate) from viable normal tissue and cancer following therapy. The results of current MRSI studies also provide evidence that the magnitude of metabolic changes in regions of cancer before therapy as well as the magnitude and time course of metabolic changes after therapy can improve our understanding of cancer aggressiveness and mechanisms of therapeutic response. Clinically, combined MRI/MRSI has already demonstrated the potential for improved diagnosis, staging, and treatment planning of brain and prostate cancer (Kurhanewicz, Vigneron, and Nelson, 2000).

11.9.2.2 ^{18}FDG-PET

PET is a non-invasive molecular imaging technique, which is able to detect functional changes. The most common application of PET in oncology is the measurement of glucose metabolism using 2′-deoxy-2′-[^{18}F]fluoro-D-glucose (FDG) (Shields, 2006). This is based on the observation by Warburg, Wind, and Negalein (1927) that glucose metabolism is increased in many tumors. To date, FDG-PET is widely used in staging, as the method has unsurpassed sensitivity for detecting occult metastases. More recently, evidence is emerging that the method can also be very useful in monitoring response to treatment (Shields, 2006). It is routinely used in the staging of lung cancer at many centers, and has found regular use in the assessment of colon, breast, head and neck, esophageal, melanoma, and lymphoma (Valk *et al.*, 2003).

A limitation of FDG, however, is the fact that enhanced glucose metabolism is not specific for tumors alone. Indeed, enhanced FDG uptake has also been observed in muscle, brain, and inflammatory tissue (Direcks *et al.*, 2008). At present, FDG is the only PET agent approved for routine clinical use in oncology, while, experimentally, tumor biosynthesis can be measured with a variety of tracers. Protein synthesis has been widely studied, especially with ^{11}C-methionine, which also measures transmethylation reactions (Shields, 2006). A number of amino acids and their analogs have been developed for use of PET, such as O-(2-^{18}F-fluoroethyl)-L-tyrosine, although none has yet found regular clinical use (Shields, 2006). Membrane biosynthesis has been examined with ^{11}C-acetate, ^{11}C-choline, and ^{18}F-fluorocholine [27Y30]. Thymidine and its analogs have been extensively explored as agents to image DNA synthesis and cell proliferation (see below).

11.9.3 Imaging tumor cell proliferation

11.9.3.1 ^{18}FLT-PET

Thymine is the only nucleotide that is exclusively incorporated into DNA and not RNA, making it and its nucleoside thymidine appropriate for studying DNA metabolism. This understanding led to the synthesis of thymidine labeled with ^{11}C by Christman *et al.* (1972). Thymidine analogs also were used in the treatment of cancer, after the initial (FLT), and 18F-1-(20-deoxy-20-fluoro-beta-D-arabinofuranosyl)-thymine (FMAU) (Shields, 2006). Thymidine and FLT are trapped in cells as part of the DNA synthetic pathway. Labeled thymidine's advantage is that it is the native compound and is readily taken up by cells, phosphorylated by thymidine kinase 1 (TK), and incorporated into DNA (Shields, 2006). Pilot studies in patients have demonstrated that it can be used to produce images of tumors and response to therapy (Shields, 2006). Unfortunately, labeled thymidine is not a practical tracer for routine use in clinical studies because it has a complex synthesis and can only be labeled with ^{11}C, which makes widespread distribution impossible and is relatively unstable. The placement development of 5-fluorouracil (5-FU) by Heidelberger *et al.* (1957), as they can interfere with DNA synthesis, 5-FU and, subsequently, 5-fluorodeoxyuridine were labeled with ^{18}F for PET imaging. This allows for the *in vivo* pharmacokinetics and tumor uptake of both these agents. This has led to the search for other pyrimidines that can be labeled for PET and provide images of proliferation. At this point, among the most extensively studied agents are ^{11}C-thymidine and FLT (^{18}F-30-deoxy-30-fluorothymidine). FLT is phosphorylated by the cytosolic cell cycle dependent enzyme TK-1, the activity of which is upregulated by a factor of 10 in the S phase. Thymidine is degraded by the enzyme thymidine phosphorylase (TP), but FLT is resistant to catabolism by this enzyme. Therefore, FLT is considered to be a proliferation marker and, as such, could be a more specific tracer of tumor 'activity' (i.e., cell proliferation) (Direcks *et al.*, 2008). Initial imaging studies with FLT demonstrated that it can produce high-contrast images of tumor, such as lung cancer (Shields *et al.*, 1998). In addition, physiologic uptake is observed in the bone marrow, a highly proliferative organ, and in the liver, kidneys, and bladder as a result of glucuronidation and clearance. FLT retention through phosphorylation by TK is very analogous to FDG retention via phosphorylation by hexokinase. Overall, FLT appears to produce images with less contrast than FDG, except in the brain. Nevertheless, FLT is most likely to find use in measuring treatment response (Shields, 2006).

11.10 Optimizing therapeutic intervention using molecular imaging

The study of MR markers over the past decade has provided evidence that the tumor microenvironment and hemodynamics play a major role in determining therapy outcome. Therefore, the identification of relevant non-invasive imaging endpoints is of crucial importance in the management of cancer patients in order to improve the therapeutic index. This is evidenced by numerous preclinical studies involving multimodal imaging, some of which are illustrated in the following section.

For the last then years, efforts have been concentrated on tumor radiosensitization and chemosensitization by transiently increasing tumor oxygenation and/or blood flow, respectively. The addition of such cotreatments and the monitoring of their effects using multi-modal imaging for timely administration in combination with radiation therapy or chemotherapy, allowed to improve the therapeutic outcome of experimental tumors. The co-reatments included vasodilators, antiantiogenic agents in their normalization phase, inhibitors of oxygen consumption, and modulators of the nitric oxide pathway. By computing data acquired on 10 different cotreatments in experimental tumors, we were able to determine which MR markers are robust in terms of predictive value of therapy outcome. Imaging endpoints under study were the following: tumor pO_2 using EPR oximetry and BOLD MRI, tumor blood flow and permeability using DCE-MRI, and tumor oxygen consumption using *ex vivo* EPR oximetry. pO_2 assessed by EPR oximetry appeared to be the most relevant marker to evaluate the potential use as a cotreatment for radiation therapy. DCE-MRI, as well as BOLD MRI were successful with regard to chemosensitization but, importantly, they failed to be predictive for the same purpose when a consumption effect was involved. Our work illustrates the usefulness of combining different MR markers in order to increase the predictive value of each technique, and thereby increase the therapeutic index on an individual basis.

From our analysis, we can reasonably suggest that: (i) the determinant factor of tumor response to radiation therapy is tumor pO_2, with blood flow being only one minor contributor besides a more important and less studied contribution of oxygen consumption; (ii) the determinant factor of tumor response to chemotherapy is vascular changes (blood flow changes), with pO_2 being less predictive in terms of response. Therefore, the assessment of absolute values of tumor pO_2 *in vivo* is of critical importance for radiation therapy planning, monitoring, and identification of resistant zones using imaging techniques (oxygen mapping). The pO_2 biomarker can be evaluated quantitatively *in vivo* using EPR oximetry (spectroscopy and imaging), ^{19}F-MRI, and DNP oximetry. The assessment of relative changes in blood flow is determinant for tumor response to chemotherapy. Those changes can be assessed *in vivo* using DCE-MRI, or 'non MR' methods, including CT with iodinated CA, and nuclear medicine (PET/SPECT), as illustrated above.

With regard to previously observed data on the predictivity potential of EPR spectroscopy, Pogue *et al.* (2002) evaluated the effect of PDT with verteporfin on radiation-induced fibrosarcoma. They observed an increase in pO_2 up to 15.2 mmHg, which was correlated to the growth delay assay. Also, Hou *et al.* (2007) established a correlation between the effect of the allosteric modifier of hemoglobin efaproxiral, on tumor oxygenation and response to 4 Gy radiation therapy. Finally, Elas *et al.* (2008) also found data arguing in favor of EPR oximetry as being a good predictor of tumor cure after radiation therapy in FSa tumors under both normal (air breathing) and clamped tumor conditions.

The BOLD signal was also assessed by different groups in the preclinical setting in order to establish correlations with the therapeutic outcome. The effect of (2-[4-[[(3,5-dimethylanilino)carbonyl]methyl]phenoxyl]-2-methylpropionic acid) (RSR13), an allosteric modifier of hemoglobin, was studied on NCI-H460 xenograft tumor response using BOLD-MRI and regrowth delay assays (Amorino *et al.*, 2001). RSR13 increased the signal intensity ratio in a dose-dependent manner that was correlated with an enhancement of radiation-induced growth delay of 2.8 while RSR13 was administered 30 minutes before a 10 Gy dose of radiation. Robinson's group further tested the prognostic potential of tumor R_2* with respect to radiotherapeutic outcome on G3H prolactinomas and radiation-induced fibrosarcoma (RIF)-1 with animals breathing either air or carbogen during radiation (Rodrigues *et al.*, 2004). When the animals breathed carbogen during radiation, the growth delay was enhanced in the G3H prolactinomas, which also exhibited a large ΔR2* in response to carbogen. In contrast, the effect of 15 Gy on the RIF-1 fibrosarcomas, which gives a negligible ΔR2* in response to carbogen, showed a much smaller growth inhibition.

DCE-MRI, using gadopentetate dimeglumine is the marker that is probably the best characterized and the more translated to the clinic. Indeed, DCE-MRI provides a powerful tool for the rapid evaluation of the acute pharmacodynamic effect of a variety of treatments such as chemotherapy, hormonal manipulation, radiotherapy, and novel therapeutic approaches (antivascular agents, immunotherapy), most notably in the case of mechanisms that affect tumor perfusion. Evidence is mounting that DCE-MRI measurements correlate with immunohistochemical surrogates of tumor angiogenesis. Many excellent articles reviewing preclinical studies have been published (Knopp *et al.*, 2001; Padhani, 2002).

We describe here some relevant studies with regard to therapy outcome. DCE-MRI with Gd-DTPA was used to monitor acute effects on tumor vascular permeability following inhibition of vascular endothelial growth factor-A (VEGF-A) signal transduction in phosphocholine-3 human prostate adenocarcinoma xenografts (Checkley *et al.*, 2003). Dose-related reductions in K_{trans} were evident following acute ZD6474 treatment and a correlation between this dose response and the growth inhibitory effect of ZD6474 following chronic treatment was also observed. This is quite different from what was observed in our studies using the same drug; however, the timing and tumor models are also different (Ansiaux *et al.*, 2009). Similarly, DCE MRI parameter maps of Dunning rat prostate cancers showed increased vascularization, k_{ep} and microvessel density, that were correlated with the slowing of tumor growth after irradiation (Kiessling *et al.*, 2004). The antiangiogenic effect of SU6668 was assessed in HT29 human colon carcinoma in mice using DCE-MRI with large molecular CA. SU6668 inhibited tumor growth, with 60% inhibition at 14 days of treatment, which was assessed early with DCE-MRI in terms of vascular permeability and fractional plasma volume at 24 hours, and 3, 7, and 14 day time points (Marzola *et al.*, 2004). DCE-MRI with macromolecular contrast media has further been used in HT-29 experimental tumors in order to assess the efficacy of an hypoxia inducible factor (HIF)-1α inhibitor and correlate with radiation therapy outcome. A dramatic reduction in tumor blood vessel permeability was observed 2 hours after treatment, and this was correlated with an enhancement in radiation response. Although the diffusion marker used in this study was even 'earlier,' the extent of response of the DCE-marker was more pronounced (Jordan *et al.*, 2005). More recently, Dewhirst's group also showed that DCE-MRI parameters (assessed with classic CA) were predictive of thermoradiotherapy outcome in dogs with soft tissue sarcoma (Viglianti *et al.*, 2009).

On a clinical scale, Mayr *et al.* had earlier suggested in 1996 that MR perfusion (DCE-MRI) imaging could be able to predict outcome in patients with advanced cervical cancer. They suggested that tumors with a higher tissue perfusion in the pretherapy study had a lower incidence of local recurrence (Mayr *et al.*, 1996, 1998). They further demonstrated that the quantity of low enhancement regions significantly predicted subsequent tumor recurrence (Mayr *et al.*, 2000). Yamashita *et al.* (2000) showed that radiation therapy was more effective in tumors with higher tissue permeability than in those with lower tissue permeability. Perfusion indices were also found to be of predictive value for therapy outcome of preoperative therapy in patients with primary rectal carcinoma (DeVries *et al.*, 2003). Similarly, George *et al.* (2001) found that responsive tumors had higher pretreatment permeability values than non-responsive tumors. Prediction of radiotherapy outcome of carcinoma of the cervix was further evaluated on 50 patients (Loncaster *et al.*, 2002): dynamic data correlated with patient outcome and patients with poorly enhancing tumors had significantly worse disease-specific survival. The predictive value of DCE-MRI in terms of chemosensitivity in breast cancer patients was evaluated by Nagashima *et al.* (2006). They demonstrated the significant correlation between pretreatment MRI data and tumor reduction by chemotherapy in breast cancer patients and suggested the possibility of defining good and non-responders prior to treatment. On the other hand, Heerschap *et al.* (2007) recently investigated the predictive value of DCE-MRI with Gd-DTPA for tumor response to first-line chemotherapy in patients with liver metastases of colorectal cancer. None of the kinetic parameters (k_{ep}, K^{trans}, V_e) was able to predict tumor response after 2 months, suggesting that the delivery of chemotherapy by tumor vasculature is not a major factor determining response in first-line treatment in liver metastasis.

Numerous works illustrate the ability of tumor ADCw to detect early changes after treatments such as chemotherapy, radiation therapy, PDT, and gene therapy in experimental tumors (reviewed in Kauppinen, 2002). A key common finding in the follow-up of different therapies with DW imaging is that tumor volume as determined from T_1 or T_2-weighted or CA-enhanced MRI starts to reduce upon cytotoxic drug treatment only days after diffusion MRI shows positive treatment response. Hence, diffusion MRI has high potential as an early and quantitative biomarker of response in experimental oncology where novel therapies for solid malignant tumors are being investigated. Gillies' group investigated the combination chemotherapy response of human breast cancer tumor xenografts sensitive or resistant to paclitaxel, an inhibitor of microtubule depolymerization, by monitoring changes in the ADC (Galons *et al.*, 1999) results indicated a substantial, and early increase in the ADC after successful therapy in drug-sensitive tumors, while no change could be observed in the ADC in *p*-glycoprotein-positive resistant tumors. Similarly, paclitaxel arrested growth in all MCF7/S tumors and had no effect on the tumors from MCF7/D40 cells. These changes in the ADC were apparent 2 days after commencement of chemotherapy, with no significant changes in tumor volumes. They further characterized the utility of DW-MRI to predict the response of prostate cancer xenografts to docetaxel, a drug that belongs to the taxoid family, in the preclinical setting (Jennings *et al.*, 2002). Subsequently, the efficacy of inhibition of the HIF-1α pathway using DW-MRI on HT-29 (a tumorigenic non-metastatic human colon carcinoma cell line) tumor-bearing severe combined immunodeficient mice (Jordan *et al.*, 2005) was evaluated. A substantial increase in mean relative tumor ADC and a right shift in ADC histograms of tumors were observed for the treated groups at 24 and 36 hours post-treatment (the increase in tumor ADC was correlated with a decrease in vessel permeability consistent with cell death and modification of the intracellular to extracellular water populations ratio) (Figure 11.6).

Largely because of the strong preclinical findings, DW-MRI monitoring of therapy is being investigated in a clinical setting (reviewed in Jordan and Gillies, 2008). Gillies' group tested the hypothesis that changes in water mobility will quantitatively presage tumor responses in patients with metastatic liver lesions from breast cancer (Theilmann *et al.*, 2004). A total of 13 patients with metastatic breast cancer and 60 measurable liver lesions were monitored by diffusion MRI after initiation of new courses of chemotherapy (taxane, vinorelbine, capecitabine, paclitaxel, and trastuzumab). The data indicate that diffusion MRI can predict response by 4 or 11

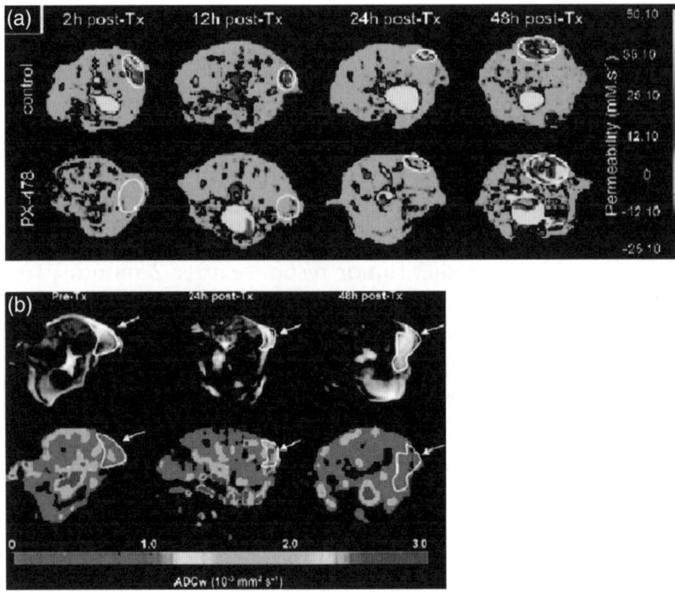

Figure 11.6 (a) Permeability maps of tumors at 2, 12, 24, and 48 hours following injection of vehicle (control) or PX-478 (drug) injection. Each image represents an axial slice of the mouse with the tumor area encircled. A substantial decrease in tumor permeability was observed as early as 2 hours following treatment and continuing until 24 hours in comparison to controls. (b) DW images at a b-value (top row) and corresponding diffusion maps (bottom row) of an HT-29 tumor-bearing mouse at 0, 24, and 48 hours following PX-478 injection. Decrease in tumor cellularity was noted at 24 and 36 hours following treatment as indicated with an increase in ADCw values. Each image represents an axial slice of the mouse with the tumor area encircled and indicated by an arrow. (Adapted from Jordan *et al.* (2005).) A full color version of this figure can be found in the color plate section.

days after commencement of therapy, depending on the analytic method, suggesting that diffusion MRI can be useful in predicting the response of liver metastases to effective chemotherapy.

Recent studies addressed tumor IFP as an early marker for anticancer therapeutics in experimental tumors (Ferretti *et al.*, 2009). Among the 13 experimental tumor models under study, most chemotherapeutics sooner (2 or 3 days) or later (6 or 7 days) lowered tumor IFP significantly, and the cytotoxic patupilone caused the greatest decrease in IFP. In rat mammary orthotopic BN472 tumors, significant drug-induced decreases in IFP and relative tumor blood volume correlated positively with each other for both patupilone and the cytostatic vatalanib. In the two orthotopic models studied, early decreases in IFP were significantly correlated with late changes in tumor volume. Thus, drug-induced decreases in tumor IFP are an early marker of response to therapy, which could aid clinical development.

11.11 Conclusions

The impact of non-invasive imaging in oncology extends from guiding preclinical development of targeted biomarkers and therapeutic agents, to assisting in the diagnosis and staging of tumors in the clinic, as well as the monitoring of therapeutic response. Even though most of the non-invasive imaging techniques may still be under development, each modality allowed to improve *in vivo* characterization of physiologic and molecular tumor properties. Multimodal non invasive imaging will allow treatment individualization for optimal therapeutic response.

References

Aboagye, E.O., Bhujwalla, Z.M., Shungu, D.C., and Glickson, J.D. (1998) Detection of tumor response to chemotherapy by 1H nuclear magnetic resonance spectroscopy: effect of 5-fluorouracil on lactate levels in radiation-induced fibrosarcoma 1 tumors. *Cancer Research*, **58**, 1063–1067.

Abramovitch, R., Dafni, H., Smouha, E. *et al.* (1999) In vivo prediction of vascular susceptibility to vascular susceptibility endothelial growth factor withdrawal: magnetic resonance imaging of C6 rat glioma in nude mice. *Cancer Research*, **59**, 5012–5016.

Amorino, G.P., Lee, H., Holburn, G.E. *et al.* (2001) Enhancement of tumor oxygenation and radiation response by the allosteric effector of hemoglobin, RSR13. *Radiation Research*, **156**, 294–300.

Ansiaux, R., Baudelet, C., Jordan, B.F. *et al.* (2006) Mechanism of reoxygenation after antiangiogenic therapy using SU5416 and its importance for guiding combined antitumor therapy. *Cancer Research*, **66**, 9698–9704.

Ansiaux, R., Dewever, J., Gregoire, V. *et al.* (2009) Decrease in tumor cell oxygen consumption after treatment by vandetanib (ZD6474) and its effects on response to radiotherapy. *Radiation Research*, **172**, 584–591.

Bammer, R. (2003) Basic principles of diffusion-weighted imaging. *European Journal of Radiology*, **45**, 169–184.

Baudelet, C. and Gallez, B. (2002) How does blood oxygen level-dependent (BOLD) contrast correlate with oxygen partial pressure (pO_2) inside tumors? *Magnetic Resonance in Medicine*, **48**, 980–986.

Baudelet, C., Ansiaux, R., Jordan, B.F. *et al.* (2004) Physiological noise in murine solid tumours using T2*-weighted gradient-echo imaging: a marker of tumour acute hypoxia? *Physics in Medicine and Biology*, **49**, 3389–3411.

Baudelet, C., Cron, G.O., and Gallez, B. (2006) Determination of the maturity and functionality of tumor vasculature by MRI: correlation between BOLD-MRI and DCE-MRI using P792 in experimental fibrosarcoma tumors. *Magnetic Resonance in Medicine*, **56**, 1041–1049.

Bentzen, L., Vestergaard-Poulsen, P., Nielsen, T. *et al.* (2005) Intravascular contrast agent-enhanced MRI measuring contrast clearance and tumor blood volume and the effects of vascular modifiers in an experimental tumor. *International Journal of Radiation Oncology, Biology, Physics*, **61**, 1208–1215.

Berliner, J.L. and Fujii, H. (1985) Magnetic resonance imaging of biological specimens by electron paramagnetic resonance of nitroxide spin labels. *Science*, **227**, 517–519.

Blankenberg, F.G. (2003) Molecular imaging: the latest generation of contrast agents and tissue characterization techniques. *Journal of Cellular Biochemistry*, **90**, 443–453.

Bobko, A.A., Dhimitruka, I., Zweier, J.L., and Khramtsov, V.V. (2007) Trityl radicals as persistent dual function pH and oxygen probes for in vivo electron paramagnetic resonance spectroscopy and imaging: concept and experiment. *Journal of the American Chemical Society*, **129**, 7240–7241.

Brasch, R.C. (1983) Work in progress: methods of contrast enhancement for NMR imaging and potential applications. A subject review. *Radiology*, **147**, 781–788.

Brasch, R.C., Li, K.C., Husband, J.E. *et al.* (2000) In vivo monitoring of tumor angiogenesis with MR imaging. *Academic Radiology*, **7**, 812–823.

Bhujwalla, Z.M. and Glickson, J.D. (1996) Detection of tumor response to radiation therapy by in vivo proton MR spectroscopy. *International Journal of Radiation Oncology, Biology, Physics*, **36**, 635–639.

Checkley, D., Tessier, J.J., Kendrew, J. *et al.* (2003) Use of dynamic contrast-enhanced MRI to evaluate acute treatment with ZD6474, a VEGF signalling inhibitor, in PC-3 prostate tumours. *British Journal of Cancer*, **89**, 1889–1895.

Cho, Z.H., Jones, J.P., and Singh, M. (1993) *Foundations of Medical Imaging*, John Wiley & Sons, Inc., NewYork.

Choyke, P.L., Dwyer, A.J., and Knopp, M.V. (2003) Functional tumor imaging with dynamic contrast-enhanced magnetic resonance imaging. *Journal of Magnetic Resonance Imaging*, **17**, 509–520.

Christman, D., Crawford, E.J., Friedkin, M., and Wolf, A.P. (1972) Detection of DNA synthesis in intact organisms with positron-emitting [methyl- 11C]thymidine. *Proceedings of the National Academy of Sciences of the United States of America*, **69**, 4.

Crokart, N., Jordan, B.F., Baudelet, C. *et al.* (2007) Glucocorticoids modulate tumor radiation response through a decrease in tumor oxygen consumption. *Clinical Cancer Research*, **13**, 630–635.

Crokart, N., Radermacher, K., Jordan, B.F. *et al.* (2005) Tumor radiosensitization by anti-inflammatory drugs: evidence for a new mechanism involving the oxygen effect. *Cancer Research*, **65**, 7911–7916.

Dadiani, M., Margalit, R., Sela, N., and Degani, H. (2004) High resolution magnetic resonance imaging of disparities in the transcapillary transfer rates in orthotopically inoculated invasive breast tumors. *Cancer Research*, **64**, 3155–3161.

Daldrup, H., Shames, D.M., Wendland, M. *et al.* (1998) Correlation of dynamic contrast-enhanced MR imaging with histologic tumor grade: comparison of macromolecular and small-molecular contrast media. *AJR American Journal of Roentgenology*, **171**, 941–949.

Dennie, J., Mandeville, J.B., Boxerman, J.L. *et al.* (1998) NMR imaging of changes in vascular morphology due to tumor angiogenesis. *Magnetic Resonance in Medicine*, **40**, 793–799.

De Saint-Hubert, M., Prinsen, K., Mortelmans, L. *et al.* (2009) Molecular imaging of cell death. *Methods*, **48**, 178–187.

DeVries, A.F., Kremser, C., Hein, P.A. *et al.* (2003) Tumor microcirculation and diffusion predict therapy outcome for primary rectal carcinoma. *International Journal of Radiation Oncology, Biology, Physics*, **56**, 958–965.

Diepart, C., Jordan, B.F., and Gallez, B. (2009) A new EPR oximetry protocol to estimate the tissue oxygen consumption in vivo. *Radiation Research*, **172**, 220–225.

Diepart, C., Verrax, J., Calderon, P.B. *et al.* (2010) Comparison of methods for measuring oxygen consumption in tumor cells in vitro. *Analytical Biochemistry*, **396**, 250–256.

Direcks, W.G., Berndsen, S.C., Proost, N. *et al.* (2008) [18F]FDG and [18F]FLT uptake in human breast cancer cells in relation to the effects of chemotherapy: an in vitro study. *British Journal of Cancer*, **99**, 481–487.

Elas, M., Bell, R., Hleihel, D. *et al.* (2008) Electron paramagnetic resonance oxygen image hypoxic fraction plus radiation dose strongly correlates with tumor cure in FSa fibrosarcomas. *International Journal of Radiation Oncology, Biology, Physics*, **71**, 542–549.

Enochs, W.S., Harsh, G., Hochberg, F. *et al.* (1999) Improved delineation of human brain tumors on MR images using a long-circulating, superparamagnetic iron oxide agent. *Journal of Magnetic Resonance Imaging*, **9**, 228–232.

Ferretti, S., Allegrini, P.R., Becquet, M.M., and McSheehy, P.M. (2009) Tumor interstitial fluid pressure as an early-response marker for anticancer therapeutics. *Neoplasia*, **11**, 874–881.

Ferretti, S., Allegrini, P.R., O'Reilly, T. *et al.* (2005) Patupilone induced vascular disruption in orthotopic rodent tumor models detected by magnetic resonance imaging and interstitial fluid pressure. *Clinical Cancer Research*, **11**, 7773–7784.

Frackowiak, R.S., Herold, S., Petty, R.K., and Morgan-Hughes, J.A. (1988) The cerebral metabolism of glucose and oxygen measured with positron tomography in patients with mitochondrial diseases. *Brain*, **111**, 1009–1024.

Frackowiak, R.S., Lenzi, G.L., Jones, T., and Heather, J.D. (1980) Quantitative measurement of regional cerebral blood flow and oxygen metabolism in man using 15O and positron emission tomography: theory, procedure, and normal values. *Journal of Computed Assisted Tomography*, **4**, 727–736.

Gallagher, F.A., Kettunen, M.I., Day, S.E. *et al.* (2008) Magnetic resonance imaging of pH in vivo using hyperpolarized 13C-labelled bicarbonate. *Nature*, **453**, 940–943.

Gallez, B., Bacic, G., Goda, F. *et al.* (1996) Use of nitroxides for assessing perfusion, oxygenation, and viability of tissues: in vivo EPR and MRI studies. *Magnetic Resonance in Medicine*, **35**, 97–106.

Gallez, B., Baudelet, C., and Jordan, B.F. (2004) Assessment of tumor oxygenation by electron paramagnetic resonance: principles and applications. *NMR in Biomedicine*, **17**, 240–262.

Gallez, B. and Swartz, H.M. (2004) In vivo EPR: when, how and why? *NMR in Biomedicine*, **17**, 223–225.

Galons, J.P., Altbach, M.I., Paine-Murrieta, C. *et al.* (1999) Early increases in breast tumor xenograft water mobility in response to paclitaxel therapy detected by non-invasive diffusion magnetic resonance imaging. *Neoplasia*, **1**, 113–117.

Garcia-Martin, M.L., Martinez, G.V., Raghunand, N. *et al.* (2006) High resolution pH(e) imaging of rat glioma using pH-dependent relaxivity. *Magnetic Resonance in Medicine*, **55**, 309–315.

George, M.L., Dzik-Jurasz, A.S., Padhani, A.R. *et al.* (2001) Non-invasive methods of assessing angiogenesis and their value in predicting response to treatment in colorectal cancer. *British Journal of Surgery*, **88**, 1628–1636.

Geraldesa, C.F.G.C. and Laurent, S. (2009) Classification and basic properties of contrast agents for magnetic resonance imaging. *Contrast Media and Molecular Imaging*, **4**, 1–23.

Gilead, A., Meir, G., and Neeman, M. (2004) The role of angiogenesis, vascular maturation, regression and stroma infiltration in dormancy and growth of implanted MLS ovarian carcinoma spheroids. *International Journal of Cancer*, **108**, 524–531.

Gillies, R.J., and Morse, D.L. (2005) In vivo magnetic resonance spectroscopy in cancer. *Annual Review of Biomedical Engineering*, **7**, 287–326.

Gillies, R.J., Raghunand, N., Garcia-Martin, M.L., and Gatenby, R.A. (2004) pH imaging. A review of pH measurement methods and applications in cancers. *EEE Engineering in Medicine and Biology Magazine*, **23**, 57–64.

Gross, D.J., Reibstein, I., Weiss, L. *et al.* (1999) The antiangiogenic agent linomide inhibits the growth rate of von Hippel–Lindau paraganglioma xenografts to mice. *Clinical Cancer Research*, **5**, 3669–3675.

Gu, Y., Bourke, V.A., Kim, J.G. *et al.* (2003) Dynamic response of breast tumor oxygenation to hyperoxic respiratory challenge monitored with three oxygen-sensitive parameters. *Applied Optics*, **42**, 2960–2967.

Gulliksrud, K., Brurberg, K.G., and Rofstad, E.K. (2009) Dynamic contrast-enhanced magnetic resonance imaging of tumor interstitial fluid pressure. *Radiotherapy and Oncology*, **91**, 107–113.

Gulsen, G., Yu, H., Wang, J. *et al.* (2002) Congruent MRI and near-infrared spectroscopy for functional and structural imaging of tumors. *Technology in Cancer Research Treatment*, **1**, 497–505.

Haider, M.A., Sitartchouk, I., Roberts, T.P.L. *et al.* (2007) Correlations between dynamic contrast-enhanced magnetic resonance imaging-derived measures of tumor microvasculature and interstitial fluid pressure in patients with cervical cancer. *Journal of Magnetic Resonance Imaging*, **25**, 153–159.

Hassid, Y., Furman-Haran, E., Margalit, R. *et al.* (2006) Noninvasive magnetic resonance imaging of transport and interstitial fluid pressure in ectopic human lung tumors. *Cancer Research*, **66**, 4159–4166.

He, X. and Yablonskiy, D.A. (2007) Quantitative BOLD: mapping of human cerebral deoxygenated blood volume and oxygen extraction fraction: default state. *Magnetic Resonance in Medicine*, **57**, 115–126.

He, X., Zhu, M., and Yablonskiy, D.A. (2008) Validation of oxygen extraction fraction measurement by qBOLD technique. *Magnetic Resonance in Medicine*, **60**, 882–888.

Heidelberger, C., Chaudhuri, N.K., Danneberg, P. *et al.* (1957) Fluorinated pyrimidines, a new class of tumor-inhibitory compounds. *Nature*, **179**, 663.

Heldin, C.H., Rubin, K., Pietras, K., and Ostman, A. (2004) High interstitial fluid pressure: an obstacle in cancer therapy. *Nature Reviews in Cancer*, **4**, 806–813.

Hou, H., Khan, N., Grinberg, O.Y. *et al.* (2007) The effects of Efaproxyn (efaproxiral) on subcutaneous RIF-1 tumor oxygenation and enhancement of radiotherapy-mediated inhibition of tumor growth in mice. *Radiation Research*, **168**, 218–225.

Howe, F.A., McPhail, L.D., Griffiths, J.R. *et al.* (2008) Vessel size index magnetic resonance imaging to monitor the effect of antivascular treatment in a rodent tumor model. *International Journal of Radiation Oncology, Biology, Physics*, **71**, 1470–1476.

Howe, F.A., Robinson, S.P., McIntyre, D.J. *et al.* (2001) Issues in flow and oxygenation dependent contrast (FLOOD) imaging of tumours. *NMR in Biomedicine*, **14**, 497–506.

Hunjan, S., Zhao, D., Constantinescu, A. *et al.* (2001) Tumor oximetry: demonstration of an enhanced dynamic mapping procedure using fluorine-19 echo planar magnetic resonance imaging in the Dunning prostate R3327-AT1 rat tumor. *International Journal of Radiation Oncology, Biology, Physics*, **49**, 1097–1108.

Hyodo, F., Soule, B.P., Matsumoto, K. *et al.* (2008) Assessment of tissue redox status using metabolic responsive contrast agents and magnetic resonance imaging. *Journal of Pharmaceutics and Pharmacology*, **60**, 1049–1060.

Ito, M., Lammertsma, A.A., Wise, R.J. *et al.* (1982) Measurement of regional cerebral blood flow and oxygen utilisation in patients with cerebral tumours using ^{15}O and positron emission tomography: analytical techniques and preliminary results. *Neuroradiology*, **23**, 63–74.

Jain, R.K. (1987) Transport of molecules in the tumor interstitium: a review. *Cancer Research*, **47**, 3039–3051.

Jennings, D., Hatton, B.N., Guo, J. *et al.* (2002) Early response of prostate carcinoma xenografts to docetaxel chemotherapy monitored with diffusion MRI. *Neoplasia*, **4**, 255–262.

Jennings, D., Raghunand, N., and Gillies, R.J. (2008) Imaging hemodynamics. *Cancer Metastasis Review*, **27**, 589–613.

Jordan, B.F., Christian, N., Crokart, N. *et al.* (2007) Thyroid status is a key modulator of tumor oxygenation: implication for radiation therapy. *Radiation Research*, **168**, 428–432.

Jordan, B.F., Crokart, N., Baudelet, C. *et al.* (2006) Complex relationship between changes in oxygenation status and changes in R*2: the case of insulin and NS-398, two inhibitors of oxygen consumption. *Magnetic Resonance in Medicine*, **56**, 637–643.

Jordan, B.F., Cron, G.O., and Gallez, B. (2009) Rapid monitoring of oxygenation by 19F magnetic resonance imaging: simultaneous comparison with fluorescence quenching. *Magnetic Resonance in Medicine*, **61**, 634–638.

Jordan, B.F. and Gillies, R.J. (2008) Post-treatment changes in tumor microenvironment: Dynamic ontrast-enhanced and diffusion-weighted magnetic resonance imaging, in *Cancer Imaging*, Instrumentations and Applications, vol. 2. (ed. M.A. Hayat), Academic Press, London, pp. 235–246.

Jordan, B.F., Grégoire, V., Demeure, R.J. *et al.* (2002) Insulin increases the sensitivity of tumors to irradiation: involvement of an increase in tumor oxygenation mediated by a nitric oxide-dependent decrease of the tumor cells oxygen consumption. *Cancer Research*, **62**, 3555–3561.

Jordan, B.F., Misson, P., Demeure, R. *et al.* (2000) Changes in tumor oxygenation/perfusion induced by the no donor, isosorbide dinitrate, in comparison with carbogen: monitoring by EPR and MRI. *International Journal of Radiation Oncology, Biology, Physics*, **48**, 565–570.

Jordan, B.F., Runquist, M., Raghunand, N. *et al.* (2005) Dynamic contrast-enhanced and diffusion MRI show rapid and dramatic changes in tumor microenvironment in response to inhibition of HIF-1alpha using PX-478. *Neoplasia*, **7**, 475–485.

Kauppinen, R.A. (2002) Monitoring cytotoxic tumour treatment response by diffusion magnetic resonance imaging and proton spectroscopy. *NMR in Biomedicine*, **15**, 6–17.

Khramtsov, V.V. (2005) Biological imaging and spectroscopy of pH. *Current Organic Chemistry*, **9**, 909–923.

Khramtsov, V.V., Marsh, D., Weiner, L., and Reznikov, V.A. (1992) The application of pH-sensitive spin labels to studies of surface potential and polarity of phospholipid membranes and proteins. *Biochimica et Biophysica Acta*, **1104**, 317–324.

Khramtsov, V.V., Panteleev, M.V., and Weiner, L.M. (1989) ESR study of proton transport across phospholipid vesicle membranes. *Journal of Biochemical and Biophysical Methods*, **18**, 237–246.

Kiessling, F., Huber, P.E., Grobholz, R. *et al.* (2004) Dynamic magnetic resonance tomography and proton magnetic resonance spectroscopy of prostate cancers in rats treated by radiotherapy. *Investigative Radiology*, **39**, 34–44.

Kim, J.G., Zhao, D., Song, Y. *et al.* (2003) Interplay of tumor vascular oxygenation and tumor pO_2 observed using near-infrared spectroscopy, an oxygen needle electrode, and ^{19}F MR pO_2 mapping. *Journal of Biomedical Optics*, **8**, 53–62.

Knopp, M.V., Giesel, F.L., Marcos, H. *et al.* (2001) Dynamic contrast-enhanced magnetic resonance imaging in oncology. *Topics in Magnetic Resonance Imaging*, **12**, 301–308.

Koh, W.J., Bergman, K.S., Rasey, J.S. *et al.* (1995) Evaluation of oxygenation status during fractionated radiotherapy in human nonsmall cell lung cancers using [F18]fluoromisonidazole positron emission tomography. *International Journal of Radiation Oncology, Biology, Physics*, **33**, 391–398.

Kondepati, V.R., Heise, H.M., and Backhaus, J. (2008) Recent applications of near-infrared spectroscopy in cancer diagnosis and therapy. *Analytical and Bioanalytical Chemistry*, **390**, 125–139.

Kondepati, V.R., Heise, H.M., and Backhaus, J. (2008) Recent applications of near-infrared spectroscopy in cancer diagnosis and therapy. *Analytical and Bioanalytical Chemistry*, **390**, 125–139.

Kragh, M., Quistorff, B., Lund, E.L., and Kristjansen, P.E. (2001) Quantitative estimates of vascularity in solid tumors by non-invasive near-infrared spectroscopy. *Neoplasia*, **3**, 324–330.

Krishnan, A.S., Neves, A.A., de Backer, M.M. *et al.* (2008) Detection of cell death in tumors by using MR imaging and a gadolinium-based targeted contrast agent. *Radiology*, **246**, 854–862.

Kroll, C., Hermann, W., Stösser, R. *et al.* (2001) Influence of drug treatment on the microacidity in rat and human skin--an in vitro electron spin resonance imaging study. *Pharmaceutical Research*, **18**, 525–530.

Kuppusamy, P., Chzhan, M., Vij, K. *et al.* (1994) Three-dimensional spectral-spatial EPR imaging of free radicals in the heart: a technique for imaging tissue metabolism and oxygenation. *Proceedings of the National Academy of Sciences of the United States of America*, **91**, 3388–3392.

Kuppusamy, P., Chzhan, M., and Zweier, J.L. (2003) Principles of imaging in vivo EPR (ESR), in *Theory and Applications: Biological Magnetic Resonance*, vol. 18 (ed. L.J. Berliner), Kluwer Academic/Plenum, New York, pp. 99–152.

Kuppusamy, P., Li, H., Ilangovan, A.J. *et al.* (2002) Noninvasive imaging of tumor redox status and its modification by tissue glutathione levels. *Cancer Research*, **62**,307–312.

Kurhanewicz, J., Vigneron, D.B., and Nelson, S.J. (2000) Three-dimensional magnetic resonance spectroscopic imaging of brain and prostate cancer. *Neoplasia*, **2**, 166–189.

Lahorte, C.M., Vanderheyden, J.L., Steinmetz, N. *et al.* (2004) Apoptosis-detecting radioligands: current state of the art and future perspectives. *European Journal of Nuclear Medicine in Molecular Imaging*, **31**, 887–919.

Laumonier, C., Segers, J., Laurent, S. *et al.* (2006) A new peptidic vector for molecular imaging of apoptosis, identified by phage display technology. *Journal of Biomolecular Screening*, **11**, 537–545.

Li, X., Vigneron, D.B., Cha, S. *et al.* (2005) Relationship of MR-derived lactate, mobile lipids and relative blood volume for in vivo gliomas. *American Journal of Neuroradiology*, **26**, 760–769.

Loncaster, J.A., Carrington, B.M., Sykes, J.R. *et al.* (2002) Prediction of radiotherapy outcome using dynamic contrast enhanced MRI of carcinoma of the cervix. *International Journal of Radiation Oncology, Biology, Physics*, **54**, 759–767.

Lutz, N.W., Tome, M.E., and Cozzone, P.J. (2003) Early changes in glucose and phospholipid metabolism following apoptosis induction by IFN-gamma/TNF-alpha inHT-29 cells. *FEBS Letters*, **544**, 123–128.

Lyng, H., Tufto, I., Skretting, A., and Rofstad, E.K. (1997) Proton relaxation times and interstitial fluid pressure in human melanoma xenografts. *British Journal of Cancer*, **75**, 180–183.

Mäder, K., Bittner, B., Li, Y. *et al.* (1998) Monitoring microviscosity and microacidity of the albumin microenvironment inside degrading microparticles from poly(lactide-co-glycolide) (PLG) or ABA-triblock polymers containing hydrophobic poly(lactide-co-glycolide) A blocks and hydrophilic poly(ethyleneoxide) B blocks. *Pharmaceutical Research*, **15**, 787–793.

Mäder, K., Gallez, B., Liu, K.J., and Swartz, H.M. (1996) Non-invasive in vivo characterization of release processes in biodegradable polymers by low-frequency electron paramagnetic resonance spectroscopy. *Biomaterials*, **17**, 457–461.

Mandl, S.J., Mari, C., Edinger, M. *et al.* (2004) Multi-modality imaging identifies key times for annexin V imaging as an early predictor of therapeutic outcome. *Molecular Imaging*, **3**, 1–8.

Margolis, D.J., Hoffman, J.M., Herfkens, R.J. *et al.* (2007) Molecular imaging techniques in body imaging. *Radiology*, **245**, 333–356.

Marzola, P., Degrassi, A., Calderan, L. *et al.* (2004) In vivo assessment of antiangiogenic activity of SU6668 in an experimental colon carcinoma model. *Clinical Cancer Research*, **10**, 739–750.

Mason, R.P. (1999) Transmembrane pH gradients in vivo: measurements using fluorinated vitamin B6 derivatives. *Current Medicinal Chemistry*, **6**, 481–499.

Mason, R.P., Hunjan, S., Constantinescu, A. *et al.* (2003) Tumor oximetry: comparison of 19F MR EPI and electrodes. *Advances in Expermintal Medicine and Biology*, **530**, 19–27.

Mason, R.P., Rodbumrung, W., and Antich, P.P. (1996) Hexafluorobenzene: a sensitive 19F NMR indicator of tumor oxygenation. *NMR in Biomedicine*, **9**, 125–134.

Mayr, N.A., Yuh, W.T., Arnholt, J.C. *et al.* (2000) Pixel analysis of MR perfusion imaging in predicting radiation therapy outcome in cervical cancer. *Journal of Magnetic Resonance Imaging*, **12**, 1027–1033.

Mayr, N.A., Yuh, W.T., Magnotta, V.A. *et al.* (1996) Tumor perfusion studies using fast magnetic resonance imaging technique in advanced cervical cancer: a new noninvasive predictive assay. *International Journal Radiation Oncology, Biology, Physics*, **36**, 623–633.

Mayr, N.A., Yuh, W.T., Zheng, J. *et al.* (1998) Prediction of tumor control in patients with cervical cancer: analysis of combined volume and dynamic enhancement pattern by MR imaging. *American Journal of Roentgenology*, **170**, 177–182.

Moffat, B.A., Hall, D.E., Stojanovska, J. *et al.* (2004) Diffusion imaging for evaluation of tumor therapies in preclinical animal models. *MAGMA*, **17**, 249–259.

Nagashima, T., Sakakibara, M., Nakamura, R. *et al.* (2006) Dynamic enhanced MRI predicts chemosensitivity in breast cancer patients. *European Journal of Radiology*, **60**, 270–274.

Neeman, M., Dafni, H., Bukhari, O. *et al.* (2001) In vivo BOLD contrast MRI mapping of subcutaneous vascular function and maturation: validation by intravital microscopy. *Magnetic Resonance in Medicine*, **45**, 887–898.

Negendank, W. (1992) Studies of human tumors by MRS: a review. *NMR in Biomedicine*, **5**, 303–324.

Ng, P., Rajendran, J.G., Schwartz, D.L. *et al.* (2003) Can [F-18] fluoromisonidazole PET imaging predict treatment response in head and neck cancer? *Journal of Nuclear Medicine*, **44**, 128P.

Norris, D.G. (2001) The effects of microscopic tissue parameters on the diffusion weighted magnetic resonance imaging experiment. *NMR in Biomedicine*, **14**, 77–93.

Ojugo, A.S., McSheehy, P.M., McIntyre, D.J. *et al.* (1999) Measurement of the extracellular pH of solid tumours in mice by magnetic resonance spectroscopy: a comparison of exogenous (19)F and (31)P probes. *NMR in Biomedicine*, **12**, 495–504.

Ogawa, S., Lee, T.M., Kay, A.R., and Tank, D.W. (1990) Brain magnetic resonance imaging with contrast dependent on blood oxygenation. *Proceedings of the National Academy of Sciences of the United States of America*, **87**, 9868–9872.

Padhani, A.R. (2002) Dynamic contrast-enhanced MRI in clinical oncology: current status and future directions. *Journal of Magnetic Resonance Imaging*, **16**, 407–422.

Padhani, A. (2006) PET imaging of tumour hypoxia. *Cancer Imaging*, **31**, S117–S121.

Padhani, A.R. and Husband, J.E. (2001) Dynamic contrast-enhanced MRI studies in oncology with an emphasis on quantification, validation and human studies. *Clinical Radiology*, **56**, 607–620.

Padhani, A.R., Krohn, K.A., Lewis, J.S. and Alber, M. (2007) Imaging oxygenation of human tumours. *European Radiology*, **17** (4), 861–872.

Pham, T.H., Hornung, R., Berns, M.W. *et al.* (2001) Monitoring tumor response during photodynamic therapy using near-infrared photon-migration spectroscopy. *Photochemistry and Photobiology*, **73**, 669–677.

Pogue, B.W., O'Hara, J.A., Goodwin, I.A. *et al.* (2002) Tumor pO(2) changes during photodynamic therapy depend upon photosensitizer type and time after injection. *Comparative Biochemistry and Physiology Part A: Molecular and Integrative Physiology*, **132**, 177–184.

Potapenko, D.I., Foster, M.A., Lurie, D.J. *et al.* (2006) Real-time monitoring of drug-induced changes in the stomach acidity of living rats using improved pH-sensitive nitroxides and low-field EPR techniques. *Journal of Magnetic Resonance*, **182**, 1–11.

Rajendran, J.G., Mankoff, D.A., O'Sullivan, F. *et al.* (2004) Hypoxia and glucose metabolism in malignant tumors: evaluation by [18F] fluoromisonidazole and [18F]fluorodeoxyglucose positron emission tomography imaging. *Clinical Cancer Research*, **10**, 2245–2252.

Rajendran, J.G., Wilson, D.C., Conrad, E.U. *et al.* (2003) [(18)F] FMISO and [(18)F]FDG PET imaging in soft tissue sarcomas: correlation of hypoxia, metabolism and VEGF expression. *European Journal of Nuclear Medicine and Molecular Imaging*, **30**, 695–704.

Robinson, S.P. and Griffiths, J.R. (2004) Current issues in the utility of 19F nuclear magnetic resonance methodologies for the assessment of tumour hypoxia. *Philosophical Transactions of the Royal Society B: Biological Sciences*, **359**, 987–996.

Robinson, S.P., Howe, F.A., Griffiths, J.R. *et al.* (2007) Susceptibility contrast magnetic resonance imaging determination of fractional tumour blood volume: a non-invasive imaging biomarker of response to the vascular disrupting agent ZD6126. *International Journal of Radiation Oncology, Biology, Physics*, **69**, 872–879.

Robinson, S.P., Kalber, T.L., Howe, F.A. *et al.* (2005) Acute tumor response to ZD6126 assessed by intrinsic susceptibility magnetic resonance imaging. *Neoplasia*, **7**, 466–474.

Rodrigues, L.M., Howe, F.A., Griffiths, J.R., and Robinson, S.P. (2004) Tumor R2* is a prognostic indicator of acute radiotherapeutic response in rodent tumors. *Journal of Magnetic Resonance Imaging*, **19**, 482–488.

Ronen, I., Merkle, H., Ugurbil, K., and Navon, G. (1998) Imaging of H217O distribution in the brain of a live rat by using proton-detected 17O MRI. *Proceedings of the National Academy of Sciences of the United States of America*, **95**, 12934–12939.

Ross, B.D., Moffat, B.A., Lawrence, T.S. *et al.* (2003) Evaluation of cancer therapy using diffusion magnetic resonance imaging. *Molecular Cancer Therapeutics*, **2**, 581–587.

Schellenberger, E.A., Bogdanov, A. Jr., Högemann, D. *et al.* (2002) Annexin V-CLIO: a nanoparticle for detecting apoptosis by MRI. *Molecular Imaging*, **1**, 102–107.

Secomb, T.W., Hsu, R., and Dewhirst, M.W. (2004) Synergistic effects of hyperoxic gas breathing and reduced oxygen consumption on tumor oxygenation: a theoretical model. *International Journal of Radiation Oncology, Biology, Physics*, **59**, 572–578.

Sheppard, G., Gruzelier, J., Manchanda, R. *et al.* (1983) ^{15}O positron emission tomographic scanning in predominantly never-treated acute schizophrenic patients. *Lancet*, **2**, 1448–1452.

Shields, A.F. (2006) Positron emission tomography measurement of tumor metabolism and growth: its expanding role in oncology. *Molecular Imaging in Biology*, **8**, 141–150.

Shields, A., Grierson, J., Stayanoff, J. *et al.* (1998) F-18 FLT can be used to image cell proliferation in vivo. *Journal of Nuclear Medicine*, **39**, 228P.

Sijens, P.E., Levendag, P.C., Vecht, C.J. *et al.* (1996) 1H MR spectroscopy detection of lipids and lactate in metastatic brain tumors. *NMR in Biomedicine*, **9**, 65–71.

Sonveaux, P., Dessy, C., Martinive, P. *et al.* (2004) Endothelin-1 is a critical mediator of myogenic tone in tumor arterioles: implications for cancer treatment. *Cancer Research*, **64**, 3209–3214.

Stubbs, M., Bhujwalla, Z.M., Tozer, G.M. *et al.* (1992) An assessment of 31P MRS as a method of measuring pH in rat tumours. *NMR in Biomedicine*, **5**, 351–359.

Subramanian, S., Matsumoto, K., Mitchell, J.B., and Krishna, M.C. (2004) Radio frequency continuous-wave and timedomain EPR imaging and Overhauser-enhanced magnetic resonance imaging of small animals: instrumental developments and comparison of relative merits for functional imaging. *NMR in Biomedicine*, **17**, 263–294.

Swartz, H.M. (1990) Principles of the metabolism of nitroxides and their implications for spin trapping. *Free Radical Research Communications*, **9**, 399–405.

Swartz, H.M., Khan, N., Buckey, J. *et al.* (2004) Clinical applications of EPR: overview and perspectives. *NMR in Biomedicine*, **17**, 335–351.

Tatum, J.L., Kelloff, G.J., Gillies, R.J. *et al.* (2006) Hypoxia: importance in tumor biology, noninvasive measurement by imaging, and value of its measurement in the management of cancer therapy. *International Journal of Radiation Biology*, **82**, 699–757.

Tatum, J.L., Kelloff, G.J., Gillies, R.J. (2006) Hypoxia: importance in tumor biology, noninvasive measurement by imaging, and value of its measurement in the management of cancer therapy. *International Journal of Radiation Biology*, **82** (10), 699–757.

Ter-Pogossian, M.M. and Herscovitch, P. (1985) Radioactive oxygen-15 in the study of cerebral blood flow, blood volume, and oxygen metabolism. *Seminars in Nuclear Medicine*, **15**, 377–394.

Theilmann, R.J., Borders, R., Trouard, T.P. *et al.* (2004) Changes in water mobility measured by diffusion MRI predict response of metastatic breast cancer to chemotherapy. *Neoplasia*, **6** (6), 831–837.

Tofts, P.S., Brix, G., Buckley, D.L. *et al.* (1999) Estimating kinetic parameters from dynamic contrast-enhanced T(1)-weighted MRI of a diffusable tracer: standardized quantities and symbols. *Journal of Magnetic Resonance Imaging*, **10**, 223–232.

Torigian, D.A., Huang, S.S., Houseni, M., and Alavi, A. (2007) Functional imaging of cancer with emphasis on molecular techniques. *CA: A Cancer Journal for Clinicians*, **57**, 206–224.

Tropres, I., Grimault, S., Vaeth, A. *et al.* (2001) Vessel size imaging. *Magnetic Resonance in Medicine*, **45**, 397–408.

Valk, P.E., Bailey, D.L., Townsend, D.W., and Maisey, M.N. (2003) *Positron Emission Tomography*, Springer-Verlag, London.

Viglianti, B.L., Lora-Michiels, M., Poulson, J.M. *et al.* (2009) Dynamic contrast-enhanced magnetic resonance imaging as a predictor of clinical outcome in canine spontaneous soft tissue sarcomas treated with thermoradiotherapy. *Clinical Cancer Research*, **15**, 4993–5001.

Warburg, O., Wind, F., and Negalein, E. (1927) The metabolism of tumours in the body. *Journal of Physiology*, **8**, 519–530.

Xia, M., Kodibagkar, V., Liu, H., and Mason, R.P. (2006) Tumour oxygen dynamics measured simultaneously by near-infrared spectroscopy and 19F magnetic resonance imaging in rats. *Physics in Medicine and Biology*, **51**, 45–60.

Yablonskiy, D.A. (1998) Quantitation of intrinsic magnetic susceptibility-related effects in a tissue matrix. Phantom study. *Magnetic Resonance in Medicine*, **39**, 417–428.

Yablonskiy, D.A. and Haacke, E.M. (1994) Theory of NMR signal behavior in magnetically inhomogeneous tissues: the static dephasing regimen. *Magnetic Resonance in Medicine*, **32**, 749–763.

Yamada, K.I., Kuppusamy, P., English, S. *et al.* (2002) Feasibility and assessment of non-invasive in vivo redox status using electron paramagnetic resonance imaging. *Acta Radiology*, **43**, 433–440.

Yamashita, Y., Baba, T., Baba, Y. *et al.* (2000) Dynamic contrast-enhanced MR imaging of uterine cervical cancer: pharmacokinetic analysis with histopathologic correlation and its importance in predicting the outcome of radiation therapy. *Radiology*, **216**, 803–809.

Zhao, D., Jiang, L., and Mason, R.P. (2004) Measuring changes in tumor oxygenation. *Methods in Enzymology*, **386**, 378–418.

Zhao, M., Beauregard, D.A., Loizou, L. *et al.* (2001) Non-invasive detection of apoptosis using magnetic resonance imaging and a targeted contrast agent. *Nature in Medicine*, **7**, 1241–1244.

Zhao, D., Constantinescu, A., Chang, C.H. *et al.* (2003) Correlation of tumor oxygen dynamics with radiation response of the dunning prostate R3327-HI tumor. *Radiation Research*, **159**, 621–631.

Zhu, X.H., Zhang, Y., Tian, R.X. *et al.* (2002) Development of (17)O NMR approach for fast imaging of cerebral metabolic rate of oxygen in rat brain at high field. *Proceedings of the National Academy Sciences of the United States of America*, **99**, 13194–13199.

Zhu, X.H., Zhang, N., Zhang, Y. *et al.* (2005) In vivo 17O NMR approaches for brain study at high field. *NMR in Biomedicine*, **18**, 83–103.

Further reading

Koo, V., Hamilton, P.W., and Williamson, K. (2006) Non-invasive in vivo imaging in small animal research. *Cellular Oncology*, **28**, 127–139.

van Laarhoven, H.W., Klomp, D.W., Rijpkema, M. *et al.* (2007) Prediction of chemotherapeutic response of colorectal liver metastases with dynamic gadolinium-DTPA-enhanced MRI and localized 19F MRS pharmacokinetic studies of 5-fluorouracil. *NMR in Biomedicine*, **20**, 128–140.

Sotgiu, A., Mäder, K., Placidi, G. *et al.* (1998) pH-sensitive imaging by low-frequency EPR: a model study for biological applications. *Physics in Medicine and Biology*, **43**, 1921–1930.

Zhang, N., Zhu, X.H., Lei, H. *et al.* (2004) Simplified methods for calculating cerebral metabolic rate of oxygen based on 17O magnetic resonance spectroscopic imaging measurement during a short 17O2 inhalation. *Journal of Cerebral Blood Flow and Metabolism*, **24**, 840–848.

12

Hypoxia-Inducible Factor 1 (HIF-1) Mediated Adaptive Responses in the Solid Tumor

Tereza Goliasova and Nicholas C. Denko
Division of Radiation and Cancer Biology, Department of Radiation Oncology, Stanford School of Medicine, Stanford, CA, USA

12.1 Introduction

In order to maintain tissue homeostasis, it is necessary to maintain a tight control over the rate of cell division and cell loss. A stable number of cells in tissues also requires a stable blood supply to perfuse it adequately. This delicate balance is disturbed in tumors. By acquiring mutations, cancer cells escape regulatory mechanisms and proliferate uncontrollably. As the tumor mass enlarges and outgrows adjacent vasculature, the delivery of oxygen and nutrients is unable to meet the demand of the tissue. Therefore areas that are poorly perfused suffer from low oxygen tension (hypoxia), low glucose (hypoglycemia), and increased waste products (acidosis). The rate limiting "nutrient" is oxygen, as this is consumed most rapidly by the tissue as it is being delivered. Studies of tumor architecture revealed more than half a century ago that hypoxic regions existed in human tumors (Thomlinson and Gray, 1955). The hypoxic regions neighbored necrotic areas that were localized at a great distance from the nearest blood vessel. Tumor cells located at 70–120 μm from a blood vessel are inadequately supplied with oxygen, because portion of the oxygen becomes metabolized by the cells closer to the blood vessel (Vaupel and Harrison, 2004). Hypoxia caused by oxygen diffusion limitations is termed chronic hypoxia. Cells located more than 150–180 μm from blood vessels are anoxic, and die by either apoptosis or necrosis. Acute fluctuations in blood flow result in perfusion-limited hypoxia, which is a consequence of transient inhomogeneities in the microcirculation caused

Tumor Microenvironment Edited by Dietmar W. Siemann

by abnormal tumor vessels (Dewhirst, Cao, and Moeller, 2008). Oxygen concentrations in human tumors are therefore highly heterogeneous with many regions at much lower values than the normal tissues from which they arose (Vaupel *et al.*, 1991).

It was also recognized that hypoxic cells were resistant to killing by ionizing radiation *in vitro* (Gray *et al.*, 1953). When the technology was available, it was later determined that in patients, low tumor oxygenation status had a negative impact on response to radiotherapy (Brizel *et al.*, 1999; Hockel *et al.*, 1996; Nordsmark, Overgaard, and Overgaard, 1996). Surprisingly, these clinical studies also determined that tumor hypoxia was also predictive of poor response to chemotherapy, and even surgery. The hypoxic tumors showed signs of increased aggressiveness (Hockel *et al.*, 1999), local invasion (Hockel *et al.*, 1996), and even distant metastasis (Brizel *et al.*, 1996). In the 50 years that followed the original observations of radiologists, molecular biology has stepped in to help explain the molecular mechanisms governing the "hypoxic tumor phenotype."

12.2 Molecular consequences of tumor hypoxia

Hypoxia is a condition generally unfavorable to cell growth, with more severe hypoxia being more infavorable. Therefore, the cell triggers a cascade of adaptive physiological responses all aimed at eliminating hypoxic stress and facilitating cellular survival. These adaptive responses are designed to bring the oxygen demand of the tissue back in agreement with the oxygen supplied by the vasculature. Re-establishing demand to meet supply by definition relieves hypoxia, and is accomplished by a combination of decreasing demand and increasing supply. One major mechanism by which cells reduce their demand for oxygen and energy is to reduce macromolecular synthesis. At moderately severe hypoxia ($<0.5\%$ oxygen), cells have been shown to reduce proliferation (DNA and lipid synthesis) (Papandreou *et al.*, 2005), reduce RNA synthesis (Johnson, Denko, and Barton, 2008), and decrease protein translation (Wouters *et al.*, 2004). In the background of this overall reduction in macromolecular synthesis, there are a group of proteins that are actively synthesized that mediate the adaptive response. The transcription factors nuclear factor (NF)-κB, activator protein-1 (AP-1), early growth response protein-1, activating transcription factor-3/4 as well as others have been found to be induced to some degree by hypoxia. However, the transcription factor with the most robust and specific response to hypoxia is the hypoxia-inducible factor 1, or HIF-1. This factor and its downstream target genes will be the focus of this chapter. These adaptive physiological changes are thought to allow for cellular survival. However, these molecular changes inadvertently contribute to the biologic observation that hypoxic tumors have a fundamentally worse clinical outcome (Harris, 2002).

12.3 Hypoxia inducible factor 1

HIF-1 is a transcription factor protein complex that is a major mediator of the hypoxic response. First characterized by Semenza and co-workers in the 1990s, it has been widely studied and shown to play important roles in embryogenesis, wound response, physiologic response to exercise or reduced oxygenation (journey to high altitudes), and tumor progression. A direct link between HIF-1 activity and tumor formation was established in 1999, when it was determined that loss of the tumor suppressor von Hippel–Lindau (VHL) resulted in constitutive HIF-1 protein and activity (Maxwell *et al.*, 1999), and this contributed to VHL's role as a tumor suppressor (Kondo *et al.*, 2002). HIF-1 generates a transcriptional response to hypoxia by binding to hypoxia-response elements (HREs) in the promoter/enhancer regions of target genes. The core HRE DNA motif contains a consensus recognition sequence 5′-G/ACGTG-3′ (Schofield and Ratcliffe, 2004).

12.4 HIF-1 subunits and domain structure

HIF-1 is a heterodimer composed of a constitutively expressed HIF-1β subunit, and a HIF-1α subunit, whose stability and transcriptional activity are precisely regulated by the cellular oxygen concentrations. The HIF-1α mRNA is also constitutively expressed mainly through the action of the SP1 transcription factor. Other transcription factors, such as AP-1 and 2, NF-1, and NF-κB also have binding sites in the promoter region of the HIF-1α gene, however the regulation of HIF-1α is at the level of the protein and not the mRNA. Under normoxic conditions, HIF-1α protein is rapidly targeted for ubiquitin-mediated degradation by the proteasome and has a half-life of less than 5 minutes (Berra *et al.*, 2001). During hypoxia, HIF-1α protein evades degradation and becomes stabilized, with a half life of 30–60 minutes. In hypoxia, effective protein translation of HIF-1α is maintained by the presence of an internal ribosome entry site (IRES) found in the 5′-untranslated region of its mRNA (Lang, Kappel, and Goodall, 2002).

Both HIF-1α and HIF-1β belong to the basic-helix-loop-helix-Per/Arnt/Sim (bHLH-PAS) transcription factor family (Wang *et al.*, 1995). The bHLH and PAS domains at the amino terminus are required for heterodimer formation between the two subunits and for DNA binding (Jiang *et al.*, 1996). HIF-1β subunit is identical to aryl hydrocarbon receptor nuclear translocator (ARNT), and can heterodimerize with either HIF-1α or aromatic hydrocarbon receptor (AHR), which is a component of the xenobiotic signaling pathway (Wang *et al.*, 1995). Upon dimerization, the HIF-1 complex translocates into the nucleus where it binds to the HREs of target genes.

Apart from the bHLH-PAS domains, HIF-1α contains a unique domain that is not present within the β subunit. The oxygen-dependent degradation domain (ODDD, aa 401–603) is responsible for normoxic degradation and hypoxic stabilization of the α subunit (Huang *et al.*, 1998). The function of the domain and the mechanism of stabilization will be described below. HIF-1α also includes two transactivation domains, namely, the N-terminal transactivation domain and C-terminal transactivation domain (N-TAD, aa 481–603 and C-TAD, aa 776–826), which mediate transcriptional activation and interaction with co-activators (Jiang *et al.*, 1997). Hypoxic regulation of these transactivation domains will also be described next.

12.5 Regulation of HIF-1α protein stability and activity by post-translational modifications

12.5.1 Prolyl hydroxylation and ubiquitination

HIF-1 transcriptional activity is dependent upon the availability of the HIF-1α subunit, which in turn is normally regulated by cellular oxygen levels. At the biochemical level, oxygen partial pressure (pO_2) is sensed by a family of protein hydroxylases that are able to transmit the intracellular pO_2 information directly to HIF-1α in the form of covalently attached hydroxy groups. The fact that the proline hydroxylases are dioxygenases and require molecular oxygen for activity connects available molecular oxygen to the regulation of the HIF-1α protein (Semenza, 2001; Wenger, 2006).

The HIF-1α ODD domain possesses two proline residues (Pro^{402} and Pro^{564}) that are hydroxylated in the presence of oxygen by one of three enzymes with prolyl hydroxylase domains (PHDs 1–3) (Epstein *et al.*, 2001; Ivan *et al.*, 2001; Masson *et al.*, 2001). PHDs are dioxygenases that require oxygen, 2-oxoglutarate, ferrous iron, and ascorbate as cofactors. In the hydroxylation reaction, they transfer one oxygen atom to the proline residue, and the second oxygen atom reacts with 2-oxoglutarate generating succinate and CO_2. The hydroxyproline residues are then recognized by, and become buried within, the hydrophobic core of von Hippel–Lindau tumor suppressor protein (pVHL). The distinction between hydroxylated and non-hydroxylated residues is mediated by two optimized hydrogen bonds, which are formed between alcohol groups of the hydroxylated prolines and two residues of pVHL (Ser^{111} and His^{115}) (Hon *et al.*, 2002). pVHL is the recognition element of a multicomponent ubiquitin ligase (pVHL-elonginB-elonginC-Cul2-Rbx) that ubiquinates and targets HIF-1α subunit for proteolysis by the ubiquitin–proteasome pathway (Schofield and Ratcliffe, 2004).

Three functional 2-oxoglutarate (2-OG)-dependent prolyl hydroxylases have been identified: PHD1, PHD2, PHD3. All three require oxygen, Fe^{2+}, and 2-OG as cofactors. PHD2 seems to be the most ubiquitously expressed, and therefore most important in the regulation of HIF-1α (Berra *et al.*, 2003). However, biochemical characterization has determined that the different PHDs have different K_ms for the various cofactors, and this could help to regulate HIF-1α protein stability under different oxygen or other environmental conditions (Appelhoff *et al.*, 2004). In addition, PHD3 is a direct target gene that is induced by HIF-1 under hypoxia (Aprelikova *et al.*, 2004), suggesting a possible feedback mechanism by which HIF-1 activity is downregulated in long-term hypoxia. Therefore, under hypoxic conditions, or by displacement of Fe^{2+} (by hypoxia mimetics), or by alterations in some 2-oxoacids (by certain mitochondrial mutations), prolyl hydroxylation and subsequent HIF-1α degradation are suppressed (Schofield and Ratcliffe, 2004).

12.5.2 Asparagine hydroxylation

A second oxygen-dependent hydroxylation checkpoint involves hydroxylation of an asparagine residue (Asn^{803}) in the C-TAD of HIF-1α (Lando *et al.*, 2002a, b). This reaction is facilitated by a 2-OG-dependent asparaginyl hydroxylase called factor inhibiting HIF-1 (FIH-1). Utilization of oxygen as a substrate allows FIH-1 to serve as a second oxygen sensor. The hydroxy-modification of Asn^{803} blocks the binding of the coactivator proteins CREB binding protein (CBP) and p300 and thus inhibits HIF-1 transactivation activity in normoxia (Lando *et al.*, 2002b). In hypoxia, C-TAD binds to the cysteine/histidine-rich 1 (CH1) domain in p300 and the transcriptional adapter zinc-binding 1 (TAZ1) domain in CBP. CBP and p300 function to physically link other sequence-specific DNA-binding proteins to the transcription initiation complex and as histone acetyltransferases, which perform chromatin remodeling that is required for gene transcription (Semenza, 2002).

In addition, independently of its hydroxylase activity, FIH-1 acts as a corepressor by interacting with pVHL. Together they recruit histone deacetylases (HDACs) that interfere with the N-TAD function (Mahon, Hirota, and Semenza, 2001; Semenza, 2002). Finally, published reports have found that the HIF-1α protein can be either acetylated (Jeong *et al.*, 2002), or nitrosylated (Li *et al.*, 2007), although the functional consequences of these modifications are still being determined.

12.6 HIF isoforms

Besides HIF-1α, there are two closely related hypoxia-inducible members of the bHLH-PAS superfamily: HIF-2α, also referred to as endothelial Per/Arnt/Sim domain protein 1 (EPAS1) and HIF-3α (Tian *et al.*, 1998). HIF-1α and HIF-2α have similar domain structure and are regulated in a similar manner, whereas HIF-3α is less closely related. HIF-1α and HIF-2α undergo proteolytic degradation that is dependent on the presence of ODDD, and they also possess two transactivation domains, N-TAD and C-TAD (Maxwell *et al.*, 1999). HIF-3α can dimerize with HIF-1β, and bind to HREs. The production of its alternatively spliced form (composed of the bHLH-PAS domains, and omitting ODDD and TAD) is striking. This protein (known as inhibitory Per/Arnt/Sim (IPAS) protein) forms transcriptionally inactive heterodimers with HIF-1α (Makino *et al.*, 2001). Moreover, IPAS can be upregulated by HIF-1 via a functional HRE under hypoxic conditions, creating a negative feedback loop (Makino *et al.*, 2001). It was reported that hypoxia does not affect mRNA levels of HIF-1α, HIF-1β, and HIF-2α, but HIF-3α mRNA increases after 2 hours of hypoxia (Heidbreder *et al.*, 2003).

There is considerable debate in the literature as to the functional consequences of having seemingly redundant HIF-α isoforms. Lower metazoans (*Caenorhabditis elegaqns*, *Drosophila melanogaster*) have only a single HIF-α gene, while higher vertebrates have two. Differences also exist in the phenotype of the respective knockout mice. While deletion of HIF-1α is universally lethal at an early embryonic age (Ryan, Lo, and Johnson, 1998), deletion of HIF-2α in some mouse backgrounds is tolerated, albeit with significant multiorgan pathology (Scortegagna *et al.*, 2003).

Two major functional differences between HIF-1α and HIF-2α have been reported: differences in their respective target genes, and differences in the kinetics and oxygen-dependence of their induction. With respect to differences in target genes, HIF-1 comprised of HIF-1α binds to the same consensus HRE as that of HIF-1 comprised of HIF-2α by biochemical criteria. However, the core HRE is small (CGTG), so the adjoining sequences certainly play a role in the binding of a large dimeric protein. To date, no bioinformatic or biochemical characteristics have been reported that can generate an HRE that discriminates between HIF-1α and HIF-2α binding. However, cellular assays using HIF-α knockdowns and knockouts have identified genes that appear to be preferentially regulated by one isoform or the other. The gene DMT1, involved in iron metabolism, is the best characterized 'HIF-2α-specific gene' found to date (Mastrogiannaki *et al.*, 2009). With respect to kinetics and oxygen dependency of induction, HIF-2α has been shown to be stabilized in neuroblastomas by more moderate oxygen concentrations, but to have a more persistent stability in cells (Holmquist-Mengelbier *et al.*, 2006). Regardless of the mechanism, there appear to be significant differences in the role that HIF-1α and HIF-2α play in tumorigenesis. Analysis of primary renal cancers was able to determine that there are two subtypes in which VHL had been lost. One set had intact HIF-1α and HIF-2α genes being expressed, and these tumors showed increased akt and extracellular-signal-related kinase (ERK) signaling. The second set had lost HIF-1α expression but still maintained HIF-2α, and were characterized with elevated myc expression (Gordan *et al.*, 2008). The significance of these findings on prognosis and therapeutic response is still being investigated.

12.7 Oxygen-independent HIF signaling

In addition to intratumor hypoxia, oncogene activation and loss of tumor suppressors may lead to induction of HIF in an oxygen-independent manner. Increased HIF-1 transcriptional activity can be mediated by the activation of the Ras oncogene and the subsequent mitogen-activated protein kinase (MAPK) pathway. Eight serine residues that may serve as putative consensus targets for MAPK family exist within HIF-1α (Richard *et al.*, 1999). One mechanism to explain the increased transcriptional activity suggests that HIF-1β binds preferentially to the phosphorylated form of HIF-1α (Suzuki, Tomida, and Tsuruo, 2001). The phosphorylation of Thr^{796} also enhances the transcriptional response in hypoxia and also prevents the hydroxylation of Asn^{803} by FIH-1 (Lancaster *et al.*, 2004). In addition, activation of the phosphoinositide 3-kinase (PI3K) pathway (for example, by activation of the HER2/neu oncogene or loss of the phosphatease and tensin homolog tumor suppressor) can result in increased translation of HIF-1α through the Akt protein kinase-dependent activation of mammalian target of rapamycin (mTOR) (Laughner *et al.*, 2001). HIF signaling is also promoted by decreasing HIF-1α ubiquitination and thus increasing its accumulation, such as in renal cell carcinoma with VHL tumor suppressor loss of function (Maxwell *et al.*, 1999). The loss of VHL function is caused by mutations that render VHL defective in respect to binding to elongin C or HIF-1α (Kondo and Kaelin, 2001).

The loss of function of a group of mitochondrial tumor suppressors also leads to increased HIF signaling by inhibiting PHD-mediated hydroxylation of HIF-1α. As has already been mentioned, PHDs are 2-OG-dependent dioxygenases, which catalyze the conversion of a prolyl residue, molecular oxygen, and 2-OG to hydroxyprolyl, carbon dioxide, and succinate using ferrous iron as cofactor. Yet succinate is not only a product of PHDs in the cytosol, but also a substrate for succinate dehydrogenase (SDH) in the mitochondria (King, Selak, and Gottlieb, 2006). SDH is a tricarboxylic acid (TCA) cycle enzyme that converts succinate to fumarate. SDH dysfunction in cells raises the levels of succinate, which then accumulates in the mitochondrial matrix and leaks out into the cytosol (Selak *et al.*, 2005). The accumulated succinate, by feedback inhibition of PHDs, leads to HIF-α stabilization, and activation of the HIF complex. A similar situation arises in fumarate hydratase (FH) deficiency. FH is the following TCA cycle enzyme after SDH that converts fumarate to malate. Fumarate is not a product of PHDs but it is chemically similar to succinate. Like succinate, it accumulates in FH-deficient cells and inhibits PHD activity in the cytosol (Isaacs *et al.*, 2005). Although SDH and FH are housekeeping genes with key bioenergetic roles, mutations in these genes predispose patients to cancer (King, Selak, and Gottlieb, 2006). SDH possesses two catalytic subunits (A, B), which are anchored in the inner mitochondrial membrane by subunits C and D. Heterozygous germline mutations in subunits B, C, or D of SDH lead to the development of paraganglioma or pheochromocytoma (Baysal, 2003). SDH subunit A mutation carriers are unlikely to develop hereditary paraganglioma (HPGL) because two types of genes exist for this subunit (paraganglia express both), and one can stand in for the other (Briere *et al.*, 2005). Unlike SDH, FH is a homotetrameric enzyme. Heterozygous germline FH mutations have been implicated in hereditary leiomyomatosis renal cell carcinoma, which is associated with skin and uterine leiomyomata (smooth muscle tumors) and papillary renal cancer (Isaacs *et al.*, 2005).

12.8 HIF target genes

Hypoxia inducible factor is central to physiological and pathological processes involved in adaptation to decreased oxygen availability. About 70 direct HIF target genes have been identified, and expression of dozens more genes are either directly or indirectly regulated by HIF. The evolutionary pressure to regulate these genes by hypoxia is to allow the organism or tissues (and tumors, too) to survive and reestablish a normoxic state. The adaptive mechanism can be through decreasing demand for oxygen and increasing supply in the hypoxic tissue. Evolutionarily, this was probably most important in wound healing, where vascular/tissue damage led to regions of hypoxia that would eventually heal and return to normoxia. In the tumor, the normoxic state is not reached, and the adaptive changes continue indefinitely, leading to much of the 'hypoxic tumor phenotype.' HIF target genes are involved in angiogenesis, erythropoiesis, iron metabolism, vascular tone regulation (all of which increase oxygen delivery), in glucose uptake, and glycolysis (for energy production and availability of biosynthetic substrates), in pH regulation (to neutralize acidic intracellular pH that results from anaerobic metabolism), as well as in other yet unidentified processes (Figure 12.1).

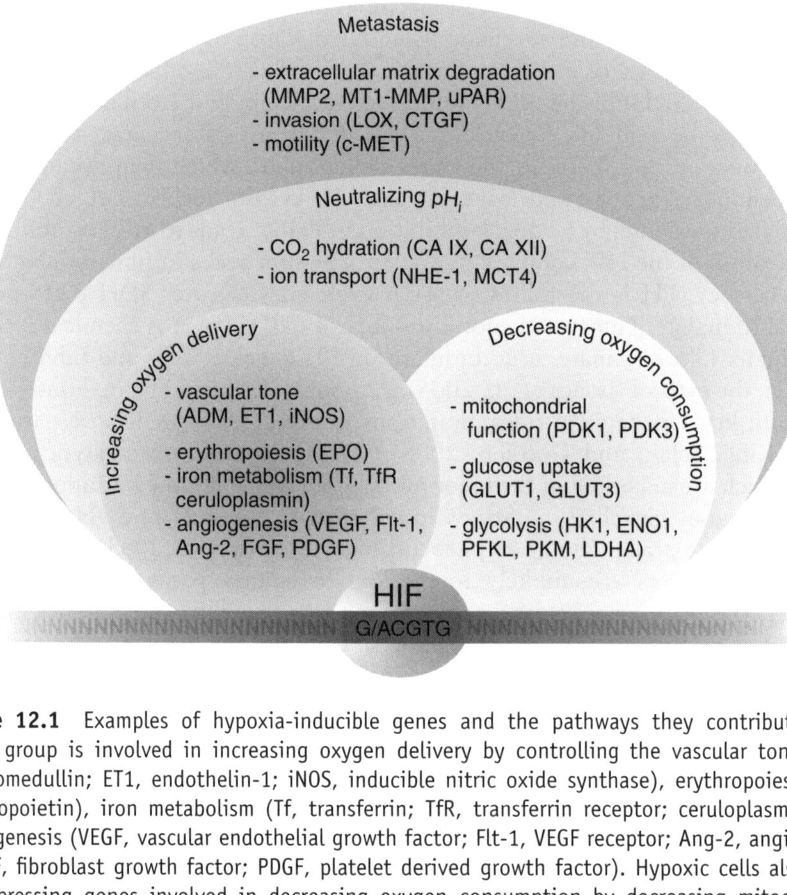

Figure 12.1 Examples of hypoxia-inducible genes and the pathways they contribute to. A major group is involved in increasing oxygen delivery by controlling the vascular tone (ADM, adrenomedullin; ET1, endothelin-1; iNOS, inducible nitric oxide synthase), erythropoiesis (EPO, erythropoietin), iron metabolism (Tf, transferrin; TfR, transferrin receptor; ceruloplasmin), and angiogenesis (VEGF, vascular endothelial growth factor; Flt-1, VEGF receptor; Ang-2, angiopoietin 2; FGF, fibroblast growth factor; PDGF, platelet derived growth factor). Hypoxic cells also adapt by expressing genes involved in decreasing oxygen consumption by decreasing mitochondrial function (PDK1 and 3, pyruvate dehydrogenase kinase 1 and 3), increasing glucose uptake (GLUT1 and 3, glucose transporter 1 and 3) and glycolysis (HK1, hexokinase 1; ENO1, enolase 1; PFKL, phosphofructokinase L; PKM, pyruvate kinase M; LDHA, lactate dehydrogenase A). Because of inefficient tumor vasculature and glycolytic metabolism, hypoxic cells also express genes involved in neutralizing pH$_i$ by CO_2 hydration (CA IX and XII, carbonic anhydrase IX and XII) or ion transport (NHE-1, Na^+/H^+ exchanger; MCT4, monocarboxylate transporter 4). All of the previously mentioned genes finally contribute to metastasis, along with hypoxia-regulated genes involved in extracellular matrix degradation (MMP2, matrix metalloproteinase 2; MT1-MMP, membrane type 1 MMP; uPAR, urokinase plasminogen activator receptor), invasion (LOX, lysyl oxidase; CTGF, connective tissue growth factor), and motility (c-Met, HGF receptor tyrosine kinase).

12.9 Hypoxia and oxygen delivery

In response to reduced oxygen availability, the organism can increase the capacity of red blood cells to transport oxygen by the increased expression of genes involved in erythropoiesis and iron-metabolism. Hypoxia increases the expression of erythropoietin (EPO), which is required for the development of red blood cells (Semenza *et al.*, 1991). An increase in the number of erythrocytes (hematocrit) increases the total oxygen carrying capacity of the blood. Products of iron-metabolizing genes control the major erythropoietic rate-limiting step of heme production. Hypoxia upregulates transferrin (Tf), which transports Fe^{3+} into cells; the Tf receptor, which binds Tf and enables cellular Tf uptake; and ceruloplasmin (also known as ferroxidase), which is required to oxidize ferrous (Fe^{2+}) to ferric (Fe^{3+}) iron (Ke and Costa, 2006). Increasing these genes supports iron supply to bone marrow for the production of functional erythrocytes.

To aid in the delivery of the erythrocytes to the hypoxic tissue, HIF-1 regulates genes involved in governing the vascular tone and heart rate. Increases of tyrosine hydroxylases in the carotid body stimulate heart rate, and increased expression of inducible nitric oxide synthase (iNOS), heme oxygenase 1, endothelin 1, adrenomedullin, and the α_{1B}-adrenergic receptor, result in increased blood flow to the hypoxic tissues (Brown and Wilson, 2004).

A response to chronic hypoxia at the tissue level is the formation of new blood vessels. Angiogenesis is a complex process that involves multiple gene products expressed by different cell types. It is best characterized as an integrated sequence of events, involving endothelial cells, stromal and tumor cells, and extracellular matrix (ECM). Consistent with a major role of hypoxia in angiogenesis, many genes involved in different steps of this process have been shown to be hypoxia-responsive, for example, vascular endothelial growth factor (VEGF), placental endothelial growth factor, angiopoietin 2, fibroblast growth factors (FGF), and their receptors, transforming growth factor β1, platelet-derived growth factor (PDGF), genes involved in ECM metabolism, including matrix metalloproteinases (MMPs) and their inhibitors, receptors, and inhibitors of plasminogen activator (Hirota and Semenza, 2006; Pugh and Ratcliffe, 2003). To form new capillaries, endothelial cells sprout and migrate from pre-existing blood vessels. VEGF is the most potent endothelial-specific mitogen, and it directly participates in angiogenesis by recruiting endothelial cells into hypoxic and avascular areas and stimulates their proliferation.

However, this complicated process that generates a functional vascular bed during embryogenesis is not executed well in the disorganized tumor. Tumor vessels show a broad spectrum of structural and functional abnormalities: they are irregularly shaped, dilated, tortuous, with blind ends, hemorrhagic (due to overproduced VEGF, also known as vascular permeability factor), and lack smooth muscle and innervation (Bergers and Benjamin, 2003; Brown, 1999). They are not organized into arterioles, venules, and capillaries, but share chaotic features of all of them. As a result, blood flow in these neovessels is highly irregular and does not eliminate tumor hypoxia. Rather, the inefficient, leaky tumor vessels provide a means for the metastatic spread of highly aggressive hypoxic cells into circulation.

12.10 Hypoxia and glucose metabolism

Under conditions of limited oxygen supply, cells switch their methods of ATP generation from the oxygen-dependent mitochondrial oxidative phosphorylation (OXPHOS) to glycolysis, the oxygen-independent metabolic pathway in the cytoplasm (Pasteur effect). The net outcome of hypoxic HIF-1 activation is likewise to shift energy production by decreasing mitochondrial function and increasing glycolysis (Denko, 2008). Nobel Prize winner Otto Warburg showed in the 1930s that tumors, in contrast to normal tissues, had a very high capacity to produce lactic acid, which was the result of a switch to glycolytic metabolism. He hypothesized that cancer cells were defective in the mitochondrial electron transport chain. We know today that HIF-1 activation in tumor cells (as well as other oncogenic changes such as activation of akt) actively shifts metabolism from oxidative to glycolytic in both hypoxic as well as oxygenated regions of the tumor. Hence, the term aerobic glycolysis describing the Warburg effect.

Glycolysis provides only two ATP molecules for each glucose molecule, in contrast to the TCA cycle, which provides up to 38 ATP molecules. In order to generate sufficient ATP by glycolysis, hypoxic cells increase glucose uptake and the activity of glycolytic enzymes through HIF-1 transactivation. HIF-1 has been shown to regulate expression of all the enzymes in the glycolytic pathway (e.g., aldolase A and C, enolase 1, hexokinase 1 and 2, pyruvate kinase M (PKM), phosphofructokinase, etc.), as well as expression of the glucose transporters GLUT1 and GLUT3, which mediate cellular glucose uptake (Semenza *et al.*, 1994). HIF-1 also actively downregulates mitochondrial oxygen consumption by inducing expression of pyruvate dehydrogenase kinase (PDK) 1 and 3 (PDK1, PDK3), which phosphorylate and inhibit pyruvate dehydrogenase (PDH) (Kim *et al.*, 2006; Lu *et al.*, 2008; Papandreou *et al.*, 2006). PDH converts pyruvate (the end product of glycolysis) to acetyl-CoA in mitochondria, the first committed step in carbohydrate oxidation. PDK activation therefore excludes pyruvate from mitochondrial OXPHOS and leads to its accumulation in the cytoplasm. Pyruvate itself has been shown to increase HIF-1α levels, promote HIF-1 DNA binding and increase mRNA levels of several HIF-target genes even in the presence of oxygen (Dalgard *et al.*, 2004). In the hypoxic cells, accumulated pyruvate is converted into lactate via lactate dehydrogenase, another HIF-1-upregulated enzyme (Semenza *et al.*, 1994). This reaction regenerates NAD^+ from NADH, which is necessary for further glycolysis (Denko, 2008). Lactate has also been reported to regulate hypoxia-inducible expression independently of hypoxia by stimulating the accumulation of HIF-1α (Lu, Forbes, and Verma, 2002). Excess lactate (lactic acid) poses a threat to the neutral intracellular milieu, so its removal is facilitated by monocarboxylate transporterss (MCTs), one of which, MCT4, is upregulated by HIF-1 (Ullah, Davies, and Halestrap, 2006).

Even though glycolysis produces far less ATP than the TCA cycle per glucose molecule, it has a much higher throughput. In addition, proliferating cells use glucose and byproducts of its metabolism as sources of carbon to produce nucleotides and lipids for the replication of cellular components (Vander Heiden, Cantley, and Thompson, 2009). Moreover, reduced mitochondrial function results in the accumulation of another biosynthetic molecule, oxygen. Molecular oxygen (O_2)

is used enzymatically in the production of sterols and unsaturated fatty acids, and as a substrate for oxygenases, hydroxylases, and demethylases (Denko, 2008). The use of oxygen in non-energy producing pathways may be responsible for the tumor growth advantage of aerobic glycolysis. Reduced mitochondrial function will conserve oxygen for cells further away from blood vessels, and benefit the tumor as a whole (Chen *et al.*, 2009). Reduced mitochondrial function therefore yields increased anabolic substrates for the rapidly dividing tumor cells (Denko, 2008).

12.11 Hypoxia and acidosis

The production of lactic acid by hypoxic tumor cells, combined with a vasculature that is inadequate to remove it, gives rise to tumor acidosis. Because lactic acid has a pK_a of 3.9, it is dissociated into a lactate anion and a proton at physiological pH (Tannock and Rotin, 1989). In most tumors, intracellular pH (pH_i) lies between 7.0 and 7.4, similar to non-tumor cells, whereas extracellular pH (pH_e) is typically between 6.9 and 7.0, although values as low as 6.0 have been reported (Swietach, Vaughan-Jones, and Harris, 2007). Intracellular acidosis poses a threat to cell survival, and maintenance of a normal pH_i is a key cellular strategy to protect against apoptotic death and permit cell proliferation (Swietach, Vaughan-Jones, and Harris, 2007).

The realization that pH_i was alkaline required a mechanism for lactic acid extrusion to prevent its cytosolic accumulation. As mentioned earlier, MCTs are such transporters that cotransport H^+ ions with monocarboxylate anions like lactate or pyruvate to extracellular space. MCT4 is a HIF-target gene (Ullah, Davies, and Halestrap, 2006).

Although it is clear that lactate produced by glycolysis generates an acidic microenvironment, cells that express defective forms of lactate dehydrogenase (LDH) still maintain a low pH_e (Yamagata *et al.*, 1998). One consequence of rapid metabolism in any tissue is the formation of protons that must be removed from the cell if the internal milieu is to maintain its normal pH. Short-term homeostasis of pH_i in cells involves the recruitment of rapid H^+-consuming mechanisms including physicochemical buffering (bicarbonate), H^+-consuming metabolic reactions, transfer of acids from the cytosol into organelles, and membrane-based ion transport mechanisms (Tannock and Rotin, 1989). In cancer cells, the protons formed during glycolysis leave the cell not only with lactate via MCTs but are also exported by the Na^+/H^+ exchanger (NHE1), using the energy of the Na^+ gradient. This antiporter, yet another HIF-1 target gene (Shimoda *et al.*, 2006), elevates cytosolic Na^+, which is subsequently pumped out by the Na^+/K^+ ATPase. H^+ might also be exported by vacuolar type H^+ ATPases in the plasma membrane (Tannock and Rotin, 1989).

A comparison of the metabolic profiles of glycolysis-impaired and parental cells revealed that cell-generated CO_2 is also a significant source of extracellular acidity in tumors (Helmlinger *et al.*, 2002). The effectiveness of extracellular acid-trapping depends on the ability of CO_2 to be hydrated to HCO_3^- and H^+ outside the cell (Swietach, Vaughan-Jones, and Harris, 2007). Carbonic anhydrases (CAs) catalyze this reaction – intracellular isoforms in the cytoplasm and transmembrane isoforms extracellularly. Two transmembrane CA isoforms, namely CA IX and XII, are associated with tumors. CA XII is overexpressed in some tumor types, whereas CA IX is almost exclusively related to cancer. They are both induced by hypoxia with CA IX as one of the most hypoxia-responsive HIF-1 target genes (Wykoff *et al.*, 2000). Expression of CA IX is linked to poor prognosis (Potter and Harris, 2003), and the protein appears to be functionally implicated in tumorigenesis – as part of regulatory mechanisms that control pH and cell adhesion (Svastova *et al.*, 2003, 2004).

It is generally accepted that CAs function in so-called metabolons. Metabolons are group of proteins all working together and in close proximity enhancing a certain metabolic effect. By a similar mechanism, CA IX could modulate pH at both sides of the plasma membrane (Pastorekova, Parkkila, and Zavada, 2006). As an extracellular component, CA IX hydrates CO_2 that diffuses from inside of the cell, to bicarbonate, which is then taken up by anion exchanger (AE) and transported to the cytosol in exchange for Cl^-. CA IX is capable of directly interacting with AEs and improve their bicarbonate transport activity (Morgan *et al.*, 2007). In the cytoplasm, CA II (an intracellular component) could use the bicarbonate to neutralize protons. Thus created carbon dioxide freely diffuses to extracellular space where CA IX can once again convert it to bicarbonate. This reaction facilitated by CA IX generates protons in the extracellular space that contribute to acidification of the tumor microenvironment, but at the same time, it prevents acidification of the intracellular milieu of cancer cells (Pastorekova, Parkkila, and Zavada, 2006).

12.12 Hypoxia and metastasis

One mechanism by which a cell could respond (or adapt) to a hypoxic environment would be to move to a more well-oxygenated one. In order to migrate to a secondary site, cells from a primary tumor need to first overcome cell-to-cell adhesion contacts, break through the ECM, and penetrate into the circulatory system. Once there, they must overcome anchorage-dependent growth (anoikis) and be able to escape host immunity, and again penetrate from the vessels to a location, where they could implant and proliferate (Chan and Giaccia, 2007). Cell migration therefore depends on the migratory machinery located intracellularly and on the interaction of the cell with the ECM (Stock and Schwab, 2009). Many of the hypoxia-regulated processes described above can positively contribute to these metastatic activities.

Standard neutral intracellular pH combined with low extracellular pH promote an aggressive phenotype. Cellular pH regulators enhance the invasiveness of cancer cells and even promote directional cell migration by their linkage to the cytoskeleton and their ion-translocating activity (Stock and Schwab, 2009). For example, the hypoxia-inducible NHE1 accumulates at the leading edge of migrating cells, where in the thin, mitochondria-free glycolytic lamellipodium (actin-rich structure that protrudes into the ECM) it helps shuttle protons out of a site of their massive local production (Stock and Schwab, 2009). NHE1 together with AE 2 also regulate local osmotic swelling supporting the leading edge outgrowth due to their function in transporting Na^+ and Cl^- ions, respectively (Stock and Schwab, 2009). CAs constitute another portion of pH-regulating cellular components. Apart from pH regulation, which is a common function of all active CAs, CA IX is unique because it is also capable of actively functioning in cell deadhesion. CA IX is localized on the cell surface, especially in the regions of cell-to-cell contacts, and overlaps with E-cadherin in adhering lateral membranes (Svastova *et al.*, 2003). E-cadherin and β-catenin are central cell adhesion molecules that generate tight intercellular contacts linked to the actin cytoskeleton. CA IX decreases the level of cytoskeleton-linked E-cadherin by competitively binding to β-catenin, and thus destabilizes cell contacts (Svastova *et al.*, 2003). Hypoxia has been also shown to downregulate E-cadherin by increasing the expression of its transcriptional repressors Snail and SIP (Evans *et al.*, 2007).

The more acidic pericellular microenvironment at the cell front also stimulates pH-sensitive proteases in the immediate vicinity of the migrating tumor cells. The cleavage of most ECM components is executed by MMPs, which are converted from pro-MMPs to active MMPs by proteolytic cascades activated at low pH_e (Stock and Schwab, 2009). MMP-2 and membrane type-1 MMP are also HIF target genes (Munoz-Najar *et al.*, 2006; Petrella, Lohi, and Brinckerhoff, 2005). Urokinase-type plasminogen activator (uPA) activated following binding to its cell surface receptor (uPAR, another HIF-1 target gene), converts plasminogen into plasmin, which is involved in the degradation of various ECM proteins directly at the advancing cell front, and in the activation of MMPs (Sullivan and Graham, 2007). Lysyl oxidase (LOX), that catalyzes collagen and elastin crosslinking and regulates focal adhesion kinase activity, and connective tissue growth factor (CTGF) are yet additional hypoxia-inducible ECM proteins implicated in invasion (Chan and Giaccia, 2007; Erler *et al.*, 2006).

A very potent activator of cell motility is the hepatocyte growth factor (HGF), a stromal-cell derived cytokine that signals through binding to the HGF receptor tyrosine kinase encoded by the Met proto-oncogene. MET (HGF receptor tyrosine kinase) signaling has been shown to alter the expression, topographical localization and activity of cadherins, integrins, and MMPs, supporting invasion through the stroma (Sullivan and Graham, 2007). Hypoxia has been shown to induce the MET gene and enhance HGF-MET signaling (Pennacchietti *et al.*, 2003).

Apart from increasing angiogenesis and lymphangiogenesis, which provide routes for tumor dissemination, hypoxic upregulation of VEGF also further enhances metastatic spread by directly increasing vascular permeability (Sullivan and Graham, 2007).

12.13 Therapeutic implications

HIF-1's role in promoting tumorigenesis has made it a potential target for anticancer therapy, and this will be discussed in a separate chapter within this book. Here we will discuss possible anticancer interventions that are based on an understanding of the specific downstream HIF-1 target genes and how these gene products contribute to an adaptive response in the tumor. It may be that development of a drug that can inhibit a specific HIF target gene may be more effective than a drug that inhibits all HIF target genes.

Even though tumor hypoxia has been shown clinically to lead to resistance to conventional treatment modalities, it also represents a unique tumor characteristic that could be exploited either for its physical parameters or its biological consequences or both. From the physical point of view, there is a group of bioreductive prodrugs that are designed to be reduced by biological enzymes to their toxic, active metabolites only or preferentially in the absence of oxygen (Wouters *et al.*, 2002). This selectivity is achieved by a fast free-radical reaction in which the drug radical in normoxia reduces oxygen to form superoxide radicals, and thus the drug returns to its inactive prodrug form. In the absence of oxygen, the highly reactive radical is not oxidized back to the inactive prodrug form, and therefore is capable of causing cell death by producing various types of DNA damage (Brown, 1999; Wouters *et al.*, 2002). Tirapazamine was the first selectively hypoxia-toxic drug described (Zeman *et al.*, 1986). It has shown promising results when used in combination with cisplatin and/or radiotherapy (Wouters *et al.*, 2002).

The biological consequences of tumor hypoxia lead to activation of genes resulting in adaptation of the tumor to the adverse microenvironment. Another approach is therefore to directly target these hypoxia-activated pathways in hope of blocking the tumor of its ability to adapt and survive. This could be done by blocking HIF as the master regulator of hypoxic gene transcription either by inhibiting its translation (cetuximab, rapamycin) or promoting its degradation (trichostatin A) or decreasing its transcriptional activity or DNA binding (echinomycin) (Semenza, 2009). However, simply targeting HIF-1 may result in additional effects from unknown target genes that could result in unwanted growth promoting effects (ie decrease in HIF-1-dependent pro-death processes) (Denko, 2008). In fact, preclinical models have shown that even though inhibition of HIF leads to tumor growth delay, eventually tumor growth resumes in a HIF-independent manner (Leo, Giaccia, and Denko, 2004). An alternative approach may be to target specific HIF downstream products, perhaps several at the same time. Clinical efficacy has been shown in the targeting of hypoxia-inducible protein VEGFA (bevacizumab), or VEGFA receptor tyrosine kinase (sorafenib), or epidermal growth factor receptor (gefitinib). In addition, drugs are being developed that target carbonic anhydrase IX (mAbG250, sulfonamides), or PDKs (dichloroacetate).

In addition, it may be possible to design a rational combination of drugs that take advantage of HIF-1 target gene inactivation (and its physiologic effects) with other therapies that could benefit from these consequences. One such combination has been reported in the literature in which bioreductive drugs are combined with agents interfering with the HIF-inducible mitochondrial pathways. As mentioned above, HIF-1 regulates the change from oxidative to glycolytic metabolism by upregulation of PDK1 and inactivation of PDH (Papandreou *et al.*, 2006). By decreasing mitochondrial function, tumor cells conserve oxygen and anabolic substrates. By inhibiting this adaptive response, it is possible to increase oxygen consumption of tumor cells and thereby make them even more hypoxic. This 'pharmacologic hypoxia' can then increase the efficacy and specificity of hypoxia-specific cytotoxins. Biochemical inhibition of HIF-1 or PDK1 by echinomycin or dichloroacetate respectively, increases the oxygen consumption rate of solid tumors resulting in increased tumor hypoxia, which sensitizes tumors to treatment with tirapazamine (Cairns *et al.*, 2007) or PR-104 (Cairns *et al.*, 2009). This strategy provides a high degree of tumor specificity because normal cells rely on OXPHOS for energy production. Normal cells' PDH is therefore in the active, unphosphorylated state so treatment with dichloroacetate does not have an effect on normal cell PDH activity.

As we learn more about how tumor cells adapt to hypoxia, we will discover more potential targets for therapeutic interventions. Using HIF-1 and HIF-1-regulated genes will allow for the targeted therapies to be truly targeted to the tumor, as there is little HIF-1 expression in normal adult tissues. It will be necessary to understand all the downstream consequences of these interventions in order to best schedule or combine these novel drugs with each other or with standard therapies.

References

Appelhoff, R.J., Tian, Y.M., Raval, R.R. *et al.* (2004) Differential function of the prolyl hydroxylases PHD1, PHD2, and PHD3 in the regulation of hypoxia-inducible factor. *Journal of Biological Chemistry*, **279** (37), 38458–38465.

Aprelikova, O., Chandramouli, G.V., Wood, M. *et al.* (2004) Regulation of HIF prolyl hydroxylases by hypoxia-inducible factors. *Journal of Cellular Biochemistry*, **92** (3), 491–501.

Baysal, B.E. (2003) On the association of succinate dehydrogenase mutations with hereditary paraganglioma. *Trends in Endocrinology and Metabolism*, **14** (10), 453–459.

Bergers, G. and Benjamin, L.E. (2003) Tumorigenesis and the angiogenic switch. *Nature Reviews Cancer*, **3** (6), 401–410.

Berra, E., Richard, D.E., Gothie, E., and Pouyssegur, J. (2001) HIF-1-dependent transcriptional activity is required for oxygen-mediated HIF-1alpha degradation. *FEBS Letters*, **491** (1-2), 85–90.

Berra, E., Benizri, E., Ginouves, A. *et al.* (2003) HIF prolyl-hydroxylase 2 is the key oxygen sensor setting low steady-state levels of HIF-1alpha in normoxia. *EMBO Journal*, **22** (16), 4082–4090.

Briere, J.J., Favier, J., Benit, P. *et al.* (2005) Mitochondrial succinate is instrumental for HIF1alpha nuclear translocation in SDHA-mutant fibroblasts under normoxic conditions. *Human Molecular Genetics* **14** (21) 3263–3269.

Brizel, D.M., Scully, S.P., Harrelson, J.M. *et al.* (1996) Tumor oxygenation predicts for the likelihood of distant metastases in human soft tissue sarcoma. *Cancer Research*, **56** (5), 941–943.

Brizel, D.M., Dodge, R.K., Clough, R.W., and Dewhirst, M.W. (1999) Oxygenation of head and neck cancer: changes during radiotherapy and impact on treatment outcome. *Radiotherapy and Oncology*, **53** (2), 113–117.

Brown, J.M. (1999) The hypoxic cell: a target for selective cancer therapy--eighteenth Bruce F. Cain Memorial Award lecture. *Cancer Research*, **59** (23), 5863–5870.

Brown, J.M. and Wilson, W.R. (2004) Exploiting tumour hypoxia in cancer treatment. *Nature Reviews Cancer*, **4** (6), 437–447.

Cairns, R.A., Papandreou, I., Sutphin, P.D. and Denko, N.C. (2007) Metabolic targeting of hypoxia and HIF1 in solid tumors can enhance cytotoxic chemotherapy. *Proceedings of the National Academic Sciences of the United States of America*, **104** (22), 9445–9450.

Cairns, R.A., Bennewith, K.L., Graves, E.E. *et al.* (2009) Pharmacologically increased tumor hypoxia can be measured by 18F-Fluoroazomycin arabinoside positron emission tomography and enhances tumor response to hypoxic cytotoxin PR-104. *Clinical Cancer Research*, **15** (23), 7170–7174.

Chan, D.A. and Giaccia, A.J. (2007) Hypoxia, gene expression, and metastasis. *Cancer and Metastasis Reviews*, **26** (2), 333–339.

Chen, Y., Cairns, R., Papandreou, I. *et al.* (2009) Oxygen consumption can regulate the growth of tumors, a new perspective on the Warburg effect. *PLoS One*, **4** (9), e7033.

Dalgard, C.L., Lu, H., Mohyeldin, A., and Verma, A. (2004) Endogenous 2-oxoacids differentially regulate expression of oxygen sensors. *Biochemistry Journal*, **380** (Pt. 2), 419–424.

Denko, N.C. (2008) Hypoxia, HIF1 and glucose metabolism in the solid tumour. *Natural Reviews Cancer*, **8** (9), 705–713.

Dewhirst, M.W., Cao, Y., and Moeller, B. (2008) Cycling hypoxia and free radicals regulate angiogenesis and radiotherapy response. *Natural Reviews Cancer*, **8** (6), 425–437.

Epstein, A.C., Gleadle, J.M., McNeill, L.A. *et al.* (2001) *C. elegans* EGL-9 and mammalian homologs define a family of dioxygenases that regulate HIF by prolyl hydroxylation. *Cell*, **107** (1), 43–54.

Erler, J.T., Bennewith, K.L., Nicolau, M. *et al.* (2006) Lysyl oxidase is essential for hypoxia-induced metastasis. *Nature*, **440** (7088), 1222–1226.

Evans, A.J., Russell, R.C., Roche, O. *et al.* (2007) VHL promotes E2 box-dependent E-cadherin transcription by HIF-mediated regulation of SIP1 and snail. *Molecular and Cellular Biology*, **27** (1), 157–169.

Gordan, J.D., Lal, P., Dondeti, V.R. *et al.* (2008) HIF-alpha effects on c-Myc distinguish two subtypes of sporadic VHL-deficient clear cell renal carcinoma. *Cancer Cell*, **14** (6), 435–446.

Gray, L.H., Conger, A.D., Ebert, M. *et al.* (1953) The concentration of oxygen dissolved in tissues at the time of irradiation as a factor in radiotherapy. *British Journal of Radiology*, **26** (312), 638–648.

Harris, A.L. (2002) Hypoxia – a key regulatory factor in tumour growth. *Nature Reviews Cancer*, **2** (1), 38–47.

Heidbreder, M., Frohlich, F., Johren, O. *et al.* (2003) Hypoxia rapidly activates HIF-3alpha mRNA expression. *FASEB Journal*, **17** (11), 1541–1543.

Helmlinger, G., Sckell, A., Dellian, M. *et al.* (2002) Acid production in glycolysis-impaired tumors provides new insights into tumor metabolism. *Clinical Cancer Research*, **8** (4), 1284–1291.

Hirota, K. and Semenza, G.L. (2006) Regulation of angiogenesis by hypoxia-inducible factor 1. *Critical Reviews in Oncology/Hematology*, **59** (1), 15–26.

Hockel, M., Schlenger, K., Aral, B. *et al.* (1996) Association between tumor hypoxia and malignant progression in advanced cancer of the uterine cervix. *Cancer Research*, **56** (19), 4509–4515.

Hockel, M., Schlenger, K., Hockel, S., and Vaupel, P. (1999) Hypoxic cervical cancers with low apoptotic index are highly aggressive. *Cancer Research*, **59** (18), 4525–4528.

Holmquist-Mengelbier, L, Fredlund, E., Lofstedt, T. *et al.* (2006) Recruitment of HIF-1alpha and HIF-2alpha to common target genes is differentially regulated in neuroblastoma: HIF-2alpha promotes an aggressive phenotype. *Cancer Cell*, **10** (5), 413–423.

Hon, W.C., Wilson, M.I., Harlos, K. *et al.* (2002) Structural basis for the recognition of hydroxyproline in HIF-1 alpha by pVHL. *Nature*, **417** (6892), 975–978.

Huang, L.E., Gu, J., Schau, M., and Bunn, H.F. (1998) Regulation of hypoxia-inducible factor 1alpha is mediated by an O2-dependent degradation domain via the ubiquitin-proteasome pathway. *Proceedings of the National Academic Science of the United States of America*, **95** (14), 7987–7992.

Isaacs, J.S., Jung, Y.J., Mole, D.R. *et al.* (2005) HIF overexpression correlates with biallelic loss of fumarate hydratase in renal cancer: novel role of fumarate in regulation of HIF stability. *Cancer Cell* **8** (2) 143–153.

Ivan, M., Kondo, K., Yang, H. *et al.* (2001) HIFalpha targeted for VHL-mediated destruction by proline hydroxylation: implications for O_2 sensing. *Science*, **292** (5516), 464–468.

Jeong, J.W., Bae, M.K., Ahn, M.Y. *et al.* (2002) Regulation and destabilization of HIF-1alpha by ARD1-mediated acetylation. *Cell*, **111** (5), 709–720.

Jiang, B.H., Rue, E., Wang, G.L. *et al.* (1996) Dimerization, DNA binding, and transactivation properties of hypoxia-inducible factor 1. *Journal of Biological Chemistry*, **271** (30), 17771–17778.

Jiang, B.H., Zheng, J.Z., Leung, S.W. *et al.* (1997) Transactivation and inhibitory domains of hypoxia-inducible factor 1alpha. Modulation of transcriptional activity by oxygen tension. *Journal of Biological Chemistry*, **272** (31), 19253–19260.

Johnson, A.B., Denko, N., and Barton, M.C. (2008) Hypoxia induces a novel signature of chromatin modifications and global repression of transcription. *Mutation Research*, **640** (1-2), 174–179.

Ke, Q. and Costa, M. (2006) Hypoxia-inducible factor-1 (HIF-1). *Molecular Pharmacology*, **70** (5), 1469–1480.

Kim, J.W., Tchernyshyov, I., Semenza, G.L., and Dang, C.V. (2006) HIF-1-mediated expression of pyruvate dehydrogenase kinase: a metabolic switch required for cellular adaptation to hypoxia. *Cell Metabolism*, **3** (3), 177–185.

King, A., Selak, M.A., and Gottlieb, E. (2006) Succinate dehydrogenase and fumarate hydratase: linking mitochondrial dysfunction and cancer. *Oncogene*, **25** (34), 4675–4682.

Kondo, K. and Kaelin, W.G. Jr. (2001) The von Hippel–Lindau tumor suppressor gene. *Experimental Cell Research*, **264** (1), 117–125.

Kondo, K., Klco, J., Nakamura, E. *et al.* (2002) Inhibition of HIF is necessary for tumor suppression by the von Hippel–Lindau protein. *Cancer Cell*, **1** (3), 237–246.

Lancaster, D.E., McNeill, L.A., McDonough, M.A. *et al.* (2004) Disruption of dimerization and substrate phosphorylation inhibit factor inhibiting hypoxia-inducible factor (FIH) activity. *Biochemistry Journal*, **383** (Pt. 3), 429–437.

Lando, D., Peet, D.J., Gorman, J.J. *et al.* (2002a) FIH-1 is an asparaginyl hydroxylase enzyme that regulates the transcriptional activity of hypoxia-inducible factor. *Genes and Development*, **16** (12), 1466–1471.

Lando, D., Peet, D.J., Whelan, D.A. *et al.* (2002b) Asparagine hydroxylation of the HIF transactivation domain a hypoxic switch. *Science*, **295** (5556), 858–861.

Lang, K.J., Kappel, A., and Goodall, G.J. (2002) Hypoxia-inducible factor-1alpha mRNA contains an internal ribosome entry site that allows efficient translation during normoxia and hypoxia. *Molecular Biology of the Cell*, **13** (5), 1792–1801.

Laughner, E., Taghavi, P., Chiles, K. *et al.* (2001) HER2 (neu) signaling increases the rate of hypoxia-inducible factor 1alpha (HIF-1alpha) synthesis: novel mechanism for HIF-1-mediated vascular endothelial growth factor expression. *Molecular and Cellular Biology*, **21** (12), 3995–4004.

Leo, C., Giaccia, A.J., and Denko, N.C. (2004) The hypoxic tumor microenvironment and gene expression. *Seminars in Radiation Oncology*, **14** (3), 207–214.

Li, F., Sonveaux, P., Rabbani, Z.N. *et al.* (2007) Regulation of HIF-1alpha stability through *S*-nitrosylation. *Molecular Cell*, **26** (1), 63–74.

Lu, C.W., Lin, S.C., Chen, K.F. *et al.* (2008) Induction of pyruvate dehydrogenase kinase-3 by hypoxia-inducible factor-1 promotes metabolic switch and drug resistance. *Journal of Biological Chemistry*, **283** (42), 28106–28114.

Lu, H., Forbes, R.A., and Verma, A. (2002) Hypoxia-inducible factor 1 activation by aerobic glycolysis implicates the Warburg effect in carcinogenesis. *Journal of Biological Chemistry*, **277** (26), 23111–23115.

Mahon, P.C., Hirota, K., and Semenza, G.L. (2001) FIH-1: a novel protein that interacts with HIF-1alpha and VHL to mediate repression of HIF-1 transcriptional activity. *Genes and Development*, **15** (20), 2675–2686.

Makino, Y., Cao, R., Svensson, K. *et al.* (2001) Inhibitory PAS domain protein is a negative regulator of hypoxia-inducible gene expression. *Nature*, **414** (6863), 550–554.

Masson, N., Willam, C., Maxwell, P.H. *et al.* (2001) Independent function of two destruction domains in hypoxia-inducible factor-alpha chains activated by prolyl hydroxylation. *EMBO Journal*, **20** (18), 5197–5206.

Mastrogiannaki, M., Matak, P., Keith, B. *et al.* (2009) HIF-2alpha, but not HIF-1alpha, promotes iron absorption in mice. *Journal of Clinical Investigation*, **119** (5), 1159–1166.

Maxwell, P.H., Wiesener, M.S., Chang, G.W. *et al.* (1999) The tumour suppressor protein VHL targets hypoxia-inducible factors for oxygen-dependent proteolysis. *Nature*, **399** (6733), 271–275.

Morgan, P.E., Pastorekova, S., Stuart-Tilley, AK *et al.* (2007) Interactions of transmembrane carbonic anhydrase, CAIX, with bicarbonate transporters. *American Journal of Cell Physiology*, **293** (2), C738–748.

Munoz-Najar, U.M., Neurath, K.M., Vumbaca, F. *et al.* (2006) Hypoxia stimulates breast carcinoma cell invasion through MT1-MMP and MMP-2 activation. *Oncogene*, **25** (16), 2379–2392.

Nordsmark, M., Overgaard, M., and Overgaard, J. (1996) Pretreatment oxygenation predicts radiation response in advanced squamous cell carcinoma of the head and neck. *Radiotherapy Oncology*, **41** (1), 31–39.

Papandreou, I., Cairns, R.A., Fontana, L. *et al.* (2006) HIF-1 mediates adaptation to hypoxia by actively downregulating mitochondrial oxygen consumption. *Cell Metabolism*, **3** (3), 187–197.

Papandreou, I., Krishna, C., Kaper, F. *et al.* (2005) Anoxia is necessary for tumor cell toxicity caused by a low-oxygen environment. *Cancer Research*, **65** (8), 3171–3178.

Pastorekova, S., Parkkila, S., and Zavada, J. (2006) Tumor-associated carbonic anhydrases and their clinical significance. *Advances in Clinical Chemistry*, **42**, 167–216.

Pennacchietti, S., Michieli, P., Galluzzo, M. *et al.* (2003) Hypoxia promotes invasive growth by transcriptional activation of the met protooncogene. *Cancer Cell*, **3** (4), 347–361.

Petrella, B.L., Lohi, J., and Brinckerhoff, C.E. (2005) Identification of membrane type-1 matrix metalloproteinase as a target of hypoxia-inducible factor-2 alpha in von Hippel-Lindau renal cell carcinoma. *Oncogene*, **24** (6), 1043–1052.

Potter, C.P. and Harris, A.L. (2003) Diagnostic, prognostic and therapeutic implications of carbonic anhydrases in cancer. *British Journal of Cancer*, **89** (1), 2–7.

Pugh, C.W. and Ratcliffe, P.J. (2003) Regulation of angiogenesis by hypoxia: role of the HIF system. *Natural Medicine*, **9** (6), 677–684.

Richard, D.E., Berra, E., Gothie, E. *et al.* (1999) p42/p44 mitogen-activated protein kinases phosphorylate hypoxia-inducible factor 1alpha (HIF-1alpha) and enhance the transcriptional activity of HIF-1. *Journal of Biological Chemistry*, **274** (46), 32631–32637.

Ryan, H.E., Lo, J., and Johnson, R.S. (1998) HIF-1 alpha is required for solid tumor formation and embryonic vascularization. *EMBO Journal*, **17** (11), 3005–3015.

Schofield, C.J. and Ratcliffe, P.J. (2004) Oxygen sensing by HIF hydroxylases. *Natural Reviews Molecular Cell Biology*, **5** (5), 343–354.

Scortegagna, M., Ding, K., Oktay, Y. *et al.* (2003) Multiple organ pathology, metabolic abnormalities and impaired homeostasis of reactive oxygen species in Epas1$^{-/-}$ mice. *Nature Genetics*, **35** (4), 331–340.

Selak, M.A., Armour, S.M., MacKenzie, E.D. *et al.* (2005) Succinate links TCA cycle dysfunction to oncogenesis by inhibiting HIF-alpha prolyl hydroxylase. *Cancer Cell*, **7** (1), 77–85.

Semenza, G.L. (2001) HIF-1, O(2), and the 3 PHDs: how animal cells signal hypoxia to the nucleus. *Cell*, **107** (1), 1–3.

Semenza, G.L. (2002) Physiology meets biophysics: visualizing the interaction of hypoxia-inducible factor 1 alpha with p300 and CBP. *Proceedings of the National Academic Sciences of United States of America*, **99** (18), 11570–11572.

Semenza, G.L. (2010) Defining the role of hypoxia-inducible factor 1 in cancer biology and therapeutics. *Oncogene*, **29** (5), 625–634.

Semenza, G.L., Nejfelt, M.K., Chi, S.M., and Antonarakis, S.E. (1991) Hypoxia-inducible nuclear factors bind to an enhancer element located 3′ to the human erythropoietin gene. *Proceedings of the National Academic Sciences of United States of America*, **88** (13), 5680–5684.

Semenza, G.L., Roth, P.H., Fang, H.M., and Wang, G.L. (1994) Transcriptional regulation of genes encoding glycolytic enzymes by hypoxia-inducible factor 1. *Journal of Biological Chemistry*, **269** (38), 23757–23763.

Shimoda, L.A., Fallon, M., Pisarcik, S. *et al.* (2006) HIF-1 regulates hypoxic induction of NHE1 expression and alkalinization of intracellular pH in pulmonary arterial myocytes. *American Journal of Physiology. Lung Cell Molecular Physiology*, **291** (5), 941–949.L

Stock, C. and Schwab, A. (2009) Protons make tumor cells move like clockwork. *Pflugers Archiv*, **458** (5), 981–992.

Sullivan, R. and Graham, C.H. (2007) Hypoxia-driven selection of the metastatic phenotype. *Cancer Metastasis Reviews*, **26** (2), 319–331.

Suzuki, H., Tomida, A., and Tsuruo, T. (2001) Dephosphorylated hypoxia-inducible factor 1alpha as a mediator of p53-dependent apoptosis during hypoxia. *Oncogene*, **20** (41), 5779–5788.

Svastova, E., Zilka, N., Zat'ovicova, M. *et al.* (2003) Carbonic anhydrase IX reduces E-cadherin-mediated adhesion of MDCK cells via interaction with beta-catenin. *Experimantal Cell Research*, **290** (2), 332–345.

Svastova, E., Hulikova, A., Rafajova, M. *et al.* (2004) Hypoxia activates the capacity of tumor-associated carbonic anhydrase IX to acidify extracellular pH. *FEBS Letters*, **577** (3), 439–445.

Swietach, P., Vaughan-Jones, R.D., and Harris, A.L. (2007) Regulation of tumor pH and the role of carbonic anhydrase 9. *Cancer and Metastasis Reviews*, **26** (2), 299–310.

Tannock, I.F. and Rotin, D. (1989) Acid pH in tumors and its potential for therapeutic exploitation. *Cancer Research*, **49** (16), 4373–4384.

Thomlinson, R.H. and Gray, L.H. (1955) The histological structure of some human lung cancers and the possible implications for radiotherapy. *British Journal of Cancer*, **9** (4), 539–549.

Tian, H., Hammer, R.E., Matsumoto, A.M. *et al.* (1998) The hypoxia-responsive transcription factor EPAS1 is essential for catecholamine homeostasis and protection against heart failure during embryonic development. *Genes and Development*, **12** (21), 3320–3324.

Ullah, M.S., Davies, A.J., and Halestrap, A.P. (2006) The plasma membrane lactate transporter MCT4, but not MCT1, is up-regulated by hypoxia through a HIF-1alpha-dependent mechanism. *Journal of Biological Chemistry*, **281** (14), 9030–9037.

Vander Heiden, M.G., Cantley, L.C., and Thompson, C.B. (2009) Understanding the Warburg effect: the metabolic requirements of cell proliferation. *Science*, **324** (5930), 1029–1033.

Vaupel, P. and Harrison, L. (2004) Tumor hypoxia: causative factors, compensatory mechanisms, and cellular response. *Oncologist*, **9** (Suppl. 5), 4–9.

Vaupel, P., Schlenger, K., Knoop, C., and Hockel, M. (1991) Oxygenation of human tumors: evaluation of tissue oxygen distribution in breast cancers by computerized O_2 tension measurements. *Cancer Research*, **51** (12), 3316–3322.

Wang, G.L., Jiang, B.H., Rue, E.A., and Semenza, G.L. (1995) Hypoxia-inducible factor 1 is a basic-helix-loop-helix-PAS heterodimer regulated by cellular O_2 tension. *Proceedings of the National Academic Sciences of United States of America*, **92** (12), 5510–5514.

Wenger, R.H. (2006) Mitochondria: oxygen sinks rather than sensors? *Medical Hypotheses*, **66** (2), 380–383.

Wouters, B.G., Weppler, S.A., Koritzinsky, M. *et al.* (2002) Hypoxia as a target for combined modality treatments. *European Journal of Cancer*, **38** (2), 240–257.

Wouters, B.G., van den Beucken, T., Magagnin, M.G. *et al.* (2004) Targeting hypoxia tolerance in cancer. *Drug Resistance Updates*, **7** (1), 25–40.

Wykoff, C.C., Beasley, N.J., Watson, P.H. *et al.* (2000) Hypoxia-inducible expression of tumor-associated carbonic anhydrases. *Cancer Research*, **60** (24), 7075–7083.

Yamagata, M., Hasuda, K., Stamato, T., and Tannock, I.F. (1998) The contribution of lactic acid to acidification of tumours: studies of variant cells lacking lactate dehydrogenase. *British Journal of Cancer*, **77** (11), 1726–1731.

Zeman, E.M., Brown, J.M., Lemmon, M.J. *et al.* (1986) SR-4233: a new bioreductive agent with high selective toxicity for hypoxic mammalian cells. *International Journal of Radiation Oncology Biology Physics*, **12** (7), 1239–1242.

13

Regulation of the Unfolded Protein Response in Cancer

Jing Zhang and Albert C. Koong

Department of Radiation Oncology, Stanford University School of Medicine, Stanford, CA, USA

13.1 Introduction

As tumors develop, they encounter stresses in the tumor microenvironment, including hypoxia, nutrient deprivation, and glucose fluctuation. Tumors must adapt to these adverse conditions in order to survive and proliferate. The endoplasmic reticulum (ER) is a key organelle that can sense and trigger an appropriate response to these extracellular stresses. By regulating protein synthesis and secretion, the ER maintains cellular homeostasis and facilitates adaptation to these microenvironmental conditions. In solid tumors, hypoxia is often accompanied by nutrient deprivation due to inadequate vascularization. Hypoxia perturbs ER function and adversely affects protein folding. As a result, misfolded proteins accumulate, causing ER stress, and leading to the activation of the unfolded protein response (UPR). An important function of the UPR is to promote cell survival by halting protein translation, increasing expression of ER chaperones and protein folding enzymes, and targeting misfolded proteins for degradation (Zhang and Kaufman, 2006). Therefore, within the context of the tumor microenvironment, the UPR is an important adaptive signaling pathway essential for the survival and proliferation of tumor cells (Claudio, 2009). However, under prolonged ER stress, the UPR can also serve a paradoxical function through its regulation of apoptosis. Indeed, dysregulation of the UPR has been reported in many types of human cancers (Feldman, Chauhan, and Koong, 2005; Moenner *et al.*, 2007). In this chapter we describe how the UPR responds to changes in the protein folding environment of the ER, transmits stress signals to the nucleus, and alters the global gene expression profile. We also discuss

Tumor Microenvironment Edited by Dietmar W. Siemann

the role of UPR in cancer and how targeting various components of the UPR may be a promising cancer therapeutic strategy.

13.2 The UPR signaling cascade

The ER is an intracellular membranous organelle presents in eukaryotic cells and is responsible for the assembly of all secreted and transmembrane proteins. Before translocation to the extracellular space or cytoplasmic membrane, these proteins undergo a series of folding and modification steps, such as disulfide bond formation and glycosylation. ER resident folding enzymes and chaperones, including protein disulfide isomerase (PDI), glucose regulated protein 78 (GRP78)/Bip and endoplasmic reticulum oxidase 1 (ERO1), bind to nascent peptides and facilitate the conformational modification of client proteins (Federovitch, Ron, and Hampton, 2005). This process is extremely sensitive to alterations in the extracellular environment, such as glucose deprivation or hypoxia. Perturbations in this process leads to the accumulation of misfolded or unfolded proteins in the ER lumen (ER stress) which, if prolonged without correction, can lead to cell death. To ensure that only correctly folded proteins are exported and that misfolded proteins are targeted for degradation, the ER has evolved a highly regulated feedback system. The UPR serves as the proximal sensor coupling the protein folding status in the ER with the activation of target genes to restore homeostasis whenever the system becomes unbalanced (Schröder and Kaufman, 2005a). For example, one critical function of the UPR is to reduce the total amount of nascent proteins entering the ER lumen by inhibiting protein translation. Activation of the UPR also results in the enhancement of ER protein folding capacity by increasing expression of genes encoding foldases and chaperones. While these processes are occurring, the protein load on the ER can be further reduced by protein degradation through a process known as (endoplasmic reticulum-associated degradation (ERAD) (Ron, 2002; Harding *et al.*, 2002). Through the integration of these cellular functions, the UPR reduces the protein load on the ER, ultimately promoting cellular adaptation and survival. Thus, it is not surprising that the UPR plays a vital role in the physiological activities of high-ER-demand professional secretory cells, including plasma cells, pancreatic β cells, and hepatocytes (Zhang and Kaufman, 2006).

Paradoxically, the UPR also instigates proapoptotic pathways leading to apoptosis under prolonged ER stress, particularly if the excessive accumulation of misfolded proteins cannot be ameliorated (Kim *et al.*, 2006; Fribley, Zhang, and Kaufman, 2009). Therefore, divergent cell fates depend upon a balance between UPR-mediated survival and pro-apoptotic pathways. Dysregulation of UPR effectors that offsets this balance has been implicated in various diseases, such as diabetes, neurodegenerative diseases, and cancer.

In mammalian cells, there are three major distinct branches of ER-resident transmembrane proteins that sense ER stress and trigger the UPR signaling cascades: inositol requiring kinase 1 (IRE1), double-stranded RNA activated protein kinase (PKR)-like endoplasmic reticulum kinase (PERK), and activating transcription factor 6 (ATF6). These signaling branches of the UPR are shown schematically in Figure 13.1.

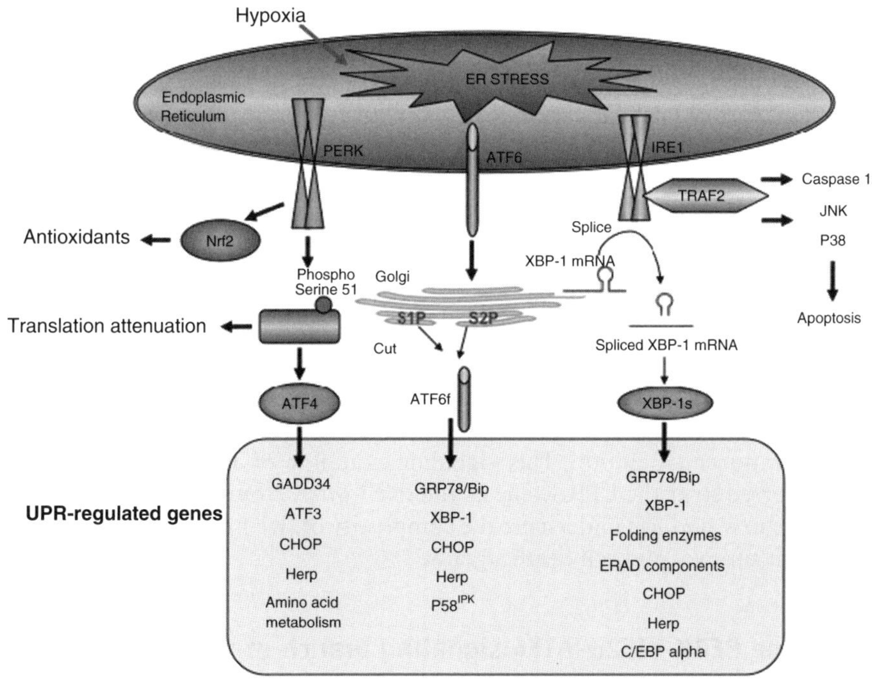

Figure 13.1 Schematic representation of UPR signaling pathways.

13.2.1 The IRE1-XBP1 signaling branch of the UPR

IRE1 is the only ER stress 'sensor' that is evolutionarily conserved throughout all eukaryotic cells. There are two isoforms of IRE1: namely, IRE1α and IRE1β, expressed in mammals. IRE1α, the human ortholog to yeast IRE1, is ubiquitously expressed in all cell types and IRE1β expression is limited to intestinal tissue. IRE1 is a type I transmembrane protein that has dual kinase activities, containing a serine/threonine kinase domain and an endoribonuclease domain at the cytoplasmic terminal. During unstressed conditions, ER chaperone protein Grp78/BiP directly binds to the luminal side of IRE1, suppressing its dimerization/oligomerization and activation. Under ER stress, Grp78/BiP is competitively titrated away by the accumulation of misfolded proteins in the ER lumen. IRE1 thereafter dimerizes and undergoes autophosphorylation, leading to the activation of the endoribonuclease domain. The active endoribonuclease excises a 26-nucleotide intron from the mRNA encoding a basic leucine zipper (bZIP) transcription factor, X-box-binding protein (XBP1). The removal of this intron causes a translational frame shift and results in the production of a larger, spliced form of XBP1 (XBP1s), which is a potent transcriptional activator of UPR targeted genes. XBP1s binds to specific promoter elements of genes encoding folding enzymes, chaperones, and ERAD machinery. ERAD reduces protein load upon the ER through a process consisting of retrograde transport of misfolded/unfolded proteins from the ER lumen to the cytosol and subsequent proteasome-mediated degradation (Vembar and Brodsky, 2008). Many ERAD components, including ER degradation enhancing α-mannosidase-like protein (EDEM), are preferentially regulated by the IRE1–XBP1 branch of the UPR (Back *et al.*, 2005; Heath-Engel, Chang, and Shore, 2008).

The IRE1–XBP1 cascade is an evolutionarily conserved signaling arm of the UPR. In yeast, IRE1-mediated gene regulation is entirely dependent on the splicing and activation of HAC1, the ortholog of mammalian XBP1 (Niwa *et al.*, 2004). However, the activity of IRE1 in metazoans is more complex. Recent studies have shown that IRE1α also functions through XBP1-independent mechanisms (Hetz and Glimcher, 2009). For example, the endoribonuclease activity of IRE1α has been shown to be responsible for the decay of selected mRNAs encoding a variety of secreted proteins (Hollien and Weissman, 2006). Independent of this endoribonuclease function, IRE1α can also interact with adaptor proteins, activating apoptosis and autophagy pathways through mitogen-activated protein kinase (MAPK) family of kinases and caspases (Urano *et al.*, 2000). This signaling complex of IRE1α-adaptor-protein kinase is referred to as the UPRosome or the IRE1 *interactome* (Hetz and Glimcher, 2009). These new findings underscore the complexity of IRE1 function in metazoans, eliciting both survival and cell death signals.

13.2.2 The PERK-eIF2α-ATF4 signaling branch of the UPR

Similar to IRE1α, PERK is also a type I transmembrane protein residing in the ER membrane. In the absence of ER stress, GRP78/BiP associates with the luminal domain of PERK, preventing the dimerization, phosphorylation, and activation of this conventional serine/threonine protein kinase. Under ER stress, GRP78/BiP is sequestrated by the accumulated unfolded proteins and dissociates from PERK, allowing dimerization of PERK and phosphorylation of its kinase domain localized to the cytosolic side. Activated PERK phosphorylates eukaryotic initiation factor 2α (eIF2α) on serine 51, resulting in attenuation of protein translation by blocking formation of the eukaryotic translation initiation complex eIF2/GTP/Met-tRNA, thereby leading to a global inhibition of mRNA translation (Wek, Jiang, and Anthony, 2006). Thus, activation of PERK has an immediate effect on reducing protein biosynthesis and decreasing the load of nascent proteins in the ER lumen.

In addition to decreasing global protein synthesis, phosphorylation of eIF2α by PERK selectively increases translation of a specific group of mRNAs encoding UPR-associated signaling molecules. These mRNAs contain multiple upstream open reading frames (uORFs) in their 5′-untranslated region (UTR), which prevents translation under non-ER stress conditions, but allows increased translation when eIF2α is phosphorylated. ATF4, a bZIP family transcription factor, is regulated in this manner under ER stress. ATF4 controls the expression of downstream transcriptional regulators, including ATF3 and C/EBP homologous protein (CHOP). This pathway is also highly regulated as CHOP induces the expression of growth arrest and DNA damage-inducible protein-34 (GADD34), which dephosphorylates eIF2α by directing the protein phosphatase PP1 to eIF2α. This degree of feedback regulation allows tight control of protein synthesis that is exquisitely responsive to ER stress. CHOP modulates the transcription of genes involved in protein folding, such as ERO1 and genes mediating UPR-associated apoptosis, such as Bcl-2 (McCullough *et al.*, 2001). In addition to its role as a key regulator of the UPR, PERK also links the UPR and oxidative stress response (Cullinan and Diehl, 2006). PERK directly phosphorylates and activates transcription factor nuclear factor-E2-related factor-2 (Nrf2), an important regulator of antioxidants and detoxifying enzymes. Activation of Nrf2 by PERK serves a cytoprotective function against oxidative stress. PERK knock-out cells or cells expressing a mutant non-phosphorylatable eIF2α^{51A}, are hypersensitive to ER stress and mouse models harboring these genetic mutations develop early onset of diabetes caused by degeneration of pancreatic β-cells (Zhang *et al.*, 2002).

13.2.3 The ATF6 signaling branch of the UPR

ATF6, a bZIP transcription factor, was identified as another important ER stress sensor in a promoter binding screen under ER stress. This class of ER stress regulators also includes ATF6 homologs, such as cAMP response element binding (CREB)H, CREB3, CREB4, and old astrocyte specifically induced substance (OASIS) (Kondo *et al.*, 2005). In unstressed cells, ATF6 is retained in the ER membrane in an inactive form by its association with GRP78/BiP (Sommer and Jarosch, 2002). During ER stress, GRP78/BiP is titrated away by the misfolded/unfolded proteins and ATF6 is released allowing transit to the Golgi compartment where it is cleaved by site-specific proteases S1P and S2P (Ye *et al.*, 2000). Cleaved or activated ATF6 is a potent transcriptional inducer of key UPR target genes such as GRP78/BiP and XBP1.

To ensure redundancy of critical UPR target genes such as GRP78/BiP, there is significant signaling overlap between the three major branches of the UPR. However, each pathway is also responsible for regulating its own distinct set of genes (Table 13.1). Expression profiling studies in which various components of the UPR were either inhibited or deleted have helped to define the primary mechanism of regulation for specific sets of genes (Yoshida *et al.*, 2000; Lee, Iwakoshi, and Glimcher, 2003b). Each major branch of ER stress sensors and their downstream transcriptional targets provide an integrated signaling network that coordinates the overall protein folding environment in the ER and serves to maintain a balance between cell survival and cell death during ER stress.

Targeting various components of the UPR for therapeutic purposes and shifting the balance of this coordinated cellular adaptation program toward cell death is an attractive anticancer approach. Understanding how key aspects of the tumor microenvironment affects UPR signaling is critical to the design of these targeted therapeutics.

13.3 Hypoxia activates UPR

Hypoxia, a common feature of the microenvironment in solid tumors, develops because of aberrantly regulated tumor vasculature and tumor growth kinetics that outpace the formation of new vessels. As a result, low and fluctuating oxygen

Table 13.1 Summary of UPR signaling-regulated genes.

UPR signaling branch	Protein	Function	Reference
IRE1-XBP1	GRP78/Bip	Chaperone, ER stress sensor	Lee, Iwakoshi, and Glimcher (2003b), Ni and Lee (2007)
	HRDJ	Co-chaperone	Lee, Iwakoshi, and Glimcher (2003b), Yu, Haslam, and Haslam (2000)
	Herp	ERAD component, ubiquitin-like protein	Lee, Iwakoshi, and Glimcher (2003b), Ma and Hendershot (2004)
	EDEM1-3	ERAD component, ER degradation enhancing, mannosidase alpha-like	Lee, Iwakoshi, and Glimcher (2003b), Yamamoto *et al.* (2004a)
	TRAP	ERAD component, translocon-associated protein	Nagasawa *et al.* (2007)
	XBP1	Transcription factor	Lee, Iwakoshi, and Glimcher (2003b), Ogawa and Mori (2004)
	CHOP	Transcription factor	Lee, Iwakoshi, and Glimcher (2003b), Endo *et al.* (2005)
	C/EBPα	Transcription factor	Sha *et al.* (2009)
	PDI	Folding enzyme, disulfide bond formation	Lee, Iwakoshi, and Glimcher (2003b), Schröder and Kaufman (2005b)
	P58 IPK	Co-chaperone, negative regulator of eIF2α kinase	Lee, Iwakoshi, and Glimcher(2003b), Rutkowski *et al.* (2007)
	Calrectulin	Lectin chaperone, retention of mono-glucosylated glycoproteins	Lee *et al.* (2007)
	GRP170/ ORP150	Chaperone	Schröder and Kaufman (2005b), Easton, Kaneko, and Subjeck (2000)
PERK-eIF2α-ATF4	CHOP	Transcription factor	Mungrue *et al.* (2009), Wek, Jiang, and Anthony (2006)
	ATF3	Transcription factor	Mungrue *et al.* (2009), Wek, Jiang, and Anthony (2006)
	GADD34	Co-chaperone, negative regulator of eIF2α kinase	Ma and Hendershot (2003), Lee, Cevallos, and Jan (2009)
	ANAS	Amino acid metabolism, asparagine synthetase	Gjymishka, Su, and Kilberg (2009)
	Herp	ERAD component, ubiquitin-like protein	Ma and Hendershot (2004)

Table 13.1 (*continued*)

UPR signaling branch	Protein	Function	Reference
ATF6	CHOP	Transcription factor	Nakanishi, Sudo, and Morishima (2005), Yoshida *et al.* (2000)
	GRP78/Bip	Chaperone, facilitate protein folding	Baumeister *et al.* (2005)
	XBP1	Transcription factor	Yoshida *et al.* (2000), Yoshida *et al.* (2001)
	P58IPK	Co-chaperone, negative regulator of eIF2α kinase	van Huizen *et al.* (2003)
	Herp	ERAD component, ubiquitin-like protein	Ma and Hendershot (2004)

levels within tumors play a crucial role in cell proliferation, malignant progression, and response to anticancer therapies. Numerous studies have correlated hypoxia with tumor growth, angiogenesis, metastasis, and resistance to therapeutic agents. These effects of hypoxia can broadly be categorized as hypoxia-inducible factor-1 (HIF-1) dependent and HIF-independent (i.e., UPR) signaling pathways (Wouters and Koritzinsky, 2008).

Although it is well-documented that hypoxia activates the UPR, the exact molecular basis for hypoxic induction of ER stress has yet to be defined. One potential mechanism is that molecular oxygen serves as the ultimate electron acceptor of the disulfide bond formation reaction mediated by PDI and ERO1 (Tu and Weissman, 2004). These results predict that proteins with a high disulfide bond content would be more sensitive to misfolding in the absence of oxygen and indicate that the accumulation of misfolded proteins in the ER is an expected consequence of hypoxia.

An immediate cellular response to hypoxia is a rapid reduction to the ATP intensive process of protein synthesis. The underlying mechanism for attenuation of protein synthesis during ER stress is activation of PERK and subsequent phosphorylation of eIF2α to hinder the assembly of mRNA translation machinery. Mouse embryonic fibroblasts (MEFs) with a PERK deletion or a mutation of eIF2α at serine 51 that prevents phosphorylation undergo apoptosis when exposed to severe hypoxia (Bi *et al.*, 2005; Koumenis *et al.*, 2002). Moreover, HT29 colorectal carcinoma cells with impaired PERK function grew significantly slower when implanted as tumor xenografts. Importantly, the tumors demonstrated increased apoptosis in hypoxic regions compared to tumor with an intact PERK pathway. Furthermore, downstream transcriptional regulator ATF4 was overexpressed and colocalized with hypoxic areas in human cancer tissues, suggesting a functional contribution to hypoxia tolerance of the tumor cells (Feldman, Chauhan, and Koong, 2005; Ameri *et al.*, 2004). These findings strongly implicate PERK-eIF2α-ATF4 signaling as an important requirement for cell survival and tumor growth under hypoxic stress.

The IRE1–XBP1 signaling arm of the UPR is also implicated in cell hypoxia tolerance *in vitro*, as well as tumorigenic ability of transformed cells in tumor xenograft models (Koong, Chauhan, and Romero-Ramirez, 2006). The first direct demonstration of the requirement of the UPR for tumor growth came from studies investigating the regulation and function of IRE1–XBP1 signaling branch under hypoxia. Transformed MEFs with deletion of XBP1 did not form tumors and HT1080 (human fibrosarcoma cells) expressing an XBP1 shRNA construct to inhibit XBP1 expression demonstrated impaired tumor growth (Chen *et al.*, 2005) when implanted into immune-deficient mice, suggesting an essential role for XBP1 in tumor development. Mechanistically, XBP1 mediated survival under hypoxia, suggesting that the impairment of tumor growth was at least partially explained by increased sensitivity of these cells to hypoxia (Romero-Ramirez *et al.*, 2004). These experiments provide strong evidence that IRE1–XBP1 signaling is essential for tumor survival and growth in a hypoxic microenvironment.

ATF6, another major signaling pathway of the UPR, is also activated by hypoxia and shares some regulatory features in common with the IRE1 and PERK. Immunoblot analysis indicated that ATF6 was cleaved in a time-dependent manner by hypoxia and the cleaved ATF6 translocated from the cytosol to the nucleus after prolonged hypoxia exposure (Koumenis *et al.*, 2007). This mechanism of activation under hypoxia is similar to activation of ATF6 by other ER stresses.

As discussed above, each of the three major branches of the UPR (IRE1, PERK, ATF6) are capable of inducing GRP78/BiP under ER stress. This level of redundancy suggests an essential role for this chaperone under these conditions. Indeed, GRP78/Bip has been shown to be upregulated in many types of human tumors and essential for hypoxia tolerance *in vitro* (Lee, 2007; Koong *et al.*, 1994).

13.4 UPR and expression of UPR-targeted genes in cancer

Fluctuating changes of hypoxia within the tumor microenvironment impose a constant challenge to malignant cells. Activation of UPR signaling pathways and their downstream genes are indispensible for hypoxia tolerance and tumor progression. Clinical studies have correlated increased expression of UPR pathway genes with angiogenesis, metastasis, and resistance to chemotherapy in multiple cancer types.

13.4.1 Activation of UPR signaling branches in cancer

Biochemical analyses using human tumor samples and biological experiments in cancer cell lines and mouse models have strongly implicated involvement of the UPR in various cancer types. In hepatocellular carcinoma (HCC), IRE1-mediated splicing of XBP1 was observed in tumor cells. Moreover, as histological grade increased, the degree of XBP1 splicing also increased (Masahiro *et al.*, 2003). XBP1 was also reported to be overexpressed in colorectal adenomas and adenocarcinomas (Fujimoto *et al.*, 2007). In addition, other investigators reported increased XBP1 mRNA and protein in human primary breast carcinomas, as well as in breast cancer cell lines, but not in the corresponding non-cancer tissues (Scriven *et al.*, 2009). Davies and colleagues correlated the expression and splicing of XBP1 with significantly worse disease-free survival of hormonally treated breast cancer patients (Davies *et al.*, 2008).

Activation XBP1 within tumors is further supported by data from human tumor xenografts and transgenic mouse models expressing a XBP–luciferase fusion protein (Feldman, Chauhan, and Koong, 2005). In these models, when XBP1 is activated, there is a frame shift in the coding mRNA, leading to translation of an XBP1–luciferase fusion protein. Thus, luciferase activity is detectable only when XBP1 is spliced into its active form. In a tumor xenograft model, luciferase activity was detectable throughout tumor growth, indicating that XBP1 activation occurs even at small tumor sizes. In transgenic mice harboring the XBP1–luciferase gene, basal luciferase activity was observed to be highest in the pancreas where ER stress occurs physiologically because of high ER demands in protein folding and secretion. When the XBP1–luciferase transgenic mouse was crossed with a genetic breast tumor mouse model (FVB/N-Tg MMTV-PyVT mice, The Jackson Laboratory, Bar Harbor, ME, USA), a surprising range of heterogeneity of XBP1 activity was observed in these breast carcinomas. The degree of XBP1 activation correlated with tumor growth rates, decreased glucose uptake, and tumor hypoxia. However, there is no correlation between XBP1 activation and tumor size (Spiotto *et al.*, 2010). These findings highlight the complicated relationship between the development of ER stress and tumor growth (Figure 13.2). Targeting XBP1 in cancer is a promising cancer therapeutic strategy and should be pursued in the context of our understanding of the underlying biology. Finally, spliced XBP1 was shown to be overexpressed in human pancreatic adenocarcinoma resection specimens and activation of XBP1 correlated with areas of increased angiogenesis (Romero-Ramirez *et al.*, 2009).

The PERK-eIF2α-ATF4 signaling branch has also been shown to be important for tumor cell survival under hypoxia and during tumorigenesis in mouse models. MEFs and HT29 colon carcinoma cells expressing a dominant-negative PERK exhibited significant attenuation of tumor growth in xenograft models (Blais *et al.*, 2006; Bi *et al.*, 2005). The hypoxic regions in PERK-null tumors co-localized with regions of increased apoptosis, suggesting one potential mechanism for PERK-mediated tumor growth. ATF4, the downstream transcriptional effecter of this pathway, has been shown to be overexpressed in human cancer with the highest level of ATF4 observed primarily in areas bordering regions of necrosis (Ameri *et al.*, 2004; Ye and Koumenis, 2009). Several genes involved in angiogenesis and tumor metastasis were reported to be translated selectively in a PERK-dependent manner, (i.e., vascular endothelial growth factor and type I collagen inducible protein (vCIP), matrix metalloproteinase-13 (MMP13)) indicating additional functions of PERK in tumors (Blais *et al.*, 2006).

The significance of ATF6 within tumors is less well characterized. However, in human HCCs, ATF6 was found to be overexpressed and activated along with increased XBP1 and GRP78/Bip (Masahiro *et al.*, 2003). Additionally, Schewe *et al.* identified ATF6 as a pivotal survival factor for quiescent but not proliferative squamous carcinoma cells through GTPase Rheb and mammalian target of rapamycin

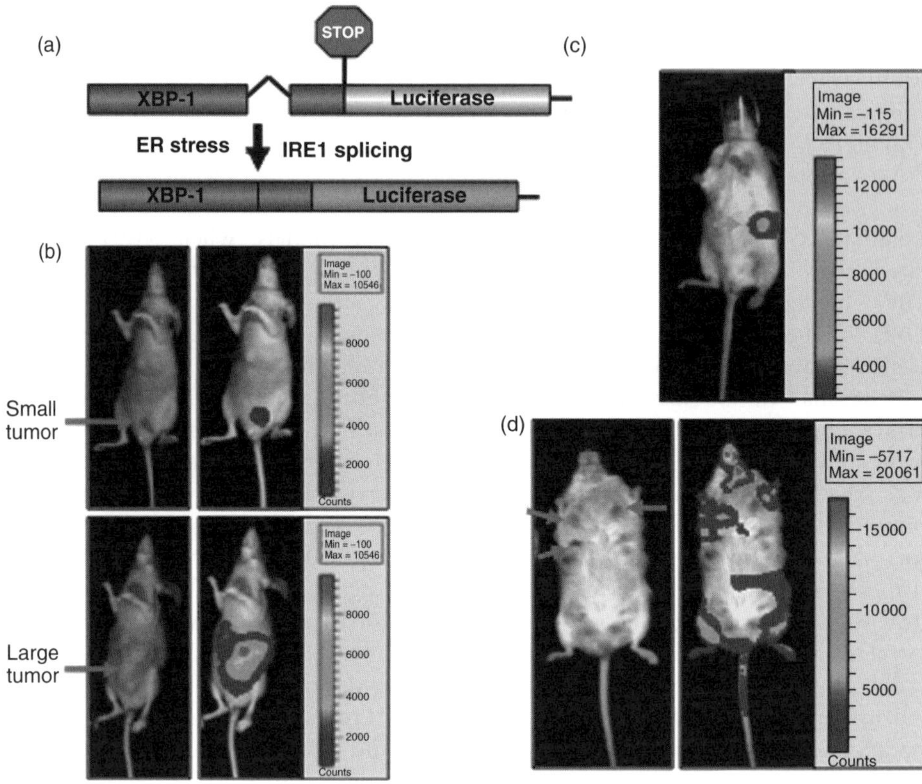

Figure 13.2 Activation of Ire1/XBP-1 pathway in implanted tumor xenografts and transgenic mouse model. (a) Schematic representation of XBP-luciferase reporter construct. (b) HT1 080 human fibrosarcoma cells stably expressing the XBP-luciferase reporter were implanted subcutaneously into nude mice. Luciferase activity was detected in both small tumor and large tumor (red arrow). (c) Transgenic mouse harboring the XBP–luciferase reporter gene. Highest basal luciferase activity was detected in the pancreas (green arrow). (d) Representative of double transgenic mouse harboring both the XBP–luciferase reporter gene and the MMTV-PyVT gene. Luciferase activity was detected in multiple spontaneous breast tumors. A full color version of this figure can be found in the color plate section.

(mTOR) (Schewe and Aguirre-Ghiso, 2008). Other investigators reported that activation of ATF6 stimulated the expression of antiapoptotic factor Mcl-1 in melanoma cells, leading to apoptosis resistance of these cells (Jiang *et al.*, 2008).

13.4.2 Altered expression of UPR-targeted genes in human cancers

A number of UPR-regulated genes display aberrant expression in human cancer and have been correlated with several aspects of tumor malignancy, including survival, angiogenesis, metastasis, and resistance to therapy. A list of these genes and the specific cancers in which they are implicated are summarized in Table 13.2.

Table 13.2 UPR-regulated genes implicated in human cancers.

Category	Gene	Cancer type	Reference
ER chaperones	GRP78/BiP	Lung, breast, colon, liver, prostate, gastric, head and neck, esophageal cancer, melanoma, malignant glioma	Hidetaka *et al.* (2005), Pootrakul *et al.* (2006), Zhang *et al.* (2006), Papalas *et al.* (2009), Pyrko *et al.* (2006), Masahiro *et al.* (2003), Lee *et al.* (2006), Chena and Change (2007), Langer *et al.* (2008), Xing *et al.* (2006), Zheng *et al.* (2008)
	GRP94/gp96	Esophagus, pancreatic, gastric, prostate, lung	Langer *et al.* (2008), Pan *et al.* (2009), Zheng *et al.* (2008), Howard *et al.* (2008), Qi *et al.* (2005)
	GRP170/ ORP150	Glioma, breast, gastric	Ozawa *et al.* (2001b), Tsukamoto *et al.* (1998), Namba *et al.* (2007)
	Calrectulin	Hepatocellular carcinoma	Nicolas *et al.* (2006)
Transcription factors	XBP1	Breast, colon, melanoma, hepatocellular carcinoma	Davies *et al.* (2008), Fujimoto *et al.* (2007), Lee *et al.* (2003a), Masahiro *et al.* (2003), Scriven *et al.* (2009)
	ATF6,	Hepatocellular carcinoma	Masahiro *et al.* (2003)
	ATF4	Breast cancer	Ameri *et al.* (2004)
	CREB-H	Hepatocellular carcinoma	Chin *et al.* (2005)
	BBF2H7	Fibromyxoid sarcoma	Storlazzi *et al.* (2003)
ERAD components	valosin-containing protein p97	Hepatocellular carcinoma, esophagus, pancreatic, prostate	Yamamoto *et al.* (2003), Yamamoto *et al.* (2004b), Yamamoto *et al.* (2004d), Tsujimoto *et al.* (2004)
	SEC61p	Glioblastoma	Lu *et al.* (2009)
Folding enzymes	PDI	Hepatocellular carcinoma	Nicolas *et al.* (2006)
	ERp29	Basal-cell carcinoma	Cheretis *et al.* (2006)

Perhaps the most well-documented UPR-regulated gene that plays an essential role in many human cancers is the ER chaperone GRP78/BiP, considered indispensable in the regulation of all three major branches of the UPR. Functionally, GRP78/BiP is involved in tumor cell proliferation, apoptosis, metastasis, and resistance to chemotherapy (Lee, 2007). Lee and coinvestigators reported that GRP78/BiP expression was predictive of resistance to Adriamycin-based chemotherapy in breast cancer patients (Lee *et al.*, 2006). Inhibition of GRP78/BiP expression through an antisense strategy sensitized tumor cells to hypoxia *in vitro* and suppressed tumor formation *in vivo* (Jamora, Dennert, and Lee, 1996). Additionally, expression of GRP78/BiP in tumors increased resistance to several DNA-damaging chemotherapeutic agents, such as etoposide and cisplatin (Lee *et al.*, 2008). Wang *et al.* also reported that blocking GRP78BiP expression sensitized breast cancer cells to microtubule-interfering agents (Wang *et al.*, 2009). Mechanistically, the cytoprotective function of GRP78/BiP has, in part, been attributed to alleviation of ER stress and inhibiting BIK/BAK-mediated and caspase-mediated proapoptotic pathway (Fu, Li, and Lee, 2007).

Another ER chaperone, GRP94, shares the same transcriptional regulatory elements with the GRP78/Bip promoter. GRP94 expression is also increased in many types of human cancers, often concurrently with the overexpression of GRP78/BiP. Furthermore, GRP94 has been implicated in mediating chemoresistance in cancer cells, most likely by inhibiting apoptosis (Fu and Lee, 2006).

Another mechanism by which ER chaperones may also influence tumor growth is through the regulation of protein secretion. For example, GRP170/ORP150 (oxygen-related protein 150) promotes angiogenesis by mediating the extracellular release of vascular endothelial growth factor A (VEGF-A) under hypoxia (Ozawa *et al.*, 2001a). This level of regulation demonstrates both the direct and indirect effects by which ER chaperones affect tumor growth.

In addition to ER chaperones other UPR target genes has also been demonstrated to be important in cancer. ERO1 and the PDI-like protein ERp29 are both components of the protein folding machinery and have been reported to be overexpressed in breast carcinomas and basal skin carcinomas, respectively (Moenner *et al.*, 2007). ERp29 has been implicated in mesenchymal–epithelial transition and functions to suppress xenograft tumor growth (Bambang *et al.*, 2009). ER translocon component SEC61γ exhibited a high frequency of gene amplification in glioblastoma multiforme (GBM) and was overexpressed primarily in the high grade gliomas (Lu *et al.*, 2009). Phenotypically, inhibition of SEC61γ expression in tumor cells led to growth suppression and apoptosis. Other genes including calreticulin, a lectin-like chaperone, were reported to be overexpressed in HCC (Nicolas *et al.*, 2006). And finally, ERAD-associated regulator, valosin-containing protein p97, has been associated with the progression of several cancers, including lung, HCC, and pancreatic ductal adenocarcinoma (Yamamoto *et al.*, 2003, 2004d, 2004c).

13.4.3 UPR as therapeutic target in cancer

Taken together, a variety of UPR-targeted genes show abnormal expression patterns in human cancers and have been correlated with tumor malignancy. These data provide convincing evidence of the critical role of UPR in cancer and form the basis for further investigation into the underlying molecular mechanisms for the development of ER stress in tumors. Targeting the various aspects of the UPR has emerged as a promising cancer therapeutic strategy (Kim, Xu, and Reed, 2008).

The therapeutic potential for pharmacological agents targeting the UPR was initially illustrated by the discovery of compounds that blocked the expression or function of GRP78/BiP. One of these compounds, versipelostatin, functioned as an inhibitor of UPR activator-induced transcription of GRP78/BiP and demonstrated antitumor activity in xenograft models (Park *et al.*, 2004). Another approach to inhibiting GRP78/BiP was utilizing interleukin 24 (IL-24), a cytokine that associates with GRP78. In these studies, IL-24 suppressed tumor activity in a breast tumor xenograft model (Gupta *et al.*, 2006). Other investigators have used a cytotoxin delivery strategy using Sub A, a bacterial toxin whose only known substrate is GRP78/BiP, into tumor cells. They found the toxin-tagged reporter protein significantly inhibited the growth of human breast and prostate tumor xenografts (Backer *et al.*, 2009).

Induction of ER stress is recognized as another mechanism of cell killing by some chemotherapeutic agents. For example, bortezomib (Velcade), a proteasome inhibitor approved for the treatment of multiple myeloma, induces ER stress by blocking protein degradation causing a net accumulation of proteins in the ER. Bortezomib has also been shown to trigger the proapoptotic effects of PERK, through the induction of ER stress (Dong *et al.*, 2009; Fribley, Zeng, and Wang, 2004). Because of its effects on increasing ER stress, bortezomib preferentially sensitizes hypoxic tumor cells to increased cell death suggesting that this is a viable therapeutic strategy to target hypoxic cells within solid tumors (Fels *et al.*, 2008). Another class of drugs originally developed as human immunodeficiency virus protease inhibitors (nelfinavir) induced ER stress in gliomas (Gupta *et al.*, 2007). Nelfinavir-activated apoptosis was potentiated by inhibition of GRP78/BiP, further substantiating the hypothesis that enhancement of tumor cell killing can be achieved through modulation of ER stress. Inhibition of caspase-4 blocked apoptosis, suggesting that nelfinavir-mediated cancer cell killing was mediated through a UPR-associated apoptotic pathway (Pyrko *et al.*, 2007). And finally, PDI inhibitors are being developed as anticancer agents that function to suppress UPR-mediated survival pathways (Lovat *et al.*, 2008).

More recently, several groups are working to develop small molecule inhibitors of PERK–ATF4 or IRE1–XBP1 as anticancer agents. In addition to their genetic validation as novel cancer targets, high throughput screens for specific inhibitors of these pathways will identify novel compounds to validate the development of targeted chemotherapeutics for these pathways. Our group has developed specific screens for compounds that modulate activity of both of these pathways and to date, have identified several promising classes of small molecules that are undergoing additional preclinical evaluation with the ultimate goal of developing a UPR-targeted cancer therapy.

13.5 Concluding remarks

In rapidly growing solid tumors, heterogeneous hypoxia regions develop because of insufficient and dysregulated tumor vasculature. The dynamic fluctuations in tumor hypoxia have a profound impact on cell proliferation, survival, invasiveness, and metastasis. Tumor hypoxia has also been associated with resistance to chemotherapy and radiation therapy resulting in poor clinical outcomes. Intensive studies on the mechanisms of hypoxia tolerance and tumor progression have identified several signaling pathways that respond to changes in oxygen and initiate an adaptive response at the translational and transcriptional levels. Besides the well-characterized HIF-1 mediated transcriptional response, the UPR has been characterized as a HIF-1-independent signaling pathway that can promote cell survival or trigger cell death, depending on the cellular context. Many UPR components, indeed, have demonstrated elevated expression and/or activity in human cancers and in animal tumor models.

The UPR is an important survival mechanism for tumor cells under hypoxic and ER stress. Adaptation to this stress is essential for the growth and proliferation of tumors. Further understanding of how the UPR exerts its effects will require a detailed examination and identification of additional components of this signaling pathway. The use of small interfering RNA libraries that allow for genome-wide screening of novel genetic modulators of the UPR will also be an effective approach to advance our understanding of this evolutionarily conserved ER stress response pathway. Such investigations are currently underway in several laboratories and will undoubtedly accelerate the development of novel targeted cancer therapeutics.

References

Ameri, K., Lewis, C.E., Raida, M. *et al.* (2004) Anoxic induction of ATF-4 through HIF-1-independent pathways of protein stabilization in human cancer cells. *Blood*, **103**, 1876–1882.

Back, S., Schröder, M., Lee, K. *et al.* (2005) ER stress signaling by regulated splicing: IRE1/HAC1/XBP1. *Methods*, **35**, 22.

Backer, J., Krivoshein, A., Hamby, C. *et al.* (2009) Chaperone-targeting cytotoxin and ER stress-inducing drug synergize to kill cancer cells. *Neoplasia*, **11**, 9.

Bambang, I.F., Xu, S., Zhou, J. *et al.* (2009) Overexpression of endoplasmic reticulum protein 29 regulates mesenchymal-epithelial transition and suppresses xenograft tumor growth of invasive breast cancer cells. *Laboratory Investigation*, **89**, 1229–1242.

Baumeister, P., Luo, S., Skarnes, W.C. *et al.* (2005) Endoplasmic reticulum stress induction of the Grp78/BiP promoter: activating mechanisms mediated by YY1 and its interactive chromatin modifiers. *Molecular and Cellular Biology*, **25**, 4529–4540.

Bi, M., Naczki, C., Koritzinsky, M. *et al.* (2005) ER stress-regulated translation increases tolerance to extreme hypoxia and promotes tumor growth. *EMBO Journal*, **24**, 3470–3481.

Blais, J.D., Addison, C.L., Edge, R. *et al.* (2006) Perk-dependent translational regulation promotes tumor cell adaptation and angiogenesis in response to hypoxic stress. *Molecular and Cell Biology*, **26**, 9517–9532.

Chen, Y., Feldman, D., Deng, C. *et al.* (2005) Identification of mitogen-activated protein kinase signaling pathways that confer resistance to endoplasmic reticulum stress in *Saccharomyces cerevisiae*. *Molecular Cancer Research*, **3**, 669–677.

Chena, A.J. and Change, C.J.T. (2007) GRP78 is over-expressed in head and neck cancer. *Radiotherapy and Oncology: Journal of the European Society for Therapeutic Radiology and Oncology*, **82**, S24–S25.

Cheretis, C., Dietrich, F., Chatzistamou, I. *et al.* (2006) Expression of ERp29, an endoplasmic reticulum secretion factor in basal-cell carcinoma. *American Journal of Dermatopathology*, **28**, 410–412.

Chin, K.-T., Zhou, H.-J., Wong, C.-M. *et al.* (2005) The liver-enriched transcription factor CREB-H is a growth suppressor protein underexpressed in hepatocellular carcinoma. *Nucleic Acids Research*, **33**, 1859–1873.

Claudio, H. (2009) The UPR as a survival factor of cancer cells: more than folding proteins? *Leukemia Research*, **33**, 880–882.

Cullinan, S.B. and Diehl, J.A. (2006) Coordination of ER and oxidative stress signaling: the PERK/Nrf2 signaling pathway. *International Journal of Biochemistry and Cell Biology*, **38**, 317–332.

Davies, M.P.A., Barraclough, D.L., Stewart, C. *et al.* (2008) Expression and splicing of the unfolded protein response gene XBP-1 are significantly associated with clinical outcome of endocrine-treated breast cancer. *International Journal of Cancer*, **123**, 85–88.

Dong, H., Chen, L., Chen, X. *et al.* (2009) Dysregulation of unfolded protein response partially underlies proapoptotic activity of bortezomib in multiple myeloma cells. *Leukemia and Lymphoma*, **50**, 974–984.

Easton, D., Kaneko, Y., and Subjeck, J. (2000) The hsp110 and Grp1 70 stress proteins: newly recognized relatives of the Hsp70s. *Cell Stress Chaperones*, **5**, 276–290.

Endo, M., Oyadomari, S., Suga, M. *et al.* (2005) The ER stress pathway involving CHOP is activated in the lungs of LPS-treated mice. *Journal of Biochemistry*, **138**, 501–507.

Federovitch, C., Ron, D., and Hampton, R. (2005) The dynamic ER: experimental approaches and current questions. *Current Opinion in Cell Biology*, **17**, 6.

Feldman, D.E., Chauhan, V., and Koong, A.C. (2005) The unfolded protein response: a novel component of the hypoxic stress response in tumors. *Molecular Cancer Research*, **3**, 597–605.

Fels, D., Ye, J., Segan, A. *et al.* (2008) Preferential cytotoxicity of bortezomib toward hypoxic tumor cells via overactivation of endoplasmic reticulum stress pathways. *Cancer Research*, **68**, 9323–9330.

Fribley, A., Zeng, Q., and Wang, C.-Y. (2004) Proteasome inhibitor PS-341 induces apoptosis through induction of endoplasmic reticulum stress-reactive oxygen species in head and neck squamous cell carcinoma cells. *Molecular and Cellular Biology*, **24**, 9695–9704.

Fribley, A., Zhang, K., and Kaufman, R.J. (2009) Regulation of apoptosis by the folded protein response. *Methods in Molecular Biology*, **559**, 191–204.

Fu, Y. and Lee, A. (2006) Glucose regulated proteins in cancer progression, drug resistance and immunotherapy. *Cancer Biology and Therapy*, **5**, 4.

Fu, Y., Li, J., and Lee, A.S. (2007) GRP78/BiP inhibits endoplasmic reticulum BIK and protects human breast cancer cells against estrogen starvation-induced apoptosis. *Cancer Research*, **67**, 3734–3740.

Fujimoto, T., Yoshimatsu, K., Watanabe, K. *et al.* (2007) Overexpression of human X-box binding protein 1 (XBP-1) in colorectal adenomas and adenocarcinomas. *Anticancer Research*, **27**, 127–131.

Gjymishka, A., Su, N., and Kilberg, M. (2009) Transcriptional induction of the human asparagine synthetase gene during the unfolded protein response does not require the ATF6 and IRE1/XBP1 arms of the pathway. *Biochemical Journal*, **417**, 695–703.

Gupta, A., Li, B., Cerniglia, G. *et al.* (2007) The HIV protease inhibitor nelfinavir downregulates Akt phosphorylation by inhibiting proteasomal activity and inducing the unfolded protein response. *Neoplasia*, **9**, 8.

Gupta, P., Walter, M.R., Su, Z.-Z. *et al.* (2006) BiP/GRP78 is an intracellular target for MDA-7/IL-24 induction of cancer-specific apoptosis. *Cancer Research*, **66**, 8182–8191.

Harding, H.P., Calfon, M., Urano, F. *et al.* (2002) Translational and translational control in th the mammalian unfolded protein response. *Annual Review of Cell and Developmental Biology*, **18**, 575–599.

Heath-Engel, H.M., Chang, N.C., and Shore, G.C. (2008) The endoplasmic reticulum in apoptosis and autophagy: role of the BCL-2 protein family. *Oncogene*, **27**, 6419–6433.

Hetz, C. and Glimcher, L.H. (2009) Fine-tuning of the unfolded protein response: assembling the IRE1[alpha] interactome. *Molecular Cell*, **35**, 551–561.

Hidetaka, U., Kenji, S., Tsunehiro, O. *et al.* (2005) Expression of endoplasmic reticulum molecular chaperone Grp78 in human lung cancer and its clinical significance. *Lung Cancer (Amsterdam, Netherlands)*, **49**, 55–62.

Hollien, J. and Weissman, J.S. (2006) Decay of endoplasmic reticulum-localized mRNAs during the unfolded protein response. *Science*, **313**, 104–107.

Howard, E., Leung, S., Yuen, H. *et al.* (2008) Decreased adhesiveness, resistance to anoikis and suppression of GRP94 are integral to the survival of circulating tumor cells in prostate cancer. *Clinical and Experimental Metastasis*, **25**, 497–508.

van Huizen, R., Martindale, J.L., Gorospe, M., and Holbrook, N.J. (2003) P58IPK, a novel endoplasmic reticulum stress-inducible protein and potential negative regulator of eIF2α signaling. *Journal of Biological Chemistry*, **278**, 15558–15564.

Jamora, C., Dennert, G., and Lee, A. (1996) Inhibition of tumor progression by suppression of stress protein GRP78/BiP induction in fibrosarcoma B/C10ME. *Proceedings of the National Academy of Sciences*, **93**, 5.

Jiang, C.C., Lucas, K., Avery-Kiejda, K.A. *et al.* (2008) Up-regulation of Mcl-1 is critical for survival of human melanoma cells upon endoplasmic reticulum stress. *Cancer Research*, **68**, 6708–6717.

Kim, R., Emi, M., Tanabe, K., and Murakami, S. (2006) Role of the unfolded protein response in cell death. *Apoptosis*, **11**, 5–13.

Kim, I., Xu, W., and Reed, J.C. (2008) Cell death and endoplasmic reticulum stress: disease relevance and therapeutic opportunities. *Nature Reviews Drug Discovery*, **7**, 1013–1030.

Kondo, S., Murakami, T., Tatsumi, K. *et al.* (2005) OASIS, a CREB/ATF-family member, modulates UPR signalling in astrocytes. *Nature Cell Biology*, **7**, 186–194.

Koong, A., Chauhan, V., and Romero-Ramirez, L. (2006) Targeting XBP-1 as a novel anti-cancer strategy. *Cancer Biology and Therapy*, **5**, 4.

Koong, A., Chen, E., Lee, A. *et al.* (1994) Increased cytotoxicity of chronic hypoxic cells by molecular inhibition of GRP78 induction. *International Journal of Radiation Oncology Biology Physics*, **28**, 661–666.

Koumenis, C., Naczki, C., Koritzinsky, M. *et al.* (2002) Regulation of protein synthesis by hypoxia via activation of the endoplasmic reticulum kinase PERK and phosphorylation of the translation initiation factor eIF2{alpha}. *Molecular and Cellular Biology*, **22**, 7405–7416.

Koumenis, C., Bi, M., Ye, J., *et al.* (2007) Hypoxia and the unfolded protein response, in *Methods in Enzymology*, vol 435 (eds S. Helmut and B. Bernhard), Academic Press, London, pp. 275–293.

Langer, R., Feith, M., Siewert, J. *et al.* (2008) Expression and clinical significance of glucose regulated proteins GRP78 (BiP) and GRP94 (GP96) in human adenocarcinomas of the esophagus. *BMC Cancer*, **8**, 70.

Lee, A.S. (2007) GRP78 induction in cancer: therapeutic and prognostic implications. *Cancer Research*, **67**, 3496–3499.

Lee, A.-H., Iwakoshi, N.N., Anderson, K.C., and Glimcher, L.H. (2003a) Proteasome inhibitors disrupt the unfolded protein response in myeloma cells. *Proceedings of the National Academy of Sciences of the United States of America*, **100**, 9946–9951.

Lee, A.-H., Iwakoshi, N.N., and Glimcher, L.H. (2003b) XBP-1 regulates a subset of endoplasmic reticulum resident chaperone genes in the unfolded protein response. *Molecular and Cellular Biology*, **23**, 7448–7459.

Lee, D., Singaravelu, G., Park, B., and Ahnn, J. (2007) Differential requirement of unfolded protein response pathway for calreticulin expression in Caenorhabditis elegans. *Journal of Molecular Biology*, **372**, 331–340.

Lee, E., Nichols, P., Spicer, D. *et al.* (2006) GRP78 as a novel predictor of responsiveness to chemotherapy in breast cancer. *Cancer Research*, **66**, 7849–7853.

Lee, H.K., Xiang, C., Cazacu, S. *et al.* (2008) GRP78 is overexpressed in glioblastomas and regulates glioma cell growth and apoptosis. *Neuro-Oncology*, **10**, 236–243.

Lee, Y.-Y., Cevallos, R.C., and Jan, E. (2009) An upstream open reading frame regulates translation of GADD34 during cellular stresses that induce eIF2α phosphorylation. *Journal of Biological Chemistry*, **284**, 6661–6673.

Lovat, P.E., Corazzari, M., Armstrong, J.L. *et al.* (2008) Increasing melanoma cell death using inhibitors of protein disulfide isomerases to abrogate survival responses to endoplasmic reticulum stress. *Cancer Research*, **68**, 5363–5369.

Lu, Z., Zhou, L., Killela, P. *et al.* (2009) Glioblastoma proto-oncogene SEC61gamma is required for tumor cell survival and response to endoplasmic reticulum stress. *Cancer Research*, **69**, 9105–9111.

Ma, Y. and Hendershot, L.M. (2003) Delineation of a negative feedback regulatory loop that controls protein translation during endoplasmic reticulum stress. *Journal of Biological Chemistry*, **278**, 34864–34873.

Ma, Y. and Hendershot, L.M. (2004) Herp is dually regulated by both the endoplasmic reticulum stress-specific branch of the unfolded protein response and a branch that is shared with other cellular stress pathways. *Journal of Biological Chemistry*, **279**, 13792–13799.

Mann, M. and Hendershot, L. (2006) UPR activation alters chemosensitivity of tumor cells. *Cancer Biology and Therapy*, **5**, 736–740.

Masahiro, S., Nobuo, K., Nobuo, I. *et al.* (2003) Activation of the ATF6, XBP1 and grp78 genes in human hepatocellular carcinoma: a possible involvement of the ER stress pathway in hepatocarcinogenesis. *Journal of Hepatology*, **38**, 605–614.

McCullough, K.D., Martindale, J.L., Klotz, L.-O. *et al.* (2001) Gadd153 sensitizes cells to endoplasmic reticulum stress by down-regulating Bcl2 and perturbing the cellular redox state. *Molecular and Cellular Biology*, **21**, 1249–1259.

Moenner, M., Pluquet, O., Bouchecareilh, M., and Chevet, E. (2007) Integrated endoplasmic reticulum stress responses in cancer. *Cancer Research*, **67**, 10631–10634.

Mungrue, I.N., Pagnon, J., Kohannim, O. *et al.* (2009) CHAC1/MGC4504 is a novel proapoptotic component of the unfolded protein response, downstream of the ATF4-ATF3-CHOP cascade. *Journal of Immunology*, **182**, 466–476.

Nagasawa, K., Higashi, T., Hosokawa, N. *et al.* (2007) Simultaneous induction of the four subunits of the TRAP complex by ER stress accelerates ER degradation. *EMBO Reports*, **8**, 483–489.

Nakanishi, K., Sudo, T., and Morishima, N. (2005) Endoplasmic reticulum stress signaling transmitted by ATF6 mediates apoptosis during muscle development. *Journal of Cell Biology*, **169**, 555–560.

Namba, T., Hoshino, T., Tanaka, K.-I. *et al.* (2007) Up-regulation of 150-kDa oxygen-regulated protein by celecoxib in human gastric carcinoma cells. *Molecular Pharmacology*, **71**, 860–870.

Ni, M. and Lee, A.S. (2007) ER chaperones in mammalian development and human diseases. *FEBS Letters*, **581**, 3641–3651.

Nicolas, C., Sufen, S., Hong, W. *et al.* (2006) Cleavage of endoplasmic reticulum proteins in hepatocellular carcinoma: detection of generated fragments in patient sera. *Gastroenterology*, **130**, 2010–2022.

Niwa, M., Patil, C., DeRisi, J., and Walter, P. (2004) Genome-scale approaches for discovering novel nonconventional splicing substrates of the Ire1 nuclease. *Genome Biology*, **6**, R3.

Ogawa, N. and Mori, K. (2004) Autoregulation of the HAC1 gene is required for sustained activation of the yeast unfolded protein response. *Genes to Cells*, **9**, 95–104.

Ozawa, K., Kondo, T., Hori, O. *et al.* (2001a) Expression of the oxygen-regulated protein ORP150 accelerates wound healing by modulating intracellular VEGF transport. *Journal of Clinical Investigation*, **108**, 41–50.

Ozawa, K., Tsukamoto, Y., Hori, O. *et al.* (2001b) Regulation of tumor angiogenesis by oxygen-regulated protein 150, an inducible endoplasmic reticulum chaperone. *Cancer Research*, **61**, 4206–4213.

Pan, Z., Erkan, M., Streit, S. *et al.* (2009) Silencing of GRP94 expression promotes apoptosis in pancreatic cancer cells. *International Journal of Oncology*, **35**, 823–828.

Papalas, J.A., Vollmer, R.T., Gonzalez-Gronow, M. *et al.* (2009) Patterns of GRP78 and MTJ1 expression in primary cutaneous malignant melanoma. *Modern Pathologyl*, **23**, 134–143.

Park, H.-R., Tomida, A., Sato, S. *et al.* (2004) Effect on tumor cells of blocking survival response to glucose deprivation. *Journal of National Cancer Institute*, **96**, 1300–1310.

Pootrakul, L., Datar, R.H., Shi, S.-R. *et al.* (2006) Expression of stress response protein Grp78 is associated with the development of castration-resistant prostate cancer. *Clinical Cancer Research*, **12**, 5987–5993.

Pyrko, P., Kardosh, A., Wang, W. *et al.* (2007) HIV-1 protease inhibitors nelfinavir and atazanavir induce malignant glioma death by triggering endoplasmic reticulum stress. *Cancer Research*, **67**, 10920–10928.

Pyrko, P., Lee, A.S., Schonthal, A. *et al.* (2006) Reduction of GRP78 expression enhances chemosensitivity of malignant gliomas. *AACR Meeting Abstracts*, **47**, 555b.

Qi, W., Zhongzhou, H., Jinhui, Z. *et al.* (2005) Overexpression of endoplasmic reticulum molecular chaperone GRP94 and GRP78 in human lung cancer tissues and its significance. *Cancer Detection and Prevention*, **29**, 544–551.

Romero-Ramirez, L., Cao, H., Nelson, D. *et al.* (2004) XBP1 is essential for survival under hypoxic conditions and is required for tumor growth. *Cancer Research*, **64**, 5943–5947.

Romero-Ramirez, L., Cao, H., Regalado, M. *et al.* (2009) X box-binding protein 1 regulates angiogenesis in human pancreatic adenocarcinomas. *Translational Oncology*, **2**, 8.

Ron, D. (2002) Translational control in the endoplasmic reticulum stress response. *Journal of Clinical Investigation*, **110**, 1383–1388.

Rutkowski, D.T., Kang, S.-W., Goodman, A.G. *et al.* (2007) The Role of p58IPK in protecting the stressed endoplasmic reticulum. *Molecular Biology of the Cell*, **18**, 3681–3691.

Schewe, D.M. and Aguirre-Ghiso, J.A. (2008) ATF6α-Rheb-mTOR signaling promotes survival of dormant tumor cells in vivo. *Proceedings of the National Academy of Sciences*, **105**, 10519–10524.

Schröder, M. and Kaufman, R. (2005a) ER stress and the unfolded protein response. *Mutation Research*, **569**, 29–63.

Schröder, M. and Kaufman, R. (2005b) The mammalian unfolded protein response. *Annual Review of Biochemistry*, **74**, 739–789.

Scriven, P., Coulson, S., Haines, R. *et al.* (2009) Activation and clinical significance of the unfolded protein response in breast cancer. *British Journal of Cancer*, **101**, 1692–1698.

Sha, H., He, Y., Chen, H. *et al.* (2009) The IRE1alpha-XBP1 pathway of the unfolded protein response is required for adipogenesis. *Cell Metabolism*, **9**, 556–564.

Sommer, T. and Jarosch, E. (2002) BiP binding keeps ATF6 at bay. *Developmental Cell*, **3**, 1–2.

Spiotto, M., Banh, A., Papandreou, I. *et al.* (2010) Visualization of the unfolded protein response in primary tumors reveals unique microenvironments that reflect metabolic variations and predict for tumor growth. *Cancer Research*, **70**, 78–88.

Storlazzi, C.T., Mertens, F., Nascimento, A. *et al.* (2003) Fusion of the FUS and BBF2H7 genes in low grade fibromyxoid sarcoma. *Human Molecular Genetics*, **12**, 2349–2358.

Tsukamoto, Y., Kuwabara, K., Hirota, S. *et al.* (1998) Expression of the 150-kd oxygen-regulated protein in human breast cancer. *Laboratory Investigation*, **78**, 699–706.

Tsujimoto, Y., Tomita, Y., Hoshida, Y. *et al.* (2004) Elevated expression of valosin-containing protein (p97) is associated with poor prognosis of prostate cancer. *Clinical Cancer Research*, **10**, 3007–3012.

Tu, B.P. and Weissman, J.S. (2004) Oxidative protein folding in eukaryotes: mechanisms and consequences. *Journal of Cell Biology*, **164**, 341–346.

Urano, F., Wang, X., Bertolotti, A. *et al.* (2000) Coupling of stress in the ER to activation of JNK protein kinases by transmembrane protein kinase IRE1. *Science*, **287**, 664–666.

Vembar, S.S. and Brodsky, J.L. (2008) One step at a time: endoplasmic reticulum-associated degradation. *Nature Reviews Molecular Cell Biology*, **9**, 944–957.

Wang, J., Yin, Y., Hua, H. *et al.* (2009) Blockade of GRP78 sensitizes breast cancer cells to microtubules-interfering agents that induce the unfolded protein response. *Journal of Cellular and Molecular Medicine*, **13**, 3888–3897.

Wek, R.C., Jiang, H.Y., and Anthony, T.G. (2006) Coping with stress: eIF2 kinases and translational control. *Biochemical Society Transactions*, **34**, 7–11.

Wouters, B.G. and Koritzinsky, M. (2008) Hypoxia signalling through mTOR and the unfolded protein response in cancer. *Nature Reviews Cancer*, **8**, 851–864.

Xing, X., Lai, M., Wang, Y. *et al.* (2006) Overexpression of glucose-regulated protein 78 in colon cancer. *Clinica Chimica Acta*, **364**, 308–315.

Yamamoto, K., Yoshida, H., Kokame, K. *et al.* (2004a) Differential contributions of ATF6 and XBP1 to the activation of endoplasmic reticulum stress-responsive cis-acting elements ERSE, UPRE and ERSE-II. *Journal of Biochemistry*, **136**, 343–350.

Yamamoto, S., Tomita, Y., Hoshida, Y. *et al.* (2004b) Expression level of valosin-containing protein (p97) is associated with prognosis of esophageal carcinoma. *Clinical Cancer Research*, **10**, 5558–5565.

Yamamoto, S., Tomita, Y., Hoshida, Y. *et al.* (2004c) Expression level of valosin-containing protein (p97) is correlated with progression and prognosis of non-small-cell lung carcinoma. *Annals of Surgical Oncology*, **11**, 697–704.

Yamamoto, S., Tomita, Y., Hoshida, Y. *et al.* (2004d) Increased expression of valosin-containing protein (p97) is associated with lymph node metastasis and prognosis of pancreatic ductal adenocarcinoma. *Annals of Surgical Oncology*, **11**, 165–172.

Yamamoto, S., Tomita, Y., Nakamori, S. *et al.* (2003) Elevated expression of valosin-containing protein (p97) in hepatocellular carcinoma is correlated with increased incidence of tumor recurrence. *Journal of Clinical Oncology*, **21**, 447–452.

Ye, J. and Koumenis, C. (2009) ATF4, an ER stress and hypoxia-inducible transcription factor and its potential role in hypoxia tolerance and tumorigenesis. *Current Molecular Medicine*, **9**, 6.

Ye, J., Rawson, R., Dave, U. *et al.* (2000) Site-1 protease (S1P) and site-2 protease (S2P), the two proteases that cleave membrane-bound SREBPs, also cleave ATF6 upon ER stress. *Molecular Biology of the Cell*, **11**, 1511.

Yoshida, H., Matsui, T., Yamamoto, A. *et al.* (2001) XBP1 mRNA is induced by ATF6 and spliced by IRE1 in response to ER stress to produce a highly active transcription factor. *Cell*, **107**, 881–891.

Yoshida, H., Okada, T., Haze, K. *et al.* (2000) ATF6 Activated by proteolysis binds in the presence of NF-Y (CBF) directly to the *cis*-acting element responsible for the mammalian unfolded protein response. *Molecular and Cell Biology*, **20**, 6755–6767.

Yu, M., Haslam, R.H.A., and Haslam, D.B. (2000) HEDJ, an Hsp40 Co-chaperone localized to the endoplasmic reticulum of human cells. *Journal of Biological Chemistry*, **275**, 24984–24992.

Zhang, J., Jiang, Y., Jia, Z. *et al.* (2006) Association of elevated GRP78 expression with increased lymph node metastasis and poor prognosis in patients with gastric cancer. *Clinical and Experimental Metastasis*, **23**, 401–410.

Zhang, K. and Kaufman, R.J. (2006) The unfolded protein response: a stress signaling pathway critical for health and disease. *Neurology*, **66**, S102–S109.

Zhang, P., McGrath, B., Li, S. *et al.* (2002) The PERK eukaryotic initiation factor 2{alpha} kinase is required for the development of the skeletal system, postnatal growth, and the function and viability of the pancreas. *Molecular and Cellular Biology*, **22**, 3864–3874.

Zheng, H., Takahashi, H., Li, X. *et al.* (2008) Overexpression of GRP78 and GRP94 are markers for aggressive behavior and poor prognosis in gastric carcinomas. *Human Pathology*, **39**, 1042–1049.

14

Influence of Hypoxia on Metastatic Spread

Richard P. Hill[1,2] and Naz Chaudary[1]

[1]*Ontario Cancer Institute/Princess Margaret Hospital and The Campbell Family Institute for Cancer Research, Toronto, Ontario, Canada*

[2]*Departments of Medical Biophysics and Radiation Oncology, University of Toronto, Toronto, Ontario, Canada*

14.1 Introduction

Metastasis is a process by which tumor cells establish new growths at sites in the body distinct from the primary tumor. This is a major cause of treatment failure and death in cancer patients. The development of metastasis is complex, requiring multiple individual stages to successfully establish a tumor at a secondary site. Tissue structure and function is intimately connected to and controlled by tumor cell–cell and cell–extracellular matrix (ECM) interactions. The metastatic process involves multiple changes at a molecular level that disrupt and modify these interactions. These include signaling through cell adhesion molecules (CAMs), such as integrins and cadherins, and tissue remodeling through the action of proteinases, such as plasmin or metalloproteinases (MMPs), as well as apoptotic machinery, chemokines, and growth factors, all of which act together to control processes such as proliferation, survival/apoptosis, migration, and invasion. These interactions are dysregulated in the tumor microenvironment and there is significant heterogeneity in these different interactions and functions between tumor types.

The microenvironment of cells in tumors involves both local interactions with surrounding cells and ECM, and exposure to pathophysiological conditions, such as low oxygen tension (hypoxia), low glucose concentrations, high lactate concentrations, low extracellular pH (acidity), and high interstitial fluid pressure (IFP), all of which can vary between different tumor regions. The expression of many of the genes involved in the various metastatic processes can be affected by exposure

Tumor Microenvironment Edited by Dietmar W. Siemann

to conditions induced by the pathophysiologic environment of tumors, particularly hypoxia. Hypoxia-inducible factors-1 and 2 (HIF-1/2) are key regulators of gene expression under hypoxic conditions and immunohistochemical analysis has found overexpression of HIF-1α in many human cancers and their metastases. High levels of both HIF-1α and HIF-2α have been positively correlated with tumor progression and poor prognosis in patients with a variety of cancers (HIF-1α – brain, non small cell lung carcinoma, breast, ovarian, uterine, and cervical tumors; HIF-2α – non-small-cell lung cancer, head and neck squamous cell, renal cell carcinoma). The acute (transient) hypoxia that occurs in tumors may induce reactive oxygen species (ROS), which in turn can activate HIF-1, promoting persistent oxidative stress and further amplifying HIF-1 activation, with downstream effects on gene expression.

In contrast to normal cells where a damaged genome is usually deleted through apoptosis or restored through repair pathways, tumor cells commonly harbor multiple mutations, including a variety of chromosomal aberrations and smaller alterations such as insertions, deletions, and base changes. Hypoxia and nutrient deprivation of cells *in vitro* have been shown to increase the number of mutations and repeated rounds of hypoxia treatment *in vitro* can generate a mutation spectrum resembling that found in solid tumors. In the absence of adequate antioxidant responses to ROS-mediated oxidative stress, various DNA adducts can be formed by ROS interaction with DNA. One of the best-studied lesions, 8-hydroxy-2′-deoxyguanosine (8-OHdG), is mutagenic and causes G–C to T–A transversions by preventing accurate DNA template reading by polymerases. ROS may also attack polyunsaturated membrane lipids causing an oxidising chain reaction and the generation of secondary mutagenic end products such as malondialdehyde (MDA), which in turn can form adducts with DNA. ROS can also contribute to tumor progression by inactivating tumor suppressor genes such as p53 or by activating oncogenes, such as c-FOS, c-JUN, and c-MYC, or transcription factors such as nuclear factor kappaB (NF-κB), activator protein 1 (AP-1) and, as noted above, HIF-1. ROS can induce expression of growth factors, such as vascular endothelial growth factor (VEGF) and interleukin-8 (IL-8), or activate proteinases, such as MMP-2. Alternatively, ROS and acidic pH may contribute to tumor progression by reacting with and inactivating proteins, including DNA repair proteins. Diminished repair of DNA damage has been demonstrated in cells exposed to hypoxic and acidic culture conditions, and hypoxia has been reported to decrease the expression of a number of DNA repair proteins.

The role of oxidative stress in tumor progression is supported by clinical studies demonstrating decreased antioxidant responses in certain human tumors relative to normal tissues, increased plasma MDA concentration in patients with metastatic versus non-metastatic disease and increased hydroxyl radical-induced DNA damage in metastatic versus non-metastatic breast tumors. The tumor microenvironment may also enhance genomic instability and promote tumor progression by selecting for genetic cell variants that have a survival advantage under the adverse microenvironmental conditions. Hypoxia has been identified as a selective pressure for mismatch DNA repair-deficient colorectal cell populations that had increased mutation frequencies. It has also been demonstrated that exposure of some transformed cells to hypoxia *in vitro* induces apoptosis, but that repeated cycles of hypoxia and reoxygenation can select for cells with mutated p53 that are resistant to apoptosis. Apoptotic cell death has also been reported to depend on p53 under acidic conditions *in vitro*, which may suggest a growth advantage for p53 deficient cells under acidic conditions *in vivo*.

14.2 The metastatic process

The spread of tumor cells can occur either via lymphatic drainage of the primary site leading to local lymph node metastases or via the blood vasculature leading to metastases in distant organs. Further spread of tumor cells from these sites may also occur, including spread from the lymphatic system into the blood vasculature. These processes have traditionally been divided into a number of specific stages, which include escape of the tumor cell from the primary site, survival in the circulation, arrest in the capillaries at a new location, migration from the capillary into the interstitial space, and establishment of a new growth at this new location. These stages represent potential barriers, all of which a tumor cell must be able to surmount if it is to be successful in forming a metastasis. These stages of the metastatic process are illustrated in Figure 14.1 as stages 1–5. Since these stages are sequential and dependent on each other in order for a successful metastatic event to occur, only a few cells may be successful in completing all of them. Thus, an important aspect of metastasis is that it is a very inefficient process. As illustrated in Figure 14.1, similar molecules or classes of molecules may be involved in more than one stage of this process.

14.2.1 Intravasation

The initial stages of the metastatic process involve detachment of cells from the primary tumor and their (intra)vasation into the lymphatic or systemic circulation. Since there appear to be few functional lymphatics in most tumors, cells at or near the invasive edge of the tumor are likely to be those which escape into the lymphatic system. Tumor cells have to migrate through the ECM and transverse the basement membrane to access these two circulatory systems. Tumor blood vessels are often poorly formed, because of the rapid angiogenesis that can occur during tumor growth. This may provide for easier access of some tumor cells into the bloodstream. This stage involves the processes of detachment, mediated by adhesion proteins responsible for cell–cell and cell–matrix interactions such as cadherins and integrins, migration, and degradation of the ECM, mediated by a number of proteinases and their inhibitors. This may be enhanced by an epithelial to mesenchymal transition (EMT), which is a conserved process critical for embryonic development, and involves the release of epithelial cells from the surrounding tissue through alterations in integrin and cadherin expression and subsequent reorganization of the ECM to permit migration. EMT has been implicated in the metastatic process, although there is also the possibility that cells may migrate via

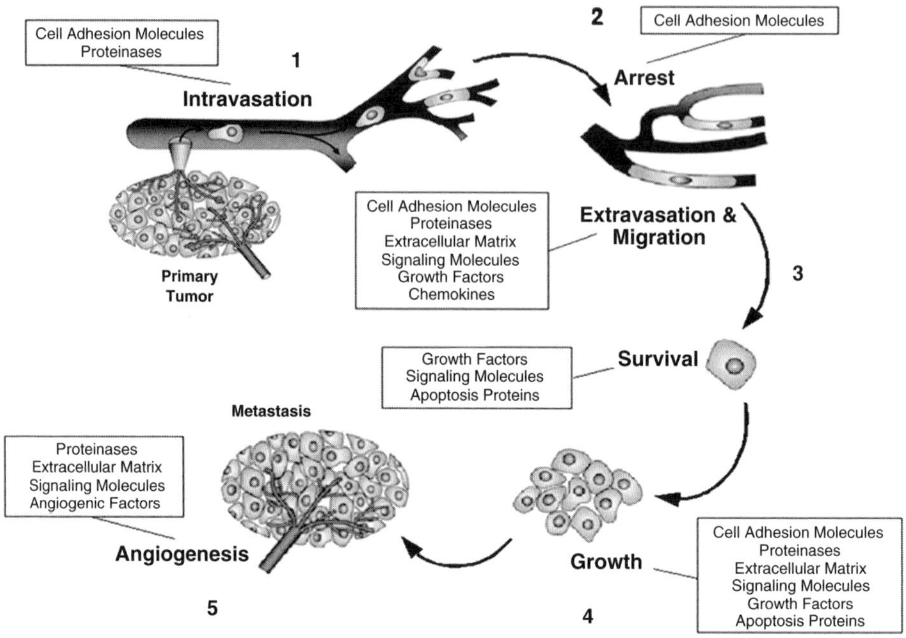

Figure 14.1 The process of metastasis. Schematic diagram showing the stages of metastasis. (labeled 1–5) and the classes of molecules known to be involved at each stage. (Figure adapted from Cairns, Khokha, and Hill, (2003) and Chambers, Groom, and Macdonald, (2002).)

an ameboid-type process that may not involve adhesion proteins. A number of genes that have been implicated in invasion and metastasis, have been found to be regulated by the hypoxic or acidic microenvironment, suggesting a direct role of the microenvironment in invasion (Table 14.1).

14.2.2 Survival, extravasation, and growth

Once in the circulatory system, the cells are likely to be arrested in the first capillary bed that they encounter owing to size constraints and/or to an increased possibility of interacting with adhesion molecules expressed on the surface of the endothelial cells (or exposed basement membrane) at lower flow rates. Depending on the location of the primary tumor, the first (blood) capillary bed will be either the liver (for intestinal tumors growing in organs whose blood drains into the portal system) or the lung (for tumors in organs in other parts of the body). Some of the arrested cells may successfully transit through this initial 'first-pass' organ and move on to other sites, but studies with radiolabeled cells have suggested that the proportion of cells which do so is low and cell type dependent. Cells may be destroyed in the circulation either by the immune system, by hemodynamic forces, or stress-induced apoptosis. Such apoptosis may also occur at the arrest site. A very small number of tumor cells may be found in the bloodstream.

It has been generally thought that the ability of tumor cells to extravasate out of (blood) vessels is an important limiting step in metastasis formation. Proteolytic enzymes similar to those discussed above are likely to be involved in this extravasation stage. Some tumor cells can apparently bypass this stage by initiating growth while still in the blood vessel. Furthermore, intravital microscopy studies have shown that, for some tumor cells, extravasation can be efficient, allowing for most of the cells (>80%) that enter the circulation to move from the vasculature into the interstitial space. On the other hand, only a small proportion of these cells were found to initiate cell division to form micrometastases, and only a proportion of these became vascularized and grew into macroscopic metastases (Chambers, Groom and Macdonald, 2002). These data indicate that growth after extravasation can be a critical limiting factor in metastatic spread.

Initiation of proliferation requires that the cells can either respond to locally available growth factors or can make sufficient growth factors themselves to achieve autocrine stimulation of growth. Meanwhile, they must also be able to avoid or ignore any negative growth signals, including apoptotic signals. Thus, metastatic growth is a result of a positive interaction between the tumor cell phenotype and the stromal tissue in which the metastasis is growing. Proteolytic enzymes are usually regarded as being involved in the intravasation and extravasation stages, but they may also be involved in the growth stimulation stage, since some growth factors that are sequestered in the interstitial matrix (e.g., VEGF) or are bound to binding proteins in the circulation (e.g., insulin-derived growth factor (IGF)) can be released by proteolytic enzyme activity. Similarly, adhesion molecules may be involved not only in the arrest but also in the other stages (e.g., apoptosis or early induction of growth) by virtue of their function in signaling cascades. Expression of molecules by both the tumor cells and the stromal cells can play a role in these processes.

14.2.3 Angiogenesis

Finally, if progressive growth to a size greater than about 1 mm diameter is to occur, the tumor cells must be able to initiate angiogenesis to stimulate the formation of new blood vessels. Angiogenesis is important for metastasis both in terms of new vessel formation in the primary tumor, allowing tumor cells to more easily enter the circulation, and for the ability of tumor cells to grow into macroscopic nodules at the metastatic site. A number of angiogenic factors have been shown to be induced by the tumor microenvironment, including VEGF, IL-8, angiogenin (ANG), basic fibroblast growth factor (bFGF), and platelet-derived endothelial cell growth factor (PD-ECGF) (Table 14.1). It has been reported in clinical studies that regions of high vascular density (hot spots) are associated with more aggressive disease and experimentally, vascular hot spots, induced in hypoxic foci, have been found to be associated with subsequent spontaneous lung metastasis formation. A direct role of microenvironmentally-induced angiogenesis at the secondary site in enhancing the initial metastasis formation has not been demonstrated, although experimental studies with human melanoma cells have implicated the expression of such molecules in metastasis formation (see below).

14.3 The tumor microenvironment and metastasis

14.3.1 Tumor metastasis and hypoxia

14.3.1.1 Tumor hypoxia

Regions of low oxygen tension (pO_2), or hypoxia, are found in most solid tumors. A proportion of tumor cells are in hypoxic regions beyond the maximum diffusion distance of oxygen from a capillary. These cells may be exposed chronically to low oxygen tensions (chronic hypoxia) for hours to days. Tumor hypoxia can also occur transiently due to the substantial instability in microregional blood flow and tissue oxygenation that can occur in animal and human tumors. These fluctuations are thought to be due to transient occlusion and narrowing of vessels and to arteriolar vasomotion. Also, the abnormal architecture of the vascular system itself may produce variations in red cell flow. High IFP may further exacerbate the situation. This blood flow instability, in the context of an already poorly organized and regulated vascular system, can produce short-term (5–60 minutes) fluctuations in oxygenation (acute transient hypoxia). Thus tumor cells adjacent to vasculature may be exposed to short-term hypoxia; however, the actual distance from blood vessels at which hypoxia occurs likely varies widely in different tumors because of the unstable delivery of oxygen within tumor blood vessels and the variable oxygen consumption of tumor cells. The extent of hypoxic regions is heterogeneous even amongst tumors of identical histopathological type, and does not correlate with standard prognostic factors such as tumor size, stage, and grade. Although the definition of hypoxia depends on the effect being studied and varies between different studies, a pO_2 level <10–15 mmHg ($<2\%$ O_2 in the gas phase) is generally considered to be associated with changes in the expression of a number of genes and has also been associated with poor prognosis in a number of clinical studies. It is important to note however that certain hypoxia-responsive genes (e.g., carbonic anhydrase-9 (CA-9), glucose transporter-1 (Glut-1), VEGF, and urokinase plasminogen activator receptor (uPAR)) are turned on (albeit to a reduced extent) at higher levels of oxygen (2–3% O_2) than other hypoxia-responsive genes (e.g., cathepsin D, glyceraldehyde phosphate dehydrogenase) that may not change until lower levels of oxygenation ($<1.0\%$) occur (Dayan *et al.*, 2006).

14.3.1.2 Clinical studies of metastasis

Multiple clinical studies using the Eppendorf pO_2 histograph have demonstrated a connection between tumor hypoxia and disease progression in a variety of human tumors including carcinomas of the cervix, prostate and the head and neck and soft tissue sarcomas. Early studies (Hockel *et al.*, 1996) demonstrated that tumor hypoxia was a poor prognostic factor in cervical carcinoma regardless of whether patients were treated with radiotherapy or with surgery alone, and suggested that hypoxia might be associated with lymph node metastases. A number of other studies have also linked hypoxia directly to distant metastases for both cervix carcinoma (Fyles *et al.*, 2006) and soft tissue sarcoma (Brizel *et al.*, 1996). The nature of the connection is difficult to explore in the clinic, although several studies have demonstrated that there are alterations within the tumor microenvironment as a consequence of hypoxia.

14.3.1.3 Experimental studies of metastasis

A large number of experimental studies provide support for the link between hypoxia and metastatic disease, as discussed in a number of recent reviews (Chan and Giaccia, 2007; Sullivan and Graham, 2007; Lunt, Chaudary, and Hill, 2008). For example, a relation between pO_2 levels of small primary tumors and micrometastases was found in the highly metastatic KHT murine fibrosarcoma when grown intramuscularly in mice. Furthermore, exposure of a squamous cell carcinoma and a melanoma cell line to hypoxia *in vitro* for 18–24 hours followed by 18 hours reoxygenation resulted in an enhanced number of lung metastases when the cells were injected intravenously into mice. Notably this effect was transient, with longer periods of reoxygenation resulting in a return to base line levels of lung metastases, suggesting a potential role for transient changes in gene expression. The mouse double minute 2 (Mdm2) gene is one of the genes transiently upregulated by hypoxic exposure in KHT cells, and it was found that transient overexpression of Mdm2 protein in oxygenated KHT cells can increase their metastatic potential. This was thought to be to the result of increased resistance to stress-induced apoptosis due to the downregulation of p53. An increase in the number of metastases was also observed in nude mice when cells of a human melanoma line were injected intravenously after a 24-hour hypoxia treatment *in vitro*. In this study the increase in metastases was observed without any period of reoxygenation and was associated with a simultaneous induction of VEGF secretion. The increase in metastases could be blocked by treatment of the mice with an anti-VEGF antibody. No effect of hypoxic exposure was observed in another human melanoma cell line, which had a constitutively high VEGF level (Rofstad and Halsor, 2000).

Currently there is no method to routinely measure and discriminate between chronic and acute hypoxia in tumors. Consequently, the extent of chronic versus acute hypoxia in human tumors and their respective importance for clinical outcome is unknown. Clinical studies have suggested a correlation between proteins (e.g., CA-9) that are regarded as markers of chronic hypoxia and metastases in a variety of cancers. Studies exposing mice bearing KHT murine fibrosarcomas or ME180 human cervix carcinomas to cycles of low oxygen breathing every day during tumor growth resulted in an enhanced number of microscopic lung metastases or lymph node metastasis respectively (Cairns, Kalliomaki, and Hill, 2001; Cairns and Hill, 2004), while no such effect was seen after daily chronic treatment of KHT-bearing mice. Since the tumors became severely hypoxic during the low oxygen part of these breathing cycles, these data suggest a role for acute hypoxia in the development of both blood-borne and lymphatic metastasis.

14.3.1.4 Hypoxia-induced gene expression and metastasis

There are cell line specific effects of graded oxygen levels on invasive potential and enhanced metastatic efficiency indicating model specific heterogeneity in response to

Table 14.1 Hypoxia regulated genes with a role in metastatic progression.

Gene symbol	Gene name	Hypoxia: ↑ or ↓ regulated	Function: stage of metastatic process	Tumor type
E-Cadherin	Epithelial-cadherin	↓	1, 2	Ovarian, renal cell, pancreatic, colon
VEGF-A	Vascular endothelial growth factor-A	↑	1, 2, 5	Melanoma, glioblastoma, hepatoma, colon, pancreatic, lung
VEGF-C	Vascular endothelial growth factor-C	↑	1, 5	Uterine, breast, ovarian
VEGFR	Vascular endothelial growth factor receptor	↑	1, 5	Pancreatic
bFGF	Basic fibroblast growth factor	↑	5	Breast, lung
IL-8	Interleukin-8	↑	5	Breast, melanoma, ovarian
PD-ECGF/TP	Platelet-derived endothelial cell growth factor/thymidine phosphorylase	↑	5	Breast, lung
uPAR	Urokinase plasminogen activator receptor	↑	1, 2	Endometrial, breast, melanoma, colon
MMP-9	Matrix-metalloproteinase-9	↑	1, 2	Breast, lung, melanoma
MMP-2	Matrix-metalloproteinase-2	↑	1, 2	Lung, melanoma, breast, colon
TIMP1	Tissue inhibitor of metalloproteinase-1	↓	1, 2	Breast
OPN	Osteopontin	↑	1, 3, 4, 5	Breast, colon, lung, head and neck, prostate
CXCR4	Chemokine receptor-4	↑	1, 2, 3	Breast, renal cell, lung

Table 14.1 *(continued)*

Gene symbol	Gene name	Hypoxia: ↑ or ↓ regulated	Function: stage of metastatic process	Tumor type
c-MET	Cell motility factor receptor	↑	1, 2, 3	Cervix, ovarian, breast, lung, osteosarcoma, pancreatic, liver, osteosarcoma
HGF	Hepatocyte growth factor	↑	1, 2, 3	Pancreatic, gastric, lung, ovarian, cervix, liver, osteosarcoma
LOX	Lysyl oxidase	↑	1, 2, 3	Breast, Prostate, Renal
Hdm2	Transformed 3T3 double minute 2	↑	4	Fibrosarcoma, brain, colon, bone
TWIST	TWIST	↑	1, 3	Head and neck, lung
SLUG	SNAI 2	↑	1, 3	Ovarian, renal cell, pancreatic, colon, breast
SNAIL	SNAI 1	↑	1, 3	Ovarian, renal cell, pancreatic, colon
AMF	Autocrine motility factor	↑	1, 3	Breast, pancreas, colon

The table lists a selection of hypoxia regulated genes with roles in the metastatic process. Whether the gene is up (↑) or down (↓) regulated in response to hypoxia is indicated. The stage(s) in the metastatic process, where each gene is believed to be involved, is indicated by a number (1–5) pertaining to Figure 14.1. The tumor type/cell lines that have demonstrated hypoxia mediated expression of the listed gene are shown. This table is modified from Lunt, Chaudary, and Hill (2009), where all the individual references are listed.

low oxygen concentrations. Most studies of the effects of hypoxia on expression of metastasis-related genes reported to date have exposed cells to prolonged periods at a fixed level of oxygen rather than fluctuating exposures. There is also a high degree of variability in the level of hypoxia used in different studies, with oxygen concentrations ranging from anoxic ($<0.1\%$ O_2) to severe hypoxia ($\sim0.2\%$ O_2) or levels approaching normoxia ($3-5\%$ O_2). Furthermore, the actual levels of hypoxia induced in these cultures (which depend critically on factors such as the cell density and the surface area to volume ratio of the media) have not often been measured.

Nevertheless, the duration and different level of hypoxia used and the resultant gene profile have highlighted both the heterogeneity and intricacy of gene expression changes in hypoxia.

The tumor cell–cell and cell–ECM interactions are vital in maintaining tissue structure and integrity. The expression of many of the genes involved in these processes can be affected by exposure to conditions induced by hypoxia (Table 14.1). A key step in tumor metastasis is postulated to involve the dysregulation of these interactions through EMT. In tumor tissue, this provides a mechanism for metastatic dissemination. Epithelial-cadherin (E-cadherin) is a transmembrane protein that is integral in maintaining epithelial cell–cell adhesion through the formation of cadherin–catenin complexes. Its loss has been demonstrated in multiple tumor cell lines including ovarian, renal cell, pancreatic, and colon. The tumor microenvironment appears to play a prominent role in effecting EMT changes. For example, the E-cadherin transcriptional repressor TWIST is positively regulated by HIF-1α (Yang *et al.*, 2008) and Notch signaling enhances EMT under hypoxic conditions by the activation of the transcriptional factors SNAI 1/2 (Sahlgren *et al.*, 2008). In a recent study (Cannito *et al.*, 2008), a number of different cancer cells were exposed to carefully controled hypoxic conditions and investigated for EMT changes. All the cancer cells were found to respond to the hypoxic exposure within 72 hours by classic EMT changes (fibroblastoid phenotype, SNAI and β-catenin nuclear translocation, and changes in E-cadherin), and they showed increased migration and invasiveness. Interestingly, the early EMT-related events induced by hypoxia were dependent on transient intracellular generation of ROS, whereas increased invasiveness was sustained by late and HIF-1α-dependent release of VEGF. These findings suggest that the early EMT-related events may be triggered by the acute transient hypoxia associated with "perfusion limitations" in tumors. The later stages might then be maintained by VEGF also produced in the cells exposed to such fluctuating hypoxia or in cells exposed to more chronic levels of hypoxia.

Important adhesion molecules mediated by hypoxia include the beta-1 integrins. The level of constitutive activity of different beta-1 integrins has been found to correlate with invasive capacity, and the use of monoclonal antibodies to inhibit their expression can block invasion *in vitro*. For example, HIF-1 increased expression of beta-1 integrin in pancreatic cancer cell lines and the use of antisense HIF-1 inhibited its expression and reduced metastases presentation *in vivo* (Chang *et al.*, 2006). Survivin, which is an antiapoptotic protein, was also found to be regulated by HIF-1 expression in this study, and the authors postulated that the combined reduction of survivin and beta-1 integrin were responsible for the reduced metastases observed. In a study with human fibrosarcoma cells hypoxic exposure *in vitro* was found to enhance apoptosis resistance and increase lung colonization but this effect was blocked by treatment of the cells with a Ras inhibitor (farnesyl thiosalicylic acid) (Zhang and Hill, 2007). VEGF-A expression was shown to be inhibited following the treatment but increased HIF-1α expression associated with the hypoxic exposure was not inhibited. These data emphasize that different cell types/lines can respond differently to similar stimuli.

During metastatic dissemination, degradation of the ECM and basement membrane can facilitate migration of the tumor cells through the stroma and into/out of the vasculature. The uPA and its receptor (uPAR) regulate the expression of the serine protease, plasmin. uPA is a proenzyme that is converted into its active form on binding to its receptor, enabling the conversion of plasminogen to plasmin, which in turn allows the localized degradation of the ECM. In addition, plasmin can activate latent forms of MMP-2 and MMP-9, further facilitating degradation of the ECM. It has been shown that hypoxia induces the expression of uPA, uPAR, MMP-2, and MMP-9 in a variety of cell lines. Furthermore, several studies have shown that this hypoxia-mediated expression translates to an increase in metastatic invasion both *in vitro* and *in vivo*. For example, the use of a monoclonal antibody against uPAR resulted in an almost complete absence of lymph node metastases in mice bearing certain human melanoma xenografts (Rofstad *et al.*, 2002a).

The balance between proteinases and their inhibitors is perturbed during tumor cell invasion. Hypoxic exposure of a human breast carcinoma line *in vitro* resulted in the downregulation of tissue inhibitor of metalloproteinase (TIMP-1) and concomitant upregulation of MMP-9. These changes were found to increase the invasive capacity (*in vitro*) of this cell line, an effect that could be blocked using an inhibitor of MMPs (Canning *et al.*, 2001). As with most metastases-promoting molecules identified to date, the importance of specific MMPs in hypoxia mediated invasion appears to be cell-line specific; the aforementioned study found no change in MMP-2 expression despite changes in MMP-9 levels, whereas no changes were observed in MMP-2, -9, or -14 levels in two cell lines (human fibrosarcoma and human breast cancer) with divergent invasive capacities under low oxygen conditions (Subarsky and Hill, 2008). In contrast to these data, other studies have demonstrated hypoxia-mediated upregulation of MMP-2 activity and a positive correlation with metastatic ability in lung and melanoma tumor models.

VEGF-A, originally identified as vascular permeability factor, is a potent inducer of tumor angiogenesis that is upregulated in response to hypoxia by both HIF-1-dependent and -independent mechanisms. Its receptors, VEGFR-1 and VEGFR-2, are also induced under hypoxic conditions. The role of this protein in metastatic disease has been extensively examined, with some studies suggesting a link between VEGF-A expression and metastatic disease and others not. In driving the development of neovasculature VEGF-A may provide a mechanism of transport for the tumor cells as well as possibly enhancing the intravasation and extravasation stages of the metastatic process due to increased vascular permeability. VEGF-A also plays a role in macrophage migration and tumor associated macrophages have been reported to localize in hypoxic regions aiding in tumor cell migration. Perhaps because of the importance of angiogenesis in tumor survival, there are several other hypoxia inducible pro-angiogenic proteins that provide a potential alternative mechanism for hypoxia and angiogenesis mediated secondary disease, including IL-8, PD-ECGF, and bFGF. A study examining the effect of neutralizing antibodies against VEGF, IL-8, PD-ECGF, and bFGF in different human melanoma lines demonstrated the effects of these proteins on tumor metastasis as well as the tumor specific effect apparent in the differing response of the tumor lines to the antibodies (Rofstad and Halsor, 2000).

There is also evidence that hypoxia might have a role in enhancing cell motility. Hepatocyte growth factor (HGF) is a cytokine with roles in both normal tissue and tumor cell growth which signals through its receptor c-MET. HGF was originally identified as a factor (scatter factor) that could induced random migration of certain cell populations *in vitro*. Hypoxia has been shown to increase transcription of c-MET thereby sensitizing the cells to HGF, significantly increasing the cell migration and ECM invasion *in vitro*. The importance of HGF/c-MET signaling in metastatic disease has also been demonstrated using RNAi lentiviral inducible system that enabled the specific inhibition of c-MET in transfected tumor cells. Using this system, it was shown that inhibition of c-MET in a human gastric carcinoma cell line not only prevented the development of experimental metastases *in vivo*, but also facilitated the regression of established metastatic lesions, clearly demonstrating an important role in the maintenance, as well as the development, of metastatic disease. Also, autocrine motility factor (AMF) has been reported to be up-regulated by hypoxia, resulting in enhanced random motility of pancreatic cancer cells.

The stromal-cell derived factor (SDF-1)/chemokine receptor-4 (CXCR4) signaling complex plays a key role in tumor cell motility and homing. In normal tissue SDF-1 signals through CXCR4 to regulate the homing of hematopoetic stem cells to the bone marrow. *In vivo* studies using a human breast cancer line showed that the formation of both spontaneous and experimental lung metastases could be significantly reduced using a monoclonal antibody against CXCR4, demonstrating its importance in the metastatic process, a role that has been supported in additional studies. Both genes have been shown to be upregulated by tumor hypoxia, facilitating the development of metastatic disease. It has been suggested that the presence of SDF-1 at secondary sites, such as lymph nodes, lungs, liver, and bone, concomitant with the expression of its receptor on circulating tumor cells enables cell adhesion and extravasation at the secondary site (Ben-Baruch, 2007).

A hypoxia-induced ECM protein, lysyl oxidase (LOX), has recently been identified as a key member of the hypoxia-regulated genes involved in metastatic disease (Erler *et al*., 2006). Clinically, LOX was found to be positively correlated with tumor hypoxia in breast cancer patients, and there was a significant relationship between LOX expression and distant metastases. When a breast cancer line with repressed LOX expression was generated and grown orthotopically in the mammary fat pad, no liver metastases and significantly fewer lung metastases were formed, consistent with a requirement for LOX in the metastatic process in this model. Similar results were observed using a monoclonal antibody against LOX and recent studies have suggested that it plays a role in the establishment of premetastatic niches. A series of *in vitro* studies also demonstrated an apparent role for LOX in invasion and migration through regulation of focal adhesion kinase (FAK) activity, suggesting multiple roles for hypoxia induced LOX in the formation of metastatic disease. In addition to its response to LOX, FAK, which is expressed at focal adhesions and activated by integrin binding, may be linked to hypoxia-mediated increases in tumor cell migration through EMT. FAK is upregulated in response to VEGF and a reduction in VEGF expression through inhibition of HIF-1 has been shown to inhibit phosphorylation of FAK.

Another important ECM protein that can be upregulated by hypoxic exposure is the secreted glycophosphoprotein osteopontin (OPN), which is expressed by multiple different cell types. OPN has roles in cell adhesion, angiogenesis, prevention of apoptosis and the anchorage-independent proliferation of tumor cells, and its expression has been found to correlate with increased metastatic potential in breast, prostate, colon, head and neck cancer, and soft tissue sarcoma (Rittling and Chambers, 2004). Clinically, OPN levels were found to be elevated in the serum of head and neck patients with the more hypoxic tumors, and this was correlated with an adverse outcome. The fact that OPN levels are elevated in the plasma or serum, and that this may correlate with outcome, provides the possibility for a simple diagnostic test that could enable stratification of high risk patients.

In summary the multiple and complex processes required for the development of metastatic disease involves the expression of many hypoxia responsive genes (Table 14.1). Taken together all these data confirm a role of hypoxia in enhancing the metastatic potential of tumor cells, but they also indicate that this effect may occur through different mechanisms in different cell types.

14.3.1.5 Tumor metastasis and pH

Another feature of the pathophysiological tumor microenvironment that arises, at least partially, as a consequence of the abnormal tumor vasculature, is low extracellular pH. Tumor acidity is a complex and multifactorial process. Warburg described high rates of glycolysis in tumor cells more than 50 years ago. This results in the accumulation of lactic acid and there is increasing evidence of complex links between the poor supply of oxygen from the inadequate vasculature and tumor acidity (Gatenby and Gillies, 2004). These authors proposed that, as early carcinogenesis often occurs in a hypoxic microenvironment, glycolysis is a critical and early adaptive mechanism of cancer cells against hypoxia, that persists because it offers resistance to apoptosis and increased availability of metabolites for growth (due to changes in mitochondria). Mechanisms through which lactic acidosis may facilitate tumor growth include breakdown of ECM allowing expansion, increased cell mobility/metastatic potential and (along with HIF) activation of angiogenesis.

Tumor acidity is not as well studied as tumor hypoxia, and the overlap between the two conditions can confound interpretation of the available data. Upregulation of HIF-1 is known to regulate the expression of various genes involved in the regulation of pH, including CA-IX and the glucose transporters Glut-1 and Glut-3, as well as key enzymes in the glycolytic pathway. Similar to tumor hypoxia, there have been several studies demonstrating a correlation between a low extracellular pH and increased invasion/metastatic spread. Studies have shown that culturing different tumor cell lines, both murine and human, under acidic conditions *in vitro* can enhance the formation of experimental lung metastases when injected *in vivo*. It was demonstrated with human melanoma cell lines that acidity induced upregulation of the proteolysis enzymes MMP-2 and MMP-9 and the angiogenic factors VEGF and IL8 *in vitro* and that blocking the activity of these enzymes prevented the increase in metastases (Rofstad *et al.*, 2002a). More recently, it has been suggested that exposure to acidic conditions selects for clonal populations with enhanced invasive capacity, as human melanoma cells exposed to acidic conditions were found to demonstrate an increase in invasion *in vitro* following reacclimatization to normal pH (Moellering *et al.*, 2008). However, an *in vivo* study, where increased tumor acidity was experimentally induced in murine tumors, failed to demonstrate an enhanced development of spontaneous lung metastases (Kalliomaki and Hill, 2004). In contrast a recent study in which increased alkalinity was induced in mice resulted in reduced spontaneous metastases in a human breast cancer xenograft model (Robey *et al.*, 2009).

There appears to be a close relationship between pH- and hypoxia-mediated invasion. In the experimental metastases studies, the genes found to be upregulated in response to an acidic microenvironment are all genes that have been shown to respond to a hypoxic environment. Walenta *et al.* used bioluminescent imaging to show a correlation between hypoxia and lactate levels in an experimental rat tumor model, suggesting that these two parameters may act in conjunction with one another (Walenta *et al.*, 2001). Furthermore, they showed that high lactate levels in head and neck cancers were associated with increased metastasis (Brizel *et al.*, 2001). Both hypoxia and an acidic extracellular pH were reported to enhance expression of the pro-angiogenic factor VEGF in brain tumor xenografts independently of one another (Fukumura *et al.*, 2001).

14.3.1.6 Metastasis and nutrient deprivation

Low glucose levels in tumors have been linked to metastasis both clinically and experimentally but there is quite limited data. Increased glucose uptake and reduced glucose concentration in tumors have been shown to be poor prognostic factors in patients with carcinomas of the head and neck, oral cavity, breast, and rectum, but not of the cervix. Glut-1 correlates with hypoxia in cervical carcinoma, but other clinical and experimental studies have not found consistent correlations between levels of hypoxia and glucose uptake or Glut-1 expression in tumors. Experimental studies with murine tumor cells cultured in glucose-deficient medium for 1–2 days followed by recovery in glucose-supplemented medium for 0–2 days demonstrated a significantly increased lung colonization ability when the cells were injected intravenously into mice (Schlappack, Zimmermann, and Hill, 1991). Similarly, serum deprivation has been reported to increase motility and invasiveness of human tumor cells through PI-3 kinase-dependent inhibition of the (pH_i-regulating) Na^+/H^+ exchanger (Reshkin *et al.*, 2000).

14.3.2 Metastasis and interstitial fluid pressure

The pathophysiologic tumor microenvironment is also characterized by elevated IFP with values ranging from 10 to 100 mmHg. Studies have demonstrated that inhibition of PDGF signaling can reduce tumor IFP values, consistent with a dynamic role in increasing tumor IFP (Heldin *et al.*, 2004). However, data on the relationship between IFP *per se* and disease progression is limited. A clinical study (Fyles *et al.*, 2006) demonstrated a significant relationship between a high tumor IFP value and risk of local recurrence and distant metastases in cervical carcinoma patients following treatment with radiotherapy alone, although there was no correlation between tumor IFP and metastatic disease at the time of diagnosis similar to the findings of a smaller study in breast cancer patients. There was no correlation between tumor hypoxia (Eppendorf pO_2 measurements) and IFP in the cervix study, and similarly no correlation has been observed between such measurements in experimental tumors. Experimentally, in one study a positive relationship was demonstrated between IFP in oxic melanoma tumors and the development of lymph node metastases (Rofstad *et al.*, 2002b). In contrast, in another no relationship was found between IFP and metastases presentation in a cervical xenograft model, murine fibrosarcoma model, or spontaneous mammary tumor model (Lunt *et al.*, 2008).

There are no known IFP mediated changes in gene expression, however, it has been reported that myofibroblast contraction and integrin interactions can result in release of active TGF-beta from the ECM (Wipff and Hinz, 2008). There are genes that may either elevate tumor IFP or act in conjunction with elevated IFP levels to increase metastatic potential. VEGF-A is a potent proangiogenic factor, as mentioned above, and VEGF-C and VEGF-D, both of which signal through VEGFR3, are lymphangiogenic factors that may increase the number of lymphatic vessels at the periphery of the tumor. Expression of these genes could provide an increased avenue for tumor cell migration into the lymphatic system. Elevated tumor IFP may aid in this process irrespective of gene expression. It has been shown that when an equilibrium exists between the capillary pressure and the interstitial pressure, the movement of fluid into the interstitial space is equal to the movement of fluid into the surrounding normal tissue, potentially forcing the tumor cells into the surrounding normal tissue where they can invade into the lymphatic system.

It is also possible that a high IFP may act as a marker of alternative factors that predispose to metastatic disease. For example, if IFP is primarily, or largely, determined by the vasculature, a high IFP may be a marker of vascular persistence. One possibility is that tumors with a higher IFP are characterized by increased vascular leakiness, potentially due to higher levels of VEGF-A expression, and decreased pericyte coverage of the blood vessels, both of which may facilitate the intravasation process. Since VEGF-A is up-regulated by tumor hypoxia, the strong role of the tumor vasculature in the development of both tumor hypoxia and IFP might imply a relationship between these two parameters of the tumor microenvironment. However, the literature to date has shown no evidence for such a correlation, either clinically or experimentally. The lack of data does not necessarily signify that no association exists, however, since none of the methods used to assess tumor hypoxia were able to distinguish between regions of chronic or acute transient hypoxia. It is possible that, as the microvascular pressure and the interstitial pressure reach equilibrium, temporary drops in vascular pressure may result in transient perturbations in flow and subsequent occlusion of the vessels. Furthermore, the packing density of the tumor cells may impact IFP and this may be related to increased vascular compression by the tumor mass. Both of these situations could result in regions of transiently hypoxic tumor cells, which may play a role in the relationship between IFP and metastases. A related issue is ECM "stiffening" that can occur in tumors due to collagen cross-linking and increased focal adhesions associated with integrin signaling. Recent work suggests that this may be related to LOX activity and induce increased invasiveness in a breast cancer model (Levental *et al.*, 2009).

14.4 Summary

The tumor microenvironment is complex, involving pathophysiological features such as hypoxia, low pH, and increased IFP caused by poorly structured vascular supply. In particular, cells can experience a wide range of oxygen levels from near arterial pO_2 levels (approximately equivalent to 100 mmHg) to average tissue pO_2 levels equivalent to 5% O_2 (~40 mmHg) to pO_2 levels close to anoxia depending on their distance from functional blood vessels. Many of the cells will experience fluctuating levels of hypoxia due to variations in the red cell flux in the blood vessels. Mechanisms of metastasis involve a complex array of genetic and epigenetic changes many of which appear to be specific both for different types of tumors and for different sites of metastasis. The pathophysiological features of tumors can cause changes in gene expression that disrupt or modify the ability of the tumor cells to interact with the cell–cell and cell–ECM signaling interactions in the tumor that may themselves be aberrant relative to normal tissues, thereby facilitating key stages of the metastatic process. Recent studies of stem cells in tumors have suggested that tumors may contain only a small fraction of cells capable of regenerating a tumor and it is presumably these cells which have metastatic ability. There is recent evidence that signaling pathways in those cells that maintain their stem cell properties may be preferentially affected by hypoxia (Barnhart and Simon, 2007). If this is the case, studies focusing on this particular subpopulation of cells in tumors would be particularly relevant to the metastatic process.

References

Barnhart, B.C. and Simon, M.C. (2007) Metastasis and stem cell pathways. *Cancer and Metastasis Reviews*, **26** (2), 261–271.

Ben-Baruch, A. (2007) Organ selectivity in metastasis: regulation by chemokines and their receptors. *Clinical and Experimental Metastasis*, **25** (4), 45–56.

Brizel, D.M., Schroeder, T., Scher, R.L. *et al.* (2001) Elevated tumor lactate concentrations predict for an increased risk of metastases in head-and-neck cancer. *International Journal of Radiation Oncology, Biology, Physics*, **51** (2), 349–353.

Brizel, D.M., Scully, S.P., Harrelson, J.M. *et al.* (1996) Tumor oxygenation predicts for likelihood of distant metastases in human soft tissue sarcoma. *Cancer Research*, **56**, 941–943.

Cairns, R.A. and Hill, R.P. (2004) Acute hypoxia enhances spontaneous lymph node metastasis in an orthotopic murine model of human cervical carcinoma. *Cancer Research*, **64** (6), 2054–2061.

Cairns, R.A., Kalliomaki, T., and Hill, R.P. (2001) Acute (cyclic) hypoxia enhances spontaneous metastasis of KHT murine tumors. *Cancer Research*, **61** (24), 8903–8908.

Cairns, R.A., Khokha, R. and Hill, R.P. (2003) Molecular mechanisms of tumor invasion and metastasis: an integrated view. *Current Molecular Medicine*, **3** (7), 659–671.

Canning, M.T., Postovit, L.M., Clarke, S.H., and Graham C.H. (2001) Oxygen-mediated regulation of gelatinase and tissue inhibitor of metalloproteinases-1 expression by invasive cells. *Experimental Cell Research*, **267** (1), 88–94.

Cannito, S., Novo, E., Compagnone, A. *et al.* (2008) Redox mechanisms switch on hypoxia-dependent epithelial–mesenchymal transition in cancer cells. *Carcinogenesis*, **29** (12), 2267–2278.

Chambers, A.F., Groom, A.C., and Macdonald, I.C. (2002) Dissemination and growth of cancer cells in metastatic sites. *Nature Reviews Cancer*, **2** (8), 563–572.

Chan, D.A. and Giaccia, A.J. (2007) Hypoxia, gene expression, and metastasis. *Cancer and Metastasis Reviews*, **26** (2), 333–339.

Chang, Q., Qin, R., Huang, T., Gao, J., and Feng, Y. (2006) Effect of antisense hypoxia-inducible factor 1alpha on progression, metastasis, and chemosensitivity of pancreatic cancer. *Pancreas*, **32** (3), 297–305.

Dayan, F., Roux, D., Brahimi-Horn, M.C., Pouyssegur, J., and Mazure, N.M. (2006) The oxygen sensor factor-inhibiting hypoxia-inducible factor-1 controls expression of distinct genes through the bifunctional transcriptional character of hypoxia-inducible factor-1alpha. *Cancer Research*, **66** (7), 3688–3698.

Erler, J.T., Bennewith, K.L., Nicolau, M. *et al.* (2006) Lysyl oxidase is essential for hypoxia-induced metastasis. *Nature*, **440** (7088), 1222–1226.

Fukumura, D., Xu, L., Chen, Y., Gohongi, T., Seed, B., and Jain, R.K. (2001) Hypoxia and acidosis independently up-regulate vascular endothelial growth factor transcription in brain tumors in vivo. *Cancer Research*, **61** (16), 6020–6024.

Fyles, A., Milosevic, M., Pintilie, M. *et al.* (2006) Long-term performance of interstial fluid pressure and hypoxia as prognostic factors in cervix cancer. *Radiotherapy and Oncology*, **80** (2), 132–137.

Gatenby, R.A. and Gillies, R.J. (2004) Why do cancers have high aerobic glycolysis? *Nature Reviews Cancer*, **4** (11), 891–899.

Heldin, C.H., Rubin, K., Pietras, K., Ostman, A. (2004) High interstitial fluid pressure: an obstacle in cancer therapy. *Nature Reviews Cancer*, **4** (10), 806–813.

Hockel, M., Schlenger, K., Aral, B., Mitze, M., Schaffer, U., and Vaupel, P. (1996) Association between tumor hypoxia and malignant progression in advanced cancer of the uterine cervix. *Cancer Research*, **56**, 4509–4515.

Kalliomaki, T. and Hill, R.P. (2004) Effects of tumour acidification with glucose+MIBG on the spontaneous metastatic potential of two murine cell lines. *British Journal of Cancer*, **90** (9), 1842–1849.

Levental, K.R., Yu, H., Kass, L. *et al.* (2009) Matrix crosslinking forces tumor progression by enhancing integrin signaling. *Cell*, **139** (5), 891–906.

Lunt, S.J., Kalliomaki, T.M., Brown, A., Yang, V.X., Milosevic, M., and Hill, R.P. (2008) Interstitial fluid pressure, vascularity and metastasis in ectopic, orthotopic and spontaneous tumours. *BMC Cancer*, **8** (1), 2.

Lunt, S.J., Chaudary, N., and Hill, R.P. (2009) The tumor microenvironment and metastatic disease. *Clinical and Experimental Metastasis*, **26** (1), 19–34.

Moellering, R.E., Black, K.C., Krishnamurty, C. *et al.* (2008) Acid treatment of melanoma cells selects for invasive phenotypes. *Clinical and Experimental Metastasis*, **25** (4), 411–425.

Reshkin, S.J., Bellizzi, A., Albarani, V. *et al.* (2000) Phosphoinositide 3-kinase is involved in the tumor-specific activation of human breast cancer cell Na(+)/H(+) exchange, motility, and invasion induced by serum deprivation. *Journal of Biological Chemistry*, **275** (8), 5361–5369.

Rittling, S.R. and Chambers, A.F. (2004) Role of osteopontin in tumour progression. *British Journal of Cancer*, **90** (10), 1877–1881.

Robey, I.F., Baggett, B.K., Kirkpatrick, N.D. *et al.* (2009) Bicarbonate increases tumor pH and inhibits spontaneous metastases. *Cancer Research*, **69** (6), 2260–2268.

Rofstad, E.K. and Halsor, E.F. (2000) Vascular endothelial growth factor, interleukin 8, platelet-derived endothelial cell growth factor, and basic fibroblast growth factor promote angiogenesis and metastasis in human melanoma xenografts. *Cancer Research*, **60** (17), 4932–4938.

Rofstad, E.K., Rasmussen, H., Galappathi, K., Mathiesen, B., Nilsen, K., and Graff, B.A. (2002a) Hypoxia promotes lymph node metastasis in human melanoma xenografts by up-regulating the urokinase-type plasminogen activator receptor. *Cancer Research*, **62** (6), 1847–1853.

Rofstad, E.K., Tunheim, S.H., Mathiesen, B. *et al.* (2002b) Pulmonary and lymph node metastasis is associated with primary tumor interstitial fluid pressure in human melanoma xenografts. *Cancer Research*, **62** (3), 661–664.

Rofstad, E.K., Mathiesen, B., Kindem, K., and Galappathi, K. (2006) Acidic extracellular pH promotes experimental metastasis of human melanoma cells in athymic nude mice. *Cancer Research*, **66** (13), 6699–6707.

Sahlgren, C., Gustafsson, M.V., Jin, S., Poellinger, L., and Lendahl, U. (2008) Notch signaling mediates hypoxia-induced tumor cell migration and invasion. *Proceedings of the National Academy of Sciences of the United States of America*, **105** (17), 6392–6397.

Schlappack, O.K., Zimmermann, A., Hill, R.P. (1991) Glucose starvation and acidosis: effect on experimental metastatic potential, DNA content and MTX resistance of murine tumour cells. *British Journal of Cancer*, **64** (4), 663–670.

Subarsky, P. and Hill, R.P. (2008) Graded hypoxia modulates the invasive potential of HT1080 fibrosarcoma and MDA MB231 carcinoma cells. *Clinical and Experimental Metastasis*.

Sullivan, R. and Graham, C.H. (2007) Hypoxia-driven selection of the metastatic phenotype. *Cancer and Metastasis Reviews*, **26** (2), 319–331.

Walenta, S., Snyder, S., Haroon, Z.A. (2001) Tissue gradients of energy metabolites mirror oxygen tension gradients in a rat mammary carcinoma model. *International Journal of Radiation Oncology, Biology, Physics*, **51** (3), 840–848.

Wipff, P.J. and Hinz, B. (2008) Integrins and the activation of latent transforming growth factor beta1 – an intimate relationship. *European Journal of Cell Biology*, **87** (8–9), 601–615.

Yang, M.H., Wu, M.Z., Chiou, S.H. *et al.* (2008) Direct regulation of TWIST by HIF-1alpha promotes metastasis. *Nature Cell Biology*, **10** (3), 295–305.

Zhang, L. and Hill, R.P. (2007) Hypoxia enhances metastatic efficiency in HT1080 fibrosarcoma cells by increasing cell survival in lungs, not cell adhesion and invasion. *Cancer Research*, **67** (16), 7789–7797.

15

Drug Penetration and Therapeutic Resistance

Andrew I. Minchinton and Alastair H. Kyle
Department of Integrative Oncology, Radiation Biology Unit, British Columbia Cancer Agency Research Centre, Vancouver, BC, Canada

15.1 Introduction

The role played by the tumor microenvironment in limiting extravascular distribution of anticancer drugs within tumor tissue has come under increasing scrutiny over the past decades. Unregulated proliferation and survival of cancer cells within solid tumors result in increased intervascular distances, producing gradients away from blood vessels, in drug concentration, oxygen tension, nutrients, proliferation and acidity, all of which can modulate the potency of cancer therapy. Through studies using experimental solid tumors and *in vitro* multicellular models it is now clear that the efficiency of extravascular distribution plays a critical epigenetic role determining the efficacy of anticancer drugs. Measurement of net tumor drug accumulation, the most commonly assessed parameter characterizing drug kinetics within solid tumors harbors a critical deficiency in that it does not differentiate between drugs that accumulate at high levels but are confined to the first few cells surrounding tumor vasculature and drugs that reach lower levels but distribute homogeneously throughout the tissue. While new techniques are being developed to assess tumor drug penetration, most chemotherapy agents in use today (and most of those currently under development) were never screened for their ability to distribute within solid tumors. The quantitative assessment of drug penetration now permits anticancer drug discovery activities to incorporate considerations of drug distribution into the development algorithm and potentially could help identification of more potent anticancer medicines.

Tumor Microenvironment Edited by Dietmar W. Siemann

15.2 Tumor microenvironment

The microenvironment of solid tumors has long been identified as a source of resistance to chemotherapy and radiotherapy. A combination of sustained expansion of the tumor cell population and aberrant neovascularization leads to the creation of a microenvironment that differentiates tumors from normal tissues. Dysregulated proliferation and survival of tumor cells result in increased separation of blood vessels (Rubin and Casarett, 1966; Thomlinson and Gray, 1955) and unstable perfusion (Brown, 1979; Chaplin, Durand and Olive, 1986; Minchinton, Durand and Chaplin, 1990; Durand and Lepard, 1995), in turn leading to a reduction in the distribution and removal of molecules supplied by the blood (Jain, 1987, 1997). Tumor cells can be located up to 15–20 cells away from the nearest blood vessel (more in some cases) while in most normal tissue cells are within a few cell layers of a vessel (Krogh, 1919). In addition to this, the process of tumor neovascularization often produces aberrant or leaky vessels, which further complicates the drug delivery process (Dvorak *et al.*, 1988; Hashizume *et al.*, 2000; Peterson and Appelgren, 1977; Dvorak, 2002). It is these limitations in diffusion and perfusion within solid tumors and resultant variable tissue oxygenation and acidification that define the tumor microenvironment (Sutherland, 1988; Jain, 1994; Tannock, 2001; Minchinton and Tannock, 2006).

15.2.1 Architecture

Examples of two tumor architectures commonly observed in experimental tumors grown in mice are shown in Figure 15.1. Sections were immunostained to reveal the relation between vasculature, proliferation, hypoxia and necrosis. In both examples the tumors are fast growing, subcutaneous tumors with little stromal component. Figure 15.1(a) shows a sparsely vascularized HCT-116 human tumor xenograft in which sheaths or 'cords' of tumor cells can be seen to surround vessels. In this example clear gradients in tissue oxygenation and proliferation are seen in relation to vasculature. Most cells are within $\sim$150 μm of a perfused vessel beyond which hypoxia is seen to occur, shown in green, followed by necrosis. Figure 15.1(b) shows an example of the experimental SCCVII mouse tumor which exhibits a dense vasculature, where most tumor tissue is within $\sim$50 μm of a vessel. In this case the hypoxic areas do not appear to be the product of diffusion limited oxygen and more likely arise from intermittent blood flow (Chaplin, Olive and Durand, 1987). Proliferating cells can be seen in both the green hypoxic areas and in the vicinity of unperfused blood vessels, consistent with the regular opening and closing of vessels.

15.2.2 Factors that define the tumor microenvironment

15.2.2.1 Diffusion limitations

The sparse or aberrant vasculature that is often observed in solid tumors leads to diffusion limited gradients of molecules supplied and removed by the blood. Early

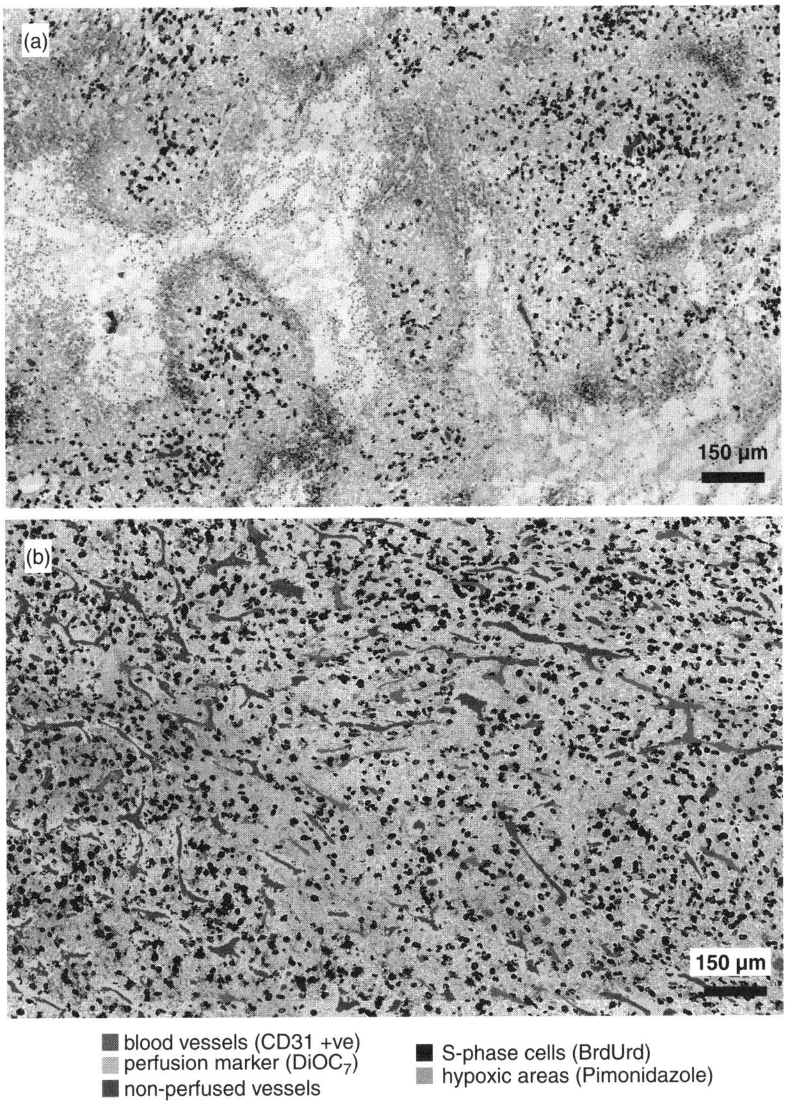

Figure 15.1 Examples of two tumor architectures commonly observed in experimental tumors grown subcutaneously in mice. (a) A human HCT-116 tumor xenograft, which exhibits a sparse vasculature where cords of tumor cells are often seen to surround individual vessels. (b) An SCCVII mouse tumor, which exhibits a dense vasculature and significant intermittent flow (Minchinton lab, unpublished). A full color version of this figure can be found in the color plate section.

evidence of a tissue diffusion limit for oxygen supplied from the blood was first obtained from observations that their often appeared to be a limit (~150 µm) from which viable tumor cells could exist from blood vessels, beyond which necrosis occurred (Thomlinson and Gray, 1955). The hypothesis of the existence of oxygen gradients in tumor tissue was later confirmed using radiolabelled misonidazole, which is reduced and forms adducts under conditions of low oxygenation (Chapman, Franko and Sharplin, 1981; Urtasun *et al.*, 1986). Oxygen gradients are now easily detected using antibodies against several molecules (e.g., pimonidazole, EF5), which are reduced and bound under conditions of low tissue oxygenation (Raleigh *et al.*, 1987). Limited diffusion also applies to other important molecules such as glucose (Groebe, Erz and Mueller-Klieser, 1994; Venkatasubramanian, Henson and Forbes, 2006) and the metabolic waste product lactic acid, build-up of which has been shown to be responsible for increasing the acidity of tumors (Tannock and Rotin, 1989; Gatenby and Gillies, 2004). All three of these factors, oxygen, glucose and lactate, can individually cause a reduction in tumor cell proliferation. Hence it is not surprising that there is also a reduction in proliferation away from blood vessels (Tannock, 1968; Hirst and Denekamp, 1979).

15.2.2.2 Variable vessel perfusion

Variable blood vessel perfusion was originally demonstrated in experimental mouse tumors grown using SCCVII cells, where it was found that at any given time on the order of 20% of the vessels exhibited intermittent flow (Chaplin, Olive and Durand, 1987). Redundancy of vessels and short periods of closure may explain why tumor cells are able to maintain division under these conditions (see Figure 15.1b). In the original study, evaluation of blood flow status was performed through mismatch experiments in which two different fluorescent markers of vessel perfusion were injected intravenously at 20-minute intervals (Chaplin, Olive and Durand, 1987). Mismatch of the two dyes seen in histological tissue sections was then used to determine the fraction of intermittently perfused vessels. Evaluation of the duration of closure and re-opening of vessels has been more recently examined using a tumor window model (Kimura *et al.*, 1996). In this assay, a small glass window is attached to the skin of mice and tumors are grown against the window. The passage of blood cells through the vessels adjacent to the window can then be recorded via video microscopy. Results from such experiments have suggested closure and opening can occur repeatedly with a cycle time as short as tens of minutes (Dewhirst, Braun and Lanzen, 1998).

15.2.2.3 Harsh conditions and aggressive phenotypes

Diffusion limited regions of chronic reduced oxygen and glucose and elevated acidity that exist in tumors are compounded by irregular flow which can lead to cyclical reoxygenation in some tumor regions, potentially creating an environment of repeated oxidative stress leading to DNA damage and mutation (Kimura *et al.*, 1996). In addition, leaky vessels in tumors allow in molecules that would not normally have regular access to extravascular space (Dvorak, 2002; Dvorak *et al.*, 1988; Hashizume *et al.*, 2000; Peterson and Appelgren, 1977). All this creates a wide range of environmental conditions which, when linked with genomic instability, may be responsible for producing aggressive phenotypes seen in later stages of tumor progression (Harris, 2002). In fact, the gradients in physiological parameters such as oxygenation, acidity and proliferation that exist away from vasculature that are observed in solid tumors could be considered the ideal system for production and selection of cell variants that will ultimately lead to tumor progression (Gatenby *et al.*, 2006; Gatenby and Gillies, 2004; Smallbone *et al.*, 2006).

15.2.3 Modulation of therapy

The tumor microenvironment, its modulation of anticancer therapy and its effect on tumor cell progression toward malignancy are topics currently under intense research. Perhaps the earliest observation of a potential for modulation of cancer therapy by the tumor microenvironment was made by Thomlinson and Gray in 1955. Upon observing sparse tumor vasculatures, as depicted in Figure 15.1a they hypothesized that reduced tissue oxygenation might exist in the tumor regions most distant from vasculature, which would result in the existence of radiation-resistant subpopulations (Thomlinson and Gray, 1955). Their observations led to efforts to modify the oxygen delivery to tumors with the purpose in mind of increasing the efficacy of radiation therapy!

Gradients in tissue oxygenation likely reduce the efficacy of radiation (Thomlinson and Gray, 1955; Moulder and Rockwell, 1987; Janssen *et al.*, 2005) as well as anticancer drugs. To attempt to counter the radioresistance of hypoxic cells, oxygen mimetic radiosensitizers, which substitute for oxygen to prevent fixation of radiation induced DNA strand breaks, were developed, for example, misonidazole and nimorazole (Overgaard *et al.*, 1989, 1998). Hypoxic cytotoxins, which are bioreductively converted to cytotoxic species under conditions of low oxygen, were later developed in an attempt to exploit reduced tissue oxygenation as an additional level of tumor selectivity, for example, tirapazamine and AQ4N (Brown, 1993; Patterson and Mckeown, 2000). Increased tumor acidity modifies the activity of many anticancer agents (Tannock and Rotin, 1989), for example, reducing the cellular uptake and efficacy of weak bases such as the anthracyclines (Mahoney *et al.*, 2003).

Parenthetically, most anticancer drugs in current use were developed to exploit the increased proliferation status of tumor cells and so one can hypothesize that reduced cellular proliferation modulates the efficacy of the mainstay of anticancer drugs currently employed. Finally, diffusion-limited penetration can result in lower drug exposure in the tumor regions that are distant from vasculature, resulting in reduced efficacy. In a sinister mechanism, the repeated exposure to subtoxic drug levels is perhaps the most facile way of cultivating drug-resistant cellular phenotypes within tumors.

In theory, the gradients in oxygen, acidity and drugs that can occur in solid tumors will effect changes in treatment response but direct causal evidence linking these issues with clinical management of cancer is complicated and therefore often and unfortunately lacking.

15.3 Drug penetration

In contrast to most normal tissues, with relatively high microvessel density, the extravascular compartment of solid tumors may pose a significant barrier to the penetration of molecules supplied by the blood though, as mentioned above, little is known of the ability of many of the most commonly used anticancer drugs to distribute within solid tumors (Tannock *et al.*, 2002). The most commonly used methods of assessing the tissue penetration of a drug are either through visualization of the drug itself, generally within tissue sections, or through characterization of drug flux through slabs of engineered tissue followed by analysis and mathematical modeling to predict drug build-up and washout within tissue. However both these approaches have certain drawbacks. Imaging techniques (e.g., autoradiography, drug derived fluorescence, or immunodetection) often cannot differentiate between parent drug and active or inactive metabolites. Drug-derived fluorescence also can be modified by cellular interactions (e.g., binding, acidic sequestration, etc.). Immunodetection may require the drug to be immobilized to avoid redistribution or washout during the staining process. Tissue sections produce discrete snapshots in time and to determine overall drug exposure requires a sequence of sections spanning the entire process of drug build-up and subsequent washout. *In vitro* flux-based studies coupled with mathematical modeling can solve many of these issues. However, successful application of flux data to predict drug build-up requires the full characterization of all relevant drug – tissue interactions. Information, that is often not available, especially for newly synthesized experimental drugs. Following these considerations estimating drug distribution must rely on a wide range of tools with each appropriate action requires a case-by-case approach and often using less than ideal techniques.

15.3.1 Factors influencing tumor drug distribution: supply and demand

Like gradients in oxygen tension the tumor distribution of any molecule supplied from the blood will be subject to the constraints of supply and demand (consumption). Supply factors are controlled by pharmacokinetic variables such as dose, metabolism, excretion and blood binding. Flux of drugs within tumor tissue is controlled by factors such as molecular weight, lipophilicity, charge and so on. Factors that determine demand of the drugs by tissue can include cellular metabolism, binding and sequestration.

15.3.2 Tumor mapping

Recent advances in immunohistochemistry and digital imaging mean that many drugs can now be visualized and mapped in relation to tumor vasculature. Key innovations were the development of antibodies to detect vasculature, for example, anti-CD31, Von Willebrand factor VIII, improvements in quantitative digital imaging, robotic microscopy and computer processing. For the last 5–10 years it has been possible to capture images of entire tumor sections at high enough resolution to map drug levels in tissue relative to tumor vasculature. Reports are now beginning to appear that examine radiation and drug effects in relation to immunodetected tumor vasculature. Examples include doxorubicin as detected by fluorescence (Primeau *et al.*, 2005), radiation via effects on proliferation (Ljungkvist *et al.*, 2006), and gemcitabine via effects on proliferation (Huxham *et al.*, 2004).

15.3.2.1 Direct imaging of drugs and their adducts

Studies examining drug distribution without immunodetected tumor vasculature information have looked at radiolabelled and fluorescent drugs and dyes. The earliest studies carried out looked at the diffusion of lissamine green and other dyes in tumors (Goldacre and Sylven, 1962). Doxorubicin is perhaps the most studied anticancer drug due to its native fluorescence (Ozols *et al.*, 1979; Lankelma *et al.*, 1999), though in general no attempt has been made to differentiate it from its less active but fluorescent metabolite doxorubicinol or between free and partially quenched bound drug. Radiolabelled drugs studied include paclitaxel (Li *et al.*, 2000) and immunotoxins (Sung *et al.*, 1993). In addition to classical detection via radiolabelling or direct fluorescence there now exist a number of drug adducts which can be detected via antibody recognition though detection of drug–adducts will not generally reflect the actual distribution of free drug. Examples of drugs that can be detected by antibodies include cisplatin–DNA adducts (Terheggen *et al.*, 1987; Tilby *et al.*, 1991; Liedert *et al.*, 2006). The thymidine analog, bromodeoxyuridine (BrdUrd), originally developed as a radiation sensitizer is incorporated into DNA during replication and can be detected via antibodies raised against DNA–BrdUrd adducts (Gratzner, 1982). Doxorubicin has also been reported to have been detected immunohistochemically (Henneberry and Aherne, 1992). Other examples include immunodetection of adducts formed by the two-nitoimidazoles such as EF5 and pimonidazole binding to thiol groups in proteins and amino acids under hypoxic conditions (Raleigh *et al.*, 1987; Lord, Harwell and Koch, 1993; Azuma, Raleigh and Thrall, 1997).

Figure 15.2 shows an example of radiolabeled paclitaxel mapped in reference to tumor vasculature. In this case a dry autoradiography technique was used in which a sheet of tritium sensitive film was clamped against the tissue section for a period of 80 days. Subsequent to this the tissue section was immunostained to detect tumor vasculature using an antibody against CD31. Both the film and tissue section were then scanned and the digital images overlaid and processed. Analysis then yields ^{3}H-paclitaxel distribution in relation to tumor vasculature (Kyle, 2007; Kyle *et al.*, 2007).

15.3.2.2 Mapping drug effect in tumors

Markers for proliferation (e.g., BrdUrd, Ki67, PCNA), apoptosis (e.g., terminal deoxynucleotidyl transferase dUTP nick end labeling, caspase 3) and double strand

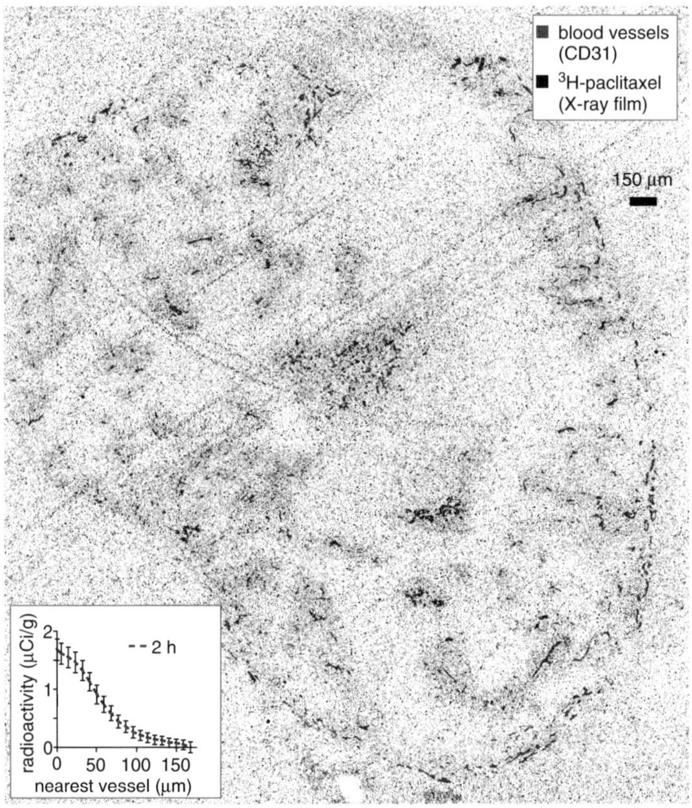

Figure 15.2 Composite image of ^{3}H-paclitaxel distribution (black) and CD31 positive vasculature (red) in an HCT-116 tumor xenograft 2 hour post i.v. injection of 10 mg/kg ^{3}H-paclitaxel. Images were obtained from tritium-sensitive film clamped against 10 μm tumor cryosection for 80 days, followed by immunohistochemical staining of CD31-positive tumor vasculature on the same section. Subsequent computer analysis of taxane distribution relative to vasculature was carried out with non-tumor areas and scratches in X-ray film cropped out. Adapted from Kyle (2007). A full color version of this figure can be found in the color plate section.

breaks (e.g., H2AX) can also be used in an attempt to map drug distributions via their effect on cells in tumors relative to vasculature. However, the relation between drug effect and drug distribution is generally obscured by changes in the intrinsic sensitivity of cells to drug with depth into tissue. Hence, these studies address more general questions of drug efficacy in solid tumors rather than questions specific to drug penetration. The earliest mapping studies were done in spheroids and tumors using either sequential trypsinization or cell-sorting techniques with cell survival as the endpoint (Giesbrecht, Wilson and Hill, 1981; Durand, 1982; Olive, Chaplin and Durand, 1985). For the cell-sorting method treated cells were dissociated and sorted based on the strength of Hoechst 33 342 fluorescence, which was allowed to diffuse into the tissue and used as an indicator of the distance of each cell from a perfused vessel/tissue edge. Work was first carried out with radiation and doxorubicin (Chaplin, Durand and Olive, 1985; Durand, 1990) and later other

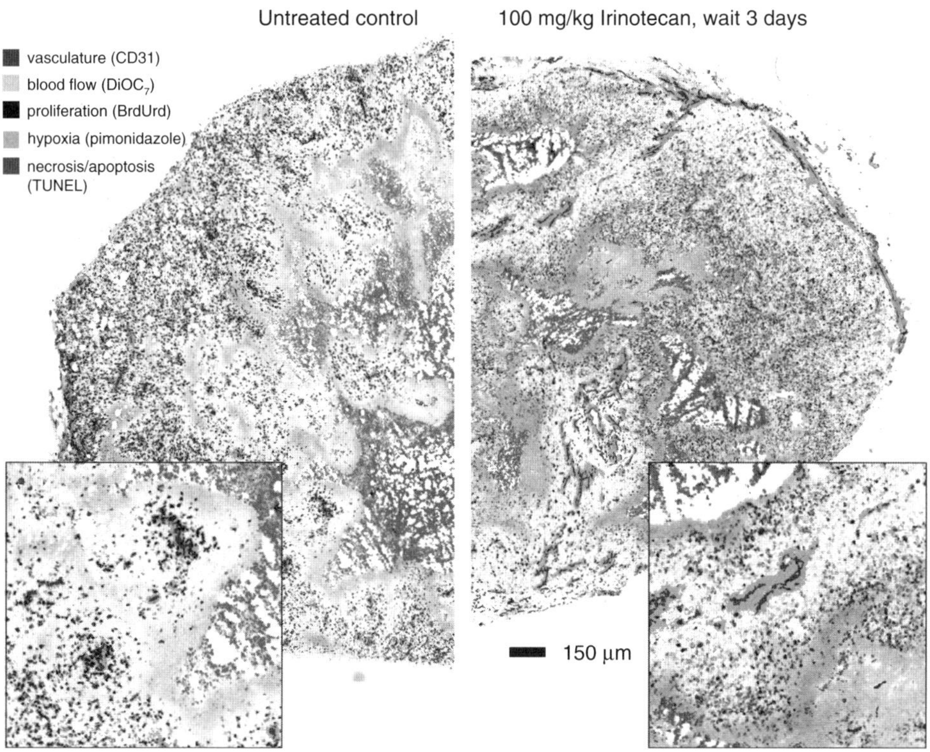

Figure 15.3 Microregional activity of 100 mg/kg Irinotecan i.v. 3 days its following treatment. Irinotecan treatment inhibits DNA synthesis in cells near vessels but results in transient stimulation in the normally quiescent cells further away (Minchinton lab, unpublished). A full color version of this figure can be found in the color plate section.

anticancer agents (Olive and Durand, 1994). New approaches involve imaging of tissue sections for markers of vasculature and drug effect. Using the tissue section-based techniques, the effects of radiation (Ljungkvist *et al.*, 2006) and drugs such as gemcitabine (Huxham *et al.*, 2004) and tirapazamine (Huxham *et al.*, 2006) have been evaluated in relation to tumor vasculature.

Figure 15.3 shows an example of using cellular markers as indicators of drug activity in tumor xenografts. In this case a tumor mapping strategy was employed to detect the effect of the topoisomerase I inhibitor Irinotecan on apoptotic and S-phase cells in relation to tumor vasculature. Irinotecan is seen to preferentially exert activity on those cells located nearer to the vascular supply. From this data it is not clear if this is the result of limited drug distribution or that the cells further from the vasculature were proliferating less.

15.3.2.3 Limitations of tumor mapping for evaluation of drug penetration

While histological based tumor mapping represents a relatively new and unexplored area of research, numerous limitations exist in terms of employing it to assess drug penetration. With regards to direct imaging, radiolabeled drugs are not always available and are generally expensive. Autoradiography-based experiments are labor intensive and getting usable data can entail waiting weeks or months while film is being exposed. In addition, the potential for the presence of radiolabeled drug metabolites means that data will often be inconclusive. Assays involving biological endpoints are prone to changes in intrinsic sensitivity of cells to drugs with changing proliferation status, oxygenation, and so on. In addition, indirect drug-derived host-toxicity may result in changes in tumor proliferation and even apoptosis via a reduction in oxygen and nutrients supplied from the blood.

15.4 *In vitro* tumor models

There exist several *in vitro* models for evaluating tumor drug diffusion, the best characterized and most widely used being multicellular spheroids (Sutherland, 1988). Other models include three-dimensional (3-D) tissue-disks (Cowan, Hicks and Wilson, 1996; Minchinton *et al.*, 1997) or chunks and slabs of tissue excised from solid tumors and cellular aggregates grown on tissue culture plates or embedded in collagen or sponges (Leighton, 1951; Swabb, Wei and Gullino, 1974; Yang *et al.*, 1979; Vescio *et al.*, 1987).

15.4.1 3D spheroidal cell culture

Multicellular spheroidal cell culture is the mainstay of *in vitro* 3-D tumor modeling. Developed over 30 years ago in the Sutherland Laboratory at the University of Western Ontario, spheroids have been shown to reproduce many of the features of the tumor microenvironment. Diffusion-limited penetration of molecules into the cultures means they can reproduce the oxygen and nutrient gradients and those of proliferation seen in solid tumors. Thicker spheroids exhibit acidified central regions, due to lactate build-up and if large enough they form a central necrotic region. In addition they have been shown to exhibit a 3-D culture cell contact effect, which modifies their response to radiation and drugs. Because they are able to integrate a number of characteristics of the tumor microenvironment they are a useful system for modeling the *in vivo* response to anticancer agents.

15.4.1.1 Modeling solid tumors

Spheroids have been applied to study relations between oxygen, glucose, serum and proliferation/necrosis (Freyer and Sutherland, 1983; Freyer *et al.*, 1984; Casciari, Sotirchos and Sutherland, 1988; Acker *et al.*, 1992; Groebe and Mueller-Klieser, 1996). Diffusion-limited penetration of molecules into the cultures means that as well as producing oxygen and nutrient gradients they can also develop an acidic central region due to lactate accumulation (Walenta *et al.*, 2000) and have been used to study relations between central pH and medium buffering capacity (Acker *et al.*, 1987). They have been grown using a number of cell types including normal and drug resistant variants (Martin *et al.*, 2003) and have been shown to create an extracellular matrix (Glimelius *et al.*, 1988).

15.4.1.2 Effect of tumor microenvironment on response

Cells located on the outer versus inner layers of spheroids have been shown to exhibit differential sensitivity to a number of drugs and also to radiation (Sutherland *et al.*, 1970; Durand and Sutherland, 1972; Sutherland and Durand, 1976; Durand, 1980; Olive and Durand, 1994). Changes in sensitivity of cells have been attributed to a number a factors including cell cycle, intercellular communication and drug penetration.

15.4.1.3 Drug penetration

The tissue penetration of drugs from several classes has been studied using spheroidal cell culture. Direct visualization of the anthracycline doxorubicin (Kerr *et al.*, 1988) and radiolabeled vinblastine (Nederman, Carlsson and Malmqvist, 1981), methotrexate (West, Weichselbaum and Little, 1980) and other biological agents (Nederman, Carlsson and Kuoppa, 1988) have been investigated using spheroids. Indirect effect-based detection has been used to study several drugs including doxorubicin, cisplatin and others (Durand, 1990; Olive and Durand, 1994). However, in these studies it was shown that the differential response of inner versus outer cells in spheroids depended on a number of factors.

15.4.2 3-D tissue-disks

Use of 3-D tissue-disks is a tissue-engineered cell culture system in which cells are grown as disks of tissue. This is often referred to as multicellular layers (MCLs) or multilayered cell culture (MCC). Cultures are grown starting from a cell suspension, which is added to a tissue culture insert and allowed to attach to form a confluent layer. Inserts are then transferred to stirred reservoirs and grown to form planar cultures typically 15–30 cell layers thick (Figure 15.4a).

15.4.2.1 Disk-based flux assay

Tissue disks were developed to allow measurement of flux of drugs through tumor-like tissue in a controlled environment. There are now numerous studies published using this model for a variety of anticancer drugs including radiosensitizers (Cowan, Hicks and Wilson, 1996; Kyle, 1999), hypoxic cytotoxins (Hicks *et al.*, 1998; Kyle and Minchinton, 1999; Phillips, Loadman and Cronin, 1998), DNA-intercalators (Hicks *et al.*, 1997) and other agents (Tannock *et al.*, 2002). Figure 15.4b shows schematics apparatus and conformations that are often used for measurement of drug

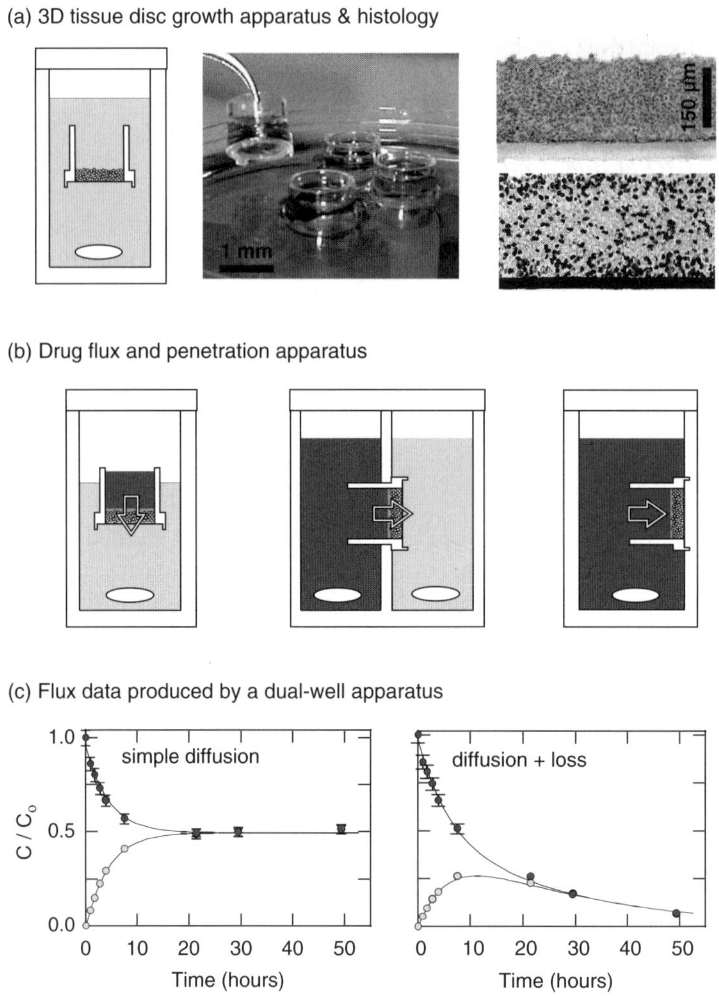

Figure 15.4 3-D tissue-disk assays. (a) Disks are grown using standard tissue culture inserts. Once cells are seeded and attached the inserts are immersed in stirred growth medium. Cryosections show hematoxylin (top) and BrdURd (bottom) showing symmetrical proliferation that occurs within the tissue. (b) Single and dual reservoir configurations used for drug flux and tissue penetration studies. For the floating disk configuration, the upper, unstirred reservoir is often filled with agar to prevent convection; samples are then taken from the lower, stirred reservoir over ime. (c) Typical data from donating and receiving reservoirs using the dual-reservoir configuration. Source Kyle (2007).

flux. Some configurations allow drug concentration from both donating and receiving reservoirs to be monitored and allow modeling of mass balance. Figure 15.4c show typical data produced using the dual-reservoir configuration. Fitting data sets from the two reservoirs simultaneously allows for maintenance of mass balance between reservoirs and monitoring of net drug sequestration/consumption in the culture.

Results from flux experiments can be used for direct comparison of relative flux between two or more drugs or experimental conditions or used to estimate rates of diffusion and sequestration/consumption within the tissue. On their own, raw data from one or both reservoirs can be used to compare the relative flux of two or more drugs. It is also possible to examine the effect of altering some experimental parameter, often the inhibition of a specific consumptive process. Examples of this include variation of tissue oxygenation to inhibit the enzymatic conversion of the hypoxic cytotoxin tirapazamine (Hicks *et al.*, 1998; Kyle and Minchinton, 1999; Phillips, Loadman and Cronin, 1998) and overexpression of Pgp to reduce cellular accumulation of doxorubicin (Tunggal *et al.*, 2000). Data from flux experiments can also be fitted to differential equations to obtain estimates of parameters representing diffusion and sequestration/consumption processes within the cultures. Examples include analysis of the DNA intercalator DAPA (Hicks *et al.*, 1997), the hypoxic cytotoxin tirapazamine (Hicks *et al.*, 1998; Kyle and Minchinton, 1999) and recently a panel of tirapazamine analogs (Hicks *et al.*, 2006).

15.4.2.2 Simulating drug distribution

In cases where the key drug–tissue interactions are understood and have been quantified it may be possible to mathematically model drug build-up in tissue. Once the initial characterization work has been carried out this approach provides flexibility in terms of exploring the effect of variations in exposure conditions such as plasma pharmacokinetics or tumor architecture on drug distribution (Hicks *et al.*, 2006; Kyle and Minchinton, 1999). However, problems may be encountered if this approach is used to model an undercharacterized system or through oversimplification of complex drug–tissue interactions, potentially leading to false predictions.

15.4.2.3 Visualizing drug fluorescence in 3-D disks

An alternative application of tissue disks to assess the distribution of drugs is to directly examine drug build-up within the tissue. One way of doing this is to close off one side of the disks and allow drug to accumulate from the other side, as shown in Figure 15.4b (right panel).

Examples of the possible range of variation in drug profiles within 150 μm thick cultures following a 1-hour drug exposure to one side are shown in Figure 15.5. Panels show drug-derived fluorescence visualized in tissue cryosections. Fluorescent profiles indicate that in some cases drug build-up can be limited to just the first few cell layers, while in others, drugs can exhibit almost uniform distribution over the full width of the tissue. Of the examples shown here the fluorescent-tagged goat IgG, with a molecular weight over 100 times greater than the other molecules, perhaps not surprisingly, exhibits the poorest distribution after the 1-hour exposure. However, ranking of the other agents does not follow molecular weight strictly. Factors that can modify the tissue penetration of a drug will include their level of

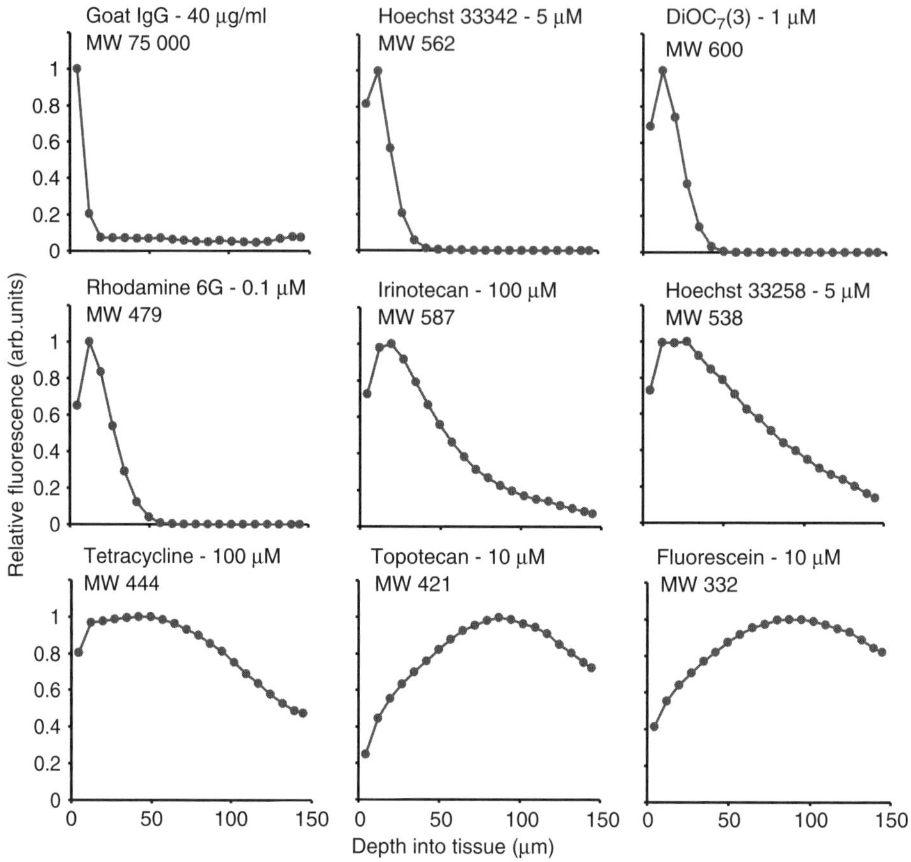

Figure 15.5 Drug derived fluorescent profiles observed in HCT-116 3D tissue-disks following 1-hour exposure using the closed-off apparatus. Drug profiles are not seen to rank solely by MW. Following drug exposure cultures were embedded in cryosectioning medium and immediately frozen. Panels show fluorescence as a function of depth into tissue as determined from epifluorescence imaging of unmounted dry-cryosections. Hoechst 33342, Hoechst 33258 (excitation 365, emission 460 nm); irinotecan, topotecan, tetracycline (ex 425 nm em 540 nm); DiOC7(3), fluorescein (ex 480 nm, em 535 nm); goat IgG–Alexa 546, rhodamine 6G (ex 540 nm, em 605 nm). Source Kyle (2007).

access to the cellular compartment and degree of sequestration/consumption within the tissue.

Comparison of the two Hoechst drugs examined here indicates that 33 258 exhibits a more uniform tissue distribution than 33 342, despite 33 342 being more lipophilic and entering cells more easily than 33 258. Both drugs accumulate in cells through binding to DNA. In the case where a drug's accumulation is negligible in comparison with its rate of flux through tissue, a more lipophilic drug will have better access to the intracellular compartment of the tissue and hence exhibit faster flux. However, if access to the intracellular compartment leads to significant accumulation then the opposite may occur and the less lipophilic drug, confined to the extracellular compartment, will exhibit a more uniform distribution. In this case it is 33 258, the less lipophilic drug, that exhibits the more uniform distribution, suggesting that drug sequestration via DNA binding may be high enough to limit tissue penetration.

Fluorescein and topotecan, which appear to distribute the most efficiently of the agents studied, both exhibit oddly shaped profiles. This is likely caused by an unusually fast washout during the freezing process, consistent with a molecule that exhibits rapid diffusion and negligible retention through cellular accumulation or binding. During freezing, the edges of the cultures come into contact with embedding gel for 1–2 minutes before it solidifies. Freezing occurs from the bottom/far edge of the cultures, which likely explains why there appears to be less washout from that side.

15.4.2.4 Assessing drug distribution using a biological endpoint

Because of the inherent difficulties in estimating drug distribution using the assays described previously in this chapter it became evident to the authors that it would be desirable to develop a drug penetration assay that used a biological endpoint in which the cells within the tissue could integrate drug exposure over time. Advantages of using a biological endpoint in which the cells within the tissue could integrate drug exposure over the period of build-up and wash-out and differentiate between active and inactivated drug but that would not be subject to confounding variations in cellular response seen *in vivo*. 3-D tissue-disks process a unique property in that they present a barrier to penetration in which there exist two populations of rapidly proliferating cells, one on each side of the culture, that are separated by a known thickness. This symmetry presents an opportunity to use the differential effect of the drugs on proliferating cells on each side as an indication of their penetration, in effect using the cells themselves as the drug detection end-point. This symmetry can be exploited by exposing cultures to drugs from one side and then comparing their effect on the cells located on the near (exposed) side versus the far side of the cultures (Huxham *et al.*, 2004; Kyle, 2007; Kyle *et al.*, 2003, 2004, 2007). In tumors and spheroids this technique is not possible because proliferation and cellular response fall off with depth into tissue and hence drug penetration is only one of several factors that will determine cellular response.

Drug exposure can be made with the far side of the cultures either open or closed-off to growth medium. Closing off the cultures models the maximum chance for drug build-up within tissue. A second 'open' situation, where a tissue disk is placed between two reservoirs, of which only one contains drug, would represent a greater barrier to drug build-up due to the constant wash-out of drug into the second reservoir. When the cultures are used in the closed-off configuration the maximum period of drug exposure will be limited to the duration that the closed-off cultures will maintain symmetrical proliferation. Following drug exposure, cultures will then be returned to the growth chambers and incubated for a period of time, of the order of days, to allow drug effect to manifest itself. During this incubation period drug may also redistribute and penetrate further into the culture. Boundary conditions during the incubation period will match the 'open' configuration to ensure that symmetrical proliferation is sustained. The incubation period must be long enough to allow manifestation of drug effect but short enough to avoid substantial regrowth of the cultures that would further complicate interpretation of the data. Following the incubation period the S-phase cell marker BrdUrd will be used to detect drug effect.

For drugs that diffuse in and wash out of tissue quickly, it may be critical that the proliferation status of the cells on the far side of the cultures is not perturbed during the exposure period. However, in other cases the proliferation status on the far side of the cultures during drug exposure may not be as important. For example if the drug is significantly retained following the exposure period, then it may only be important that the cells are proliferating during the incubation period. Alternatively, for drugs that accumulate and are retained in the first few layers of cells and later diffuse slowly into the tissue, significant exposure to the far side of the cultures may not actually occur until the incubation period, when the drug that accumulated in the first few layers of cells slowly redistributes within the tissue. In this case, it might not be necessary to close off the cultures during the exposure period.

Figure 15.6 shows an example of using 3-D tissue-disks to estimate drug penetration. In this case cultures were exposed to either vincristine or cisplatin for 1 hour and then left for 1 or 3 days to integrate the net effect of drug build-up and wash out on overall exposure/distribution. Each drug's activity on the exposed edge of the disk was then compared with its activity on the far-off side of ~150 μm thick disks. The disks were then cryosectioned and immunostained for incorporated BrdUrd to show S-phase cells *(top panels)*. The cultures showed sustained proliferation in central areas even at the highest exposure, indicating differential response of non-cycling cells. Analysis of BrdUrd immunostaining is summarized in the bottom pannels. Vincristine can be seen to consistently show a differential effect on near versus far sides, with results indicating a nearly threefold drug gradient across the culture. Cisplatin showed a much more uniform effect on near versus far sides.

A summary of the 3-D tissue-disk based tissue distribution data is shown in Table 15.1. Each drug's ability to distribute within tissue was estimated from the differential effect observed on the two sides following exposure to one side only. Drug exposure to the far sides of the cultures was seen to vary from as low as 3 to 10% of reservoir exposure for the anthracyclines to cases where no reduction in exposure could be detected on far sides, as was seen for 5-fluorouracil and cisplatin. In between were the taxanes, where far sides received 10–20% of reservoir exposures. Several drugs were observed to exhibit marginal reduction in exposure to the far side such as the vinca alkaloids, 33–100% of reservoir exposure. Cases where the assay was observed to fail included situations where 1 hour drug exposures produced no detectable effect on the BrdUrd end-point, for example, cytarabine and thiotepa both of which are normally administered as continuous infusions. Other examples of assay failure included drugs for which control experiments under conditions of reduced oxygen and glucose showed response on the far side to vary for reasons other than drug penetration, for example, bleomycin and etoposide which exhibit oxygen dependent toxicity.

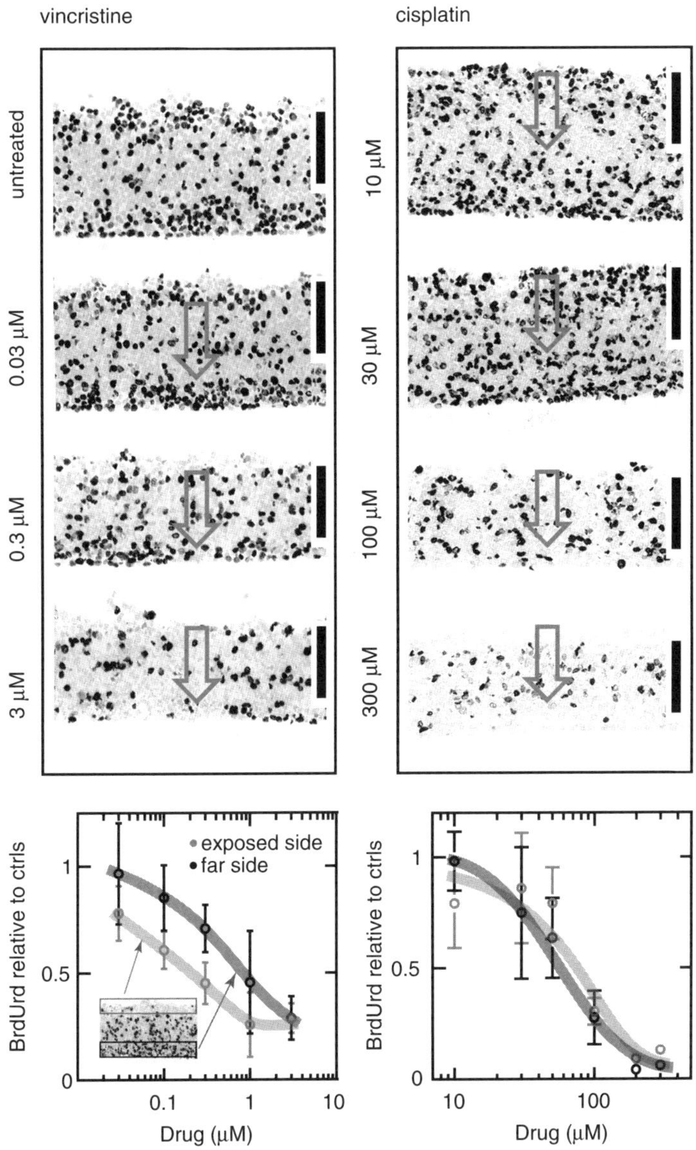

Figure 15.6 Estimating drug distribution using a biological endpoint. Top panels: distribution of S-phase cells in 3-D tissue-disks 1 or 3 days after a 1-hour exposure to increasing concentrations of vincristine or cisplatin from one side. Scale bars, 150 μm, show approximate thickness of culture at time of drug exposure. Bottom panels: comparison of drug activity in the first 30 μm of tissue on the near side and far side of the tissue-disks. Identical control experiments under normal versus low oxygen and glucose were performed to validate the findings (not shown). Data show BrdUrd immunostaining expressed as the fraction of control levels seen in untreated tissue discs. Adapted from Kyle (2007).

Table 15.1 Summary of results obtained from the closed-off 3-D tissue-disk effect-based assay for evaluation of drug distribution. Gradient refers to the fold reduction in exposure estimated on the far sides of the discs following 1 hour drug using the closed-off assay and the percentage exposure is calculated from this. Concentration refers to the concentration range that was studied and wait refers to the incubation period following completion of drug exposure before evaluating drug effect.

		MW	Concentration range (μM)	Wait (days)	Gradient (fold)	% exposure to far side
Anthracyclines	Doxorubicin	543	0.3–10	1	12	
	Epirubicin	543	0.3–10	1	10	
	Daunorubicin	530	0.3–10	1	30	
	Mitoxantrone	444	0.3–10	1	>30	
Taxanes	Paclitaxel	854	0.03–3	2	5	
	Docetaxel	862	0.03–3	2	10	
Anti-metabolites	Gemcitabine	263	3–90	1	>10	
	5-Fluorouracil	130	100–400	1	0	
	Cytarabine	243	100–1000	1	—	
Vinca alkaloids	Vincristine	825	0.03–3	1	3	
	Vinblastine	811	0.03–3	1	<3	
	Vinorelbine	980	0.03–3	1	0	
Others	Mitomycin C	334	0.3–10	3	2	
	Bleomycin	1420	1–50 U/l	1	0	
	Cisplatin	298	10–300	3	0	
	Etoposide	589	3–100	1	2	
	BCNU	215	10–90	3	0	
	Thiotepa	189	3–30	3	—	

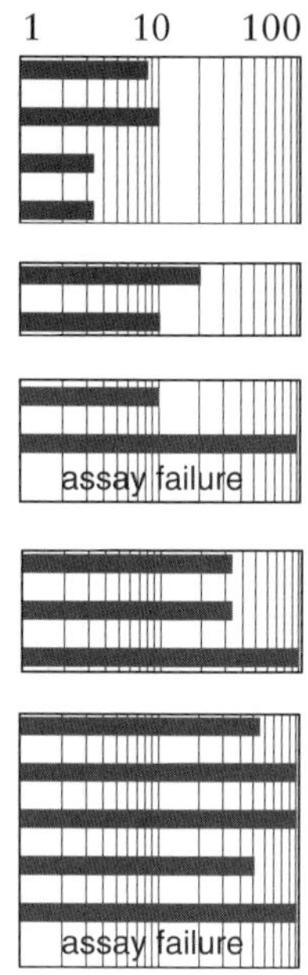

15.5 Conclusions

We currently do not know how extravascular drug distribution contributes to the efficacy (or lack of efficacy) of most anticancer drugs, but current data strongly suggests an important role. The quest for new anticancer agents would therefore benefit from quantitative knowledge of extravascular drug distribution since it would enable drugs that will fail in clinical trials to be identified early in the development process and conversely allow drugs with good distribution properties to be identified and selected for further development. While the tissue penetration of most anticancer drugs in clinical use has not been fully characterized, it is sometimes argued that repeated dosing regimes used clinically, in addition to addressing cell cycle issues also address those related to inefficient drug distribution. Under such regimes, a drug that exhibits inefficient distribution might kill tumor cells within a certain distance of the vasculature with each treatment, and work in a manner similar to peeling away layers of an onion. Though this conjecture needs experimental verification such fractionated dosing regimes might actually provoke malignancy by providing an ideal environment for the cultivation of drug resistant phenotypes since the cells residing distal to tumor vasculature would receive repeated sublethal drug exposures. Finally, more tools are needed that permit the issue of drug distribution to be addressed fully and this will become more crucial as large molecular weight drugs such as proteins are developed for clinical use.

References

Acker, H., Carlsson, J., Holtermann, G. *et al.* (1987) Influence of glucose and buffer capacity in the culture medium on growth and pH in spheroids of human thyroid carcinoma and human glioma origin. *Cancer Research*, **47**, 3504–3508.

Acker, H., Holtermann, G., Bolling, B., and Carlsson, J. (1992) Influence of glucose on metabolism and growth of rat glioma cells (C6) in multicellular spheroid culture. *International Journal of Cancer*, **52**, 279–285.

Azuma, C., Raleigh, J.A., and Thrall, D.E. (1997) Longevity of pimonidazole adducts in spontaneous canine tumors as an estimate of hypoxic cell lifetime. *Radiation Research*, **148**, 35–42.

Brown, J.M. (1979) Evidence for acutely hypoxic cells in mouse tumours and a possible mechanism of reoxygenation. *British Journal of Radiology*, **52**, 650–656.

Brown, J.M. (1993) SR 4233 (tirapazamine): a new anticancer drug exploiting hypoxia in solid tumours. *British Journal of Cancer*, **67**, 1163–1170.

Casciari, J.J., Sotirchos, S.V., and Sutherland, R.M. (1988) Glucose diffusivity in multicellular tumor spheroids. *Cancer Research*, **48**, 3905–3909.

Chaplin, D.J., Durand, R.E., and Olive, P.L. (1985) Cell selection from a murine tumour using the fluorescent probe Hoechst 33342. *British Journal of Cancer*, **51**, 569–572.

Chaplin, D.J., Durand, R.E., and Olive, P.L. (1986) Acute hypoxia in tumour: Implications for modifiers of radiation effects. *International Journal of Radiation Oncology, Biology, Physics*, **12**, 1279–1282.

Chaplin, D.J., Olive, P.L., and Durand, R.E. (1987) Intermittent blood flow in a murine tumor: radiobiological effects. *Cancer Research*, **47**, 597–601.

Chapman, J.D., Franko, A.J., and Sharplin, J. (1981) A marker for hypoxic cells in tumours with potential clinical applicability. *British Journal of Cancer*, **43**, 546–550.

Cowan, D.S.M., Hicks, K.O., and Wilson, W.R. (1996) Multicellular membranes as an in vitro model for extravascular diffusion in tumours. *British Journal of Cancer*, **74**, S28–S31.

Dewhirst, M.W., Braun, R.D., and Lanzen, J.L. (1998) Temporal changes in PO2 of R3230AC tumors in Fischer-344 rats. *International Journal of Radiation Oncology, Biology, Physics*, **42**, 723–726.

Durand, R.E. (1980) Variable radiobiological responses of spheroids. *Radiation Research*, **81**, 85–99.

Durand, R.E. (1982) Use of Hoechst 33342 for cell selection from multicell systems. *Journal of Histochemistry and Cytochemistry*, **30**, 117–122.

Durand, R.E. (1990) Slow penetration of anthracyclines into spheroids and tumors: a therapeutic advantage? *Cancer Chemotherapy and Pharmacology*, **26**, 198–204.

Durand, R.E. and Lepard, N.E. (1995) Contribution of transient blood flow to tumour hypoxia in mice. *Acta Oncologica*, **34**, 317–323.

Durand, R.E. and Sutherland, R.M. (1972) Effects of intercellular contact on repair of radiation damage. *Experimental Cell Research*, **71**, 75–80.

Dvorak, H.F. (2002) Vascular permeability factor/vascular endothelial growth factor: a critical cytokine in tumor angiogenesis and a potential target for diagnosis and therapy. *Journal of Clinical Oncology*, **20**, 4368–4380.

Dvorak, H.F., Nagy, J.A., Dvorak, J.T., and Dvorak, A.M. (1988) Identification and characterization of the blood vessels of solid tumors that are leaky to circulating macromolecules. *American Journal of Clinical Pathology*, **133**, 95–109.

Freyer, J.P. and Sutherland, R.M. (1983) Determination of diffusion constants for metabolites in multicell tumor spheroids. *Advances in Experimental Medicine and Biology*, **159**, 463–475.

Freyer, J.P., Tustanoff, E., Franko, A.J., and Sutherland, R.M. (1984) In situ oxygen consumption rates of cells in V-79 multicellular spheroids during growth. *Journal of Cellular and Comparative Physiology*, **118**, 53–61.

Gatenby, R.A., Gawlinski, E.T., Gmitro, A.F. *et al.* (2006) Acid-mediated tumor invasion: a multidisciplinary study. *Cancer Research*, **66**, 5216–5223.

Gatenby, R.A. and Gillies, R.J. (2004) Why do cancers have high aerobic glycolysis? *Nature Reviews Cancer*, **4**, 891–899.

Giesbrecht, J.L., Wilson, W.R., and Hill, R.P. (1981) Radiobiological studies of cells in multicellular spheroids using a sequential trypsinization technique. *Radiation Research*, **86**, 368–386.

Glimelius, B., Norling, B., Nederman, T., and Carlsson, J. (1988) Extracellular matrices in multicellular spheroids of human glioma origin: increased incorporation of proteoglycans and fibronectin as compared to monolayer cultures. *Apmis*, **96**, 433–444.

Goldacre, R.J. and Sylven, B. (1962) On the access of blood-borne dyes to various tumour regions. *British Journal of Cancer*, **16**, 306–322.

Gratzner, H.G. (1982) Monoclonal antibody to 5-bromo- and 5-iododeoxyuridine: A new reagent for detection of DNA replication. *Science*, **218**, 474–475.

Groebe, K., Erz, S., and Mueller-Klieser, W. (1994) Glucose diffusion coefficients determined from concentration profiles in EMT6 tumor spheroids incubated in radioactively labeled L-glucose. *Advances in Experimental Medicine and Biology*, **361**, 619–625.

Groebe, K. and Mueller-Klieser, W. (1996) On the relation between size of necrosis and diameter of tumor spheroids. *International Journal of Radiation Oncology, Biology, Physics*, **34**, 395–401.

Harris, A.L. (2002) Hypoxia – a key regulatory factor in tumour growth. *Nature Reviews Cancer*, **2**, 38–47.

Hashizume, H., Baluk, P., Morikawa, S. *et al.* (2000) Openings between defective endothelial cells explain tumor vessel leakiness. *American Journal of Clinical Pathology*, **156**, 1363–1380.

Henneberry, H.P. and Aherne, G.W. (1992) Visualisation of doxorubicin in human and animal tissues and in cell cultures by immunogold-silver staining. *British Journal of Cancer*, **65**, 82–86.

Hicks, K.O., Ohms, S.J., Van Zijl, P.L. *et al.* (1997) An experimental and mathematical model for the extravascular transport of a DNA intercalator in tumours. *British Journal of Cancer*, **76**, 894–903.

Hicks, K.O., Fleming, Y., Siim, B.G. *et al.* (1998) Extravascular diffusion of tirapazamine: effect of metabolic consumption assessed using the multicellular layer model. *International Journal of Radiation Oncology, Biology, Physics*, **42**, 641–649.

Hicks, K.O., Pruijn, F.B., Secomb, T.W. *et al.* (2006) Use of three-dimensional tissue cultures to model extravascular transport and predict in vivo activity of hypoxia-targeted anticancer drugs. *Journal of National Cancer Institute*, **98**, 1118–1128.

Hirst, D.G. and Denekamp, J. (1979) Tumour cell proliferation in relation to the vasculature. *Cell and Tissue Kinetics*, **12**, 31–42.

Huxham, L.A., Kyle, A.H., Baker, J.H. *et al.* (2006) Tirapazamine causes vascular dysfunction in HCT-116 tumour xenografts. *Radiotherapy and Oncology*, **78**, 138–145.

Huxham, L.A., Kyle, A.H., Baker, J.H. *et al.* (2004) Microregional effects of gemcitabine in HCT-116 xenografts. *Cancer Research*, **64**, 6537–6541.

Jain, R.J. (1994) Barriers to drug delivery in solid tumours. *Scientific American*, **271** (1), 58–65.

Jain, R.K. (1987) Transport of molecules in the tumor interstitium: a review. *Cancer Research*, **47**, 3039–3051.

Jain, R.K. (1997) Delivery of molecular and cellular medicine to solid tumors. *Advanced Drug Delivery Reviews*, **26**, 71–90.

Janssen, H.L., Haustermans, K.M., Balm, A.J., and Begg, A.C. (2005) Hypoxia in head and neck cancer: how much, how important? *Head and Neck*, **27**, 622–638.

Kerr, D.J., Wheldon, T.E., Hydns, S., and Kaye, S.B. (1988) Cytotoxic drug penetration studies in multicellular tumour spheroids. *Xenobiotica*, **18**, 641–648.

Kimura, H., Braun, R.D., Ong, E.T. *et al.* (1996) Fluctuations in red cell flux in tumor microvessels can lead to transient hypoxia and reoxygenation in tumor parenchyma. *Cancer Research*, **56**, 5522–5528.

Krogh, A. (1919) The number and distribution of capillaries in muscles with calculations of the oxygen pressure head necessary for supplying the tissue. *Journal of Physiology*, **52**, 409–415.

Kyle, A. (1999) Modelling extravascular drug penetration using multilayered cell culture. MSc Thesis, Department of Physics, Vancouver, University of British Columbia.

Kyle, A. (2007) Evaluating the extravascular penetration and distribution of anti-cancer drugs using a novel application of the multilayered cell culture model. PhD Thesis, Department of Experimental Medicine, Vancouver, University of British Columbia.

Kyle, A.H. and Minchinton, A.I. (1999) Measurement of delivery and metabolism of tirapazamine to tumour tissue using the multilayered cell culture model. *Cancer Chemotherapy and Pharmacology*, **43**, 213–220.

Kyle, A.H., Huxham, L.A., Baker, J.H. *et al.* (2003) Tumor distribution of bromodeoxyuridine-labeled cells is strongly dose dependent. *Cancer Research*, **63**, 5707–5711.

Kyle, A.H., Huxham, L.A., Chiam, A.S. *et al.* (2004) Direct assessment of drug penetration into tissue using a novel application of three-dimensional cell culture. *Cancer Research*, **64**, 6304–6309.

Kyle, A.H., Huxham, L.A., Yeoman, D.M., and Minchinton, A.I. (2007) Limited tissue penetration of taxanes: a mechanism for resistance in solid tumors. *Clinical Cancer Research,* **13** (9), 2804–2810.

Lankelma, J., Dekker, H., Luque, F.R. *et al.* (1999) Doxorubicin gradients in human breast cancer. *Clinical Cancer Research*, **5**, 1703–1707.

Leighton, J. (1951) A sponge matrix method for tissue culture; formation of organized aggregates of cells in vitro. *Journal of National Cancer Institute*, **12**, 545–561.

Li, C., Newman, R.A., Wu, Q.P. *et al.* (2000) Biodistribution of paclitaxel and poly(L-glutamic acid)-paclitaxel conjugate in mice with ovarian OCa-1 tumor. *Cancer Chemotherapy and Pharmacology*, **46**, 416–422.

Liedert, B., Pluim, D., Schellens, J., and Thomale, J. (2006) Adduct-specific monoclonal antibodies for the measurement of cisplatin-induced DNA lesions in individual cell nuclei. *Nucleic Acids Research*, **34**, e47.

Ljungkvist, A.S., Bussink, J., Kaanders, J.H. *et al.* (2006) Dynamics of hypoxia, proliferation and apoptosis after irradiation in a murine tumor model. *Radiation Research*, **165**, 326–336.

Lord, E.M., Harwell, L., and Koch, C.J. (1993) Detection of hypoxic cells by monoclonal antibody recognizing 2-nitroimidazole adducts. *Cancer Research*, **53**, 5721–5726.

Mahoney, B.P., Raghunand, N., Baggett, B., and Gillies, R.J. (2003) Tumor acidity, ion trapping and chemotherapeutics. I. Acid pH affects the distribution of chemotherapeutic agents in vitro. *Biochemical Pharmacology*, **66**, 1207–1218.

Martin, C., Walker, J., Rothnie, A., and Callaghan, R. (2003) The expression of P-glycoprotein does influence the distribution of novel fluorescent compounds in solid tumour models. *British Journal of Cancer*, **89**, 1581–1589.

Minchinton, A.I. and Tannock, I.F. (2006) Drug penetration in solid tumours. *Nature Reviews Cancer*, **6**, 583–592.

Minchinton, A.I., Durand, R.E., and Chaplin, D.J. (1990) Intermittent blood flow in the KHT sarcoma – flow cytometry studies using Hoechst 33342. *British Journal of Cancer*, **62**, 195–200.

Minchinton, A.I., Wendt, K.R., Clow, K.A., and Fryer, K.H. (1997) Multilayers of cells growing on a permeable support: An in vitro tumour model. *Acta Oncologica*, **36**, 13–16.

Moulder, J.E. and Rockwell, S. (1987) Tumor hypoxia: its impact on cancer therapy. *Cancer Metastasis Reviews*, **5**, 313–341.

Nederman, T., Carlsson, J., and Kuoppa, K. (1988) Penetration of substances into tumour tissue. Model studies using saccharides, thymidine and thymidine-5′-triphosphate in cellular spheroids. *Cancer Chemotherapy and Pharmacology*, **22**, 21–25.

Nederman, T., Carlsson, J., and Malmqvist, M. (1981) Penetration of substances into tumor tissue – a methodological study on cellular spheroids. *In vitro*, **17**, 290–298.

Olive, P.L. and Durand, R.E. (1994) Drug and radiation resistance in spheroids: cell contact and kinetics. *Cancer Metastasis Reviews*, **13**, 121–138.

Olive, P.L., Chaplin, D.J., and Durand, R.E. (1985) Pharmacokinetics, binding and distribution of Hoechst 33342 in spheroids and murine tumours. *British Journal of Cancer*, **52**, 739–746.

Overgaard, J., Hansen, H.S., Andersen, A.P. *et al.* (1989) Misonidazole combined with split-course radiotherapy in the treatment of invasive carcinoma of larynx and pharynx: report from the DAHANCA 2 study. *International Journal of Radiation Oncology, Biology, Physics*, **16**, 1065–1068.

Overgaard, J., Hansen, H.S., Overgaard, M. *et al.* (1998) A randomized double-blind phase III study of nimorazole as a hypoxic radiosensitizer of primary radiotherapy in supraglottic larynx and pharynx carcinoma. Results of the Danish Head and Neck Cancer Study (DAHANCA) Protocol 5-85. *Radiotherapy and Oncology*, **46**, 135–146.

Ozols, R.F., Locker, G.Y., Doroshow, J.H. *et al.* (1979) Pharmacokinetics of adriamycin and tissue penetration in murine ovarian cancer. *Cancer Research*, **39**, 3209–3214.

Patterson, L.H. and Mckeown, S.R. (2000) AQ4N: a new approach to hypoxia-activated cancer chemotherapy. *British Journal of Cancer*, **83**, 1589–1593.

Peterson, H.I. and Appelgren, L. (1977) Tumour vessel permeability and transcapillary exchange of large molecules of different size. *Bibliotheca Anatomica*, **15** (1), 262–265.

Phillips, R.M., Loadman, P.M., and Cronin, B.P. (1998) Evaluation of a novel in vitro assay for assessing drug penetration into avascular regions of tumours. *British Journal of Cancer*, **77**, 2112–2119.

Primeau, A.J., Rendon, A., Hedley, D. *et al.* (2005) The distribution of the anticancer drug Doxorubicin in relation to blood vessels in solid tumors. *Clinical Cancer Research*, **11**, 8782–8788.

Raleigh, J.A., Miller, G.G., Franko, A.J. *et al.* (1987) Fluorescence immunohistochemical detection of hypoxic cells in spheroids and tumours. *British Journal of Cancer*, **56**, 395–400.

Rubin, P. and Casarett, G. (1966) Microcirculation of tumors – Part I: Anatomy, Function and necrosis. *Clinical Radiology*, **17**, 220–229.

Smallbone, K., Gatenby, R.A., Gillies, R.J. *et al.* (2006) Metabolic changes during carcinogenesis: Potential impact on invasiveness. *Journal of Theoretical Biology*. **244** (4), 703–713

Sung, C., Dedrick, R.L., Hall, W.A. *et al.* (1993) The spatial distribution of immunotoxins in solid tumors: assessment by quantitative autoradiography. *Cancer Research*, **53**, 2092–2099.

Sutherland, R.M. (1988) Cell and environment interactions in tumor microregions: the multicell spheroid model. *Science*, **240**, 177–184.

Sutherland, R.M. and Durand, R.E. (1976) Radiation response of multicell spheroids--an in vitro tumour model. *Current Topics in Radiation Research Quarterly*, **11**, 87–139.

Sutherland, R.M., Inch, W.R., Mccredie, J.A., and Kruuv, J. (1970) A multi-component radiation survival curve using an in vitro tumour model. *International Journal of Radiation Biology and Related Studies in Physics, Chemistry, and Medicine*, **18**, 491–495.

Swabb, E.A., Wei, J., and Gullino, P.M. (1974) Diffusion and convection in normal and neoplastic tissues. *Cancer Research*, **34**, 2814–2822.

Tannock, I.F. (1968) The relation between cell proliferation and the vascular system in a transplanted mouse mammary tumour. *British Journal of Cancer*, **22**, 258–273.

Tannock, I.F. (2001) Tumor physiology and drug resistance. *Cancer Metastasis Reviews*, **20**, 123–132.

Tannock, I.F. and Rotin, D. (1989) Acid pH in tumors and its potential for therapeutic exploitation. *Cancer Research*, **49**, 4373–4384.

Tannock, I.F., Lee, C.M., Tunggal, J.K. *et al.* (2002) Limited penetration of anticancer drugs through tumor tissue: a potential cause of resistance of solid tumors to chemotherapy. *Clinical Cancer Research*, **8**, 878–884.

Terheggen, P.M., Floot, B.G., Scherer, E. *et al.* (1987) Immunocytochemical detection of interaction products of cis-diamminedichloroplatinum(II) and cis-diammine(1,1-cyclobutanedicarboxylato)platinum(II) with DNA in rodent tissue sections. *Cancer Research*, **47**, 6719–6725.

Thomlinson, R.H. and Gray, L.H. (1955) The histological structure of some human lung cancers and the possible implications for radiotherapy. *British Journal of Cancer*, **9**, 539–549.

Tilby, M.J., Johnson, C., Knox, R.J. *et al.* (1991) Sensitive detection of DNA modifications induced by cisplatin and carboplatin in vitro and in vivo using a monoclonal antibody. *Cancer Research*, **51**, 123–129.

Tunggal, J.K., Melo, T., Ballinger, J.R., and Tannock, I.F. (2000) The influence of expression of P-glycoprotein on the penetration of anticancer drugs through multicellular layers. *International Journal of Cancer*, **86**, 101–107.

Urtasun, R.C., Koch, C.J., Franko, A.J. *et al.* (1986) A novel technique for measuring human tissue pO2 at the cellular level. *British Journal of Cancer*, **54**, 453–457.

Venkatasubramanian, R., Henson, M.A., and Forbes, N.S. (2006) Incorporating energy metabolism into a growth model of multicellular tumor spheroids. *Journal of Theoretical Biology*, **242**, 440–453.

Vescio, R.A., Redfern, C.H., Nelson, T.J. *et al.* (1987) In vivo-like drug responses of human tumors growing in three-dimensional gel-supported primary culture. *Proceedings of the National Academy of Sciences of the United States of America*, **84**, 5029–5033.

Walenta, S., Doetsch, J., Mueller-Klieser, W., and Kunz-Schughart, L.A. (2000) Metabolic imaging in multicellular spheroids of oncogene-transfected fibroblasts. *Journal of Histochemistry and Cytochemistry*, **48**, 509–522.

West, G.W., Weichselbaum, R., and Little, J.B. (1980) Limited penetration of methotrexate into human osteosarcoma spheroids as a proposed model for solid tumor resistance to adjuvant chemotherapy. *Cancer Research*, **40**, 3665–3668.

Yang, J., Richards, J., Bowman, P. *et al.* (1979) Sustained growth and three-dimensional organization of primary mammary tumor epithelial cells embedded in collagen gels. *Proceedings of the National Academy of Sciences of the United States of America*, **76**, 3401–3405.

16

Impact on Radiotherapy

Michael R. Horsman[1], Jens Overgaard[1] and Dietmar W. Siemann[2]

[1]*Department of Experimental Clinical Oncology, Aarhus University Hospital, Aarhus, Denmark*

[2]*Department of Radiation Oncology, University of Florida, Gainesville, USA*

16.1 Introduction

A functional vascular supply is essential for the growth and development of solid tumors. The neovasculature that develops, from the host vascular system via the process of angiogenesis, is primitive and chaotic when compared to the normal tissue vasculature and can not meet the demands of the growing tumour mass. Thus, areas of nutrient and oxygen deprivation develop. Cells that survive the oxygen deprivation (hypoxia) are known to be resistant to radiation therapy. Numerous approaches have been developed to deal with this hypoxia-induced radioresistance. They include increasing tumour oxygen delivery, chemically radiosensitizing the hypoxic cells, or killing them with specific cytotoxins. More recent approaches have concentrated on the tumour vascular supply itself. Many of these hypoxia targeted therapies have undergone clinical evaluation and significant improvements in radiation response have been reported. In this chapter we will review the characteristics of hypoxia, how it can influence radiation response, and what approaches can be used to overcome such resistance.

16.2 The tumour vasculature and microenvironment

The successful growth and development of most solid tumors requires that they develop their own functional vascular supply (Brem *et al.*, 1976; Folkman, 1986). This they do from the normal host vascular network by the process of angiogenesis (Hahnfeldt *et al.*, 1999; Bergers and Benjamin, 2003). Only when this neovasculature

Tumor Microenvironment Edited by Dietmar W. Siemann

is established can the tumor cells obtain the oxygen and nutrients necessary for survival (Vaupel, Kallinowski, and Okunieff, 1989) and remove the toxic waste products associated with cellular metabolism (Siemann, Chaplin, and Horsman, 2004). It is also the principal route by which metastatic spread can occur (Stoeltzing and Ellis, 2006).

Although a functional tumor vascular supply is necessary, the neo-vasculature that develops is actually inadequate to meet all the demands of the growing tumor mass. Endothelial cells divide at a slower rate than tumor cells (Denekamp and Hobson, 1982). As a consequence, the developing tumor vasculature is unable to keep pace with the expanding tumor population. The tumor vasculature that is formed is also very different from that of normal tissues (Vaupel, Kallinowski, and Okunieff, 1989; Vaupel, 2004). Structurally, it is very chaotic, there is a loss of hierarchy, vascular density is abnormal, and the vessels have contour irregularities, are tortuous, and elongated. The vessels are also very primitive in nature, having incomplete or missing basement membranes and endothelial lining, and lacking pericytes, smooth muscle, and pharmacological receptors. In addition, there are numerous functional abnormalities, including unstable speed and direction of blood flow, high vascular resistance, increased vascular fragility, red blood cell sludging, leukocyte sticking, and blockage of vessels by circulating white blood cells or tumor cells. All these factors result in the development of areas within the tumor that are oxygen deprived, nutrient deficient, and highly acidic (Vaupel, Kallinowski, and Okunieff, 1989; Vaupel, 2004).

From a radiation standpoint it is cells that exist at low oxygen concentrations (hypoxia) that are the most important (see Section 16.3). The first real indication that hypoxia existed in tumors was made by Thomlinson and Gray (1955). From histological sections of carcinoma of the bronchus, they typically found viable tumor regions surrounded by vascular stroma, with regions of necrosis growing in the center of the tumor mass. The thickness of the resulting shell of viable tissue was found to be between 100 and 180 μm thick, which corresponded to the diffusion distance of oxygen in respiring tissues. They suggested that as oxygen diffused from the stroma, it was consumed by the cells, and although those beyond the diffusion distance were unable to survive, the cells immediately bordering the necrosis might be viable yet hypoxic. Later, an inverted version of the Thomlinson and Gray picture was described, with functional blood vessels surrounded by cords of viable tumor cells outside of which were areas of necrosis (Tannock, 1968). This is the more typical picture found in most solid tumors and is illustrated in Figure 16.1a. Hypoxia is shown by the green fluorescence and is distant from the blood vessels shown in red. Beyond the hypoxia areas of necrosis can be seen. This type of hypoxia is more commonly referred to as diffusion-limited chronic hypoxia and is similar to the Thomlinson and Gray model, in that it arises because of the oxygen gradient that is created as the oxygen diffuses from the blood vessel (Figure 16.1b). There are also nutrient and pH gradients thus these hypoxic cells have a tendency to be associated with nutrient deprivation and acidity (Figure 16.1b). It has been suggested that such hypoxic cells can survive under these adverse conditions for several days (Durand and Sham, 1998). Death will also occur as the hypoxic cells move further away from the blood supply as the tumor grows. The actual distance at which this hypoxia

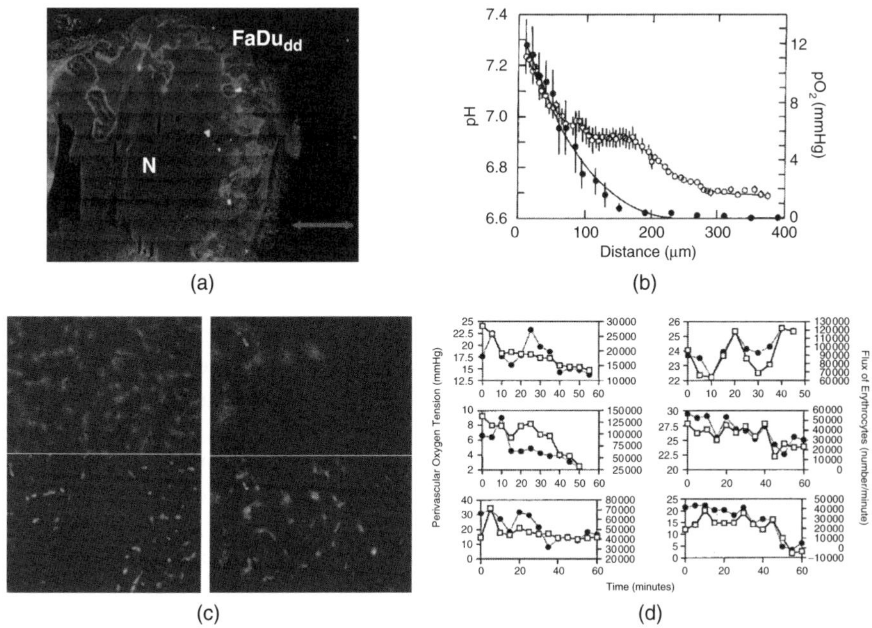

Figure 16.1 Illustrations of hypoxia in tumors. (a) Pseudocolor gray value composite image reconstructed after scanning histological sections from a head and neck xenograft ($FaDu_{dd}$) grown in nude mice. Images show hypoxia (green) and vessels (red).N denotes necrosis; red scale bar = 200 μm. (b) Combined measurements of pH and pO_2 profiles from human colon adenocarcinoma xenogafts (LS174T) grown in a window chamber in severe combined immunodeficient mice. Results show mean values of fluorescent ratio imaging for pH (○) and phosphorence quenching for pO_2 (●) for 24 profiles from seven tumors. (c) Histological sections from SCCVII carcinomas showing 'closing' (non-perfusion) of previously perfused microvessels (left) or 'opening' (perfusion) of previously non-perfused microvessels (right) during a 20-minute interval between intravenous injection of the fluorescent markers H.33342 (blue) and DiOC7 (green). (d) Representative figures showing the relationship between microvascular erythrocyte flux (●) and perivascular oxygen tensions (□) measured simultaneously in dorsal flap window chambers of Fisher-344 rats with implanted R3230Ac tumors. Erythrocytes were fluorescently labeled and observed with intravital video microscopy, while oxygenation was monitored using microelectrodes. Figures and data from Chaplin and Trotter (1991), Kimura *et al.* (1996), Helmlinger *et al.* (1997), Busk *et al.* (2008). A full color version of this figure can be found in the color plate section.

exists from the blood supply can be highly variable (Figure 16.1a) and will depend on the actual concentration of oxygen that is delivered by the blood supply. Indeed, as is shown Figure 16.1b even in the cells next to the blood vessel the oxygen concentrations can start as low as 2% (Helmlinger *et al.*, 1997) and this would certainly reduce the oxygen diffusion distance.

Diffusion-limited chronic hypoxia was the working model for hypoxia from the 1950s until around the 1980s, when it was then suggested that a second type of hypoxia could exist in tumors and one that was acute in nature (Brown, 1979). This was later confirmed and shown to be the result of transient stoppages in tumor blood flow (Chaplin, Olive, and Durand, 1987). These perfusion changes are illustrated in Figure 16.1c, where completely occluded vessels are shown either opening or closing during a 20 minute time period. The hypoxia that results has been termed perfusion limited acute hypoxia. Although complete shut-downs in flow are seen (Figure 16.1c,d) it is clear that these fluctuations can be highly variable and less severe (Figure 16.1d). Complete shut-downs would starve cells of both oxygen and nutrients, but with partial shut-downs one could expect situations where red blood cell delivery is impaired, but plasma flow and thus nutrient supply is not affected. The exact causes of these transient stoppages are not known, but suggested mechanisms include plugging of vessels by circulating blood or tumor cells, collapse of vessels in regions of high interstitial pressure, or spontaneous vasomotion in incorporated host arterioles affecting blood flow in downstream tumor vessels (Horsman, 1995). These temporary cessations in blood flow that account for this have been observed in mouse and rat tumors, as well as human tumor xenografts, with anywhere from around 4–8% of the total functional vessels involved (Horsman, 1995). Clinical studies also implicate their existence in human tumors (Powell *et al.*, 1997; Begg *et al.*, 2001; Janssen *et al.*, 2002). Whatever the cause of tumor hypoxia, its presence has now been identified in most solid animal tumor models (Moulder and Rockwell, 1984) and numerous human cancers (Vaupel, Kallinowski, and Okunieff, 1989).

16.3 Influence of tumor hypoxia on radiation therapy

The potential of microenvironmental parameters to influence outcome to radiotherapy was first suggested from experiments in which the radiation response of skin was markedly decreased if the blood flow to the irradiated area was reduced by compression (Schwarz, 1909). This was followed by a report that tissues in which blood flow was stimulated by diathermia showed a more prominent response to radiation (Müller, 1910). At that time neither study actually attributed the effects to an oxygen dependency. The latter study was also the first to show how radiation response could be improved by modifying the tumor microenvironment. Further experimental observations led Gray and coworkers to finally postulate the role of oxygen deficiency as a major source of radiation resistance (Gray *et al.*, 1953). This resistance is illustrated in Figure 16.2a for Chinese hamster V79 cells irradiated in culture. Cell surviving fraction is shown as a function of radiation dose administered under either normal aerated conditions or under hypoxia, the latter being achieved by gasing the cells with nitrogen gas for at least 1 hour. It is clear that under hypoxic conditions roughly three times as much radiation is required to kill the same number of cells as seen under normal oxygenated condition.

Figure 16.2b shows how hypoxia influences response of cells within a tumor to radiation. In this situation SCCVII tumors are excised after irradiation and the viability of the neoplastic cells within each tumor measured using a clonogenic cell survival assay. Radiation is administered either under normal air breathing conditions or when the tumors are clamped. The clamped dose–response curve represents total hypoxia and is equivalent to the nitrogen curve from Figure 16.2a. For the air breathing study a biphasic curve is obtained. The initial part of the survival curve is determined by the aerobic cells in the tumor and again is equivalent to the normal oxygenated curve in Figure 16.2a. However, at higher dose the hypoxic tumor cells begin to dominate the response and thus the curve begins to parallel the clamped result. This 'hypoxic tail' not only allows us to state that this tumor model contains radioresistant hypoxia, but since we know that the clamped curve represents 100% hypoxia, the amount that the 'hypoxic tail' of the air breathing animals is displaced allows us to actually state the percentage of hypoxia. In this example the curve is some two log units below the clamped result so the hypoxic fraction is around 1%. Two other classic assays are used to demonstrate radiobiological hypoxia in animal tumors (Moulder and Rockwell, 1984; Horsman, 1993). They are the clamped tumor growth delay assay, where measurements are made of the time taken for tumors to reach a specific size after treatment; and the clamped tumor control assay, in which the percentage of animals showing local tumor control at a certain time after treatment is recorded. For each technique it is necessary to produce full radiation dose–response curves under air breathing and fully anoxic (clamped tumor) conditions. As with the clonogenic survival assay, the hypoxic fractions can then be calculated from the displacement of the dose–response curves. Using these assays the hypoxic fractions have been estimated to range from 1% to well over 50% of the viable tumor cells in animal tumors and human tumor xenografts (Moulder and Rockwell, 1984).

None of these approaches are applicable to humans. Although numerous other methods have been developed to try and identify hypoxia in human tumors (Raleigh, Dewhirst, and Thrall, 1996; Horsman, 1993, 1998; Krohn, Link, and Mason, 2008), not all have been used to demonstrate the relationship between hypoxia and radiation response. The earliest attempts to do the latter were based on indirect approaches (Horsman, 1993, 1998), and involved estimates of tumor vascularization, using such endpoints as intercapillary distance, vascular density, and the distance from tumor cells to the nearest blood vessel (Kolstad, 1968; Lauk *et al.*, 1989; Révész *et al.*, 1989). All showed that patients with less well vascularized tumors, and presumably more hypoxic, had a poorer outcome to radiation therapy. More direct approaches have now used exogenous markers that are injected into the host and bind to regions of tumor hypoxia or endogenous markers that are genes/proteins upregulated under hypoxia. The former include nitroimidazole or copper-based derivatives, which can be identified immunohistochemically from histological sections (Raleigh, Dewhirst, and Thrall, 1996; Olive and Aquino-Parsons, 2004) or non-invasively using positron emission tomography (Rasey *et al.*, 1996; Piert *et al.*, 2005), single-photon emission computed tomography (Urtasun *et al.*, 1996), or magnetic resonance spectroscopy (Seddon *et al.*, 2003). Although such exogenous markers can be used to identify tumor hypoxia and even associated with outcome following radiation therapy in head and neck carcinoma patients (Bussink, Kaanders, and van der Kogel, 2003), there has never been a proper randomized trial. The endogenous markers include such factors as carbonic anhydrase IX, GLUT-1, hypoxia inducible factor-1, and osteopontin (Hui *et al.*, 2002; Airley *et al.*, 2003; Le *et al.*, 2003; Overgaard *et al.*, 2005). These can be measured from biopsy material using protein immunohistochemistry or as gene expression, or proteins identified from blood samples. Although endogenous markers have been correlated with outcome to radiation therapy in some studies it is not a universal finding (Bussink, Kaanders, and van der Kogel, 2003), which probably reflects the fact that many of these endogenous markers are not hypoxic specific rather than any indication that hypoxia does not play a role in influencing radiation response.

Probably the most direct method for estimating tumor hypoxia and one that has certainly been used to show that the negative influence of hypoxia on radiation response is the measurement of oxygen partial pressure (pO_2) distributions with polarographic electrodes. This latter effect is illustrated in Figure 16.2 both for animal (Figure 16.2c) and human (Figure 16.2d) tumors. In these examples the oxygen status was assessed in mouse C3H mammary carcinomas or human head and neck cancers prior to radiation treatment. They were then divided into two groups based on whether the percentage of pO_2 values less than or equal to 2.5 mmHg were above or below the median values for the whole group. Although the models are very different in terms of the tumor type, host, and radiation treatment, the results are very similar in that the groups with the greatest number of low values (more hypoxic) had the poorer local tumor control outcome. Numerous clinical studies have now clearly shown the ability of hypoxia to influence outcome in squamous cell carcinomas of the cervix and head and neck (Kolstad, 1968; Gatenby *et al.*, 1988; Hoeckel *et al.*, 1993; Brizel *et al.*, 1997; Nordsmark, Overgaard, and Overgaard, 1996).

16.4 Reducing hypoxia by increasing oxygen delivery

Numerous efforts have been made to try and find clinically applicable approaches that can eliminate the hypoxic cell population and thereby improve the therapeutic potential of radiation therapy. One of the first methods tested involved increasing the oxygen supply by simply allowing the tumor-bearing host to breathe high oxygen content gas mixtures before and during irradiation. Early experimental studies reported that both oxygen and carbogen (95% oxygen + 5% carbon dioxide) breathing could substantially enhance the response of murine tumors to radiation (Du Sault, 1963; Suit, Marshall, and Woerner, 1972). The best effect was generally seen when the gases were inspired under hyperbaric (typically 3 atmospheres) rather than normobaric conditions (Du Sault, 1963; Suit, Marshall, and Woerner, 1972), which is perhaps not surprising because hyperbaric conditions would be expected to saturate the blood with oxygen more than normobaric conditions. However, later studies indicated that the radiosensitizations produced by normabaric oxygen or carbogen are in fact quite substantial (Siemann, Hill, and Bush, 1977; Rojas, 1991; Grau, Horsman, and Overgaard, 1992). Such an effect is illustrated in Figure 16.3a.

The clinical use of high oxygen content gas breathing was introduced relatively early (Churchill-Davidson, 1968). Most trials that were performed were fairly small, and suffered from the applications of unconventional fractionation schemes, but it appeared that the effect of hyperbaric oxygen was superior to radiotherapy given in air, especially when few and large fractions were applied (Churchill-Davidson, 1968; Dische, 1979; Dische *et al.*, 1983). Significant improvements in local tumor control and subsequent survival were certainly seen in the large multicenter clinical trials conducted by the UK Medical Research Council for both uterine cervix

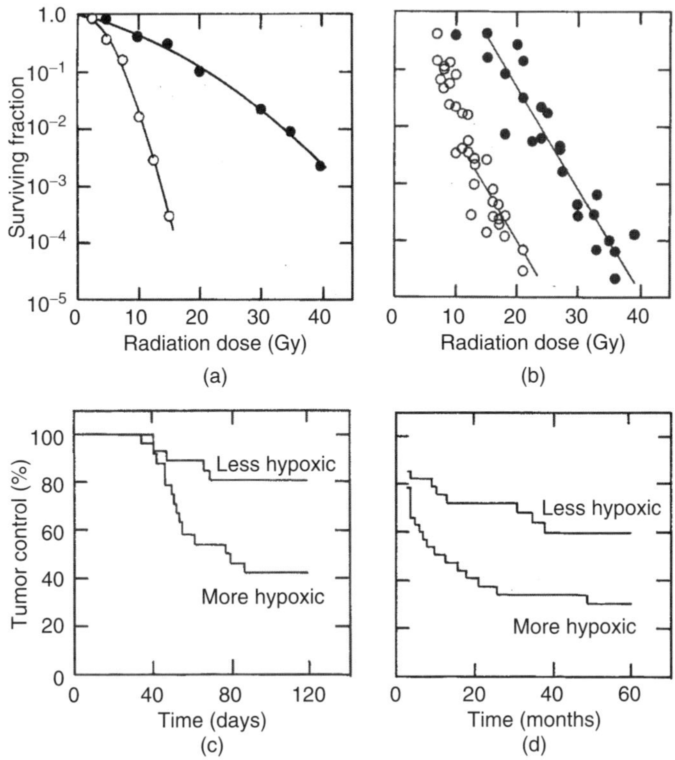

Figure 16.2 Effect of hypoxia on radiation response and tumor oxygenation. (a) Radiation response of Chinese hamster V79 cells irradiated in suspension culture at room temperature under aerobic (○) or hypoxic (●) conditions. (b) Radiation response of individual SCCVII carcinomas irradiated under normal air breathing (○) or clamped (●) conditions and tumor cell survival measured using an excision assay performed 24 hours later. (c) The oxygenation status of C3H mammary carcinomas was assessed using the Eppendorf oxygen electrode and immediately after measurement tumors were irradiated with a single dose of 55 Gy. The time to tumor recurrence was determined upto 120 days later and expressed depending on whether the percentage of pO_2 values <2.5 mmHg was above (more hypoxic) or below (less hypoxic) the median value of 48%. (d) Tumor oxygenation status was determined with the Eppendorf electrode in 67 head and neck cancer patients prior to them receiving conventional radiation therapy (66–68 Gy). Local tumor control was recorded and patients divided into two groups based on whether the percentage of pO_2 values <2.5 mmHg was above (more hypoxic) or below (less hypoxic) the median value of 22%. Figures modified from Horsman *et al.* (1994a), Nordsmark *et al.* (2007); or from Nordsmark and Horsman (unpublished observations).

(Figure 16.4) and advanced head and neck (Dische, 1979; Henk, Kunkler, and Smith, 1977; Henk and Smith, 1977; Watson *et al.*, 1978; Overgaard, 1989). The same benefit was not observed in bladder cancer; nor was it repeated in a number of smaller studies (Overgaard, 1989). In retrospect, the use of hyperbaric oxygen was stopped somewhat prematurely, partly due to the introduction of chemical hypoxic cell radiosensitizers, and partly because of patient compliance concerns.

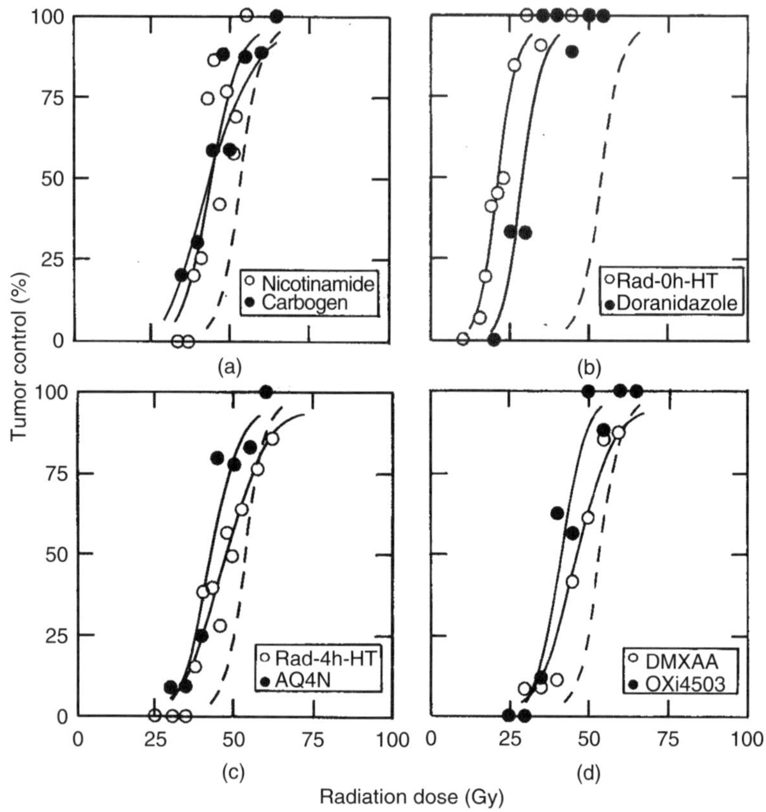

Figure 16.3 Groups of CDF1 mice implanted with a C3H mammary carcinoma had their tumors locally irradiated with single doses of radiation. The percentage of animals showing tumour control within 90 days was recorded and the dose–response curve determined following logit analysis. This is shown as the dashed line in each figure. Also shown is how this radiation response curve is modified following treatment with either (a) carbogen breathing for 5 minutes before and during irradiation (●) or an intraperitoneal (i.p.) injection with 1000 mg/kg nicotinamide 30 minutes before irradiating (○); (b) i.p. injecting 500 mg/kg doranidazole 30 minutes prior to irradiating (●) or irradiating (Rad) tumors in the middle of heating (HT) at 42.5 °C for 60 minutes (○); (c) irradiating (Rad) tumors and heating (HT) at 42.5 °C for 60 minutes 4 hours later (○) or intravenously injecting mice with 120 mg/kg AQ4N 2 hours before irradiating (●); (d) irradiating and then within 1 hour i.p. injecting 50 mg/kg OXi4503 (●) or 20 mg/kg DMXAA (○). Data modified from Horsman, Chaplin, and Overgaard (1990), Grau, Horsman, and Overgaard (1992), Murata and Horsman (2004), Murata, Tsujitani, and Horsman (2008), or unpublished observations.

Breathing normobaric high oxygen content gas has also been tried clinically but the early studies failed to show any dramatic improvement in radiation response (Bergsjø and Kolstad, 1968; Rubin *et al.*, 1979). This may have been due to the failure to achieve the optimum preirradiation gas breathing time. A number of experimental studies have shown this to be critical for the enhancement of radiation damage, and that it can vary from tumor to tumor (Suit, Marshall, and

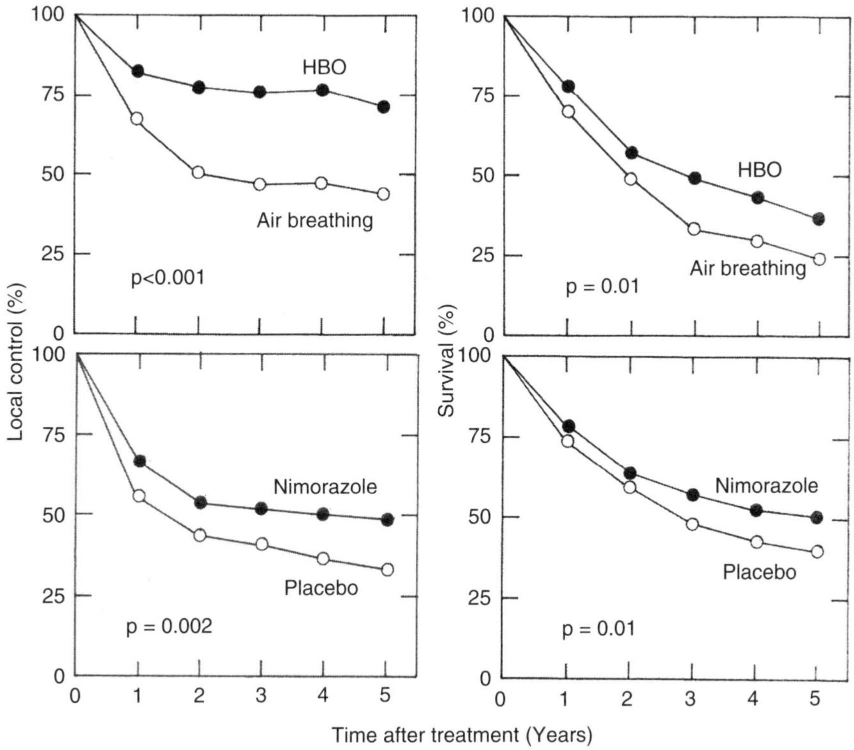

Figure 16.4 Results of clinical randomized trials in which patients received hypoxic modification in conjunction with radiotherapy. (Top panels) Results from the Medical Research Council hyperbaric oxygen (HBO) trial showing actuarial local tumor control and survival in patients with stage III carcinoma of the cervix treated with either HBO (●; 119 patients) or air breathing (○; 124 patients). Modified from Watson *et al.* (1978). (Bottom panels) Results from the DAHANCA 5 study showing actuarial estimated locoregional control and disease-specific survival rate in patients given nimorazole (●; 219 patients)or placebo (○; 195 patients). Figures modified from Overgaard *et al.* (1998).

Woerner, 1972; Siemann, Hill, and Bush, 1977; Rojas, 1991; Chaplin, Horsman, and Siemann, 1993). More recent studies with carbogen in head and neck, using a short preirradiation breathing time, have yielded conflicting findings, with both a benefit (Kaanders, Bussink, and van der Kogel, 2002a) and no improvement (Mendenhall *et al.*, 2005) in radiation response observed. The positive results in the former study may have been because carbogen was administered as part of the *a*ccelerated *r*adiation, *c*arbogen, and *n*icotinamide (ARCON) therapy. Accelerated radiotherapy was included to address the issue of accelerated repopulation that is known to occur after irradiation (Petersen *et al.*, 2001). Nicotinamide was included to deal with fluctuating hypoxia which is not likely to be influenced by high oxygen content breathing alone. Experimental studies have clearly demonstrated that the vitamin B_3 analog, nicotinamide, can enhance radiation damage in a variety of murine tumor models using both single (Figure 16.3a) and fractionated treatments (Horsman, 1995). This enhanced radiation response is dependent on the tumor type, drug dose, and time of irradiation after drug administration (Horsman, 1995). Nicotinamide can also enhance radiation damage in certain normal tissues, but in general the effects are less than those seen in tumors (Horsman, 1995). The effect of nicotinamide seems to primarily involve the drug preventing/reducing the transient changes in tumor blood flow, which normally lead to the development of fluctuating hypoxia. This finding led to the suggestion that the optimal approach would be to combine nicotinamide with treatments that specifically overcome chronic hypoxia, and was subsequently demonstrated in combination with hyperthermia (Horsman, Chaplin, and Overgaard, 1990), perfluorochemical emulsions (Chaplin, Horsman, and Aoki, 1991), and carbogen breathing (Kjellen *et al.*, 1991; Chaplin, Horsman, and Siemann, 1993; Horsman *et al.*, 1994b). This combination of nicotinamide with carbogen is now being tested in a number of European clinical studies, and the preliminary results for at least two, one in head and neck (Kaanders *et al.*, 2002b) and the other in bladder (Hoskin, Rojas, and Saunders, 2009b), already suggest that such an approach can improve radiation response.

Hemoglobin concentration is one of the major factors influencing the delivery of oxygen to tumors. It is, therefore, not surprising that low hemoglobin concentration in general has a negative impact on tumor radiation response. In a review of 51 studies involving 17 272 patients the prognostic relationship between hemoglobin concentration and local tumor control were analysed. Of these, 39 studies (14 482 patients) showed a correlation, while only 12 (2790 patients) did not (Grau and Overgaard, 1998). However, a large (357 patients) international multicenter study in head and neck failed to show any correlation between hemoglobin concentration and pretreatment pO_2 measurements questioning the relationship between hemoglobin concentration and tumor oxygenation status (Nordsmark *et al.*, 2005). Despite this latter study, the potential benefit of increasing hemoglobin by blood transfusion prior to radiotherapy has been investigated in a number of studies (Thomas, 2002). The first clinical investigation of this approach was in advanced squamous cell carcinoma of the uterine cervix (Evans and Bergsjø, 1965). Transfusion to patients with low hemoglobin level resulted in an increased tumor oxygen tension, as measured directly using oxygen electrodes. The same study was also the first to show that transfusion to a hemoglobin level of 11 g/dl or higher was able to significantly improve survival. A Canadian retrospective study of 605 cervix cancer patients showed that the negative influence of low hemoglobin on prognosis could be overcome by transfusion (Grogan *et al.*, 1999). However, these observations have not been supported by data from controlled randomized trials; a prospective phase III trial, from the Danish Head and Neck Cancer (DAHANCA) study group, failed to demonstrate any benefit of transfusion in head and neck cancer patients with low hemoglobin levels (Overgaard *et al.*, 1998).

The concentration of hemoglobin can also be increased by stimulation with the hormone erythropoietin (EPO). Several preclinical studies have shown that the induction of anemia in animals could be corrected by serial injection with EPO and that this EPO treatment also overcame the anemia-induced radiation resistance (Thews *et al.*, 1998; Stuben *et al.*, 2003). The concept of using EPO to correct for anemia has also been tested in a number of clinical trials. However, although low hemoglobin can be effectively and safely improved by EPO (Lavey and Dempsey, 1993; Hoskin *et al.*, 2009a), a number of studies in patients undergoing treatment for head and neck cancer failed to show any benefit (Henke *et al.*, 2003; Machtay *et al.*, 2007; Overgaard *et al.*, 2007; Hoskin *et al.*, 2009a). More significantly, those patients that actually received EPO in conjunction with radiation therapy did significantly worse than those patients that did not receive EPO and as a result all EPO and radiation trials have been stopped.

Several other 'hemoglobin-related' methods for improving tumor oxygenation have been investigated. These include the use of artificial blood substances, such as perfluorocarbons (Rockwell, 1985), which are small particles capable of carrying more oxygen than hemoglobin, or by manipulating the oxygen unloading capacity of blood by modifying the oxyhemoglobin dissociation curve. This can be achieved either by increasing the red blood cell 2,3-DPG content (Siemann and Macler, 1986), 2,3-DPG being one of the most important allosteric factors controlling the hemoglobin–oxygen dissociation curve, or using antilipidemic drugs (Hirst and Wood, 1991). Although each of these approaches has been shown to improve the oxygenation status of experimental tumors and/or enhance radiation damage, none of them have yet reached controlled clinical testing, thus their potential usefulness in the clinic is uncertain.

16.5 Radiosensitizing hypoxic cells

Probably the most extensively investigated method for overcoming tumor hypoxic cell radioresistance is the use of chemical agents that mimic oxygen. Unlike oxygen, these drugs are not rapidly metabolized by the tumor cells through which they diffuse and thus can penetrate further than oxygen and so reach all cells in the tumor. In the early 1960s it was found that the efficiency of radiosensitization was directly related to electron affinity (Adams and Cooke, 1969) and that ultimately led to *in vitro* studies demonstrating preferential radiosensitization of hypoxic cells by highly electron-affinic nitroaromatic compounds (Asquith *et al.*, 1974; Adams *et al.*, 1976). Several of these compounds were later shown to be effective at enhancing radiation damage in tumors (Overgaard, 1994). This is illustrated in Figure 16.3b for doranidazole, one of the latest radiosensitizers to be developed.

Clinical radiosensitizer studies were initially performed with metronidazole in brain tumors (Urtasun *et al.*, 1976) and were followed, in the latter part of the 1970s, by a series of clinical trials exploring the potential of misonidazole as a radiosensitizer (Dische, 1985; Overgaard, 1989, 1994). Most trials with misonidazole were unable to generate any significant improvement in radiation response, although a benefit was seen in some trials, especially the DAHANCA 2 study (Overgaard *et al.*, 1989). The lack of benefit with misonidazole in most studies could be attributed to clinical evaluation in tumor sites lacking convincing evidence of the existence of hypoxia, the small number of patients in many of the trials, and mostly because the drug doses necessary for effective radiosensitization also produced substantial dose-limiting clinical toxicity. As a result there followed a search for more efficient or less toxic hypoxic sensitizers. Subsequent trials evaluated pimonidazole, etanidazole, and nimorazole. The European pimonidazole trial in uterine cervix was very disappointing (Dische, Machin, and Chassagne, 1993), while the two other multicenter trials in head and neck cancer, using etanidazole, showed no benefit (Lee *et al.*, 1995; Eschwége *et al.*, 1997). On the other hand, studies with the less toxic drug nimorazole given to patients with supraglottic and pharynx carcinomas (DAHANCA 5) showed a highly significant benefit in terms of improved locoregional tumor control and disease-free survival (see Figure 16.4), thereby confirming the result of the DAHANCA 2 study (Overgaard *et al.*, 1998). A more recent International Atomic Energy Agency (IAEA) trial with the 3-nitrotriazole compound, sanazole (AK-2123), in uterine cervical cancer also demonstrated a significant improvement in both local tumor control and overall survival (Dobrowsky *et al.*, 2007).

A meta-analysis of all randomized clinical studies in which hypoxic radiosensitizers were given with primary radiotherapy (Overgaard and Horsman, unpublished observations), involving more than 11 000 patients in 91 randomized trials, showed that radiosensitizer modification of tumor hypoxia significantly improved the locoregional tumor control after radiotherapy with an odds ratio of 0.78 (0.72–0.85; 95% confidence intervals). This treatment benefit was mostly the result of an improved response in head and neck with an odds ratio of 0.74 (0.65–0.83) and to a lesser extent in bladder tumors. No significant effect was observed in other tumor sites (cervix, lung, central nervous system and esophagus). The overall survival rate was also improved with an overall odds ratio of 0.83 (0.77–0.90). In conclusion, the hypoxic radiosensitizer era has clearly demonstrated an overall significant benefit of both local control and survival in certain tumor sites. The overall observed gain was in the order of a 5–10% improvement in local control. Although small, such gains are certainly relevant to pursue, especially because they are apparently associated with a similar improvement in survival. On this basis there is no need for a feeling of disappointment at the apparent lack of a major clinical benefit. The non-significant outcome of most clinical trials is not due to a biological lack of importance of hypoxia, but in most cases is considered a consequence of poor clinical trials methodology with a too optimistic study design and an expected treatment gain which goes far beyond what is reasonable. Overall, the results with nitroimidazoles add to the general concensus that if a non-toxic hypoxic modification can be applied then such treatments may certainly be relevant as a baseline therapy together with radiotherapy for cancers like advanced head and neck. Such a strategy has been adopted in Denmark where nimorazole has become part of the standard radiotherapy treatment in head and neck.

One of the most effective radiation sensitizers is hyperthermia (Horsman and Overgaard, 2007). Numerous preclinical studies have shown both *in vitro* (Li and Kal, 1977; Sapareto, Raaphorst, and Dewey, 1979; Hahn, 1982) and *in vivo* (for review see Horsman and Overgaard, 2007) that irradiating tumors and heating with temperatures up to 43 °C at around the same time substantially enhanced radiation response (Figure 16.3b). The exact mechanism by which heat sensitizes cells to radiation is not known, but most evidence suggests that heat primarily interferes with the cells ability to deal with radiation-induced DNA damage (Kampinga and

Dikomey, 2001; Roti Roti, 2004). This large body of preclinical data showing the benefit of combining radiation and hyperthermia has resulted in the translation of this approach into a number of clinical trials. A meta-analysis of all published trials (1861 patients from 23 trials), in which patients were randomized to radiation or radiation and heat, demonstrated significant improvements in local tumor control, the most relevant endpoint for such locally applied treatments, in a number of distinct sites including chest wall, cervix, rectum, bladder, melanoma, and head and neck (Horsman and Overgaard, 2007). Some of these clinical studies even reported improvements in overall survival (Horsman and Overgaard, 2007).

16.6 Killing the resistant cell population

If an interval of several hours is introduced between the radiation and heat treatments, the sensitization effect of heat is lost. But, the heat still enhances radiation response by a small, yet significant, amount (Horsman and Overgaard, 2007). This effect is illustrated in Figure 16.3c and results from the heat simply killing the radioresistant hypoxic cell population. Several *in vitro* studies have reported that cells under hypoxic conditions are more sensitive to the lethal effects of hyperthermia than cells in a well-oxygenated environment (Gerweck, Gillette, and Dewey, 1974; Gerweck, Nygaard, and Burlett, 1979; Overgaard and Bichel, 1977; Suit and Gerweck, 1979). Under well-defined nutrient conditions, acute hypoxia alone does not have any significant influence on the cellular response to hyperthermia (Gerweck, Gillette, and Dewey, 1974; Power and Harris, 1977; Gerweck, Nygaard, and Burlett, 1979). However, prolonged oxygen deprivation or chronic hypoxia will increase cellular heat sensitivity (Gerweck, Nygaard, and Burlett, 1979; Overgaard and Nielsen, 1980; Nielsen, 1981). Prolonged hypoxia generally leads to metabolic changes, which in turn alter several other parameters especially acidity, and it is this change that is responsible for the increased cell killing by heat (Gerweck, Gillette, and Dewey, 1974; Hahn, 1974; Overgaard and Bichel, 1977; Overgaard and Nielsen, 1980).

Early pre-clinical studies with electron-affinic radiosensitizers also showed that these agents, which were relatively non-toxic to cells under normal oxygenated conditions, were reduced to a more toxic form under hypoxia (Hall and Roizin-Towle, 1975). This led to the development of various bioreductive drugs that preferentially killed the radiation resistant hypoxic tumor cell population. Basically, these drugs can be divided into three major groups: the quinones, nitroaromatics, and *N*-oxides (McKeown, Cowen, and Williams, 2007).

The prototype bioreductive drug is mitomycin C (MMC), which is activated by bioreduction to form products that cross-link DNA (Kennedy, Rockwell, and Sartorelli, 1980). It has been used for many years in patients as a chemoradiosensitizer, but there are also several studies in which it has been used to specifically overcome hypoxic cell radioresistance. In two randomized clinical trials in patients with squamous cell carcinoma of the head and neck, MMC improved radiation-induced local tumor control without any enhancement of radiation reactions in normal tissues (Weissberg *et al.*, 1989; Haffty *et al.*, 1993). However, in two other trials no major influence on response or survival was seen (Dobrowsky *et al.*, 1995; Grau *et al.*, 2003). This absence of any improved response is perhaps not surprising since the drug was only given once during the radiation schedule and also shows little differential between aerobic and hypoxic cell killing (Stratford and Stephens, 1989; Hall, 1994). Attempts to find more efficient quinones have been undertaken and to that end porfiromycin and EO9 have been developed (McKeown, Cowen, and Williams, 2007). While EO9 has only gone through preliminary phase I/II testing, porfiromycin was included in a prospective randomized trial in combination with radiation therapy in head and neck cancer, but was found to be no better than MMC (Haffty *et al.*, 2005).

With the finding that misonidazole showed preferential toxcity toward hypoxic cells numerous efforts were made to find other nitroimidazole radiosensitizers that were effective as hypoxic cell cytotoxins. To that end RSU 1069 was developed. This compound has the classic 2-nitroimidazole radiosensitizing properties, but an aziridine ring at the terminal end of the chain gives the molecule substantial potency as a hypoxic cell cytotoxin. Although the drug was found to have substantial activity in hypoxic cells *in vitro* and in tumors *in vivo* (Stratford *et al.*, 1986) large animal studies indicated dose-limiting gastrointestinal toxicity. A less toxic prodrug, RB 6145, which *in vivo* is reduced to RSU 1069, was also shown to have potent anti-tumor activity in experimental systems, but was dropped when toxicity studies revealed that it induced blindness in dogs. However, other nitroaromatic compounds are under development, including NLCQ-1, CB1954, SN-23862, and PR-104 (McKeown, Cowen, and Williams, 2007).

One of the more promising group of bioreductives are the organic nitroxides, of which the benzotriazene di-*N*-oxide, tirapazamine, is the lead compound. The parent moiety shows limited toxicity toward aerobic cells, but after reduction under hypoxic conditions a product is formed which has been shown to be highly toxic to both cells *in vitro* and can substantially enhance radiation damage in tumors *in vivo* (Zeman *et al.*, 1988). Most clinical studies have involved combining tirapazamine with chemotherapy, although there have been a few trials with radiation $\pm$ chemotherapy (McKeown, Cowen, and Williams, 2007). The results from the phase II trials generally showed promise, but in the few randomized trials that have been completed the results have been somewhat disappointing. However, it has now been suggested that a benefit of tirapazamine might be achieved if one could select patients with hypoxic tumors by positron emission tomography imaging prior to treatment. Other *N*-oxides currently under development include chlorambucil *N*-oxide and AQ4N (banoxantrone), the latter being combined with radiation in a number of phase II trials (McKeown, Cowen, and Williams, 2007). The ability of AQ4N to enhance radiation-induced local tumor control is illustrated in Figure 16.3c.

16.7 Vascular targeting approaches

The tumor vasculature not only plays a significant role in determining the microenvironment within tumors, it is also essential for the successful growth and development of those tumors (Folkman, 1986). This makes the vasculature an attractive target for therapy and two major groups of vascular targeting agents (VTAs) have emerged

(Siemann *et al.*, 2005). One approach involves preventing the development of the tumor vasculature by inhibiting various steps in the angiogenic process (angiogenesis inhibiting agents; AIs). An alternative method is to use agents that damage the already established tumor vessels (vascular disrupting agents; VDAs).

Although both AIs and VDAs show antitumor effects, they never result in tumor control, even when used in combination (Horsman and Siemann, 2006). This has led to the suggestion that their potential clinical application would be when combined with more conventional treatments, especially radiation. In fact, numerous preclinical studies have now shown that the response of tumors to radiation can be significantly improved when animals are treated with either AIs or VDAs (Horsman and Siemann, 2006). This is illustrated in Figure 16.3d for two of the leading VDAs in clinical development; the flavone derivative 5,6-dimethylxanthenone-4-acetic acid (DMXAA) and the tubulin-binding drug OXi4503.

With both AIs and VDAs, hypoxia has been implicated in the mechanism for this enhancement of radiation response. For AIs, the consensus opinion is that the improvement in radiation response found in preclinical studies is the consequence of normalization of the tumor vasculature resulting in a decrease in tumor hypoxia (Jain, 2001). While there are certainly preclinical studies showing an improved tumor oxygenation status with such treatments, there are just as many studies showing no change and even a decrease in tumor oxygenation (Horsman and Siemann, 2006). These findings not only make it unclear as to the role of vessel normalization in influencing the combination of angiogenesis inhibitors with radiation, they also suggest that timing and sequencing of the two modalities may be critical for an optimal benefit. With VDAs the situation is less controversial. VDAs damage tumor vasculature and as a result cause a severe reduction in tumor blood flow leading to ischemia and cell death (Horsman and Siemann, 2006). The cells that die first are most likely the hypoxic cells that are already oxygen and nutrient deprived, and the improved antitumor responses observed when such agents are combined with radiotherapy probably reflects an additive tumor response resulting from the VDA eliminating treatment resistant hypoxic tumor cells while the radiotherapy acts against the aerobic tumor cell population. However, there are indications that VDAs may induce hypoxia in cells that were not previously hypoxic and that such cells survive this treatment (Horsman and Siemann, 2006). This supports the concept that timing of VDA therapy relative to radiation treatment is critical, and that for the best affect the drugs should be given soon after irradiating (Wilson *et al.*, 1998; Murata *et al.*, 2001a, 2001b; Siemann and Rojiani, 2002).

16.8 Conclusions and future perspectives

One major focus of current cancer research is the role of cancer stem cells in tumorigenesis and therapy. Such cells amount to around 1–25% of the total viable tumor cell population (Hill, Marie-Egyptienne, and Hedley, 2009), but they are believed to be the cells that must be eliminated to obtain local tumor control with radiation therapy (Baumann, Krause, and Hill, 2008). Significant effort is currently being made to identify these cells and specifically target them. However, recent evidence suggests a possible link between hypoxia and cancer stem cells (Baumann, Krause, and Hill, 2008). Hypoxia may affect cancer stem cell generation and maintenance through the upregulation of hypoxia induced factors (Baumann, Krause, and Hill, 2008; Hill, Marie-Egyptienne, and Hedley, 2009). Pre-clinical studies have also shown an inverse correlation between hypoxia and local tumor control after irradiation (Horsman *et al.*, 1993; Yaromina *et al.*, 2006), suggesting that hypoxia may also actually protect the cancer stem cells from the lethal effects of radiation. Using treatments that eliminate hypoxia should, therefore, be an effective method for enhancing the radiosensitivity of cancer stem cells. This fact alone should argue for substantially more effort being made to get hypoxia modifying treatments established clinically.

It has also been proposed that 'targeted therapies' are the future for cancer therapy. In that context, hypoxia must be considered the ultimate target. It is a fundamental feature of virtually all solid tumors, whether animal or human, that can be identified by a host of clinically applicable techniques. There is definitive evidence that its existence in specific tumor types will have a significant negative impact on cancer treatment, especially radiotherapy. Numerous preclinical studies have identified a range of different approaches that can reduce/eliminate hypoxia and preferentially improve tumor radiation response, and many of these have undergone clinical testing and substantial improvements in outcome following radiation therapy obtained. However, except for Denmark in which the hypoxic cell radiosensitizer nimorazole is routinely used in the treatment of head and neck cancer, none of these hypoxic modifiers have become established as a standard therapy with radiation. Whether this is a consequence of the non-glamorous nature of this target, the lack of potential profit from using easily obtainable treatments, or some other unexplained reason is not known. Nevertheless, despite this negative situation, extensive preclinical and clinical studies are ongoing in this area, so hopefully it is simply a matter of time before the hypoxia problem is eliminated.

References

Adams, G.E. and Cooke, M.S. (1969) Electron-affinic sensitization. I. A structural basis for chemical radiosensitizers in bacteria. *International Journal of Radiation Biology*, **15**, 457–471.

Adams, G.E., Flockhart, I.R., Smithen, C.E. *et al.* (1976) Electron–affinic sensitization VII: A correlation between structures, one–electron reduction potentials and efficiencies of some nitroimidazoles as hypoxic cell radiosensitizers. *Radiation Research*, **67**, 9–20.

Airley, R.E., Loncaster, J., Raleigh, J. *et al.* (2003) GLUT-1 and CAIX as intrinsic markers of hypoxia in carcinoma of the cervix: relationship to pimonidazole binding. *International Journal of Cancer*, **104**, 85–91.

Asquith, J.C., Watts, M.E., Patel, K. *et al.* (1974) Electron–affinic sensitization V. Radiosensitization of hypoxic bacteria and mammalian cells in vitro by some nitroimidazoles and nitropyrazoles. *Radiation Research*, **60**, 108–118.

Baumann, M., Krause, M., and Hill, R. (2008) Exploring the role of cancer stem cells in radioresistance. *Nature Reviews Cancer*, **8**, 545–554.

Begg, A.C., Janssen, H., Sprong, D. *et al.* (2001) Hypoxia and perfusion measurements in human tumours: initial experience with pimonidazole and IUdR. *Acta Oncologica*, **40**, 924–928.

Bergers, G. and Benjamin, L.E. (2003) Tumorigenesis and the angiogenic switch. *Nature Reviews Cancer*, **3**, 401–410.

Bergsjø, P. and Kolstad, P. (1968) Clinical trial with atmospheric oxygen breathing during radiotherapy of cancer of the cervix. *Scandinavian Journal of Clinical Laboratory Investigations*, **106**, 167–171.

Brem, S., Brem, H., Folkman, J. *et al.* (1976) Prolonged tumor dormancy by prevention of neovascularization in the vitreous. *Cancer Research*, **36**, 2807–2812.

Brizel, D.M., Sibley, G.S., Prosnitz, L.R. *et al.* (1997) Tumor hypoxia adversely affects the prognosis of carcinoma of the head and neck. *International Journal of Radiation Oncology Biology Physics*, **38**, 285–289.

Brown, J.M. (1979) Evidence for acutely hypoxic cells in mouse tumours, and a possible mechanism of reoxygenation. *British Journal of Radiology*, **52**, 650–656.

Busk, M., Horsman, M.R., Jakobsen, S. *et al.* (2008) Imaging hypoxia in xenografted and murine tumors with ^{18}F-fluoroazomycin arabinoside: a comparative study involving microPET, autoradiography, pO_2-polarography, and fluorescent microscopy. *International Journal of Radiation Oncology Biology Physics*, **70**, 1202–1212.

Bussink, J., Kaanders, J.H.A.M., and van der Kogel, A.J. (2003) Tumor hypoxia at the micro-regional level: clinical relevance and predictive value of exogenous and endogenous hypoxic cell markers. *Radiotherapy and Oncology*, **67**, 3–15.

Chaplin, D.J., Horsman, M.R., and Aoki, D.S. (1991) Nicotinamide, fluosol DA and carbogen: a strategy to reoxygenate acutely and chronically hypoxic cells in vivo. *British Journal of Cancer*, **63**, 109–113.

Chaplin, D.J., Horsman, M.R., and Siemann, D.W. (1993) Further evaluation of nicotinamide and carbogen as a strategy to reoxygenate hypoxic cells *in vivo*: importance of nicotinamide dose and pre-irradiation breathing time. *British Journal of Cancer*, **68**, 269–273.

Chaplin, D.J., Olive, P.L., and Durand, R.E. (1987) Intermittent blood flow in a murine tumor: radiobiological effects. *Cancer Research*, **47**, 597–601.

Chaplin, D.J. and Trotter, M.J. (1991) Chemical modifiers of tumor blood flow, in *Tumor Blood Supply and Metabolic Microenvironment* (eds P. Vaupel and R.K. Jain), Gustav Fischer Verlag, Stuttgart, New York, pp. 65–85.

Churchill-Davidson, I. (1968) The oxygen effect in radiotherapy: historical review. *Frontiers of Radiation Therapeutic Oncology*, **1**, 1–15.

Denekamp, J. and Hobson, B. (1982) Endothelial-cell proliferation in experimental tumours. *British Journal of Cancer*, **46**, 711–720.

Dische, S. (1979) Hyperbaric oxygen: the medical research council trials and their clinical significance. *British Journal of Radiology*, **51**, 888–894.

Dische, S. (1985) Chemical sensitizers for hypoxic cells: a decade of experience in clinical radiotherapy. *Radiotherapy and Oncology*, **3**, 97–115.

Dische, S., Anderson, P.J., Sealy, R. *et al.* (1983) Carcinoma of the cervix: anaemia, radiotherapy and hyperbaric oxygen. *British Journal of Radiology*, **56**, 251–255.

Dische, S., Machin, D., and Chassagne, D. (1993) A trial of Ro 03–8799 (pimonidazole) in carcinoma of the uterine cervix: an interim report from the Medical Research Council Working Party on advanced carcinoma of the cervix. *Radiotherapy and Oncology*, **26**, 93–103.

Dobrowsky, W., Huigol, N.G., Jayatilake, R.S. *et al.* (2007) AK-2123 (sanazol) as a radiation sensitizer in the treatment of stage III cervical cancer: results of an IAEA multicentre randomized trial. *Radiotherapy and Oncology*, **82**, 24–29.

Dobrowsky, W., Naude, J., Dobrowsky, E. *et al.* (1995) Mitomycin C (MMC) and unconventional fractionation (V-CHART) in advanced head and neck cancer. *Acta Oncologica*, **34**, 270–272.

Durand, R.E. and Sham, E. (1998) The lifetime of hypoxic tumor cells. *International Journal of Radiation Oncology Biology Physics*, **42**, 711–715.

Du Sault, L.A. (1963) The effect of oxygen on the response of spontaneous tumours in mice to radiotherapy. *British Journal of Radiology*, **36**, 749–754.

Eschwége, F., Sancho-Garnier, H., Chassagne, D. *et al.* (1997) Results of a European randomized trial of etanidazole combined with radiotherapy in head and neck carcinomas. *International Journal of Radiation Oncology Biology Physics*, **39**, 275–281.

Evans, J.C. and Bergsjø, P. (1965) The influence of anemia on the results of radiotherapy in carcinoma of the cervix. *Radiology*, **84**, 709–717.

Folkman, J. (1986) How is blood vessel growth regulated in normal and neoplastic tissue? *Cancer Research*, **46**, 467–473.

Gatenby, R.A., Kessler, H.B., Rosenblum, J.S. *et al.* (1988) Oxygen distribution in squamous cell carcinoma metastases and its relationship to outcome of radiation therapy. *International Journal of Radiation Oncology Biology Physics*, **14**, 831–838.

Gerweck, L.E., Gillette, E.L., and Dewey, W.C. (1974) Killing of chinese hamster cells in vitro by heating under hypoxic or aerobic conditions. *European Journal of Cancer*, **10**, 691–693.

Gerweck, L.E., Nygaard, T.G., and Burlett, M. (1979) Response of cells to hyperthermia under acute and chronic hypoxic conditions. *Cancer Research*, **39**, 966–972.

Grau, C., Agarwal, J.P., Jabeen, K. *et al.* (2003) Radiotherapy with or without mitomycin c in the treatment of locally advanced head and neck cancer: results of the IAEA multicentre randomized trial. *Radiotherapy and Oncology*, **67**, 17–27.

Grau, C., Horsman, M.R., and Overgaard, J. (1992) Improving the radiation response in a C3H mouse mammary carcinoma by normobaric oxygen and carbogen breathing. *International Journal of Radiation Oncology Biology Physics*, **22**, 415–419.

Grau, C. and Overgaard, J. (1998) Significance of haemoglobin cincentration for treatment outcome, in *Medical Radiology: Blood Perfusion and Microenvironment of Human Tumours* (eds M. Molls and P. Vaupel), Springer-Verlag, Heidelberg, pp. 101–112.

Gray, L.H., Conger, A.D., Ebert, M. *et al.* (1953) The concentration of oxygen dissolved in tissues at the time of irradiation as a factor in radiotherapy. *British Journal of Radiology*, **26**, 638–648.

Grogan, M., Thomas, G.M., Melamed, I. *et al.* (1999) The importance of hemoglobin levels during radiotherapy for carcinoma of the cervix. *Cancer*, **86**, 1528–1536.

Haffty, B.G., Son, Y.H., Sasaki, C.T. *et al.* (1993) Mitomycin C as an adjunct to postoperative radiation therapy in squamous cell carcinoma of the head and neck: results from two randomized clinical trials. *International Journal of Radiation Oncology Biology Physics*, **27**, 241–250.

Haffty, B.G., Wilson, L.D., Son, Y.H. *et al.* (2005) Concurrent chemo-radiotherapy with mitomycin C compared with porfiromycin in squamous cell cancer of the head and neck: final results of a randomized clinical trial. *International Journal of Radiation Oncology Biology Physics*, **61**, 5–6.

Hahn, G.M. (1974) Metabolic aspects of the role of hyperthermia in mammalian cell inactivation and their possible relevance to cancer treatment. *Cancer Research*, **34**, 3117–3123.

Hahn, G.M. (1982) *Hyperthermia and Cancer*, Plenum Press, New York.

Hahnfeldt, P., Panigrahy, D., Folkman, J., and Hlatky, L. (1999) Tumor development under angiogenic signaling: a dynamic theory of tumor growth, treatment response, and postvascular dormancy. *Cancer Research*, **59**, 4770–4775.

Hall, E.J. (1994) *Radiobiology for the Radiobiologist*, 4th edn, JB Lippincott, Philadelphia.

Hall, E.J. and Roizin-Towle, L. (1975) Hypoxic sensitizers: radiobiological studies at the cellular level. *Radiology*, **117**, 453–457.

Helmlinger, G., Yuan, F., Dellian, M., and Jain, R.K. (1997) Interstitial pH and pO_2 gradients in solid tumors in vivo: high-resolution measurements reveal a lack of correlation. *Nature Medicine*, **3**, 177–182.

Henk, J.M., Kunkler, P.B., and Smith, C.W. (1977) Radiotherapy and hyperbaric oxygen in head and neck cancer. Final report of first controlled clinical trial. *Lancet*, **2** (8029), 101–103.

Henk, J.M. and Smith, C.W. (1977) Radiotherapy and hyperbaric oxygen in head and neck cancer. Interim report of second clinical trial. *Lancet*, **2** (8029), 104–105.

Henke, M., Laszig, R., Rube, C. *et al.* (2003) Erythropoietin to treat head and neck cancer patients with anaemia undergoing radiotherapy: randomized, double-blind, placebo-controlled trial. *Lancet*, **362**, 1255–1260.

Hill, R.P., Marie-Egyptienne, D.T., and Hedley, D.W. (2009) Cancer stem cells, hypoxia and metastasis. *Seminars in Radiation Oncology*, **19**, 106–111.

Hirst, D.G. and Wood, P.J. (1991) Could manipulation of the binding affinity of haemoglobin for oxygen be used clinically to sensitize tumours to radiation? *Radiotherapy and Oncology*, **20**, 53–57.

Hoeckel, M., Knoop, C., Schlenger, K. *et al.* (1993) Intratumoral pO2 predicts survival in advanced cancer of the uterine cervix. *Radiotherapy and Oncology*, **26**, 45–50.

Horsman, M.R. (1993) Hypoxia in tumours: its relevance, identification and modification, in *Current Topics in Clinical Radiobiology of Tumours* (ed. H. Beck-Bornholdt), Springer-Verlag, Berlin, pp. 99–112.

Horsman, M.R. (1995) Nicotinamide and other benzamide analogs as agents for overcoming hypoxic cell radiation resistance in tumours. *Acta Oncologica*, **34**, 571–587.

Horsman, M.R. (1998) Measurement of tumor oxygenation. *International Journal of Radiation Oncology Biology Physics*, **42**, 701–704.

Horsman, M.R. and Overgaard, J. (2007) Hyperthermia: a potent enhancer of radiotherapy. *Clinical Oncology*, **19**, 418–426.

Horsman, M.R. and Siemann, D.W. (2006) Pathophysiological effects of vascular targeting agents and the implications for combination therapies. *Cancer Research*, **66**, 11520–11539.

Horsman, M.R., Chaplin, D.J., and Overgaard, J. (1990) Combination of nicotinamide and hyperthermia to eliminate radioresistant chronically and acutely hypoxic tumor cells. *Cancer Research*, **50**, 7430–7436.

Horsman, M.R., Khalil, A.A., Nordsmark, M. *et al.* (1993) Relationship between radiobiological hypoxia and direct estimates of tumour oxygenation in a mouse tumour model. *Radiotherapy and Oncology*, **28**, 69–71.

Horsman, M.R., Khalil, A.A., Siemann, D.W. *et al.* (1994a) Relationship between radiobiological hypoxia in tumours and electrode measurements of tumour oxygenation. *International Journal of Radiation Oncology Biology Physics*, **29**, 439–442.

Horsman, M.R., Nordsmark, M., Khalil, A.A. *et al.* (1994b) Reducing chronic and acute hypoxia in tumours by combining nicotinamide with carbogen breathing. *Acta Oncology*, **33**, 371–376.

Hoskin, P.J., Robinson, M., Slevin, N. *et al.* (2009a) Effect of epoetin alfa on survival and cancer treatment-related anemia and fatigue in patients receiving radical radiotherapy with curative intent for head and neck cancer. *Journal of Clinical Oncology*, **27**, 5751–5756.

Hoskin, P.J., Rojas, A., and Saunders, M.I. (2009b) Accelerated radiotherapy, carbogen, and nicotinamide (ARCON) in the treatment of advanced bladder cancer: Mature results of a phase II nonrandomized study. *International Journal of Radiation Oncology Biology Physics*, 73, 1425–1431.

Hui, E.P., Chan, A.T.C., Pezzella, F. *et al.* (2002) Coexpression of hypoxia-inducible factors 1α and 2 α, carbonic anhydrase IX, and vascular endothelial growth factor in nasopharyngeal carcinoma and relationship to survival. *Clinical Cancer Research*, 8, 2595–2604.

Jain, R.K. (2001) Normalizing tumor vasculature with anti-angiogenic therapy: a new paradigm for combination therapy. *Nature Medicine*, 7, 987–989.

Janssen, H.L., Haustermans, K.M., Sprong, D. *et al.* (2002) HIF-1A, pimonidazole, and iododeoxyuridine to estimate hypoxia and perfusion in human head-and-neck tumors. *International Journal of Radiation Oncology Biology Physics*, **54**, 1537–1549.

Kaanders, J.H., Bussink, J., and van der Kogel, A.J. (2002a) ARCON: a novel biology-based approach in radiotherapy. *Lancet Oncology*, **3**, 728–737.

Kaanders, J.H.A.M., Pop, L.A.M., Marres, H.A.M. *et al.* (2002b) ARCON: experience in 215 patients with advanced head-and-neck cancer. *International Journal of Radiation Oncology Biology Physics*, **52**, 769–778.

Kampinga, H.H. and Dikomey, E. (2001) Hyperthermic radiosensitization: mode of action and clinical relevance. *International Journal of Radiation Biology*, 77, 399–408.

Kennedy, K.A., Rockwell, S., and Sartorelli, A.C. (1980) Preferential activation of mitomycin C to cytotoxic metabolites by hypoxic tumor cells. *Cancer Research*, **40**, 2356–2360.

Kimura, H., Braun, R.D., Ong, E.T. *et al.* (1996) Fluctuations in red cell flux in tumor microvessels can lead to transient hypoxia and reoxygenation in tumor parenchyma. *Cancer Research*, **56**, 5522–5528.

Kjellen, E., Joiner, M.C., Collier, J.M. *et al.* (1991) A therapuetic benefit from combining normobaric carbogen or oxygen with nicotinamide in fractionated X-ray treatments. *Radiotherapy and Oncology*, **22**, 81–91.

Kolstad, P. (1968) Intercapillary distance, oxygen tension and local recurrence in cervix cancer. *Scandinavian Journal of Clinical Laboratory Investigation*, **106**, 145–157.

Krohn, K.A., Link, J.M., and Mason, R.P. (2008) Molecular imaging of hypoxia. *Journal of Nuclear Medicine*, **49** (Suppl. 2), 129S–148S.

Lauk, S., Skates, S., Goodman, M., and Suit, H.D. (1989) Morphometric study of the vascularity of oral squamous cell carcinomas and its relation to outcome of radiation therapy. *European Journal of Cancer Clinical Oncology*, **25**, 1431–1440.

Lavey, R.S. and Dempsey, W.H. (1993) Erythropoietin increases hemoglobin in cancer patients during radiotherapy. *International Journal of Radiation Oncology Biology Physics*, 27, 1147–1152.

Le, Q.T., Sutpin, P.D., Raychaudhuri, S. *et al.* (2003) Identification of osteopontin as a prognostic plasma marker for head and neck squamous cell carcinomas. *Clinical Cancer Research*, **9**, 59–67.

Lee, D.J., Cosmatos, D., Marcial, V.A. *et al.* (1995) Results of an RTOG phase III trial (RTOG 85-27) comparing radiotherapy plus etanidazole (SR–2508) with radiotherapy alone for locally advanced head and neck carcinomas. *International Journal of Radiation Oncology Biology Physics*, **32**, 567–576.

Li, G.C. and Kal, H.B. (1977) Effect of hyperthermia on the radiation response of two mammalian cell lines. *European Journal of Cancer*, **13**, 65–69.

Machtay, M., Pajak, T., Suntharalingam, M. *et al.* (2007) Radiotherapy with or without erythropoietin for anemic patients with head and neck cancer: a randomized trial of the Radiation Therapy Oncology Group (RTOG 99-03). *International Journal of Radiation Oncology Biology Physics*, **69**, 1008–1017.

McKeown, S.R., Cowen, R.L., and Williams, K.J. (2007) Bioreductive drugs: from concept to clinic. *Clinical Oncology*, **19**, 427–442.

Mendenhall, W.M., Morris, C.G., Amdur, R.J. *et al.* (2005) Radiotherapy alone or combined with carbogen breathing for squamous cell carcinoma of the head and neck: a prospective randomized trial. *Cancer*, **104**, 333–337.

Moulder, J.E. and Rockwell, S. (1984) Hypoxic fractions of solid tumour. *International Journal of Radiation Oncology Biology Physics*, **10**, 695–712.

Müller, C. (1910) Eine neue Behandlungsmethode bösartiger Geschwülste. *Munchener Medizinische Wochenschrift*, **28**, 1490–1493.

Murata, R. and Horsman, M.R. (2004) Tumor specific enhancement of thermotherapy at mild temperatures by the vascular targeting agent 5,6-dimethylxanthenone-4-acetic acid. *International Journal of Hyperthermia*, **20**, 393–404.

Murata, R., Siemann, D.W., Overgaard, J., and Horsman, M.R. (2001a) Interaction between combretastatin A-4 disodium phosphate and radiation in murine tumours. *Radiotherapy and Oncology*, **60**, 155–161.

Murata, R., Siemann, D.W., Overgaard, J., and Horsman, M.R. (2001b) Improved tumor response by combining radiation and the vascular damaging drug 5,6-dimethylxanthenone-4-acetic acid. *Radiation Research*, **156**, 503–509.

Murata, R., Tsujitani, M., and Horsman, M.R. (2008) Enhanced local tumour control after single and fractionated radiation treatment using the hypoxic cell radiosensitizer doranidazole. *Radiotherapy and Oncology*, **87**, 331–338.

Nielsen, O.S. (1981) Effect of fractionated hyperthermia on hypoxic cells in vitro. *International Journal of Radiation Biology*, **39**, 73–80.

Nordsmark, M., Overgaard, M., and Overgaard, J. (1996) Pretreatment oxygenation predicts radiation response in advanced squamous cell carcinoma of the head and neck. *Radiotherapy and Oncology*, **41**, 31–39.

Nordsmark, M., Bentzen, S.M., Rudat, V. *et al.* (2005) Prognostic value of tumor oxygenation in 397 head and neck tumors after primary radiation therapy. An international multi-center study. *Radiotherapy and Oncology*, **77**, 18–24.

Nordsmark, M., Eriksen, J.G., Gebski, V. *et al.* (2007) Differential risk assessments from five hypoxia specific assays: the basis for biologically adapted individualized radiotherapy in advanced head and neck cancer patients. *Radiotherapy and Oncology*, **83**, 389–397.

Olive, P.L. and Aquino-Parsons, C. (2004) Measurement of tumor hypoxia using single cell methods. *Seminars in Radiation Oncology*, **14**, 241–248.

Overgaard, J. (1989) Sensitization of hypoxic tumour cells – clinical experience. *International Journal of Radiation Biology*, **56**, 801–811.

Overgaard, J. (1994) Clinical evaluation of nitroimidazoles as modifiers of hypoxia in solid tumors. *Oncology Research*, **6**, 509–518.

Overgaard, J. and Bichel, P. (1977) The influence of hypoxia and acidity on the hyperthermic response of malignant cells *in vitro*. *Radiology*, **123**, 511–514.

Overgaard, J. and Nielsen, O.S. (1980) The role of tissue environmental factors on the kinetics and morphology of tumor cells exposed to hyperthermia. *Annuals of the New York Academy of Science*, **335**, 254–280.

Overgaard, J., Hansen, H.S., Andersen, A.P. *et al.* (1989) Misonidazole combined with split-course radiotherapy in the treatment of invasive carcinoma of the larynx and pharynx: report from the DAHANCA 2 study. *International Journal of Radiation Oncology Biology Physics*, **16**, 1065–1068.

Overgaard, J., Sand Hansen, H., Overgaard, M. *et al.* (1998) A randomised double-blind phase III study of nimorazole as a hypoxic radiosensitizer of primary radiotherapy in supraglottic larynx and pharynx carcinoma. Results of the Danish head and neck cancer study (DAHANCA) protocol 5-85. *Radiotherapy and Oncology*, **46**, 135–146.

Overgaard, J., Eriksen, J.G., Nordsmark, M. *et al.* (2005) Plasma osteopontin predicts hypoxia and response to the hypoxic sensitizer nimoprazole in radiotherapy of head and neck cancer. Results from the randomized DAHANCA 5 trial. *Lancet Oncology*, **6**, 757–764.

Overgaard, J., Hoff, C., Sand Hansen, H. *et al.* (2007) Randomized study of the importance of novel erythropoiesis stimulating protein (aranesp) for the effect of radiotherapy in patients with primary squamous cell carcinoma of the head and neck (HNSCC) – the Danish Head and Neck Cancer group DAHANCA 10 randomized trial. *European Journal of Cancer*, 5, 7.

Petersen, C., Zips, D., Krause, M. *et al.* (2001) Repopulation of FaDu human squamous cell carcinoma during fractionated radiotherapy correlates with reoxygenation. *International Journal of Radiation Oncology Biology Physics*, **51**, 483–493.

Piert, M., Machulla, H.J., Picchio, M. *et al.* (2005) Hypoxia-specific tumor imaging with 18F-Fluoroazomycin arabinoside. *Journal of Nuclear Medicine*, **46**, 106–113.

Powell, M.E., Hill, S.A., Saunders, M.I. *et al.* (1997) Human tumor blood flow is enhanced by nicotinamide and carbogen breathing. *Cancer Research*, **57**, 5261–5264.

Power, J.A. and Harris, J.W. (1977) Response of extremely hypoxic cells to hyperthermia: survival and oxygen enhancement ratios. *Radiology*, **123**, 767–770.

Raleigh, J.A., Dewhirst, M.W., and Thrall, D.E. (1996) Measuring tumor hypoxia. *Seminars in Radiation Oncology*, **6**, 37–45.

Rasey, J.S., Koh, W.J., Evans, M.L. *et al.* (1996) Quantifying regional hypoxia in human tumors with positron emission tomography of 18F-fluoromisonidazole: a pretherapy study of 37 patients. *International Journal of Radiation Oncology Biology Physics*, **36**, 417–428.

Révész, L., Siracka, E., Siracky, J. *et al.* (1989) Variation of vascular density within and between tumors of the uterine cervix and its predictive value for radiotherapy. *International Journal of Radiation Oncology Biology Physics*, **16**, 1161–1163.

Rockwell, S. (1985) Use of a perfluorochemical emulsion to improve oxygenation in a solid tumor. *International Journal of Radiation Oncology Biology Physics*, **11**, 97–103.

Rojas, A. (1991) Radiosensitization with normobaric oxygen and carbogen. *Radiotherapy and Oncology*, **20** (Suppl. 1), 65–70.

Roti Roti, J.L. (2004) Introduction: radiosensitization by hyperthermia. *International Journal of Hyperthermia*, **20**, 109–114.

Rubin, P., Hanley, J., Keys, H.M. *et al.* (1979) Carbogen breathing during radiation therapy. The RTOG study. *International Journal of Radiation Oncology Biology Physics*, 5, 1963–1970.

Sapareto, S.A., Raaphorst, P., and Dewey, W.C. (1979) Cell killing and the sequencing of hyperthermia and radiation. *International Journal of Radiation Oncology Biology Physics*, 5, 343–347.

Schwarz, G. (1909) Über Desensibiliserung gegen Röntgen – und Radiumstrahlen. *Munchener Medizinische Wochenschrift*, **24**, 1–2.

Seddon, B.M., Payne, G.S., Simmons, L. *et al.* (2003) A phase I study of SR-4554 via intravenous administration for noninvasive investigation of tumor hypoxia by magnetic resonance spectroscopy in patients with malignancy. *Clinical Cancer Research*, **9**, 5101–5112.

Siemann, D.W. and Macler, L.M. (1986) Tumor radiosensitization through reductions in hemoglobin affinity. *International Journal of Radiation Oncology Biology Physics*, **12**, 1295–1297.

Siemann, D.W. and Rojiani, A.M. (2002) Enhancement of radiation therapy by the novel vascular targeting agent ZD6126. *International Journal of Radiation Oncology Biology Physics*, **53**, 164–171.

Siemann, D.W., Bibby, M.C., Dark, G. *et al.* (2005) Differentiation and definition of vascular-targeted therapies. *Clinical Cancer Research*, **11**, 416–420.

Siemann, D.W., Chaplin, D.J., and Horsman, M.R. (2004) Vascular-targeting therapies for treatment of malignant disease. *Cancer*, **100**, 2491–2499.

Siemann, D.W., Hill, R.P., and Bush, R.S. (1977) The importance of the pre–irradiation breathing times of oxygen and carbogen (5% CO2; 95% O2) on the in vivo radiation response of a murine sarcoma. *International Journal of Radiation Oncology Biology Physics*, **2**, 903–911.

Stoeltzing, O. and Ellis, L.M. (2006) The role of microvasculature in metastasis formation, in *Vascular-targeted Therapies in Oncology* (ed. D.W. Siemann), John Wiley & Sons, Ltd, Chichester, pp. 31–62.

Stratford, I.J. and Stephens, M.A. (1989) The differential hypoxic cytotoxicity of bioreductive agents determined in vitro by the MTT assay. *International Journal of Radiation Oncology Biology Physics*, **16**, 973–976.

Stratford, I.J., O'Neill, P., Sheldon, P.W. *et al.* (1986) RSU 1069, a nitroimidazole containing an aziridine group: bioreduction greatly increases cytotoxicity under hypoxic conditions. *Biochemical Pharmacology*, **35**, 105–109.

Stuben, G., Pottgen, C., Knuhmann, K. *et al.* (2003) Erythropoietin restores the anemia-induced reduction in radiosensitivity of experimental human tumors in nude mice. *International Journal of Radiation Oncology Biology Physics*, **55**, 1358–1362.

Suit, H.D. and Gerweck, L.E. (1979) Potential for hyperthermia and radiation therapy. *Cancer Research*, **39**, 2290–2298.

Suit, H.D., Marshall, N., and Woerner, D. (1972) Oxygen, oxygen plus carbondioxide, and radiation therapy of a mouse mammary carcinoma. *Cancer*, **30**, 1154–1158.

Tannock, I.F. (1968) The relationship between cell proliferation and the vascular system in a transplanted mouse mammary tumour. *British Journal of Cancer*, **22**, 258–273.

Thews, O., Koenig, R., Kelleher, D.K. *et al.* (1998) Enhanced radiosensitivity in experimental tumours following erythropoietin treatment of chemotherapy-induced anaemia. *British Journal of Cancer*, **78**, 752–756.

Thomas, G.M. (2002) Raising hemoglobin: an opportunity for increasing survival? *Oncology*, **63**, 19–28.

Thomlinson, R.H. and Gray, L.H. (1955) The histological structure of some human lung cancers and the possible implications for radiotherapy. *British Journal of Cancer*, **9**, 539–549.

Urtasun, R., Band, P., Chapman, J.D. *et al.* (1976) Radiation and high-dose metronidazole in supratentorial glioblastomas. *New England Journal of Medicine*, **294**, 1364–1367.

Urtasun, R.C., Parliament, M.B., McEwan, A.J. *et al.* (1996) Measurement of hypoxia in human tumours by non-invasive spect imaging of iodoazomycin arabinoside. *British Journal of Cancer*, **74**, S209–S212.

Vaupel, P. (2004) Tumor microenvironmental physiology and its implications for radiation oncology. *Seminars in Radiation Oncology*, **14**, 198–206.

Vaupel, P., Kallinowski, F., and Okunieff, P. (1989) Blood flow, oxygen and nutrient supply, and metabolic micro-environment of human tumors: a review. *Cancer Research*, **49**, 6449–6465.

Watson, E.R., Halnan, K.E., Dische, S. *et al.* (1978) Hyperbaric oxygen and radiotherapy: a medical research council trial in carcinoma of the cervix. *British Journal of Radiology*, **51**, 879–887.

Weissberg, J.B., Son, Y.H., Papac, R.J. *et al.* (1989) Randomized clinical trial of mitomycin C as an adjunct to radiotherapy in head and neck cancer. *International Journal of Radiation Oncology Biology Physics*, **17**, 3–9.

Wilson, W.W., Li, A.E., Cowan, D., and Siim, B.G. (1998) Enhancement of tumor radiation response by the anti-vascular agent 5,6-dimethylxanthenone-4-acetic acid. *International Journal of Radiation Oncology Biology Physics*, **42**, 905–908.

Yaromina, A., Zips, D., Thames, H.D. *et al.* (2006) Pimonidazole labeling and response to fractionated irradiation of five human squamous cell carcinoma (hSCC) lines in nude mice: the need for a multivariate approach in biomarker studies. *Radiotherapy and Oncology*, **81**, 122–129.

Zeman, E.M., Hirst, V.K., Lemmon, M.J., and Brown, J.M. (1988) Enhancement of radiation-induced tumor cell killing by the hypoxic cell toxin SR 4233. *Radiotherapy andOncology*, **12**, 209–218.

17

HIF-1 Inhibitors for Cancer Therapy

Annamaria Rapisarda and Giovanni Melillo
Tumor Hypoxia Laboratory, Developmental Therapeutics Program, SAIC-Frederick, Inc, National Cancer Institute at Frederick, Frederick, MD, USA

17.1 Introduction

Despite the heterogeneity of human cancers, which encompass a broad and diverse collection of diseases with quite distinct biological behavior, hypoxia represents a unifying feature of the microenvironment of solid tumors. Cancer cells located far from a functional blood vessel have limited oxygen availability as a result of rapid growth and increased oxygen consumption as well as structurally and functionally abnormal vasculature. Hypoxia induces an array of responses in tumor cells, such as inhibition of apoptotic and activation of prosurvival pathways; induction of a more invasive and metastatic phenotype; switch to a glycolytic metabolism and induction of angiogenesis; responses that provide adaptation to; and or escape from, the hypoxic microenvironment. Ultimately, intratumor hypoxia is associated with resistance to treatment, especially radiation therapy, and is predictive of metastasis and poor outcome in a variety of tumor types (Hockel and Vaupel, 2001; Brown and Giaccia, 1998; Brizel *et al.*, 1996; Semenza, 2007).

Many of the responses triggered by hypoxia are regulated by a family of transcription factors induced by oxygen deprivation (hypoxia inducible factor or HIF) (Semenza, 2007; Manalo *et al.*, 2005). HIF-1 is a heterodimeric protein comprised of a constitutively expressed β subunit and an α subunit which is tightly regulated by oxygen concentration (Wang *et al.*, 1995; Wang and Semenza, 1995).

National Cancer Institute, National Institutes of Health
N01-CO-12400

Tumor Microenvironment Edited by Dietmar W. Siemann

Under normoxic conditions, HIF-α protein is hydroxylated by oxygen-dependent prolyl hydroxylases (PHDs), and rapidly targeted for ubiquitination and proteasomal degradation. Furthermore, HIF-α is hydroxylated on asparagine 803 by factor inhibiting HIF-1 (FIH), which inhibits the recruitment of the cofactor p300/CREB binding protein (CBP) and blocks HIF-1-dependent transcriptional activation under normoxic conditions. Conversely, hypoxia stabilizes HIF-α allowing it to translocate to the nucleus and dimerize with HIF-1β (also known as aryl hydrocarbon receptor nuclear translocator, ARNT). By binding to hypoxia responsive elements (HREs), HIF-1 activates transcription of numerous target genes involved in key steps of tumorigenesis, including angiogenesis, metabolism, proliferation, metastasis, and differentiation (Figure 17.1) (Semenza, 2008). Three isoforms of the HIF-α subunit have been identified so far, although HIF-1α and HIF-2α have been better characterized and for the most part implicated in human cancers. Overwhelming evidence indicates that HIF-1α and/or HIF-2α are indeed over-expressed in the majority of human cancers (Zhong *et al.*, 1999; Talks *et al.*, 2000; Bos *et al.*, 2001), where they are associated with patient mortality and poor response to treatment (Table 17.1) (Aebersold *et al.*, 2001; Birner *et al.*, 2000, 2001b; Bos *et al.*, 2003; Koukourakis *et al.*, 2006). However, the relative contribution of HIF-1α and/or HIF-2α to tumor progression is still poorly understood and may be tumor type dependent. Notably, increased HIF-1α expression is associated with a variety of genetic alterations that inactivate tumor suppressor genes or activate oncogenic pathways, including but not limited to: von Hippel–Lindau (*VHL*); *PTEN* (phosphatase and tensin homolog); *CDKN2A* (cyclin-dependent kinase inhibitor 2A, which encodes ARF (ADP-ribosylation factor) loss of function; *ERBB2* (v-erb-b2 erythroblastic leukemia viral oncogene homolog 2); *SRC* (v-src sarcoma (Schmidt–Ruppin A-2) viral oncogene homolog) gain of function (Semenza, 2003). In addition, the contribution of receptor tyrosine kinase (RTK)-dependent signaling pathways, such as endothelial growth factor receptor (EGFR), Her2/Neu (v-erb-b2 erythroblastic leukemia viral oncogene homolog 2), and PI3K (phosphoinositide-3-kinase)/AKT (v-akt murine thymoma viral oncogene homolog 1)/mammalian target of rapamycin (mTOR), to the induction of HIF-1α also suggests that many of these pathways converge on or implicate HIF-1 in mediating cell survival and growth (Pouyssegur, Dayan and Mazure, 2006; Majumder *et al.*, 2004; Pore *et al.*, 2006b; Hudson *et al.*, 2002).

Hypoxia has long been postulated to be not only a strength but also a potential weakness of cancer cells, in that it may offer unique opportunities for the development of therapeutics. The discovery of HIF-1 has finally led to the identification of a definable molecular target associated with intratumor hypoxia that can be used for the development of therapeutic agents. Despite the challenges associated with attempts to target transcription factors, several HIF-1 inhibitors have been identified so far and few of them are being tested in the clinical setting (Onnis, Rapisarda, and Melillo, 2009; Melillo, 2006, 2007; Welsh, Koh, and Powis, 2006).

17.2 Small molecule inhibitors of HIF-1

The discovery of small molecules targeting HIF-1 has mainly been carried out using cell-based high throughput screens based on reporter gene assays (Rapisarda

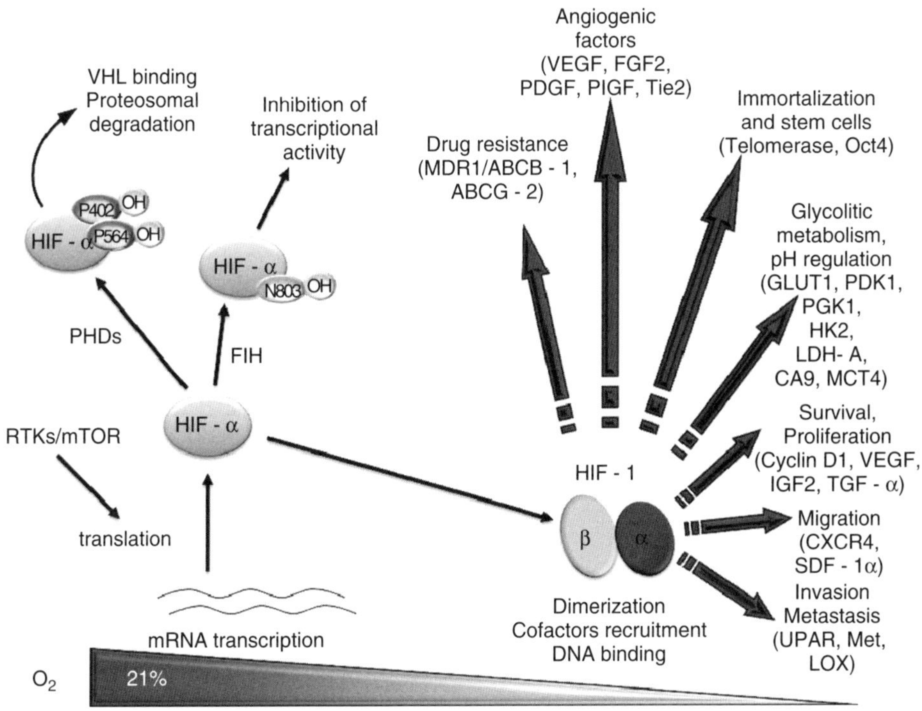

Figure 17.1 Regulation of hypoxia inducible factor-α (HIF-α)- and HIF-1-dependent gene expression. Under normal oxygen conditions, HIF-α is continuously transcribed and translated, but it is rapidly degraded through the VHL-proteasome pathway following hydroxylation of proline residues by prolyl hydroxylases (PHDs). Factor inhibiting HIF-1 (FIH) mediates the hydroxylation of an asparagine residue, thus inhibiting the recruitment of the cofactor p300/CBP and HIF-1-dependent transcriptional activity. These processes are inhibited when oxygen concentration decreases, hence the HIF-α subunit is stabilized, translocates to the nucleus where it dimerized with HIF-1β, recruits co-factors and induces the transcription of a large number of genes involved in several crucial aspects of cancer biology. In addition, receptors tyrosine kinases (RTKs) can boost HIF-1 translation through the activation of the AKT/mTOR pathway.

et al., 2002). Potential advantage of this strategy is the identification of previously unrecognized signaling pathways involved in the activation of HIF-1α. Foremost, drawbacks are the challenges associated with elucidation of the molecular mechanism of hits identified and the overall lack of selectivity of the inhibitors identified. Indeed, the majority of HIF-1 inhibitors discovered to date, target signaling pathways that are not only involved in HIF-1 regulation but also affect distinct cellular functions. Examples of HIF-1 inhibitors identified include agents that target chaperone proteins (Isaacs *et al.*, 2002; Mabjeesh *et al.*, 2002), microtubules (Mabjeesh *et al.*, 2003), topoisomerase I and topoisomerase II (Rapisarda *et al.*, 2004a; Creighton-Gutteridge *et al.*, 2007), soluble guanylate-cyclase (sGC) (Yeo *et al.*, 2003), farnesyltransferase (Han *et al.*, 2005), thioredoxin (Welsh *et al.*, 2003), histone deacetylase (Kong *et al.*, 2006) and p53/HDM2 (Mdm2 p53 binding protein homolog) (Yang *et al.*, 2009;

Table 17.1 HIF-1α overexpression and patient mortality.

Cancer type	Reference
Astrocytoma (diffuse)	Korkolopoulou *et al.* (2004)
Bladder (superficial urothelial)[a]	Theodoropoulos *et al.* (2004)
Bladder (transitional cells)	Theodoropoulos *et al.* (2005)
Breast	Vleugel *et al.* (2005), Dales *et al.* (2005)
Breast (c-Erb2 positive)	Giatromanolaki *et al.* (2007)
Breast (positive linfonode)	Schindl *et al.* (2002)
Breast (negative linfonode)	Bos *et al.* (2003)
Cervix (early stage)	Birner *et al.* (2000)
Cervix (radiation therapy)	Burri *et al.* (2003)
Cervix (grade IB-IIIB, radiation therapy)	Bachtiary *et al.* (2003)
Endometrial	Sivridis *et al.* (2002)
Gastric	Griffiths *et al.* (2007)
Gastrointestinal stromal tumor (stomach)	Takahashi *et al.* (2003)
Lung (non small cell lung carcinoma)	Swinson *et al.* (2004)
Melanoma (malignant)[b]	Giatromanolaki *et al.* (2003)
Oligodendrioglioma	Birner *et al.* (2001a)
Oropharynx-squamous cell carcinoma	Aebersold *et al.* (2001)
Ovarian[a]	Birner *et al.* (2001b)
Pancreas	Sun *et al.* (2007)

[a]Combination of HIF-1α over-expression and mutant p53.
[b]HIF-2α overexpression.

Lee *et al.*, 2009c) as well as a number of signaling pathways that have been associated with HIF-1α activation including mTOR (Majumder *et al.*, 2004; Hudson *et al.*, 2002), AKT (Tan *et al.*, 2005), Her2/Neu (Laughner *et al.*, 2001), EGFR (Pore *et al.*, 2006b; Koukourakis *et al.*, 2003; Zhong *et al.*, 2000), and Bcr-Abl (breakpoint cluster region-c-abl oncogene 1, receptor tyrosine kinase (Mayerhofer *et al.*, 2002) (Figure 17.2). It should also be noted that many of the HIF inhibitors described so far may indeed block both HIF-1α and HF-2α, making pharmacological validation of the relative contribution of these transcription factors to tumorigenesis very complicated. The effects on HIF-2α expression and/or activity has not been reported for many of the agents identified; conversely, only one example of a putative selective inhibitor of HIF-2α has been described (Zimmer *et al.*, 2008). However, this molecule has been recently used as a HIF-1α inhibitor as well (Zhong *et al.*, 2010).

The observation that a number of signaling pathways are involved in the regulation of HIF-1 activity emphasizes the potential role that HIF-1 may play as a common downstream mediator of cancer progression and the potential benefits of its inhibition for cancer therapy. Likewise, the overall lack of selectivity of HIF-1 inhibitors identified, which hampers the pharmacological validation of HIF as a target, may ultimately contribute to the anticancer activity observed with these agents by targeting multiple signaling pathways involved in cancer cells survival, proliferation and angiogenesis.

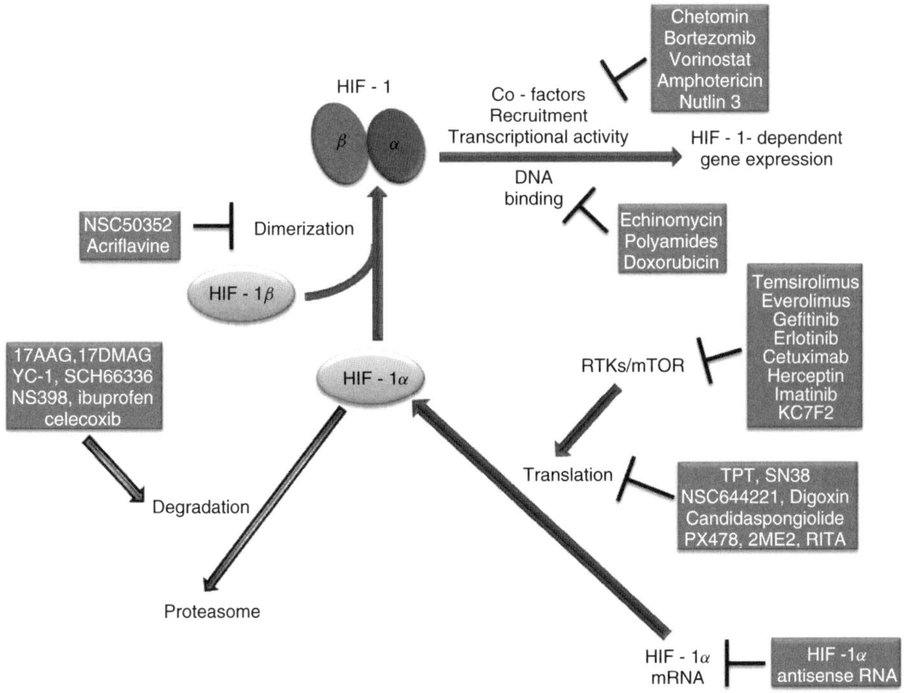

Figure 17.2 Signaling pathways targeted by small molecule inhibitors of HIF-1.

In an attempt to identify small molecule inhibitors with a higher degree of selectivity for HIF-1, assays have been developed to target either HIF-1-DNA binding or HIF-α/HIF-1β dimerization. Indeed, several small molecules that inhibit HIF-1 DNA binding have been identified, including synthetic polyamides (Olenyuk *et al.*, 2004), echinomycin (Kong *et al.*, 2005) and anthracyclines (Lee *et al.*, 2009a) (Figure 17.2) and efforts are ongoing to identify small molecule inhibitors of HIF-α/HIF-1β dimerization (Yang *et al.*, 2005 Park *et al.*, 2006; Lee *et al.*, 2009b; Scheuermann *et al.*, 2009).

17.2.1 Mechanism(s) of action of HIF-1 inhibitors

Proposed mechanisms of action of HIF-1 inhibitors identified to date are shown in Figure 17.2 and agents that are either Food and Drug Administration (FDA) approved or in early clinical development are shown in Table 17.2.

HIF-1α accumulation is controlled primarily at the level of protein degradation and protein translation. Interestingly enough, the majority of HIF-1 inhibitors identified thus far do indeed decrease HIF-1α protein levels, by a number of different mechanisms.

Table 17.2 HIF-1 inhibitors FDA approved or in clinical development with proposed mechanism(s) of action.

Agent	Target	Mechanism	Reference
EZN-2968	HIF-1α mRNA	Degradation	Greenberger *et al.* (2008)
Aminoflavone	AhR	HIF-1α mRNA Processing	Terzuoli, (submitted)
Topotecan	Topoisomerase 1	Translation	Rapisarda *et al.* (2004a)
EZN-2208	Topoisomerase 1	Translation	Sapra *et al.* (2008)
PX-478	Unknown	Translation	Welsh *et al.* (2004)
Erlotinib, gefitinib, cetuximab	EGFR	Signaling	Pore *et al.* (2006b), Luwor *et al.* (2005)
Trastuzumab	Her2/Neu	Translation	Koukourakis *et al.* (2003)
Temsirolimus, everolimus	mTOR	Translation	Del *et al.* (2006), Wan *et al.* (2006)
17AAG, 17DMAG	HSP90	Degradation	Isaacs *et al.* (2002), Mabjeesh *et al.* (2002)
Doxorubicin	DNA	DNA binding	Lee *et al.* (2009a)
Bortezomib	Proteasome	Transcriptional activity	Kaluz, Kaluzova, and Stanbridge (2006)
Vorinostat, LAQ824, FK228	HDAC	Degradation, transcriptional activity	Kong *et al.* (2006), Mie *et al.* (2003), Qian *et al.* (2006)
Acriflavine	HIF-1α PAS-B	Dimerization	Lee *et al.* (2009b)

17.2.2 Small molecules that affect HIF-1α protein accumulation

17.2.2.1 RTKs inhibitors

The levels of HIF-1α protein are primarily controlled at the level of protein degradation; however, since HIF-1α is rapidly degraded under normoxic conditions, it is postulated that ongoing protein translation is required for its accumulation. HIF-1α protein synthesis is controlled, at least in part, by the PI3K/AKT/mTOR pathway and can be induced by a variety of growth factor-mediated RTKs-dependent signaling pathways in a cell type dependent fashion.

Evidence has been provided that HIF-1α is induced by an EGFR-dependent pathway in prostate cancer cell lines (Zhong *et al.*, 2000) and that small molecule inhibitors of EGFR, such as gefitinib (Iressa) and erlotinib (Tarceva), inhibit vascular endothelial growth factor (VEGF) expression by downregulating HIF-1α protein translation in squamous cell carcinomas (Pore *et al.*, 2006b). In addition, cetuximab, a monoclonal antibody targeting EGFR, has been shown to inhibit HIF-1α protein expression in epidermoid carcinoma cells (Luwor *et al.*, 2005). Similarly, Her2/Neu-dependent signaling has been implicated in HIF-1α protein translation in breast cancer cell lines (Laughner *et al.*, 2001). Accordingly, it has been suggested that trastuzumab (Herceptin), a humanized monoclonal antibody approved for the treatment of patients with Her2/Neu positive breast cancer, may exert its antiangiogenic effects at least in part by inhibiting HIF-1α-dependent induction of VEGF (Koukourakis *et al.*, 2003).

Bcr/Abl is an oncoprotein generated by the reciprocal translocation t(9;22), which is present in chronic myelogenous leukemia. Bcr/Abl is a constitutively active tyrosine kinase activity that leads to activation of numerous signaling pathways, including PI3K. It has been shown that Bcr/Abl induces VEGF expression in a HIF-1-dependent fashion and that this effect is repressed in the presence of PI3K and mTOR inhibitors (Mayerhofer *et al.*, 2002). Imatinib, a small molecule inhibitor of Bcr/Abl, has also been shown to down-regulate HIF-1α expression in small cell lung cancer cell lines, presumably by inhibiting c-kit-dependent induction of the PI3K pathway (Litz and Krystal, 2006).

17.2.2.2 PI3K/AKT/mTOR inhibitors

Several HIF-1 inhibitors described in the literature target signaling pathways relevant to human cancer, including, but not limited to, the PI3K/AKT pathway (Cao *et al.*, 2004; Fang *et al.*, 2004, 2005; Pore *et al.*, 2006a; Tan *et al.*, 2005; Zhang *et al.*, 2005, 2006; Zhong *et al.*, 2004).

The mTOR pathway has been implicated in the growth factor-dependent induction of HIF-1α. It is then conceivable that mTOR inhibitors currently in clinical development might inhibit HIF-1 and impact on downstream pathways including angiogenesis. Indeed, inhibitors of mTOR have been shown to inhibit HIF-1α protein accumulation (temsirolimus/CCI-779 and everolimus/RAD001) (Del *et al.*, 2006; Wan *et al.*, 2006). Notably, loss of the VHL tumor suppressor gene (*VHL*) sensitized kidney cancer cells to the mTOR inhibitor, temsirolimus, both *in vitro* and in mouse models (Thomas *et al.*, 2006). Recently, screening of a natural product-like chemical compound library led to the identification of a novel HIF-1 inhibitor, KC7F2, which inhibits 4EBP1 (eukaryotic translation initiation factor 4E binding protein 1) and S6K phosphorylation, suggesting that this compound inhibits HIF-1α protein synthesis by targeting the mammalian target of rapamycin complex 1 (mTORC1) pathway (Narita *et al.*, 2009). HIF-1α might be a determinant of response in cancers in which the mTOR pathway is dysregulated and may also represent an attractive biomarker of activity of mTOR inhibitors. However, this possibility has yet to be validated in the clinical setting.

17.2.2.3 Topoisomerase inhibitors

Several small molecule inhibitors of topoisomerase I (Top 1) have been shown to block HIF-1α protein translation (Rapisarda *et al.*, 2004a; Sapra *et al.*, 2008). Topotecan, an FDA approved Top 1 poison, was one of the first small molecule inhibitors of HIF-1α identified in a cell-based high throughput screen at the NCI in Frederick (Rapisarda *et al.*, 2004a). Further evidence has indicated that virtually all of the camptothecin analogs Top 1 poisons as well as non-camptothecin Top 1 inhibitors share the ability to inhibit HIF-1α protein accumulation (Rapisarda

A, Shoemaker RH, and Melillo G, unpublished observation). Topotecan inhibits HIF-1α expression by a Top 1-dependent pathway that is independent from DNA replication or DNA damage, yet requires ongoing RNA transcription. Notably, inhibition of HIF-1α protein translation by topotecan is independent from the AKT-mTOR signaling pathway (Rapisarda *et al.*, 2004a). Daily administration of low doses of topotecan (1 mg/kg), but not higher doses (10 mg/kg) given on an intermittent schedule, inhibited tumor growth, angiogenesis and expression of HIF-1-target genes in a human glioma xenograft model, suggesting a potential association between HIF-1 inhibition and a beneficial therapeutic effect (Rapisarda *et al.*, 2004b). However, a limitation of these studies is that topotecan may also exert cytostatic/cytotoxic effects, which precludes from conclusively relating HIF-1 inhibition to tumor growth delay. Currently a pilot study to address whether oral administration of 'metronomic' topotecan modulates HIF-1 expression in tumor tissue is ongoing at NCI (http://clinicaltrials.gov/ct2/show/NCT00182676).

Interestingly, several topoisomerase II (Topo 2) inhibitors have also been shown to inhibit HIF-1α expression. In particular, NSC644221 was shown to inhibit HIF-1α protein synthesis in a Topo 2α dependent fashion in several human cancer cell lines. Interestingly, NSC644221 inhibited HIF-1α in a cell-type specific fashion and more importantly, was unable to affect HIF-1α expression in HCT116 cells in which induced DNA-damage dependent responses (i.e., p21 expression, phosphorylation of H2A.X, G2M arrest), providing further evidence of a biochemical dissociation between DNA damage and HIF-1 inhibition and suggesting the possibility of pathway-specific intrinsic resistance to HIF-1 inhibition in cancer cells (Creighton-Gutteridge *et al.*, 2007).

17.2.2.4 Microtubule inhibitors

Microtubule stabilizing (taxanes and epothilones) as well as destabilizing (*Vinka* alkaloids and colchicine binding site molecules) agents have the potential to inhibit HIF-1α expression (Escuin, Kline, and Giannakakou, 2005). In particular 2ME2, a naturally occurring estrogen metabolite that binds to the colchicine binding site of tubulin, and its analog ENMD-1198 inhibit HIF-1α protein accumulation by decreasing its translation, downstream of microtubule disruption (Mabjeesh *et al.*, 2003; Moser *et al.*, 2008). Interestingly, another study showed that microtubule depolymerizing agents induce the accumulation of a transcriptionally active HIF-1 through the induction of nuclear factor-kappaB (NFκB) (Jung *et al.*, 2003), emphasizing a potential pathway specific regulation of HIF-1 in different cell types.

17.2.2.5 p53 activators

The relationship between p53 and HIF-1α has been extensively investigated, although evidence has been conflicting regarding either a negative (Ravi *et al.*, 2000) or a positive effect of the p53/HDM2 pathway on HIF-1α protein accumulation (An *et al.*, 1998; Carroll and Ashcroft, 2008). It is conceivable then that small molecules modulating the p53 pathway may also have an effect on HIF-1α expression and/or activity.

Small-molecule antagonists of Hdm2 (nutlins) have been recently described as potential anticancer agents. Nutlins interfere with the binding of Hdm2 to p53, thus stabilizing p53 and activating p53-dependent cell death pathways. Several recent reports also demonstrated that nutlin-3 inhibited HIF-1α and, in turn, VEGF production and angiogenesis, in a p53-dependent manner (LaRusch *et al.*, 2007; Secchiero *et al.*, 2007). However, an additional mechanism of action proposed implicates inhibition of HIF-1 transcriptional activity. Indeed, it appears that, under hypoxic conditions (1% O_2), Hdm2 competes with FIH for HIF-1α–C-TAD (C-terminal transactivation domain) binding, thus preventing hydroxylation of Asn^{803}, which is a negative regulator of p300 recruitment and full transcriptional activation. In the presence of nutlin-3, the interaction between Hdm2 and C-TAD is dissociated, and hydroxylation of Asn^{803} can occur, blocking p300 recruitment and transcription of hypoxia-induced genes (Lee *et al.*, 2009c). Interestingly, inhibition of HIF-1α transcriptional activity by nutlin-3 appears to be p53 independent.

In addition, the small molecule RITA (reactivation of p53 and induction of tumor cell apoptosis, NSC652287) was found to inhibit HIF-1α protein accumulation by a novel mechanism implicating the p53-dependent phosphorylation of eIF2α (eukaryotic translation initiation factor 2, subunit 1 alpha) (Narita *et al.*, 2009). However, RITA also blocks the expression of the p53 target proteins HDM2 and p21, raising the possibility that phosphorylation of eIF2α may have a more general effect on protein synthesis.

17.2.2.6 *Cardiac glycosides*

Recently, cardiac glycosides, including digoxin, have been shown to inhibit HIF-1α protein translation by a yet undefined mechanism of action (Zhang *et al.*, 2008). In addition, glycosides isolated from natural products have also been identified in a HIF-1 cell-based assay (Klausmeyer *et al.*, 2009). However, the potential clinical application of this class of compounds as HIF-1 inhibitors remains to be established, amid concerns of potential severe toxicities.

17.2.2.7 *Heat-shock protein (Hsp90) and histone deacetylase inhibitors*

Once HIF-1α is translated, the chaperon protein Hsp90 (heat shock protein 90kDa alpha) is pivotal for its correct folding, a necessary step to prevent HIF-1α degradation. Accordingly, Hsp90 inhibitors (17-AAG, 17-DMAG) have been shown to increase HIF-1α protein degradation (Isaacs *et al.*, 2002; Mabjeesh *et al.*, 2002). Inhibition of Hsp90 has been implicated in the mechanism of action of a variety of other small molecules, such as radicicol (Hur *et al.*, 2002) and its derivative KF58333 (Kurebayashi *et al.*, 2001); the farnesyltransferase inhibitor SCH66336 (Han *et al.*, 2005); as well as histone deacetylase inhibitors. Histone deacetylase (HDAC) inhibitors (vorinostat, LAQ824, FK228) may affect HIF-1α protein levels by several mechanisms, including: (i) increased HIF-1α degradation, following hyperacetylation of Hsp90 and degradation of HIF-1α by the proteasome in a VHL- and ubiquitin-independent mechanism; (ii) direct acetylation of HIF-1α followed by polyubiquitination and proteasomal degradation (Ellis, Hammers, and Pili, 2008) and finally, (iii) inhibition of HIF-1α transcriptional activity by direct acetylation of the cofactor p300 and inhibition of its binding to HIF-1α (Ellis, Hammers, and Pili, 2008).

17.2.3 Inhibitors of HIF-1 dimerization

HIF-1 dimerization and DNA binding are mediated by the basic-helix-loop-helix and Per/ARNT/Sim (PAS) domains, PAS-A and PAS-B, present in the N-terminal end of HIF-α and HIF-1β. Protein domains that mediate dimerization of HIF-1α are attractive, yet elusive, targets for development of selective small molecule inhibitors.

17.2.3.1 NSC50352

Over-expression of the HIF-1α-PAS-A domain was found to be sufficient to inhibit endogenous HIF-1 transcriptional activity in cancer cell lines, validating its potential role as target for drug development. In an attempt to identify small molecule inhibitors of HIF-1α and HIF-1β dimerization, recombinant fragments containing the HIF-1α-PAS-A and HIF-1β PAS-A domains were used to screen chemical libraries. Interestingly, NSC50352 specifically inhibited HIF-1α and HIF-1β interaction mediated by PAS-A containing fragments, but not the interaction mediated by unrelated domains. Unfortunately, NSC50352 was inactive in cell-based assays (Park *et al.*, 2006); analogs of this compound are being generated and tested to address whether they may retain activity in intact cells.

17.2.3.2 Acriflavine

Acriflavine is a mixture of 3, 6-diamino-10-methylacridinium chloride (trypaflavin) and 3, 6-diaminoacridine (proflavine), which has trypanocidal, antibacterial, and antiviral activities. Acriflavine was recently identified by screening a library of FDA approved drugs using a cell-based split-luciferase assay. Acriflavine inhibits HIF-1 dimerization by binding to the PAS-B subdomain of HIF-1α and HIF-2α. Interestingly, acriflavine does not affect MYC (v-myc myelocytomatosis viral oncogene homolog) dependent gene expression, showing a certain degree of selectivity for the HIF-1 pathway. Consistent with inhibition of HIF-1 dimerization, acriflavine inhibited HIF-1 transcriptional activity, leading to inhibition of tumor growth and vascularization (Lee *et al.*, 2009b).

17.2.4 Inhibitors of HIF-1-DNA binding

HIF-1 binds to a specific DNA sequence known as the hypoxia-responsive element, which contains the 5′-RCGTG-3′ motif. HIF-1 binding to DNA is largely mediated by the basic domains in the N-terminal regions of HIF-α and HIF-1β. Sequence-specific inhibition of HIF-1-DNA binding has the potential to provide inhibitors selective for HIF-1 transcriptional activity and several examples of agents that interfere with HIF-1 DNA binding have been described.

17.2.4.1 Polyamides

Polyamides are synthetic oligomers containing *N*-methylpyrrole and *N*-methylimidazole amino acids. Sequence-specific DNA-binding molecules have been designed to specifically inhibit binding of the HIF-1α/HIF-1β heterodimer to its target DNA sequence to downregulate the expression of VEGF and other hypoxia-inducible genes (Olenyuk *et al.*, 2004). Notably, HIF-1 targeting polyamides prevented HIF-1-DNA binding in electrophoretic mobility shift assay (EMSA) and blocked the expression of HIF-1 target genes in a genome-wide analysis (Olenyuk *et al.*, 2004). Recent progress in the delivery of polyamides to mammalian cells raises the possibility of targeting HIF-1 transcriptional activity in a fairly selective fashion (Nickols *et al.*, 2007).

17.2.4.2 Echinomycin

One hundred twenty-eight active compounds, from a HIF-1–targeted cell-based high-throughput screen of the National Cancer Institute 140 000 small-molecule library, were tested in a 96-well plate DNA binding enzyme-linked immunosorbent assay (ELISA) for inhibition of HIF-1 activity. Echinomycin (NSC-13502), a cyclic peptide of the family of quinoxaline antibiotics, known to bind DNA in a sequence-specific fashion, was identified as a potent inhibitor of HIF-1 DNA binding in ELISA, EMSA, and chromatin immunoprecipitation experiments (Kong *et al.*, 2005). Importantly, echinomycin did not affect the binding of AP-1 or NF-κB to their cognate sequences, but it inhibited the binding of myc to the E-box sequence (5′-CACGTG-3), which shares sequence homology with the core HIF-1 DNA binding site. Echinomycin was extensively tested as an anticancer agent in phase I/II clinical trials in the late 1980s and found to be fairly inactive. It remains to be determined if schedules aimed to modulate HIF-1 activity may increase its therapeutic efficacy.

17.2.4.3 Anthracyclines

Anthracyclines (daunorubicin, epirubicin, idarubicin, and doxorubicin) were recently identified as inhibitors of HIF-1 DNA binding in a recent screen of a library of FDA-approved agents (Lee *et al.*, 2009a). Anthracyclines are DNA binding agents that preferentially bind sites of 5′-(A/T)CG-3′ and 5′-(A/T)GC-3′. Interestingly, HIF-1 binds to the sequence 5′-(A/G)CGTG-3′, which overlaps with the optimal binding site for anthracyclines. Anthracyclines have been shown to inhibit HIF-1 binding to DNA and HIF-1-dependent gene expression and angiogenesis in tumor-bearing mice. Anthracyclines have also been reported to inhibit adaptor-related protein complex 1 (AP1), GATA binding protein 4 (GATA4), and Sp1 transcription factor (Kim *et al.*, 2003; Szulawska, Gniazdowski, and Czyz, 2005; Mansilla and Portugal, 2008), indicating that they may have multiple effects on transcription.

17.2.5 Inhibitors of HIF-1α transcriptional activity

HIF-1 transcriptional activity is largely mediated by two functional domains of HIF-1α, N-TAD, and C-TAD. Hydroxylation of Asn^{803} by FIH-1 prevents HIF-1α binding to the co-activator p300/CBP and downregulates its transcriptional activity under normoxic conditions.

17.2.5.1 Chetomin

Chetomin is a member of the epidithiodiketopiperazine family of natural products previously known to have antimicrobial activity. Chetomin was identified as a potential HIF-1 inhibitor in a high-throughput screen aimed to discover small molecules capable of disrupting the interaction between HIF-1α and p300. Chetomin binds to the tertiary structure of the cysteine/histidine-rich 1 domain of p300 and disrupts its zinc-binding sites, thus precluding its interaction with HIF-1α and attenuating hypoxia-inducible transcription (Kung *et al.*, 2004; Cook *et al.*, 2009). Systemic administration of chetomin inhibited hypoxia-inducible transcription within tumors and inhibited tumor growth in HCT116 and PC3 xenograft models (Kung *et al.*, 2004). However, toxicity concerns discouraged its further development for clinical application.

17.2.5.2 Proteasome inhibitors

The proteasome inhibitor bortezomib (Velcade) has the paradoxical effect of inducing accumulation of HIF-1α protein by inhibiting its proteasomal degradation while, at the same time, interfering with the function of the carboxyl-terminal transactivation domain (C-TAD) of HIF-1α (Kaluz, Kaluzova, and Stanbridge, 2006). C-TAD is a domain essential for the recruitment of cofactors such as p300/CBP. A potential mechanism of inhibition of HIF-1 by bortezomib implicates increased HIF-1α hydroxylation on asparagine 803, which prevents binding of p300/CBP to HIF-1 (Shin *et al.*, 2008), although additional mechanisms have been suggested (Kaluz, Kaluzova, and Stanbridge, 2008).

17.2.6 Inhibitors of HIF-1α mRNA accumulation

Additional mechanisms of action of HIF-1α inhibition that have been proposed include modulation of HIF-1α mRNA levels using antisense oligonucleotides (EZN-2968) (Greenberger *et al.*, 2008). EZN-2968 is a locked nucleic acid RNA antagonist that specifically binds to and inhibits the expression of HIF-1α mRNA. *In vitro* and *in vivo* experiments confirmed that EZN-2968 induced a potent, selective, and durable inhibition of HIF-1α mRNA and protein expression associated with inhibition HIF-1-dependent gene expression. However, delivery of this therapeutic agent to cancer cells remains a main practical limitation of this approach. Ongoing phase I studies of EZN-2968 in patients with advanced malignancies will determine optimal dose and schedule for subsequent phase II studies.

17.2.7 Miscellaneous of HIF-1 inhibitors

HIF-1 inhibitors described in this section have either undefined mechanism of action or affect HIF-1 at several levels of regulation.

17.2.7.1 Soluble guanylate cyclase activators

YC-1, originally described as a pharmacological agent that activates sGC, inhibits HIF-1α protein accumulation and the expression of HIF-1-target genes by a sGC-independent mechanism (Yeo *et al.*, 2003). Evidence has been provided that the region of HIF-1α between aa 720 and 780 may be required for YC-1-dependent degradation of HIF-1α (Kim *et al.*, 2006), while more recently inhibition of HIF-1α transcriptional activity has also been implicated (Li *et al.*, 2008). Overall, the predominant mechanism of action of this compound and its potential clinical application remain to be fully determined.

17.2.7.2 PX478

PX-478 is a small molecule that inhibits HIF-1α expression and has pronounced antitumor activity in a number of xenograft models (Welsh *et al.*, 2004). Studies on the mechanism of action of PX-478 showed that this compound decreases the levels of HIF-1α mRNA and inhibits its translation. In addition, treatment with this agent also results, although to a lesser extent, in inhibition of HIF-1α deubiquitination, thus enhancing HIF-1α degradation (Koh *et al.*, 2008). PX478 is currently in a phase I clinical trial in patients with metastatic cancer and results should become available in the near future.

17.2.7.3 NSAIDs

Nonsteroidal anti-inflammatory drugs (NSAIDs), both non-selective (indometacin, ibuprofen) and cyclo-oxygenase-2 (COX-2) selective (NS-398) (Jones *et al.*, 2002; Palayoor, Tofilon, and Coleman, 2003), as well as NO-sulindac (Stewart *et al.*, 2009), have been shown to downregulate HIF-1α protein accumulation and angiogenesis. Interestingly, NO-sulindac has been shown to do so by inhibiting AKT-dependent HIF-1α translation (Stewart *et al.*, 2009).

17.2.7.4 Antioxidants

Several studies have suggested that reactive oxygen species (ROS) levels are essential for the stability of HIF-1α. Reactive oxygen scavengers, such as ascorbate and *N*-acetylcysteine, have been shown to inhibit HIF-1α protein accumulation and tumor growth (Lu *et al.*, 2005; Gao *et al.*, 2007). It is proposed that ROS inactivates PHDs, through oxidation of the ferrous ion that is essential for the catalytic activity of the enzymes, and hence stabilizes HIF-1α. Ascorbate has been shown to decrease HIF-1α levels by preventing the oxidation of the catalytic ferrous ion (Lu *et al.*, 2005).

17.2.7.5 AMPK inhibitors

Compound C, a known competitive inhibitor of AMPK (protein kinase, AMP-activated), has been shown to inhibit HIF-1α protein accumulation in hypoxia by a mechanism AMPK-independent, possibly through inhibition of mitochondrial-generated ROS (Emerling *et al.*, 2007).

17.2.7.6 Compounds that promote IRP1/IRE interaction in HIF-2α untranslated region

Using a cell-based screen to identify inhibitors of HIF-2α activity, small molecules that selectively decrease HIF-2α translation in an mTOR-independent manner have been identified (Zimmer *et al.*, 2008). These agents enhance the binding of iron-regulatory protein 1 (IRP1) to a recently identified iron-responsive element (IRE) within the 5′-untranslated region (UTR) of the HIF-2α mRNA, therefore inhibiting its translation. Knocking down the expression of IRP1 by shRNA abolished the effect of these compounds. Interestingly, these authors showed that hypoxia releases HIF-2α translation by disrupting the interaction between IRP-1/IRE, highlighting the role that IRP-1 plays in regulating HIF-2α protein synthesis.

17.2.7.7 Mitochondrial respiration inhibitors

Several recent studies have demonstrated that hypoxia-dependent production of ROS by complex III of the mitochondria is both necessary and sufficient for hypoxic induction of HIF-1α protein accumulation, suggesting that mitochondria may act as an oxygen sensor for HIF-1α regulation (Klimova and Chandel, 2008). Using a cell-based high throughput screen, alkyliminophenylacetate compounds were identified as HIF-1 inhibitors. These compounds were found to downregulate hypoxia but not deferoxamine-induced HIF-1α protein stabilization. Further analysis indicated that the alkyliminophenylacetate compounds likely inhibit the HIF-1 pathway by blocking the hypoxia-induced mitochondrial ROS production. Strikingly, all of the non-alkyliminophenylacetate HIF-1 inhibitors identified from the small-molecule library screen were found to target mitochondria, highlighting the essential role of mitochondria in HIF-1 regulation (Lin *et al.*, 2008), yet decreasing the enthusiasm for further clinical development of this class of inhibitors.

17.2.8 Natural products

During the past few years many natural products able to interfere with or inhibit HIF-1 expression and/or activity have been described. Unfortunately, many of the identified compounds are either toxic or inhibit mitochondrial respiration, decreasing the likelihood of clinical development. The main classes of compounds identified as HIF-1 inhibitors are: flavonoids (apigenin, vitexin, baicalein, deguelin, chrysin, alpinumisoflavone 4′-O-methylalpinumisoflavone, quercetin, and silibinin); benzochromenones (benzochromenones A–E); hydroxyketones (curcumin, EF24); macrolides (latrunculin A); substituted pyrroles (lamellarin O, neolamellarins); cardiac glycosides (proscillaridin A, ouabain, digoxin, strophanthidin); and aromatic compounds (benzofurans). Many of these compounds have been elegantly described elsewhere (Nagle and Zhou, 2006; Manolescu *et al.*, 2009). It is interesting to point out that several natural products inhibit HIF-1α by interfering with its translation or with protein synthesis. Candidaspongiolide, a protein synthesis inhibitor that induces eIF2α phosphorylation and apoptosis in tumor cells, potently inhibited HIF-1α protein accumulation (Trisciuoglio *et al.*, 2008). Moreover, the natural product silibinin (a polyphenolic flavonoid) was shown to downregulates HIF-1α protein translation possibly through inhibition of mTOR/p70s6K/4EBP1 signaling pathway (Garcia-Maceira and Mateo, 2009), while recent evidence also suggests that it might inhibit global protein synthesis by decreasing the formation of eIF4F complex and phosphorylation of eIF2α (Jung *et al.*, 2009).

17.3 Exploiting HIF-1 inhibitors in combination strategies

The role of HIF-1 as a potential target for development of cancer therapeutics has been addressed in a number of experimental models which have led, not unexpectedly, to somewhat controversial results that reflect, at least in part, the complexity of the involvement of HIF-1 in human cancers and the intrinsic limitations of experimental animal models (Melillo, 2006). In contrast, studies of the expression of HIF-1 in cancer patients have been overall more consistent with a positive correlation between HIF-1α and/or HIF-2α levels and tumor progression and/or patient mortality (Table 17.1) (Aebersold *et al.*, 2001; Birner *et al.*, 2000, 2001b; Bos *et al.*, 2003; Koukourakis *et al.*, 2006). However, the relative contribution of HIF-1α and HIF-2α to tumor progression is still poorly understood and may vary in a tumor-type specific fashion. In addition, it remains to be fully determined when and how to inhibit HIF-1 for cancer therapy. Pharmacological validation of HIF-1 as a therapeutic target, both in preclinical models and more so in the clinical setting, is hampered by the lack of selective inhibitors and reliable biomarkers predictive of therapeutic activity.

The focal and heterogeneous expression of HIF-1α in the majority of solid tumors also suggests that single-agent inhibition of HIF-1 may have limited clinical activity and that this approach may be more effective in a context in which survival of tumor cells may be largely and more consistently relying on HIF-1-dependent mechanisms. An example may be vascular regression caused by administration of antiangiogenic agents, which may be associated with a concomitant increase in intratumor hypoxia. This therapy-induced hypoxic stress of the tumor microenvironment may be an ideal setting in which HIF-1-targeted therapies may be more effective. Proof of principle that this may indeed be the case has been recently provided by evidence indicating that combination of anti-VEGF antibody with daily administration of topotecan exerted synergistic antitumor activity, which was associated with inhibition of HIF-1α expression and activity in tumor tissue (Rapisarda *et al.*, 2009). However, further studies are required to better understand in which tumor types and/or under what circumstances such increase in intratumor hypoxia may actually occur.

17.4 Conclusions

In the complex biology of human cancer, HIF-1 is yet another potential target for development of cancer therapeutics. Its role as a master regulator of the transcriptional response to oxygen deprivation, a pathway thought to offer unique opportunities for the development of novel therapeutics, has led to the identification of an ever-growing number of small molecule inhibitor of HIFs. While most of these agents have been tested only in cell culture experiments, several have also been validated in preclinical models, in which case inhibition of angiogenesis and tumor growth have been considered paramount of HIF-1 inhibition. However, the most difficult challenge lies ahead and is represented by validation of this therapeutic approach in the clinical setting. Whether or not HIF-1 inhibitors identified so far will indeed block HIF-1 activity in patients with cancer has yet to be determined. More importantly, the potential clinical benefit of HIF-1 inhibition in cancer patients remains elusive at this time. Despite the remarkable and fairly rapid progress, over the last two decades, regarding the role that HIFs may play in cancer biology, our understanding of the actual contribution of these factors to human cancers is still poor and fragmented. Rigorous validation and rationale development of novel therapeutic approaches targeting the hypoxic signaling pathway are instrumental to successfully translate findings from preclinical models to the clinical setting, if our goal is to benefit patients with cancer.

Acknowledgements

The authors would like to thank members of the Tumor Hypoxia Laboratory and Dr R. H. Shoemaker for helpful discussion. This project has been funded in whole or in part with Federal funds from the National Cancer Institute, National Institutes of Health, under Contract No. N01-CO-12400. The content of this publication does not necessarily reflect the views or policies of the Department of Health and Human Services, nor does mention of trade names, commercial products, or organizations imply endorsement by the US Government. This research was supported [in part] by the Developmental Therapeutics Program, DCTD, of the National Cancer Institute, NIH.

References

Aebersold, D.M., Burri, P., Beer, K.T. *et al.* (2001) Expression of hypoxia-inducible factor-1alpha: a novel predictive and prognostic parameter in the radiotherapy of oropharyngeal cancer.*Cancer Research*, **61**, 2911–2916.

An, W.G., Kanekal, M., Simon, M.C. *et al.* (1998) Stabilization of wild-type p53 by hypoxia-inducible factor 1alpha. *Nature*, **392**, 405–408.

Bachtiary, B., Schindl, M., Potter, R. *et al.* (2003) Overexpression of hypoxia-inducible factor 1alpha indicates diminished response to radiotherapy and unfavorable prognosis in patients receiving radical radiotherapy for cervical cancer. *Clinical Cancer Research*, **9**, 2234–2240.

Birner, P., Schindl, M., Obermair, A. *et al.* (2000) Overexpression of hypoxia-inducible factor 1alpha is a marker for an unfavorable prognosis in early-stage invasive cervical cancer. *Cancer Research*, **60**, 4693–4696.

Birner, P., Gatterbauer, B., Oberhuber, G. *et al.* (2001a) Expression of hypoxia-inducible factor-1 alpha in oligodendrogliomas: its impact on prognosis and on neoangiogenesis. *Cancer*, **92**, 165–171.

Birner, P., Schindl, M., Obermair, A. *et al.* (2001b) Expression of hypoxia-inducible factor 1alpha in epithelial ovarian tumors: its impact on prognosis and on response to chemotherapy. *Clinical Cancer Research*, **7**, 1661–1668.

Bos, R., van der Groep, G.P., Greijer, A.E. *et al.* (2003) Levels of hypoxia-inducible factor-1alpha independently predict prognosis in patients with lymph node negative breast carcinoma. *Cancer*, **97**, 1573–1581.

Bos, R., Zhong, H., Hanrahan, C.F. *et al.* (2001) Levels of hypoxia-inducible factor-1 alpha during breast carcinogenesis. *Journal of the National Cancer Institute*, **93**, 309–314.

Brizel, D.M., Scully, S.P., Harrelson, J.M. *et al.* (1996) Tumor oxygenation predicts for the likelihood of distant metastases in human soft tissue sarcoma. *Cancer Research*, **56**, 941–943.

Brown, J.M. and Giaccia, A.J. (1998) The unique physiology of solid tumors: opportunities (and problems) for cancer therapy. *Cancer Research*, **58**, 1408–1416.

Burri, P., Djonov, V., Aebersold, D.M. *et al.* (2003) Significant correlation of hypoxia-inducible factor-1alpha with treatment outcome in cervical cancer treated with radical radiotherapy. *International Journal of Radiation Oncology, Biology, Physics*, **56**, 494–501.

Cao, Z., Fang, J., Xia, C. *et al.* (2004) *trans*-3,4,5′-Trihydroxystibene inhibits hypoxia-inducible factor 1alpha and vascular endothelial growth factor expression in human ovarian cancer cells. *Clinical Cancer Research*, **10**, 5253–5263.

Carroll, V.A. and Ashcroft, M. (2008) Regulation of angiogenic factors by HDM2 in renal cell carcinoma. *Cancer Research*, **68**, 545–552.

Cook, K.M., Hilton, S.T., Mecinovic, J. *et al.* (2009) Epidithiodiketopiperazines block the interaction between hypoxia-inducible factor-1alpha (HIF-1alpha) and p300 by a zinc ejection mechanism. *Journal of Biological Chemistry*, **284**, 26831–26838.

Creighton-Gutteridge, M., Cardellina, J.H., Stephen, A.G. *et al.* (2007) Cell type-specific, topoisomerase II-dependent inhibition of hypoxia-inducible factor-1alpha protein accumulation by NSC 644221. *Clinical Cancer Research*, **13**, 1010–1018.

Dales, J.P., Garcia, S., Meunier-Carpentier, S. *et al.* (2005) Overexpression of hypoxia-inducible factor HIF-1alpha predicts early relapse in breast cancer: retrospective study in a series of 745 patients. *International Journal of Cancer*, **116**, 734–739.

Del, B.D., Ciuffreda, L., Trisciuoglio, D. *et al.* (2006) Antiangiogenic potential of the Mammalian target of rapamycin inhibitor temsirolimus. *Cancer Research*, **66**, 5549–5554.

Ellis, L., Hammers, H., and Pili, R. (2008) Targeting tumor angiogenesis with histone deacetylase inhibitors. *Cancer Letters*, **280 (2)**, 145–153.

Emerling, B.M., Viollet, B., Tormos, K.V. *et al.* (2007) Compound C inhibits hypoxic activation of HIF-1 independent of AMPK. *FEBS Letters*, **581**, 5727–5731.

Escuin, D., Kline, E.R., and Giannakakou, P. (2005) Both microtubule-stabilizing and microtubule-destabilizing drugs inhibit hypoxia-inducible factor-1alpha accumulation and activity by disrupting microtubule function. *Cancer Research*, **65**, 9021–9028.

Fang, J., Cao, Z., Chen, Y.C. *et al.* (2004) 9-beta-D-arabinofuranosyl-2-fluoroadenine inhibits expression of vascular endothelial growth factor through hypoxia-inducible factor-1 in human ovarian cancer cells. *Molecular Pharmacology*, **66**, 178–186.

Fang, J., Xia, C., Cao, Z. *et al.* (2005) Apigenin inhibits VEGF and HIF-1 expression via PI3K/AKT/p70S6K1 and HDM2/p53 pathways. *FASEB Journal*, **19**, 342–353.

Gao, P., Zhang, H., Dinavahi, R. *et al.* (2007) HIF-dependent antitumorigenic effect of antioxidants in vivo. *Cancer Cell*, **12**, 230–238.

Garcia-Maceira, P. and Mateo, J. (2009) Silibinin inhibits hypoxia-inducible factor-1alpha and mTOR/p70S6K/4E-BP1 signalling pathway in human cervical and hepatoma cancer cells: implications for anticancer therapy. *Oncogene*, **28**, 313–324.

Giatromanolaki, A., Sivridis, E., Kouskoukis, C. *et al.* (2003) Hypoxia-inducible factors 1alpha and 2alpha are related to vascular endothelial growth factor expression and a poorer prognosis in nodular malignant melanomas of the skin. *Melanoma Research*, **13**, 493–501.

Giatromanolaki, A., Koukourakis, M.I., Simopoulos, C. *et al.* (2007) Metastatic cancer cells from c-erbB-2 negative primary breast cancer maintain the original c-erbB-2/HIF1alpha phenotype. *Cancer Biology and Therapy*, **6**, 153–155.

Greenberger, L.M., Horak, I.D., Filpula, D. *et al.* (2008) A RNA antagonist of hypoxia-inducible factor-1alpha, EZN-2968, inhibits tumor cell growth. *Molecular Cancer Therapeutics*, **7**, 3598–3608.

Griffiths, E.A., Pritchard, S.A., Valentine, H.R. *et al.* (2007) Hypoxia-inducible factor-1alpha expression in the gastric carcinogenesis sequence and its prognostic role in gastric and gastro-oesophageal adenocarcinomas. *British Journal of Cancer*, **96**, 95–103.

Han, J.Y., Oh, S.H., Morgillo, F. *et al.* (2005) Hypoxia-inducible factor 1alpha and antiangiogenic activity of farnesyltransferase inhibitor SCH66336 in human aerodigestive tract cancer. *Journal of the National Cancer Institute*, **97**, 1272–1286.

Hockel, M. and Vaupel, P. (2001) Tumor hypoxia: definitions and current clinical, biologic, and molecular aspects. *Journal of the National Cancer Institute*, **93**, 266–276.

Hudson, C.C., Liu, M., Chiang, G.G. *et al.* (2002) Regulation of hypoxia-inducible factor 1alpha expression and function by the mammalian target of rapamycin. *Molecular and Cellular Biology*, **22**, 7004–7014.

Hur, E., Kim, H.H., Choi, S.M. *et al.* (2002) Reduction of hypoxia-induced transcription through the repression of hypoxia-inducible factor-1alpha/aryl hydrocarbon receptor nuclear translocator DNA binding by the 90-kDa heat-shock protein inhibitor radicicol. *Molecular Pharmacology*, **62**, 975–982.

Isaacs, J.S., Jung, Y.J., Mimnaugh, E.G. *et al.* (2002) Hsp90 regulates a von Hippel Lindau-independent hypoxia-inducible factor-1 alpha-degradative pathway. *The Journal of Biological Chemistry*, **277**, 29936–29944.

Jones, M.K., Szabo, I.L., Kawanaka, H. *et al.* (2002) von Hippel Lindau tumor suppressor and HIF-1alpha: new targets of NSAIDs inhibition of hypoxia-induced angiogenesis. *FASEB Journal*, **16**, 264–266.

Jung, H.J., Park, J.W., Lee, J.S. *et al.* (2009) Silibinin inhibits expression of HIF-1alpha through suppression of protein translation in prostate cancer cells. *Biochemical and Biophysical Research Communications*, **390**, 71–76.

Jung, Y.J., Isaacs, J.S., Lee, S. *et al.* (2003) Microtubule disruption utilizes an NFkappa B-dependent pathway to stabilize HIF-1alpha protein. *Journal of Biological Chemistry*, **278**, 7445–7452.

Kaluz, S., Kaluzova, M., and Stanbridge, E.J. (2006) Proteasomal inhibition attenuates transcriptional activity of hypoxia-inducible factor 1 (HIF-1) via specific effect on the HIF-1alpha C-terminal activation domain. *Molecular and Cellular Biology*, **26**, 5895–5907.

Kaluz, S., Kaluzova, M., and Stanbridge, E.J. (2008) Comment on the role of FIH in the inhibitory effect of bortezomib on hypoxia-inducible factor-1. *Blood*, **111**, 5258–5259.

Kim, H.L., Yeo, E.J., Chun, Y.S. *et al.* (2006) A domain responsible for HIF-1alpha degradation by YC-1, a novel anticancer agent. *International Journal of Oncology*, **29**, 255–260.

Kim, Y., Ma, A.G., Kitta, K. *et al.* (2003) Anthracycline-induced suppression of GATA-4 transcription factor: implication in the regulation of cardiac myocyte apoptosis. *Molecular Pharmacology*, **63**, 368–377.

Klausmeyer, P., Zhou, Q., Scudiero, D.A. *et al.* (2009) Cytotoxic and HIF-1alpha inhibitory compounds from crossosoma bigelovii. *Journal of Natural Products, April 30.*

Klimova, T. and Chandel, N.S. (2008) Mitochondrial complex III regulates hypoxic activation of HIF. *Cell Death and Differentiation*, **15**, 660–666.

Koh, M.Y., Spivak-Kroizman, T., Venturini, S. *et al.* (2008) Molecular mechanisms for the activity of PX-478, an antitumor inhibitor of the hypoxia-inducible factor-1alpha. *Molecular Cancer Therapeutics*, **7**, 90–100.

Kong, D., Park, E.J., Stephen, A.G. *et al.* (2005) Echinomycin, a small-molecule inhibitor of hypoxia-inducible factor-1 DNA-binding activity. *Cancer Research*, **65**, 9047–9055.

Kong, X., Lin, Z., Liang, D. *et al.* (2006) Histone deacetylase inhibitors induce VHL and ubiquitin-independent proteasomal degradation of hypoxia-inducible factor 1alpha. *Molecular and Cellular Biology*, **26**, 2019–2028.

Korkolopoulou, P., Patsouris, E., Konstantinidou, A.E. *et al.* (2004) Hypoxia-inducible factor 1alpha/vascular endothelial growth factor axis in astrocytomas. Associations with microvessel morphometry, proliferation and prognosis. *Neuropathology and Applied Neurobiology*, **30**, 267–278.

Koukourakis, M.I., Simopoulos, C., Polychronidis, A. *et al.* (2003) The effect of trastuzumab/docatexel combination on breast cancer angiogenesis: dichotomus effect predictable by the HIFI alpha/VEGF pre-treatment status? *Anticancer Research*, **23**, 1673–1680.

Koukourakis, M.I., Bentzen, S.M., Giatromanolaki, A. *et al.* (2006) Endogenous markers of two separate hypoxia response pathways (hypoxia inducible factor 2 alpha and carbonic anhydrase 9) are associated with radiotherapy failure in head and neck cancer patients recruited in the CHART randomized trial. *Journal of Clinical Oncology*, **24**, 727–735.

Kung, A.L., Zabludoff, S.D., France, D.S. *et al.* (2004) Small molecule blockade of transcriptional coactivation of the hypoxia-inducible factor pathway. *Cancer Cell*, **6**, 33–43.

Kurebayashi, J., Otsuki, T., Kurosumi, M. *et al.* (2001) A radicicol derivative, KF58333, inhibits expression of hypoxia-inducible factor-1alpha and vascular endothelial growth factor, angiogenesis and growth of human breast cancer xenografts. *Japanese Journal of Cancer Research*, **92**, 1342–1351.

LaRusch, G.A., Jackson, M.W., Dunbar, J.D. *et al.* (2007) Nutlin3 blocks vascular endothelial growth factor induction by preventing the interaction between hypoxia inducible factor 1alpha and Hdm2. *Cancer Research*, **67**, 450–454.

Laughner, E., Taghavi, P., Chiles, K. *et al.* (2001) HER2 (neu) signaling increases the rate of hypoxia-inducible factor 1alpha (HIF-1alpha) synthesis: novel mechanism for HIF-1-mediated vascular endothelial growth factor expression. *Molecular and Cellular Biology*, **21**, 3995–4004.

Lee, K., Qian, D.Z., Rey, S. *et al.* (2009a) Anthracycline chemotherapy inhibits HIF-1 transcriptional activity and tumor-induced mobilization of circulating angiogenic cells. *Proceedings of the National Academy of Sciences of the United States of America*, **106**, 2353–2358.

Lee, K., Zhang, H., Qian, D.Z. *et al* (2009b) Acriflavine inhibits HIF-1 dimerization, tumor growth, and vascularization. *Proceedings of the National Academy of Sciences of the United States of America*, **106**, 17910–17915.

Lee, Y.M., Lim, J.H., Chun, Y.S. *et al.* (2009c) Nutlin-3, an Hdm2 antagonist, inhibits tumor adaptation to hypoxia by stimulating the FIH-mediated inactivation of HIF-1alpha. *Carcinogenesis*, **30**, 1768–1775.

Li, S.H., Shin, D.H., Chun, Y.S. *et al.* (2008) A novel mode of action of YC-1 in HIF inhibition: stimulation of FIH-dependent p300 dissociation from HIF-1alpha. *Molecular Cancer Therapeutics*, **7**, 3729–3738.

Lin, X., David, C.A., Donnelly, J.B. *et al.* (2008) A chemical genomics screen highlights the essential role of mitochondria in HIF-1 regulation. *Proceedings of the National Academy of Sciences of the United States of America*, **105**, 174–179.

Litz, J. and Krystal, G.W. (2006) Imatinib inhibits c-Kit-induced hypoxia-inducible factor-1alpha activity and vascular endothelial growth factor expression in small cell lung cancer cells. *Molecular Cancer Therapeutics*, **5**, 1415–1422.

Lu, H., Dalgard, C.L., Mohyeldin, A. *et al.* (2005) Reversible inactivation of HIF-1 prolyl hydroxylases allows cell metabolism to control basal HIF-1. *Journal of Biological Chemistry*, **280**, 41928–41939.

Luwor, R.B., Lu, Y., Li, X. *et al.* (2005) The antiepidermal growth factor receptor monoclonal antibody cetuximab/C225 reduces hypoxia-inducible factor-1 alpha, leading to transcriptional inhibition of vascular endothelial growth factor expression. *Oncogene*, **24**, 4433–4441.

Mabjeesh, N.J., Post, D.E., Willard, M.T. *et al.* (2002) Geldanamycin induces degradation of hypoxia-inducible factor 1alpha protein via the proteosome pathway in prostate cancer cells. *Cancer Research*, **62**, 2478–2482.

Mabjeesh, N.J., Escuin, D., LaVallee, T.M. *et al.* (2003) 2ME2 inhibits tumor growth and angiogenesis by disrupting microtubules and dysregulating HIF. *Cancer Cell*, **3**, 363–375.

Majumder, P.K., Febbo, P.G., Bikoff, R. *et al.* (2004) mTOR inhibition reverses Akt-dependent prostate intraepithelial neoplasia through regulation of apoptotic and HIF-1-dependent pathways. *Nature Medicine*, **10**, 594–601.

Manalo, D.J., Rowan, A., Lavoie, T. *et al.* (2005) Transcriptional regulation of vascular endothelial cell responses to hypoxia by HIF-1. *Blood*, **105**, 659–669.

Manolescu, B., Oprea, E., Busu, C. *et al.* (2009) Natural compounds and the hypoxia-inducible factor (HIF) signalling pathway. *Biochimie*, **91**, 1347–1358.

Mansilla, S. and Portugal, J. (2008) Sp1 transcription factor as a target for anthracyclines: effects on gene transcription. *Biochimie*, **90**, 976–987.

Mayerhofer, M., Valent, P., Sperr, W.R. *et al.* (2002) BCR/ABL induces expression of vascular endothelial growth factor and its transcriptional activator, hypoxia inducible factor-1alpha, through a pathway involving phosphoinositide 3-kinase and the mammalian target of rapamycin. *Blood*, **100**, 3767–3775.

Melillo, G. (2006) Inhibiting hypoxia-inducible factor 1 for cancer therapy. *Molecular Cancer Research*, **4**, 601–605.

Melillo, G. (2007) Targeting hypoxia cell signaling for cancer therapy. *Cancer Metastasis Reviews*, **26**, 341–352.

Mie, L.Y., Kim, S.H., Kim, H.S. *et al.* (2003) Inhibition of hypoxia-induced angiogenesis by FK228, a specific histone deacetylase inhibitor, via suppression of HIF-1alpha activity. *Biochemical and Biophysical Research Communications*, **300**, 241–246.

Moser, C., Lang, S.A., Mori, A. *et al.* (2008) ENMD-1198, a novel tubulin-binding agent reduces HIF-1alpha and STAT3 activity in human hepatocellular carcinoma(HCC) cells, and inhibits growth and vascularization in vivo. *BMC Cancer*, **8**, 206.

Nagle, D.G. and Zhou, Y.D. (2006) Natural product-derived small molecule activators of hypoxia-inducible factor-1 (HIF-1). *Current Pharmaceutical Design*, **12**, 2673–2688.

Narita, T., Yin, S., Gelin, C.F. *et al.* (2009) Identification of a novel small molecule HIF-1alpha translation inhibitor. *Clinical Cancer Research*, **15**, 6128–6136.

Nickols, N.G., Jacobs, C.S., Farkas, M.E. *et al.* (2007) Improved nuclear localization of DNA-binding polyamides. *Nucleic Acids Research*, **35**, 363–370.

Olenyuk, B.Z., Zhang, G.J., Klco, J.M. *et al.* (2004) Inhibition of vascular endothelial growth factor with a sequence-specific hypoxia response element antagonist. *Proceedings of the National Academy of Sciences of the United States of America*, **101**, 16768–16773.

Onnis, B., Rapisarda, A., and Melillo, G. (2009) Development of HIF-1 inhibitors for cancer therapy. *Journal of Cellular and Molecular Medicine*, **13 (9A)**, 2780–2786.

Palayoor, S.T., Tofilon, P.J., and Coleman, C.N. (2003) Ibuprofen-mediated reduction of hypoxia-inducible factors HIF-1alpha and HIF-2alpha in prostate cancer cells. *Clinical Cancer Research*, **9**, 3150–3157.

Park, E.J., Kong, D., Fisher, R. *et al.* (2006) Targeting the PAS-A domain of HIF-1alpha for development of small molecule inhibitors of HIF-1. *Cell Cycle*, **5**, 1847–1853.

Pore, N., Gupta, A.K., Cerniglia, G.J. *et al.* (2006a) Nelfinavir down-regulates hypoxia-inducible factor 1alpha and VEGF expression and increases tumor oxygenation: implications for radiotherapy. *Cancer Research*, **66**, 9252–9259.

Pore, N., Jiang, Z., Gupta, A. *et al.* (2006b) EGFR tyrosine kinase inhibitors decrease VEGF expression by both hypoxia-inducible factor (HIF)-1-independent and HIF-1-dependent mechanisms. *Cancer Research*, **66**, 3197–3204.

Pouyssegur, J., Dayan, F., and Mazure, N.M. (2006) Hypoxia signalling in cancer and approaches to enforce tumour regression. *Nature*, **441**, 437–443.

Qian, D.Z., Kachhap, S.K., Collis, S.J. *et al.* (2006) Class II histone deacetylases are associated with VHL-independent regulation of hypoxia-inducible factor 1 alpha. *Cancer Research*, **66**, 8814–8821.

Rapisarda, A., Hollingshead, M., Uranchimeg, B. *et al.* (2009) Increased antitumor activity of bevacizumab in combination with hypoxia inducible factor-1 inhibition. *Molecular Cancer Therapeutics*, **8**, 1867–1877.

Rapisarda, A., Uranchimeg, B., Scudiero, D.A. *et al.* (2002) Identification of small molecule inhibitors of hypoxia-inducible factor 1 transcriptional activation pathway. *Cancer Research*, **62**, 4316–4324.

Rapisarda, A., Uranchimeg, B., Sordet, O. *et al.* (2004a) Topoisomerase I-mediated inhibition of hypoxia-inducible factor 1: mechanism and therapeutic implications. *Cancer Research*, **64**, 1475–1482.

Rapisarda, A., Zalek, J., Hollingshead, M. *et al.* (2004b) Schedule-dependent inhibition of hypoxia-inducible factor-1alpha protein accumulation, angiogenesis, and tumor growth by topotecan in U251-HRE glioblastoma xenografts. *Cancer Research*, **64**, 6845–6848.

Ravi, R., Mookerjee, B., Bhujwalla, Z.M. *et al.* (2000) Regulation of tumor angiogenesis by p53-induced degradation of hypoxia-inducible factor 1alpha. *Genes and Development*, **14**, 34–44.

Sapra, P., Zhao, H., Mehlig, M. *et al.* (2008) Novel delivery of SN38 markedly inhibits tumor growth in xenografts, including a camptothecin-11-refractory model. *Clinical Cancer Research*, **14**, 1888–1896.

Scheuermann, T.H., Tomchick, D.R., Machius, M. *et al.* (2009) Artificial ligand binding within the HIF2alpha PAS-B domain of the HIF2 transcription factor. *Proceedings of the National Academy of Sciences of the United States of America*, **106**, 450–455.

Schindl, M., Schoppmann, S.F., Samonigg, H. *et al.* (2002) Overexpression of hypoxia-inducible factor 1alpha is associated with an unfavorable prognosis in lymph node-positive breast cancer. *Clinical Cancer Research*, **8**, 1831–1837.

Secchiero, P., Corallini, F., Gonelli, A. *et al.* (2007) Antiangiogenic activity of the MDM2 antagonist nutlin-3. *Circulation Research*, **100**, 61–69.

Semenza, G.L. (2003) Targeting HIF-1 for cancer therapy. *Nature Reviews Cancer*, **3**, 721–732.

Semenza, G.L. (2007) Hypoxia and cancer. *Cancer Metastasis Reviews*, **26**, 223–224.

Semenza, G.L. (2008) Hypoxia-inducible factor 1 and cancer pathogenesis. *IUBMB Life*, **60**, 591–597.

Shin, D.H., Chun, Y.S., Lee, D.S. *et al.* (2008) Bortezomib inhibits tumor adaptation to hypoxia by stimulating the FIH-mediated repression of hypoxia-inducible factor-1. *Blood*, **111**, 3131–3136.

Sivridis, E., Giatromanolaki, A., Gatter, K.C. *et al.* (2002) Association of hypoxia-inducible factors 1alpha and 2alpha with activated angiogenic pathways and prognosis in patients with endometrial carcinoma. *Cancer*, **95**, 1055–1063.

Stewart, G.D., Nanda, J., Brown, D.J. *et al.* (2009) NO-sulindac inhibits the hypoxia response of PC-3 prostate cancer cells via the Akt signalling pathway. *International Journal of Cancer*, **124**, 223–232.

Sun, H.C., Qiu, Z.J., Liu, J. *et al.* (2007) Expression of hypoxia-inducible factor-1 alpha and associated proteins in pancreatic ductal adenocarcinoma and their impact on prognosis. *International Journal of Oncology*, **30**, 1359–1367.

Swinson, D.E., Jones, J.L., Cox, G. *et al.* (2004) Hypoxia-inducible factor-1 alpha in non small cell lung cancer: relation to growth factor, protease and apoptosis pathways. *International Journal of Cancer*, **111**, 43–50.

Szulawska, A., Gniazdowski, M., and Czyz, M. (2005) Sequence specificity of formaldehyde-mediated covalent binding of anthracycline derivatives to DNA. *Biochemical Pharmacology*, **69**, 7–18.

Takahashi, R., Tanaka, S., Hiyama, T. *et al.* (2003) Hypoxia-inducible factor-1alpha expression and angiogenesis in gastrointestinal stromal tumor of the stomach. *Oncology Reports*, **10**, 797–802.

Talks, K.L., Turley, H., Gatter, K.C. *et al.* (2000) The expression and distribution of the hypoxia-inducible factors HIF-1alpha and HIF-2alpha in normal human tissues, cancers, and tumor-associated macrophages. *American Journal of Pathology*, **157**, 411–421.

Tan, C., de Noronha, R.G., Roecker, A.J. *et al.* (2005) Identification of a novel small-molecule inhibitor of the hypoxia-inducible factor 1 pathway. *Cancer Research*, **65**, 605–612.

Theodoropoulos, V.E., Lazaris, A.C., Sofras, F. *et al.* (2004) Hypoxia-inducible factor 1 alpha expression correlates with angiogenesis and unfavorable prognosis in bladder cancer. *European Urology*, **46**, 200–208.

Theodoropoulos, V.E., Lazaris, A.C., Kastriotis, I. *et al.* (2005) Evaluation of hypoxia-inducible factor 1alpha overexpression as a predictor of tumour recurrence and progression in superficial urothelial bladder carcinoma. *British Journal of Urology International*, **95**, 425–431.

Thomas, G.V., Tran, C., Mellinghoff, I.K. *et al.* (2006) Hypoxia-inducible factor determines sensitivity to inhibitors of mTOR in kidney cancer. *Nature Medicine*, **12**, 122–127.

Trisciuoglio, D., Uranchimeg, B., Cardellina, J.H. *et al.* (2008) Induction of apoptosis in human cancer cells by candidaspongiolide, a novel sponge polyketide. *Journal of the National Cancer Institute*, **100**, 1233–1246.

Vleugel, M.M., Greijer, A.E., Shvarts, A. *et al.* (2005) Differential prognostic impact of hypoxia induced and diffuse HIF-1alpha expression in invasive breast cancer. *Journal of Clinical Pathology*, **58**, 172–177.

Wan, X., Shen, N., Mendoza, A. *et al.* (2006) CCI-779 inhibits rhabdomyosarcoma xenograft growth by an antiangiogenic mechanism linked to the targeting of mTOR/Hif-1alpha/VEGF signaling. *Neoplasia*, **8**, 394–401.

Wang, G.L. and Semenza, G.L. (1995) Purification and characterization of hypoxia-inducible factor 1. *The Journal of Biological Chemistry*, **270**, 1230–1237.

Wang, G.L., Jiang, B.H., Rue, E.A. *et al.* (1995) Hypoxia-inducible factor 1 is a basic-helix-loop-helix-PAS heterodimer regulated by cellular O2 tension. *Proceedings of the National Academy of Sciences of the United States of America*, **92**, 5510–5514.

Welsh, S.J., Koh, M.Y., and Powis, G. (2006) The hypoxic inducible stress response as a target for cancer drug discovery. *Seminars in Oncology*, **33**, 486–497.

Welsh, S.J., Williams, R.R., Birmingham, A. *et al.* (2003) The thioredoxin redox inhibitors 1-methylpropyl 2-imidazolyl disulfide and pleurotin inhibit hypoxia-induced factor 1alpha and vascular endothelial growth factor formation. *Molecular Cancer Therapeutics*, **2**, 235–243.

Welsh, S., Williams, R., Kirkpatrick, L. *et al.* (2004) Antitumor activity and pharmacodynamic properties of PX-478, an inhibitor of hypoxia-inducible factor-1alpha. *Molecular Cancer Therapeutics*, **3**, 233–244.

Yang, J., Zhang, L., Erbel, P.J. *et al.* (2005) Functions of the Per/ARNT/Sim domains of the hypoxia-inducible factor. *Journal of Biological Chemistry*, **280**, 36047–36054.

Yang, J., Ahmed, A., Poon, E. *et al.* (2009) Small-molecule activation of p53 blocks hypoxia-inducible factor 1alpha and vascular endothelial growth factor expression in vivo and leads to tumor cell apoptosis in normoxia and hypoxia. *Molecular and Cellular Biology*, **29**, 2243–2253.

Yeo, E.J., Chun, Y.S., Cho, Y.S. *et al.* (2003) YC-1: a potential anticancer drug targeting hypoxia-inducible factor 1. *Journal of the National Cancer Institute*, **95**, 516–525.

Zhang, H., Qian, D.Z., Tan, Y.S. *et al.* (2008) Digoxin and other cardiac glycosides inhibit HIF-1alpha synthesis and block tumor growth. *Proceedings of the National Academy of Sciences of the United States of America*, **105**, 19579–19586.

Zhang, Q., Tang, X., Lu, Q.Y. *et al.* (2005) Resveratrol inhibits hypoxia-induced accumulation of hypoxia-inducible factor-1alpha and VEGF expression in human tongue squamous cell carcinoma and hepatoma cells. *Molecular Cancer Therapeutics*, **4**, 1465–1474.

Zhang, Q., Tang, X., Lu, Q. *et al.* (2006) Green tea extract and (-)-epigallocatechin-3-gallate inhibit hypoxia- and serum-induced HIF-1alpha protein accumulation and VEGF expression in human cervical carcinoma and hepatoma cells. *Molecular Cancer Therapeutics*, **5**, 1227–1238.

Zhong, H., De Marzo, A.M., Laughner, E. *et al.* (1999) Overexpression of hypoxia-inducible factor 1alpha in common human cancers and their metastases. *Cancer Research*, **59**, 5830–5835.

Zhong, H., Chiles, K., Feldser, D. *et al.* (2000) Modulation of hypoxia-inducible factor 1alpha expression by the epidermal growth factor/phosphatidylinositol 3-kinase/PTEN/AKT/FRAP pathway in human prostate cancer cells: implications for tumor angiogenesis and therapeutics. *Cancer Research*, **60**, 1541–1545.

Zhong, L., D'Urso, A., Toiber, D. *et al.* (2010) The histone deacetylase Sirt6 regulates glucose homerostasis via Hif1a. *Cell*, **140**, 280–293.

Zhong, X.S., Zheng, J.Z., Reed, E. *et al.* (2004) SU5416 inhibited VEGF and HIF-1alpha expression through the PI3K/AKT/p70S6K1 signaling pathway. *Biochemical and Biophysical Research Communications*, **324**, 471–480.

Zimmer, M., Ebert, B.L., Neil, C. *et al.* (2008) Small-molecule inhibitors of HIF-2a translation link its 5′UTR iron-responsive element to oxygen sensing. *Molecular Cell*, **32**, 838–848.

18

Vascular-Targeted Molecular Therapy

Graeme J. Dougherty[1,2] and Shona T. Dougherty[1]

[1]*Department of Radiation Oncology, University of Arizona, Tucson, AZ, USA*
[2]*Angiogene Pharmaceuticals Ltd., Oxford, UK*

18.1 Introduction

The term 'molecular therapy' refers to the use of genetic information, in all its forms, in the treatment of disease. Included under this broad definition are approaches that employ genomic DNA, cDNA, RNA, ribozymes, siRNA, shRNA, and so on. While an understanding of the complex mechanisms that control gene expression and function in normal and malignant cells is clearly key to the success of such approaches, it is mostly difficulties related to the development of safe, efficient, and practical *in vivo* delivery systems that have slowed the more widespread introduction of potentially promising molecular therapies into the clinic (Scanlon, 2004).

For obvious reasons, attempts at the molecular therapy of cancer have tended to target the malignant cell population. In most cases, the objective has been to either directly kill tumor cells or sensitize them to the cytotoxic effects of radiation and/or chemotherapeutic agents (Palmer, Young, and Mautner, 2006). Generally speaking, such approaches have required the direct intratumoral injection or locoregional administration of viral or other vectors carrying a gene of interest (van Etten *et al.*, 2005; Li *et al.*, 2009). While impressive results have been obtained in some studies (Buller *et al.*, 2002), such approaches are limited to the relatively small number of

National Cancer Institute
CA100004
Arizona Biomedical Research Commission
Contract 0706
University of Arizona, Women's Cancers Better Than Ever Program

Tumor Microenvironment Edited by Dietmar W. Siemann

well-defined solid tumors into which a needle or canula can be readily placed, such as cutaneous melanomas or head and neck tumors (Lai *et al.*, 2009), as well as tumors localized to enclosed body cavities such as the peritoneum, pleural space or bladder, into which large volumes of vector can be instilled (Adam *et al.*, 2007; Gomella *et al.*, 2001; Vasey *et al.*, 2002). Even in such circumstances, however, good transduction efficiencies are often only seen in the immediate vicinity of the injection site (Smiley *et al.*, 1997). Presumably, such findings reflect the significant obstacles, both physical and physiological in nature, that prevent the efficient diffusion of vectors through tumor tissue. As transduction efficiencies thus tend to fall off rapidly with distance from a site of injection, the overall effect of treatment on tumor growth may be fairly poor even when the gene in question functions exactly as hoped (Rainov and Kramm, 2001; Rochlitz, 2001; Smiley *et al.*, 1997).

It is for these and other reasons that increasing thought has been given to targeting, not the malignant population itself, but instead various normal cell types upon which the survival and growth of a solid tumor depends (Aharinejad *et al.*, 2009; Chung *et al.*, 2003; Hsieh *et al.*, 2004; Jackson, Franco, and Bhowmick, 2008). Particular interest in this regard has focused on the tumor vasculature. Tumor-associated blood vessels deliver the oxygen and nutrients required to maintain tumor cell viability and if blood flow is interrupted for a sufficiently long period of time, then tumor cell death will result. As described elsewhere in this volume, studies with small molecule vascular disrupting agents (VDAs), including Combretastatin-A4 (CA4P) and ZD6126 have helped confirm the therapeutic potential of approaches that work in this way (Chaplin and Hill, 2002; Davis *et al.*, 2002; Denekamp, Dasu, and Waites, 1998; Siemann, Chaplin, and Horsman, 2004; Thorpe, 2004; Thorpe, Chaplin, and Blakey, 2003).

What makes the vasculature so especially attractive from the perspective of molecular therapy is the amplified killing that can potentially be achieved. Thus vessel occlusion produced in response to localized damage will kill not only tumor cells in the immediate vicinity of the point of blockage but also those located downstream that are supplied by the affected vessel. In this way, even the relatively poor *in vivo* transduction efficiencies that can realistically be obtained using current vector systems may cause sufficient cell death to slow or prevent tumor growth, particularly when used in conjunction with established therapies.

The fact that the endothelial cells that line tumor-associated blood vessels are in direct contact with the circulation is an additional advantage allowing the vectors used for molecular therapy to be administered intravenously. As diffusion out of the vasculature is not required to produce a benefit, the large size of the viral and other vehicles normally used to deliver molecular therapies is likely to be less of a concern. Importantly, as vascular endothelial cells are non-transformed, it also follows that they may respond more predictably than genetically unstable tumor cells to a particular therapeutic approach. Resistance to therapy is also less likely to emerge. This raises the possibility that a single strategy could potentially be applied to a broad range of histologically distinct malignancies involving various tissue sites. On the other hand, it does not seem unreasonable to assume that the exposure of endothelial cells within the tumor microenvironment to mitogenic and proinflammatory or other signals may induce unique phenotypic changes that can be exploited in the targeting of molecular therapy (Thorpe, 2004; Thorpe, Chaplin, and Blakey, 2003).

18.2 Approaches to targeting tumor vasculature *in vivo*

It goes without saying that the safe implementation of molecular therapies, whether directed against the vasculature or other normal or malignant cell types, requires that the expression and/or activity of the therapeutic genes used be restricted to a particular organ, tissue, and/or cell type. Various approaches have been explored in an effort to achieve this critical but elusive goal. While strategies involving differential transduction and/or restricted transcription have generally attracted most attention, novel approaches involving post-transcriptional and/or functional targeting have shown great promise (Figure 18.1).

18.2.1 Transductional targeting

Many viruses show tropism for particular cell types. Indeed, the symptoms of viral infection are largely a product of the cells being targeted. Although little explored, it is possible that viruses that preferentially infect endothelial cells could be engineered to generate vectors that might prove useful in vascular-targeted molecular therapy. Among the viruses that have been shown to infect and/or replicate in endothelial cells are a number that have been fairly well characterized at the molecular level and about which much is known regarding their behavior and effects *in vivo*. Examples include cytomegalovirus, hepatitis C, human immunodeficiency virus (HIV), and human T cell leukemia virus-1 (HTLV-1) (Agnello and Abel, 1997; Chi *et al.*, 2000; Dropulic and Masters, 1990; Gerna, Baldanti, and Revello, 2004; Yang *et al.*, 2000).

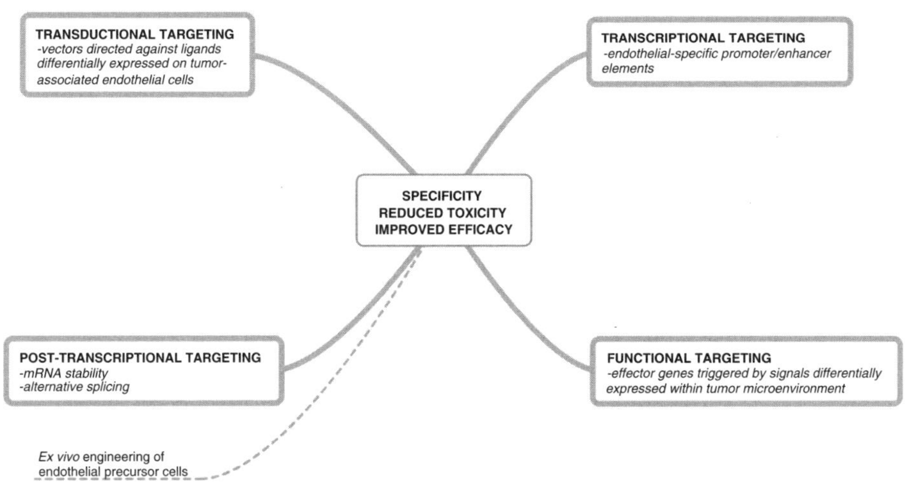

Figure 18.1 Targeting molecular therapy to tumor vasculature *in vivo*.

Ebola and certain arboviruses also infect endothelial cells and this activity may play an important role in the pathogenesis of these agents (Dropulic and Masters, 1990; Schnittler and Feldmann, 2003; Hensley and Geisbert, 2005; Sullivan, Yang, and Nabel, 2003; Solignat *et al.*, 2009; Tulman *et al.*, 2009). It is important to keep in mind, however, that tropism for endothelial cells is often relative and in most cases other cell types may also be infected, particularly later in the course of disease. While this limits the usefulness of such viruses in the delivery of therapeutic genes, an understanding of the mechanisms they employ to achieve specificity might nevertheless provide insights that could be incorporated into established vectors in order to improve the efficiency of endothelial targeting.

The viral vectors most often employed in the molecular therapy of cancer are derived from well-studied retroviral, adenoviral or adeno-associated viruses (AAVs). As the counter-receptors these viruses recognize in order to gain entry into a target cell are present at a high level on many different cell types, it follows that they lack specificity for particular cellular targets. Indeed, their broad specificity was one of the reasons why these particular viruses were chosen for this purpose. For example, the initial attachment of non-enveloped subgroup C adenoviruses to target cells is mediated by an interaction between the 'knob' portion of the viral fiber coat protein and a broadly distributed cellular receptor known as the coxsackie virus and adenovirus receptor (CAR) (Bergelson *et al.*, 1997; Tomko, Xu, and Philipson, 1997). Importantly, subsequent entry of the virus into a cell involves clathrin-dependent endocytosis, a process that is facilitated by an interaction between an RGD (Arg-Gly-Asp) motif in the viral penton base protein and one or more cell surface integrins (primarily $\alpha v\beta 3$ or $\alpha v\beta 5$) that are also widely expressed (Bergelson, 1999; Greber *et al.*, 1993; Wang *et al.*, 1998; Wickham *et al.*, 1993). Mutational changes that block both CAR and integrin binding dramatically reduce infection of target cells *in vitro*. Transduction of the liver and other tissues following *in vivo* administration is similarly reduced. Complicating the simple model outlined above in which CAR mediates initial viral attachment and integrins subsequent uptake, is the finding that mutations that disrupt only CAR binding had relatively little effect on liver transduction *in vivo* (Michael *et al.*, 1995). Most studies, however, continue to support a key role for CAR in the infection of target cells by adenoviral vectors and cells and tissues that lack the ligand are not easily transduced (Pearson *et al.*, 1999).

What makes these and other similar studies important, is that they raised the possibility that the broad tropism of adenoviral vectors could perhaps be altered and infection targeted to particular cell types if the virus could be prevented from binding to its natural ligand and instead redirected to an alternative structure present on the surface of the cell type of interest (Bilbao, Gomez-Navarro, and Curiel, 1998; Krasnykh *et al.*, 2000). Of course, when first proposed it was far from clear whether other proteins indeed possess the biological characteristic necessary to substitute for CAR. Moreover, abrogation of CAR binding had to be achieved without interfering with the key post-attachment steps involved in viral uptake and/or subsequent gene expression. It was reassuring then, when studies in which Fab fragments or chimeric scFv were used to link the knob domain of the adenoviral fiber protein to various naturally occurring cell surface molecules, demonstrated that a surprisingly large and structurally diverse collection of proteins could take the place of CAR and allow viral infection to proceed if the target cells involved express the integrins required for the subsequent uptake of bound viral particles. Even more remarkable, it appears that certain cell surface proteins can substitute for CAR and integrins allowing infection of cells that lack both molecules (Wickham *et al.*, 1997).

The next question, of course, is whether cell surface molecules with the characteristics needed to function as adenoviral receptors are differentially expressed on tumor-associated endothelial cells and if so, how might these best be targeted. It is encouraging that analysis of expressed sequence tag (EST) data suggests that endothelial cells possess a large number of unique mRNA transcripts, some of which presumably encode surface molecules (Adams *et al.*, 1995). Whether any of these potential targets are differentially expressed on tumor-associated endothelial cells is less clear although the fact that there exist obvious structural and functional differences between tumor blood vessels and those in normal tissues lends weight to the possibility that unique structures suitable for targeting purposes may be found (Denekamp, 1990).

One obvious difference between endothelial cells in normal and malignant tissues concerns their rate of proliferation. It has long been appreciated that the endothelial cells that line normal blood vessels have a very long half-life, dividing only every 1000 days or so (Denekamp, 1993). In contrast, tumor-associated endothelial cells are far more active, with some cells dividing every day or two (Denekamp, 1993). Thus, cell surface molecules upregulated in the course of endothelial cell proliferation may be differentially expressed within the tumor microenvironment and could thus serve as possible candidates for the targeting of molecular therapy (Denekamp, 1993). Examples include the receptor tyrosine kinases Tie-1 and Tie-2 (angiopoietin-1 receptor), vascular endothelial cell (VE)-cadherin, neural (N)-cadherin, and the β1 and β3 integrins. Not unexpectedly, the magnitude of induction is dependent upon the nature of the mitogenic signal used to induce endothelial cell proliferation. Thus, while integrin α5 expression increased more than fivefold when human dermal microvascular endothelial cells were stimulated with epidermal growth factor (EGF) only a small increase was triggered upon treatment with vascular endothelial cell growth factor (VEGF) (Zhang *et al.*, 1999). VEGF does, however, dramatically upregulate expression of α1 and α2 integrins on human dermal microvascular endothelial cells (Senger *et al.*, 1997). In keeping with this finding, α1β1 appears to be overexpressed on tumor blood vessels (Fabbri *et al.*, 1996). α2β1 is similarly upregulated at the sprouting tips of neonatal blood vessels (Enenstein and Kramer, 1994). These findings are not unexpected given the key role played by integrins at various stages in the angiogenic process including matrix reorganization and the regulation of endothelial cell proliferation, migration, and survival (Defilippi *et al.*, 1991). Antagonists of αvβ3, α1β1, and α2β1 inhibit VEGF-induced angiogenesis in various model systems and can slow tumor growth *in vivo* (Brooks *et al.*, 1994, 1995; Senger *et al.*, 1997).

While the ease of assay systems that allow identification of determinants associated with endothelial cell proliferation induced *in vitro* in response to treatment with proangiogenic cytokines has focused attention on this area, it is clear that the angiogenic process involves a complex and well-coordinated series of cell–cell and cell–matrix interactions that control endothelial migration and organization into functional vessels. The molecules that control these processes could also potentially serve as targets for the delivery of molecular therapy. In addition to integrins, studies on bovine aortic endothelial cells have demonstrated that the activin-binding protein follistatin, which is generally absent from quiescent endothelium, is upregulated on migrating endothelial cells (Kozian and Augustin, 1995). Osteopontin and PC4 are similarly upregulated when endothelial cells organize into tube-like structures *in vitro*, while CEC5, a novel protein with homology to calmodulin-dependent protein kinases, is downregulated (Prols, Loser, and Marx, 1998).

Accepting for the moment that there are surface structures differentially expressed on tumor-associated endothelial cells that can function as adenoviral receptors, the next challenge is to develop strategies that redirect adenovirus to these particular molecules as opposed to their widely distributed natural ligands. While bispecific antibodies and antibody fragments work well *in vitro*, they are far less practical *in vivo*, particularly in a clinical setting. From a purely practical perspective it would clearly be preferable if targeting moieties were directly incorporated into the vector itself. Of course this has to done in a manner that does not disrupt post-attachment processes essential to viral uptake and gene expression. By far the most popular approach is to incorporate short peptide sequences directly into the H1 loop of the Ad5 fiber protein (Krasnykh *et al.*, 1998). Peptide sequences that bind preferentially to tumor-associated vascular endothelial cells can be readily identified using well-established screening approaches and standard recombinant DNA techniques used to both disrupt the CAR-binding function of the penton base protein and generate chimeric fiber proteins that can be used to redirect infection to differentially expressed cognate receptors.

A good example of this approach is illustrated in a study by Nicklin *et al.* (2004) in which 12mer peptides with endothelial cell binding specificity (MTPFPTSNEANL (MTP) and MSLTTPPAVARP (MSL)) derived by phage display were used to target adenoviral vectors to endothelial cells *in vitro* while maintaining low levels of infectivity for various other cell types.

While this particular discussion has focused on adenoviral vectors it is clear that similar approaches can be used to target AAV and retroviral vectors to tumor-associated endothelial cells. Although less well developed, some progress has been made in this area.

18.2.2 Transcriptional targeting

Transcriptional targeting involves approaches that employ specific DNA sequences to preferentially drive the expression of a therapeutic gene in a cell type of interest. The promoter/enhancer elements that control the expression of a number of genes that are restricted to endothelial cells have been isolated and characterized and some have been shown to confer a reasonable degree of endothelial cell specificity when placed upstream of a gene of interest in one or more vector systems. Among the most promising are sequences derived from endoglin (Graulich *et al.*, 1999), Flt-1 (Nicklin *et al.*, 2001; Reynolds *et al.*, 2001), Tie 2 (De Palma, Venneri, and Naldini, 2003; Korhonen *et al.*, 1995), intercellular adhesion molecule-2 (ICAM-2) (Cowan *et al.*,

1996; Nicklin *et al.*, 2001; Richardson, Kaspers, and Porter, 2004; Velasco *et al.*, 2001), VE-cadherin (Dancer *et al.*, 2003), and pre-proendothelin-1 (Cho *et al.*, 2003; Greenberger *et al.*, 2004; Jager, Zhao, and Porter, 1999; Mavria, Jager, and Porter, 2000; Varda-Bloom *et al.*, 2001). 'Leakiness' or low level gene expression in cell types other than endothelial cells remains a problem. One inventive solution proposed by Savontaus and colleagues (Savontaus *et al.*, 2002) involves a binary system in which two genes required for adenoviruses to replicate in and lyse target cells (E1A and E1B) are placed under the control of two different endothelial cell-specific promoter elements derived respectively from the Flk-1 and endothelin genes.

In addition to the known and well-characterized endothelial cell-specific genes described above, techniques such as 'differential display' have lead to the identification of numerous less well-characterized genes that appear to be preferentially expressed in endothelial cells in general or in specific endothelial cell populations including those that line tumor-associated blood vessels. Although certainly worthy of further study, isolating the promoter/enhancer elements that determine the cellular specificity of such genes for use in transcriptional targeting presents some formidable challenges. In particular, the sequences involved may be located some distance from the gene itself and may include multiple interacting transcription factor binding sites distributed over a large stretch of DNA requiring extensive mutational and deletional analysis if the key regions responsible for specificity are to be identified and assembled into an element short enough to meet the size constraints dictated by current vector design.

An alternative approach that has gained some credence in recent years is to produce entirely synthetic promoters in which multiple transcription factor binding sites are simply assembled in a head-to-tail fashion forming an element that confers the desired cellular specificity. Unfortunately, analysis of DNA sequences located adjacent to genes differentially expressed in endothelial cells has failed to reveal either a particular arrangement of transcription factor binding sites or other common features that might help guide the rational design of synthetic promoters (Dai, McAninch, and Sutton, 2004). The general strategy has, however, been validated by a somewhat laborious and less directed approach in which short duplex oligonucleotides corresponding to known transcription factor binding sites were simply ligated together in a random fashion and cloned into a plasmid vector next to minimal skeletal muscle promoter upstream of an indicator gene (luciferase) (Li *et al.*, 1999). The library was introduced into muscle and other cell types and clones exhibiting high levels of muscle-specific activity isolated and sequenced. A number of promising elements have been identified in this way including some that exhibited a level of activity that rivaled that obtained for strong viral promoters such as CMV and RSV. Surprisingly, but in keeping with studies on naturally occurring promoters, evidence of shared structural features or sequence arrangements has yet to emerge from such work (Li *et al.*, 1999).

More recently, similar approaches have been used to isolate synthetic endothelial cell-specific promoters. In one such study, Dai, McAninch, and Sutton (2004) employed a system in which duplex oligonucleotides corresponding to various transcription factor-binding sites were cloned together in a random fashion next to a minimal ICAM-2 promoter upstream of an enhanced green fluorescent protein (GFP) indicator gene. The library was packaged into self-inactivating (SIN) HIV-1 vectors and introduced into an established Rhesus monkey choroidal endothelial cell line using vesicular stomatitis virus G protein pseudotyped particles. Cells that expressed high levels of fluorescence were enriched by repeated cycles of fluorescence-activated cell sorting. Ultimately, polymerase chain reaction was used to recover candidate promoter elements from GFP-positive clones and their activity confirmed using a luciferase reporter. Several promising promoters with high activity and apparent endothelial cell specificity were identified using this very efficient system (Dai, McAninch, and Sutton, 2004).

18.2.3 Post-transcriptional targeting

Improved understanding of the complex mechanisms that control the processing of mRNA in recent years has encouraged initial exploration of the potential of various post-transcriptional targeting strategies in the molecular therapy of cancer and other diseases (Anderson, 2010; Kerr and Davidson, 2010; Le Roy *et al.*, 2009).

Much of the early emphasis has focused on mRNA stability. In most cases the half-life of a transcribed message is determined by the presence within the 5′ and/or 3′ untranslated region (UTR) of defined sequences that interact with specific proteins that either protect the message from degradation causing it to accumulate within a cell, or enhance its degradation (Millevoi and Vagner, 2009). Differential expression of the proteins involved in these processes may allow particular transcripts to accumulate preferentially in certain cell types and not others. Alterations in message stability may also contribute to the dramatic changes in the level of certain mRNAs seen in response to various exogenous and/or microenvironmental stimuli.

Although little work directly concerned with the post-transcriptional targeting of endothelial cells has yet been carried out, studies focussed on hypoxia may be relevant (Ingram and Porter, 2005; Shibata, Giaccia, and Brown, 2000). Thus, while the endothelial cells lining normal blood vessels are obviously well oxygenated, the tortuous and extended nature of the tumor vasculature may generate a hypoxic microenvironment within certain tumor blood vessels, particularly toward their distal end. Indeed, cryospectroscopy studies have provided convincing evidence that a significant proportion of erythrocytes present within some tumor-associated vessels may be almost completely deoxygenated. Transient microregional interruptions in tumor blood flow produced as a result of temporary vessel occlusion may compound this problem (Chaplin and Hill, 1995).

While alterations in the expression of the hypoxia-inducible transcription factor HIF-1α, regulated via the ubiquitination and subsequent proteasome-mediated degradation of pVHL, is clearly critical in determining the rapid cellular response to hypoxia, alterations in message stability are also important. For example, under hypoxic conditions various proteins including erythropoietin mRNA binding protein bind to a sequence in the 3′-UTR of erythropoietin mRNA, stabilizing the message and increasing steady state levels (Rondon *et al.*, 1991; McGary, Rondon, and Beckman, 1997). Importantly, introduction of the relevant 3′-UTR sequence from

Epo into other genes produced a 10-fold increase in expression in various cells and tissues under hypoxic conditions (Lee *et al.*, 2006; Choi *et al.*, 2007).

VEGF RNA also accumulates under hypoxic conditions. However, in this instance, the mechanism appears to be more complicated and involves interactions between sequences present in both the coding region and the 5′- and 3′-UTRs (Dibbens *et al.*, 1999) and as such may not readily lend itself for use in molecular therapy applications (Dachs *et al.*, 1997). Although certain of the proteins that bind to the 3′-UTR of *Epo* under hypoxic conditions also bind to the 3′-UTR of *VEGF*, the sequence element involved does not appear to play an essential role in determining VEGF steady-state levels in low oxygen tensions (Shibata, Giaccia, and Brown, 2000).

Another sequence element involved in mRNA stabilization under hypoxic conditions that could potentially be exploited in the targeting of hypoxic tumor-associated endothelial cells is the iron responsive element (IRE) found in the 3′-UTR of the transferrin receptor and the 5′-UTRs of ferritin and mitochondrial aconitase (Schneider and Leibold, 2003). A sequence present in the 5′-UTR of HIF-1α has similarly been shown to play a role in the control of HIF-1α protein levels under hypoxia conditions (Gorlach *et al.*, 2000). In this case, however, the mechanism appears to involve effects on the efficiency of mRNA translation rather than stability (Gorlach *et al.*, 2000).

Finally in this connection, it has recently been suggested that microRNAs may play a role in determining gene expression under hypoxic conditions by regulating the degradation of the transcripts they interact with (Ivan *et al.*, 2008; Crosby *et al.*, 2009; Kulshreshtha *et al.*, 2007a, b). While the therapeutic potential of targeting such molecules has yet to be extensively investigated, it is certainly worthy of serious consideration (Stenvang and Kauppinen, 2008; Stenvang, Lindow, and Kauppinen, 2008; Lee, Rennie, and Jia, 2009; Spizzo *et al.*, 2009).

A second post-transcriptional process that can be exploited in order to target molecular therapy to endothelial or other cell types is alternative mRNA splicing. This is an approach that has been of particular interest to our group for some time. Specifically, we have developed a novel strategy that makes use of minigene constructs in which a leaderless but otherwise full-length cDNA encoding a therapeutic gene is cloned downstream of an engineered sequence that includes two or more exons along with their corresponding introns isolated from a gene that encodes an mRNA transcript that is alternatively spliced in the cell type being targeted. As the introns contain multiple stop codons in all three frames, a functional chimeric protein incorporating the alternatively spliced exons and the therapeutic gene is only produced in cells that are capable of accurately removing the intronic sequences from the initial pre-mRNA transcript.

In studies designed to validate this approach, we generated a series of constructs in which expression of a chimeric protein capable of activating phosphorylated prodrugs was dependent upon appropriate processing of various combinations of the three alternatively spliced exons (v8, v9, and v10) that are included in an isoform of the hyaluronan-binding CD44, designated CD44R1 (Hayes *et al.*, 2002). CD44R1 was chosen for these studies as this particular CD44 isoform is differentially expressed on metastatic tumor cells and on endothelial cells induced to proliferate by treatment with the pro-angiogenic cytokine bFGF (Griffioen *et al.*, 1997). The

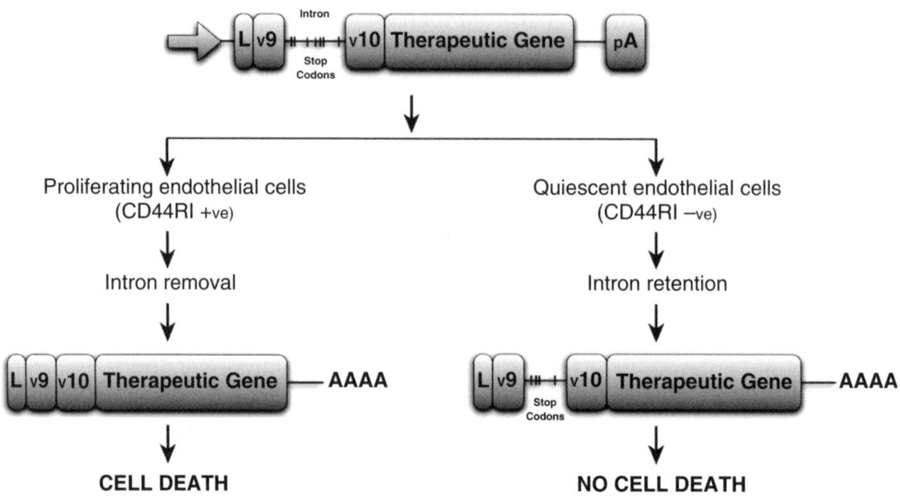

Figure 18.2 Using alternative splicing to target molecular therapy to tumor vasculature. The approach developed in our laboratory employs series of minigene constructs in which the expression of a chimeric protein incorporating the product of a therapeutic gene is dependent upon the accurate and complete removal of the intron separating two alternatively spliced exons (v9 and v10) derived from the adhesion protein CD44R1. Actively proliferating endothelial cells are CD44R1+ve and can carry out this splicing process. They are thus efficiently killed if transduced with a vector encoding the minigene construct. Resting endothelial cells express primarily CD44H and cannot mediate the required splicing event. As the intronic sequence retained in the resulting mRNA transcript contains multiple stop codons in all three frames, a functional protein product is not produced and the cells remain viable (Hayes *et al.*, 2002).

results obtained demonstrate that while the v8-v9 intron appears to be spliced irrespective of which CD44 isoform predominates in a cell, accurate and complete removal of the v9-v10 intron only occurs in cells that express CD44R1 or other v10 containing CD44 isoforms (Hayes *et al.*, 2002). These are very encouraging findings and serve to emphasize the exquisite specificity of the molecular processes that control alternative splicing (Figure 18.2).

While a systematic comparison of resting and tumor-associated endothelial cells is yet to be carried out, there is increasing evidence from the study of other cell types that a substantial number of genes are alternatively spliced in a cell- and/or activation-dependent manner (Orengo and Cooper, 2007; Pettigrew and Brown, 2008; Tazi, Bakkour, and Stamm, 2009). It is likely, therefore, that the general approach outlined above may prove broadly applicable and potentially useful in the context of vascular targeting.

18.2.4 Functional targeting

A little appreciated advantage of vascular-directed molecular therapies concerns the expanded range of genes that can be used to therapeutic effect. Thus while tumor-directed approaches generally require that the cells being targeted are either directly killed or sensitized to the cytotoxic effects of ionizing radiation or various chemotherapeutic agents, vascular-directed approaches can, in addition, take advantage of genes that trigger vessel occlusion without necessarily inducing death. Examples of potentially useful genes include those encoding proteins involved in the regulation of complement activation and/or the coagulation cascade. Genes that alter the shape or adherence properties of endothelial cells might also prove useful. Potentially, such effectors may offer a safety advantage as they may prove less harmful if inadvertently expressed in non-targeted tissues.

Building on this argument, our group have recently begun to explore a novel targeting approach that makes use of genes that produce a biological effect (i.e. the induction of cell death) only when activated by signals that occur differentially within the tumor microenvironment. The rationale underlying this functional targeting strategy is that even if the genes involved are introduced into non-targeted tissues, toxicity will not result unless the signals required to trigger the effector function are also present.

A good example of a signal that can be exploited in this way is provided by the endothelial cell mitogen, VEGF. VEGF has been shown to play a central role in the induction and regulation of angiogenesis in many solid tumor types (Birk *et al.*, 2008; Ferrara, 2009; Raica, Cimpean, and Ribatti, 2009; Verheul and Pinedo, 2000). Critically, while expression of VEGF is generally low or absent in normal tissues, it is often dramatically upregulated within tumors as a result of exposure to various microenvironmental stimuli such as hypoxia, or as a consequence of the expression of various oncogenes and/or tumor suppressor genes (Birk *et al.*, 2008; Ferrara, 2009; Raica, Cimpean, and Ribatti, 2009). Correlations have been established between VEGF levels and the induction of tumor vascularization, as well as the degree of local invasion, metastatic spread, and prognosis (Des Guetz *et al.*, 2006; Kosem *et al.*, 2009; Nieto *et al.*, 2007; Raica, Cimpean, and Ribatti, 2009). These activities of VEGF are mediated by binding to one or more differentially expressed endothelial cell surface receptors of which the receptor tyrosine kinases Flt-1 (vascular endothelial cell growth factor receptor-1,VEGFR-1) and Flk-1/KDR (VEGFR-2) appear to be the most important (Kliche and Waltenberger, 2001; Neufeld *et al.*, 1999; Shibuya, 2002, 2006).

In order to take advantage of the differential production of VEGF within tumors for targeting purposes, we generated a family of chimeric proteins in which the extracellular domains of the the VEGF receptors Flt-1 and Flk-1 were fused in-frame to the membrane spanning and cytoplasmic 'death domain' of the pro-apoptotic protein Fas (Carpenito *et al.*, 2002). Since the induction of death by wild-type Fas only occurs upon oligomerization triggered by the binding of the cognate Fas ligand (FasL), it was hypothesized that the chimeric Flk-1/Fas and Flt-1/Fas proteins would similarly only kill transduced target cells in the presence of VEGF (i.e., within tumor tissues but not normal tissues). Extensive *in vitro* studies have confirmed that this is indeed the case (Carpenito *et al.*, 2002; Dougherty and Dougherty, 2009b). Moreover, it was demonstrated that the induction of death via Flk-1/Fas only occurs when VEGF is present above a critical threshold level (Carpenito *et al.*, 2002; Dougherty and Dougherty, 2009b). This is potentially important as it suggests that when placed under the control of a promoter element that is assumed to have specificity for tumor vasculature, some 'leakiness' in the activity of the promoter resulting in low levels of transgene expression in normal endothelial cells, can be tolerated.

A similar strategy has been employed with equal success for other differentially-expressed tumor-associated ligands. Thus, a chimera in which the extracellular hyaluronan-binding of the adhesion protein CD44 is fused in-frame to the death domain of Fas (CD44/Fas) induced cell death only when crosslinked by hyaluronan or breakdown products thereof, which have been shown in previous studies to possess potent angiogenic activity (Dougherty and Dougherty, 2009a). A PDGFRα/Fas chimera was also effective. Interestingly, a chimera in which the entire extracellular domain of the human epidermal cell growth factor receptor (EGFR) was fused to the death domain of Fas did not induce the death of transduced cells even in the presence of high concentrations of EGF. Why this is the case is not yet clear. The degree of oligomerization induced upon ligand binding is likely to be key (Dougherty and Dougherty, 2009a). Thus while VEGF exists as a homodimer and the motif recognized by CD44 is repeated many times in a high molecular weight hyaluronan molecule, evidence from the study of the interaction of EGF with a soluble form of the EGFR extracellular domain, favors a model in which monomeric EGF binds to a single EGFR. Receptor dimerization, which is necessary for the triggering of downstream signal transduction events, appears to involve the subsequent 1 : 1 association of two monomeric EGF–EGFR complexes (Lemmon, 2009; Lemmon *et al.*, 1997).

18.3 Alternative targeting strategies

While most tumor vasculature appears to be generated via the proliferation, migration, and differentiation of 'mature' endothelial cells lining blood vessels in nearby normal tissues, it has become evident in recent years that there also exists a small but significant population of endothelial precursor cells in the bone marrow and blood that have the capacity under appropriate conditions to home to sites of active angiogenesis where they may contribute to the formation of new vessels (Rafii, 2000; Rafii and Lyden, 2003; Reyes *et al.*, 2002). Assuming such cells can be harvested in sufficient numbers or maintained and expanded *in vitro*, then the possibility exists that they could be genetically engineered *ex vivo* and used subsequently as vehicles to carry therapeutic genes to tumor sites. Indeed, this approach has already been explored and its potential efficacy confirmed in a number of preclinical studies (Davidoff *et al.*, 2001; De Palma *et al.*, 2003; Lyden *et al.*, 2001). Technologies that permit the isolation and genetic manipulation of hemopoietic progenitor cells are already well established and it does not seem unreasonable to expect that similar approaches could perhaps be applied to endothelial cell precursors, facilitating rapid clinical development of the strategy. It is important to keep in mind, however, that the precise phenotype of endothelial precursor cells remains the subject of some debate and much needs to be discovered concerning the mechanisms that control their recruitment to, and expansion in, the tumor vasculature.

18.4 Concluding remarks

The encouraging results obtained with small molecule VDAs such as CA4P and ZD6126 in early phase clinical studies have helped validate the vasculature as a target in cancer therapy. Clearly molecular medicines of the type described in this chapter are complex and substantial practical and regulatory challenges will need to be overcome before they can progress to the clinic (Fox, 2003). What makes such efforts worthwhile is the belief that by incorporating multiple independent targeting mechanisms into a molecular medicine it might be possible to direct an effect very specifically to tumor-associated vasculature greatly reducing the potential damage to vasculature in normal tissues.

Acknowledgements

Work from the authors' laboratory referred to in this article was supported by grants from the National Cancer Institute (CA100004), the Arizona Biomedical Research Commission (Contract 0706) and the University of Arizona, Women's Cancers Better Than Ever Program.

References

Adam, L., Black, P.C., Kassouf, W. *et al.* (2007) Adenoviral mediated interferon-alpha 2b gene therapy suppresses the pro-angiogenic effect of vascular endothelial growth factor in superficial bladder cancer. *Journal of Urology*, **177**, 1900–1906.

Adams, M.D., Kerlavage, A.R., Fleischmann, R.D. *et al.* (1995) Initial assessment of human gene diversity and expression patterns based upon 83 million nucleotides of cDNA sequence. *Nature*, **377**, 3–174.

Agnello, V. and Abel, G. (1997) Localization of hepatitis C virus in cutaneous vasculitic lesions in patients with type II cryoglobulinemia. *Arthritis and Rheumatism*, **40**, 2007–2015.

Aharinejad, S., Sioud, M., Lucas, T., and Abraham, D. (2009) Targeting stromal-cancer cell interactions with siRNAs. *Methods in Molecular Biology*, **487**, 243–266.

Anderson, P. (2010) Post-transcriptional regulons coordinate the initiation and resolution of inflammation. *Nature Reviews Immunology*, **10**, 24–35.

Bergelson, J.M. (1999) Receptors mediating adenovirus attachment and internalization. *Biochemical Pharmacology*, **57**, 975–979.

Bergelson, J.M., Cunningham, J.A., Droguett, G. *et al.* (1997) Isolation of a common receptor for Coxsackie B viruses and adenoviruses 2 and 5. *Science*, **275**, 1320–1323.

Bilbao, G., Gomez-Navarro, J., and Curiel, D.T. (1998) Targeted adenoviral vectors for cancer gene therapy. *Advances in Experimental Medicine and Biology*, **451**, 365–374.

Birk, D.M., Barbato, J., Mureebe, L., and Chaer, R.A. (2008) Current insights on the biology and clinical aspects of VEGF regulation. *Vascular and Endovascular Surgery*, **42**, 517–530.

Brooks, P.C., Montgomery, A.M., Rosenfeld, M. *et al.* (1994) Integrin alpha v beta 3 antagonists promote tumor regression by inducing apoptosis of angiogenic blood vessels. *Cell*, **79**, 1157–1164.

Brooks, P.C., Stromblad, S., Klemke, R. *et al.* (1995) Antiintegrin alpha v beta 3 blocks human breast cancer growth and angiogenesis in human skin. *The Journal of Clinical Investigation*, **96**, 1815–1822.

Buller, R.E., Runnebaum, I.B., Karlan, B.Y. *et al.* (2002) A phase I/II trial of rAd/p53 (SCH 58500) gene replacement in recurrent ovarian cancer. *Cancer Gene Therapy*, **9**, 553–566.

Carpenito, C., Davis, P.D., Dougherty, S.T., and Dougherty, G.J. (2002) Exploiting the differential production of angiogenic factors within the tumor microenvironment in the design of a novel vascular-targeted gene therapy-based approach to the treatment of cancer. *International Journal of Radiation Oncology, Biology, Physics*, **54**, 1473–1478.

Chaplin, D.J. and Hill, S.A. (1995) Temporal heterogeneity in microregional erythrocyte flux in experimental solid tumours. *British Journal of Cancer*, **71**, 1210–1213.

Chaplin, D.J. and Hill, S.A. (2002) Selective induction of tumor ischemia: development of vascular targeting agents for cancer therapy. *Current Opinion in Investigational Drugs*, **3**, 1381–1384.

Chi, D., Henry, J., Kelley, J. *et al.* (2000) The effects of HIV infection on endothelial function. *Endothelium*, **7**, 223–242.

Cho, J., Lim, W., Jang, S., and Lee, Y. (2003) Development of an efficient endothelial cell specific vector using promoter and 5′ untranslated sequences from the human preproendothelin-1 gene. *Experimental and Molecular Medicine*, **35**, 269–274.

Choi, B.H., Ha, Y., Ahn, C.H. *et al.* (2007) A hypoxia-inducible gene expression system using erythropoietin 3′ untranslated region for the gene therapy of rat spinal cord injury. *Neuroscience Letters*, **412**, 118–122.

Chung, L.W., Hsieh, C.L., Law, A. *et al.* (2003) New targets for therapy in prostate cancer: modulation of stromal–epithelial interactions. *Urology*, **62**, 44–54.

Cowan, P.J., Shinkel, T.A., Witort, E.J. *et al.* (1996) Targeting gene expression to endothelial cells in transgenic mice using the human intercellular adhesion molecule 2 promoter. *Transplantation*, **62**, 155–160.

Crosby, M.E., Devlin, C.M., Glazer, P.M. *et al.* (2009) Emerging roles of microRNAs in the molecular responses to hypoxia. *Current Pharmaceutical Design*, **15**, 3861–3866.

Dachs, G.U., Patterson, A.V., Firth, J.D. *et al.* (1997) Targeting gene expression to hypoxic tumor cells. *Nature Medicine*, **3**, 515–520.

Dai, C., McAninch, R.E., and Sutton, R.E. (2004) Identification of synthetic endothelial cell-specific promoters by use of a high-throughput screen. *Journal of Virology*, **78**, 6209–6221.

Dancer, A., Julien, S., Bouillot, S. *et al.* (2003) Expression of thymidine kinase driven by an endothelial-specific promoter inhibits tumor growth of Lewis lung carcinoma cells in transgenic mice. *Gene Therapy*, **10**, 1170–1178.

Davidoff, A.M., Ng, C.Y., Brown, P. *et al.* (2001) Bone marrow-derived cells contribute to tumor neovasculature and, when modified to express an angiogenesis inhibitor, can restrict tumor growth in mice. *Clinical Cancer Research*, **7**, 2870–2879.

Davis, P.D., Dougherty, G.J., Blakey, D.C. *et al.* (2002) ZD6126: a novel vascular-targeting agent that causes selective destruction of tumor vasculature. *Cancer Research*, **62**, 7247–7253.

De Palma, M., Venneri, M.A., and Naldini, L. (2003) In vivo targeting of tumor endothelial cells by systemic delivery of lentiviral vectors. *Human Gene Therapy*, **14**, 1193–1206.

De Palma, M., Venneri, M.A., Roca, C., and Naldini, L. (2003) Targeting exogenous genes to tumor angiogenesis by transplantation of genetically modified hematopoietic stem cells. *Nature Medicine*, **9**, 789–795.

Defilippi, P., van Hinsbergh, V., Bertolotto, A. *et al.* (1991) Differential distribution and modulation of expression of alpha 1/beta 1 integrin on human endothelial cells. *Journal of Cell Biology*, **114**, 855–863.

Denekamp, J. (1990) Vascular attack as a therapeutic strategy for cancer. *Cancer Metastasis Reviews*, **9**, 267–282.

Denekamp, J. (1993) Review article: angiogenesis, neovascular proliferation and vascular pathophysiology as targets for cancer therapy. *British Journal of Radiology*, **66**, 181–196.

Denekamp, J., Dasu, A., and Waites, A. (1998) Vasculature and microenvironmental gradients: the missing links in novel approaches to cancer therapy? *Advances in Enzyme Regulation*, **38**, 281–299.

Des Guetz, G., Uzzan, B., Nicolas, P. *et al.* (2006) Microvessel density and VEGF expression are prognostic factors in colorectal cancer. Meta-analysis of the literature. *British Journal of Cancer*, **94**, 1823–1832.

Dibbens, J.A., Miller, D.L., Damert, A. *et al.* (1999) Hypoxic regulation of vascular endothelial growth factor mRNA stability requires the cooperation of multiple RNA elements. *Molecular Biology of the Cell*, **10**, 907–919.

Dougherty, G.J. and Dougherty, S.T. (2009a) A cancer gene therapy approach that targets tumor-associated hyaluronan. *Cancer Growth Metastasis*, **2**, 29–43.

Dougherty, G.J. and Dougherty, S.T. (2009b) Exploiting the tumor microenvironment in the development of targeted cancer gene therapy. *Cancer Gene Therapy*, **16**, 279–290.

Dropulic, B. and Masters, C.L. (1990) Entry of neurotropic arboviruses into the central nervous system: an in vitro study using mouse brain endothelium. *Journal of Infectious Diseases*, **161**, 685–691.

Enenstein, J. and Kramer, R.H. (1994) Confocal microscopic analysis of integrin expression on the microvasculature and its sprouts in the neonatal foreskin. *Journal of Investigative Dermatology*, **103**, 381–386.

van Etten, B., Eggermont, A.M., van Tiel, S.T. *et al.* (2005) Gene therapy in in vivo isolated perfusion models. *Current Gene Therapy*, **5**, 195–202.

Fabbri, M., Castellani, P., Gotwals, P.J. *et al.* (1996) A functional monoclonal antibody recognizing the human alpha 1-integrin I-domain. *Tissue Antigens*, **48**, 47–51.

Ferrara, N. (2009) Vascular endothelial growth factor. *Arteriosclerosis, Thrombosis, and Vascular Biology*, **29**, 789–791.

Fox, J.L. (2003) US authorities uphold suspension of SCID gene therapy. *Nature Biotechnology*, **21**, 217.

Gerna, G., Baldanti, F., and Revello, M.G. (2004) Pathogenesis of human cytomegalovirus infection and cellular targets. *Human Immunology*, **65**, 381–386.

Gomella, L.G., Mastrangelo, M.J., McCue, P.A. *et al.* (2001) Phase i study of intravesical vaccinia virus as a vector for gene therapy of bladder cancer. *The Journal of Urology*, **166**, 1291–1295.

Gorlach, A., Camenisch, G., Kvietikova, I. *et al.* (2000) Efficient translation of mouse hypoxia-inducible factor-1alpha under normoxic and hypoxic conditions. *Biochimica et Biophysica Acta*, **1493**, 125–134.

Graulich, W., Nettelbeck, D.M., Fischer, D. *et al.* (1999) Cell type specificity of the human endoglin promoter. *Gene*, **227**, 55–62.

Greber, U.F., Willetts, M., Webster, P., and Helenius, A. (1993) Stepwise dismantling of adenovirus 2 during entry into cells. *Cell*, **75**, 477–486.

Greenberger, S., Shaish, A., Varda-Bloom, N. *et al.* (2004) Transcription-controlled gene therapy against tumor angiogenesis. *Journal of Clinical Investigation*, **113**, 1017–1024.

Griffioen, A.W., Coenen, M.J., Damen, C.A. *et al.* (1997) CD44 is involved in tumor angiogenesis; an activation antigen on human endothelial cells. *Blood*, **90**, 1150–1159.

Hayes, G.M., Carpenito, C., Davis, P.D. *et al.* (2002) Alternative splicing as a novel of means of regulating the expression of therapeutic genes. *Cancer Gene Therapy*, **9**, 133–141.

Hensley, L.E. and Geisbert, T.W. (2005) The contribution of the endothelium to the development of coagulation disorders that characterize Ebola hemorrhagic fever in primates. *Thrombosis and Haemostasis*, **94**, 254–261.

Hsieh, C.L., Gardner, T.A., Miao, L. *et al.* (2004) Cotargeting tumor and stroma in a novel chimeric tumor model involving the growth of both human prostate cancer and bone stromal cells. *Cancer Gene Therapy*, **11**, 148–155.

Ingram, N. and Porter, C.D. (2005) Transcriptional targeting of acute hypoxia in the tumour stroma is a novel and viable strategy for cancer gene therapy. *Gene Therapy*, **12**, 1058–1069.

Ivan, M., Harris, A.L., Martelli, F., and Kulshreshtha, R. (2008) Hypoxia response and microRNAs: no longer two separate worlds. *Journal of Cellular and Molecular Medicine*, **12**, 1426–1431.

Jackson, R.S., Franco, O.E., and Bhowmick, N.A. (2008) Gene targeting to the stroma of the prostate and bone. *Differentiation*, **76**, 606–623.

Jager, U., Zhao, Y., and Porter, C.D. (1999) Endothelial cell-specific transcriptional targeting from a hybrid long terminal repeat retrovirus vector containing human prepro-endothelin-1 promoter sequences. *Journal of Virology*, **73**, 9702–9709.

Kerr, T.A. and Davidson, N.O. (2010) Therapeutic RNA manipulation in liver disease. *Hepatology*. **51**, 1055–1061.

Kliche, S. and Waltenberger, J. (2001) VEGF receptor signaling and endothelial function. *IUBMB Life*, **52**, 61–66.

Korhonen, J., Lahtinen, I., Halmekyto, M. *et al.* (1995) Endothelial-specific gene expression directed by the tie gene promoter in vivo. *Blood*, **86**, 1828–1835.

Kosem, M., Tuncer, I., Kotan, C. *et al.* (2009) Significance of VEGF and microvascular density in gastric carcinoma. *Hepatogastroenterology*, **56**, 1236–1240.

Kozian, D.H. and Augustin, H.G. (1995) Rapid identification of differentially expressed endothelial cell genes by RNA display. *Biochemical and Biophysical Research Communications*, **209**, 1068–1075.

Krasnykh, V., Dmitriev, I., Mikheeva, G. *et al.* (1998) Characterization of an adenovirus vector containing a heterologous peptide epitope in the HI loop of the fiber knob. *Journal of Virology*, **72**, 1844–1852.

Krasnykh, V., Dmitriev, I., Navarro, J.G. *et al.* (2000) Advanced generation adenoviral vectors possess augmented gene transfer efficiency based upon coxsackie adenovirus receptor-independent cellular entry capacity. *Cancer Research*, **60**, 6784–6787.

Kulshreshtha, R., Ferracin, M., Negrini, M. *et al.* (2007a) Regulation of microRNA expression: the hypoxic component. *Cell Cycle*, **6**, 1426–1431.

Kulshreshtha, R., Ferracin, M., Wojcik, S.E. *et al.* (2007b) A microRNA signature of hypoxia. *Molecular and Cellular Biology*, **27**, 1859–1867.

Lai, S.Y., Koppikar, P., Thomas, S.M. *et al.* (2009) Intratumoral epidermal growth factor receptor antisense DNA therapy in head and neck cancer: first human application and potential antitumor mechanisms. *Journal of Clinical Oncology*, **27**, 1235–1242.

Lee, C.Y., Rennie, P.S., and Jia, W.W. (2009) MicroRNA regulation of oncolytic herpes simplex virus-1 for selective killing of prostate cancer cells. *Clinical Cancer Research*, **15**, 5126–5135.

Lee, M., Choi, D., Choi, M.J. *et al.* (2006) Hypoxia-inducible gene expression system using the erythropoietin enhancer and 3′-untranslated region for the VEGF gene therapy. *Journal of Controlled Release*, **115**, 113–119.

Lemmon, M.A. (2009) Ligand-induced ErbB receptor dimerization. *Experimental Cell Research*, **315**, 638–648.

Lemmon, M.A., Bu, Z., Ladbury, J.E. *et al.* (1997) Two EGF molecules contribute additively to stabilization of the EGFR dimer. *EMBO Journal*, **16**, 281–294.

Le Roy, F., Charton, K., Lorson, C.L., and Richard, I. (2009) RNA-targeting approaches for neuromuscular diseases. *Trends in Molecular Medicine*, **15**, 580–591.

Li, J.L., Liu, H.L., Zhang, X.R. *et al.* (2009) A phase I trial of intratumoral administration of recombinant oncolytic adenovirus overexpressing HSP70 in advanced solid tumor patients. *Gene Therapy*, **16**, 376–382.

Li, X., Eastman, E.M., Schwartz, R.J., and Draghia-Akli, R. (1999) Synthetic muscle promoters: activities exceeding naturally occurring regulatory sequences. *Nature Biotechnology*, **17**, 241–245.

Lyden, D., Hattori, K., Dias, S. *et al.* (2001) Impaired recruitment of bone-marrow-derived endothelial and hematopoietic precursor cells blocks tumor angiogenesis and growth. *Nature Medicine*, **7**, 1194–1201.

Mavria, G., Jager, U., and Porter, C.D. (2000) Generation of a high titre retroviral vector for endothelial cell-specific gene expression in vivo. *Gene Therapy*, **7**, 368–376.

McGary, E.C., Rondon, I.J., and Beckman, B.S. (1997) Post-transcriptional regulation of erythropoietin mRNA stability by erythropoietin mRNA-binding protein. *Journal of Biological Chemistry*, **272**, 8628–8634.

Michael, S.I., Hong, J.S., Curiel, D.T., and Engler, J.A. (1995) Addition of a short peptide ligand to the adenovirus fiber protein. *Gene Therapy*, **2**, 660–668.

Millevoi, S. and Vagner, S. (2009) Molecular mechanisms of eukaryotic pre-mRNA 3′ end processing regulation. *Nucleic Acids Research*. [Epub ahead of print]

Neufeld, G., Cohen, T., Gengrinovitch, S., and Poltorak, Z. (1999) Vascular endothelial growth factor (VEGF) and its receptors. *FASEB Journal*, **13**, 9–22.

Nicklin, S.A., Reynolds, P.N., Brosnan, M.J. *et al.* (2001) Analysis of cell-specific promoters for viral gene therapy targeted at the vascular endothelium. *Hypertension*, **38**, 65–70.

Nicklin, S.A., White, S.J., Nicol, C.G. *et al.* (2004) In vitro and in vivo characterisation of endothelial cell selective adenoviral vectors. *Journal of Gene Medicine*, **6**, 300–308.

Nieto, Y., Woods, J., Nawaz, F. *et al.* (2007) Prognostic analysis of tumour angiogenesis, determined by microvessel density and expression of vascular endothelial growth factor, in high-risk primary breast cancer patients treated with high-dose chemotherapy. *British Journal of Cancer*, **97**, 391–397.

Orengo, J.P. and Cooper, T.A. (2007) Alternative splicing in disease. *Advances in Experimental Medicine and Biology*, **623**, 212–223.

Palmer, D.H., Young, L.S., and Mautner, V. (2006) Cancer gene-therapy: clinical trials. *Trends in Biotechnology*, **24**, 76–82.

Pearson, A.S., Koch, P.E., Atkinson, N. *et al.* (1999) Factors limiting adenovirus-mediated gene transfer into human lung and pancreatic cancer cell lines. *Clinical Cancer Research*, **5**, 4208–4213.

Pettigrew, C.A. and Brown, M.A. (2008) Pre-mRNA splicing aberrations and cancer. *Frontiers in Bioscience*, **13**, 1090–1105.

Prols, F., Loser, B., and Marx, M. (1998) Differential expression of osteopontin, PC4, and CEC5, a novel mRNA species, during in vitro angiogenesis. *Experimental Cell Research*, **239**, 1–10.

Rafii, S. (2000) Circulating endothelial precursors: mystery, reality, and promise. *Journal of Clinical Investigation*, **105**, 17–19.

Rafii, S. and Lyden, D. (2003) Therapeutic stem and progenitor cell transplantation for organ vascularization and regeneration. *Nature Medicine*, **9**, 702–712.

Raica, M., Cimpean, A.M., and Ribatti, D. (2009) Angiogenesis in pre-malignant conditions. *European Journal of Cancer*, **45**, 1924–1934.

Rainov, N.G. and Kramm, C.M. (2001) Vector delivery methods and targeting strategies for gene therapy of brain tumors. *Current Gene Therapy*, **1**, 367–383.

Reyes, M., Dudek, A., Jahagirdar, B. *et al.* (2002) Origin of endothelial progenitors in human postnatal bone marrow. *Journal of Clinical Investigation*, **109**, 337–346.

Reynolds, P.N., Nicklin, S.A., Kaliberova, L. *et al.* (2001) Combined transductional and transcriptional targeting improves the specificity of transgene expression in vivo. *Nature Biotechnology* **19**, 838–842.

Richardson, T.B., Kaspers, J., and Porter, C.D. (2004) Retroviral hybrid LTR vector strategy: functional analysis of LTR elements and generation of endothelial cell specificity. *Gene Therapy*, **11**, 775–783.

Rochlitz, C.F. (2001) Gene therapy of cancer. *Swiss Medical Weekly*, **131**, 4–9.

Rondon, I.J., MacMillan, L.A., Beckman, B.S. *et al.* (1991) Hypoxia up-regulates the activity of a novel erythropoietin mRNA binding protein. *Journal of Biological Chemistry*, **266**, 16594–16598.

Savontaus, M.J., Sauter, B.V., Huang, T.G., and Woo, S.L. (2002) Transcriptional targeting of conditionally replicating adenovirus to dividing endothelial cells. *Gene Therapy*, **9**, 972–979.

Scanlon, K.J. (2004) Cancer gene therapy: challenges and opportunities. *Anticancer Research*, **24**, 501–504.

Schneider, B.D. and Leibold, E.A. (2003) Effects of iron regulatory protein regulation on iron homeostasis during hypoxia. *Blood*, **102**, 3404–3411.

Schnittler, H.J. and Feldmann, H. (2003) Viral hemorrhagic fever--a vascular disease? *Thrombosis and Haemostasis*, **89**, 967–972.

Senger, D.R., Claffey, K.P., Benes, J.E. *et al.* (1997) Angiogenesis promoted by vascular endothelial growth factor: regulation through alpha1beta1 and alpha2beta1 integrins. *Proceedings of the National Academy of Sciences of the United States of America*, **94**, 13612–13617.

Shibata, T., Giaccia, A.J., and Brown, J.M. (2000) Development of a hypoxia-responsive vector for tumor-specific gene therapy. *Gene Therapy*, **7**, 493–498.

Shibuya, M. (2002) Vascular endothelial growth factor receptor family genes: when did the three genes phylogenetically segregate? *Biological Chemistry*, **383**, 1573–1579.

Shibuya, M. (2006) Differential roles of vascular endothelial growth factor receptor-1 and receptor-2 in angiogenesis. *Journal of Biochemistry and Molecular Biology*, **39**, 469–478.

Siemann, D.W., Chaplin, D.J., and Horsman, M.R. (2004) Vascular-targeting therapies for treatment of malignant disease. *Cancer*, **100**, 2491–2499.

Smiley, W.R., Laubert, B., Howard, B.D. *et al.* (1997) Establishment of parameters for optimal transduction efficiency and antitumor effects with purified high-titer HSV-TK retroviral vector in established solid tumors. *Human Gene Therapy*, **8**, 965–977.

Solignat, M., Gay, B., Higgs, S. *et al.* (2009) Replication cycle of chikungunya: a re-emerging arbovirus. *Virology*, **393**, 183–197.

Spizzo, R., Rushworth, D., Guerrero, M., and Calin, G.A. (2009) RNA inhibition, microRNAs, and new therapeutic agents for cancer treatment. *Clinical Lymphoma and Myeloma*, **9** (Suppl 3), S313–S318.

Stenvang, J. and Kauppinen, S. (2008) MicroRNAs as targets for antisense-based therapeutics. *Expert Opinion on Biological Therapy*, **8**, 59–81.

Stenvang, J., Lindow, M., and Kauppinen, S. (2008) Targeting of microRNAs for therapeutics. *Biochemical Society Transactions*, **36**, 1197–1200.

Sullivan, N., Yang, Z.Y., and Nabel, G.J. (2003) Ebola virus pathogenesis: implications for vaccines and therapies. *Journal of Virology*, **77**, 9733–9737.

Tazi, J., Bakkour, N., and Stamm, S. (2009) Alternative splicing and disease. *Biochimica et Biophysica Acta*, **1792**, 14–26.

Thorpe, P.E. (2004) Vascular targeting agents as cancer therapeutics. *Clinical Cancer Research*, **10**, 415–427.

Thorpe, P.E., Chaplin, D.J., and Blakey, D.C. (2003) The first international conference on vascular targeting: meeting overview. *Cancer Research*, **63**, 1144–1147.

Tomko, R.P., Xu, R., and Philipson, L. (1997) HCAR and MCAR: the human and mouse cellular receptors for subgroup C adenoviruses and group B coxsackieviruses. *Proceedings of the National Academy of Sciences of the United States of America*, **94**, 3352–3356.

Tulman, E.R., Delhon, G.A., Ku, B.K., and Rock, D.L. (2009) African swine fever virus. *Current Topics in Microbiology and Immunology*, **328**, 43–87.

Varda-Bloom, N., Shaish, A., Gonen, A. *et al.* (2001) Tissue-specific gene therapy directed to tumor angiogenesis. *Gene Therapy*, **8**, 819–827.

Vasey, P.A., Shulman, L.N., Campos, S. *et al.* (2002) Phase I trial of intraperitoneal injection of the E1B-55-kd-gene-deleted adenovirus ONYX-015 (dl1520) given on days 1 through 5 every 3 weeks in patients with recurrent/refractory epithelial ovarian cancer. *Journal of Clinical Oncology*, **20**, 1562–1569.

Velasco, B., Ramirez, J.R., Relloso, M. *et al.* (2001) Vascular gene transfer driven by endoglin and ICAM-2 endothelial-specific promoters. *Gene Therapy*, **8**, 897–904.

Verheul, H.M. and Pinedo, H.M. (2000) The role of vascular endothelial growth factor (VEGF) in tumor angiogenesis and early clinical development of VEGF-receptor kinase inhibitors. *Clinical Breast Cancer*, **1** (Suppl 1), S80–S84.

Wang, K., Huang, S., Kapoor-Munshi, A., and Nemerow, G. (1998) Adenovirus internalization and infection require dynamin. *Journal of Virology*, **72**, 3455–3458.

Wickham, T.J., Lee, G.M., Titus, J.A. *et al.* (1997) Targeted adenovirus-mediated gene delivery to T cells via CD3. *Journal of Virology*, **71**, 7663–7669.

Wickham, T.J., Mathias, P., Cheresh, D.A., and Nemerow, G.R. (1993) Integrins alpha v beta 3 and alpha v beta 5 promote adenovirus internalization but not virus attachment. *Cell*, **73**, 309–319.

Yang, Z.Y., Duckers, H.J., Sullivan, N.J. *et al.* (2000) Identification of the Ebola virus glycoprotein as the main viral determinant of vascular cell cytotoxicity and injury. *Nature Medicine*, **6**, 886–889.

Zhang, H.T., Gorn, M., Smith, K. *et al.* (1999) Transcriptional profiling of human microvascular endothelial cells in the proliferative and quiescent state using cDNA arrays. *Angiogenesis*, **3**, 211–219.

Index

Tumor Microenvironment Edited by Dietmar W. Siemann